Recent Progress in Plant Physiology

Recent Progress in Plant Physiology

Edited by **Edgar Crombie**

SYRAWOOD
PUBLISHING HOUSE

New York

Published by Syrawood Publishing House,
750 Third Avenue, 9th Floor,
New York, NY 10017, USA
www.syrawoodpublishinghouse.com

Recent Progress in Plant Physiology
Edited by Edgar Crombie

International Standard Book Number: 978-1-68286-184-4 (Hardback)

Contents

Preface

Plant physiology is a discipline which studies the internal processes of plants and the factors influencing those processes. The study of this discipline has enabled mankind to develop new techniques and methods particularly useful for agriculture and horticulture. Mechanisms like photosynthesis, which take place at molecular levels to development of plants that occur at macro levels, many physiological mechanisms have been discussed in detail. The aim of this book is to provide apt references to graduate and post graduate students of biology and botany about various life processes like transpiration, germination, etc. This book traces the recent progresses in this field and will also be beneficial for research scholars.

This book has been the outcome of endless efforts put in by authors and researchers on various issues and topics within the field. The book is a comprehensive collection of significant researches that are addressed in a variety of chapters. It will surely enhance the knowledge of the field among readers across the globe.

It gives us an immense pleasure to thank our researchers and authors for their efforts to submit their piece of writing before the deadlines. Finally in the end, I would like to thank my family and colleagues who have been a great source of inspiration and support.

Editor

Betacyanin accumulation and guaiacol peroxidase activity in *Beta vulgaris* L. leaves following copper stress

Janet María León Morales, Mario Rodríguez-Monroy, Gabriela Sepúlveda-Jiménez*

Department of Biotechnology, Center of Biotic Products Development, National Polytechnic Institute, CePProBi 8, San Isidro, Yautepec, Morelos 62731, México

Abstract

The effect of copper stress on betacyanin accumulation and guaiacol peroxidase (GPOD) activity in leaves of different age was evaluated in red beet (*Beta vulgaris* L. var. Crosby Egyptian) plants. In hydroponic culture, plants were treated with 0.3 μM (control), 50 μM, 100 μM, and 250 μM of CuSO$_4$ for 6 days. Copper was taken up and accumulated in old roots but was not translocated to leaves. However in young leaves, the increase of lipid peroxidation and reduction of growth were evident from day 3 of copper exposure; whereas in old leaves, the lipid peroxidation and growth were the same from either copper-treated or control plants. In response to copper exposure, the betacyanin accumulation was evident in young leaves by day 3, and continued to increase until day 6. Betacyanin only were accumulated in old leaves until day 6, but the contents were from 4 to 5 times lower than those observed in young leaves at the same copper concentrations. GPOD activity increased 3.3- and 1.4-fold in young and old leaves from day 3 of copper treatment respectively, but only in the young leaves was sustained at the same level until day 6. Old roots shown betacyanin in the control plants, but the betacyanin level and growth were reduced with the copper exposure. In contrast, young roots emerged by copper effect also accumulated copper and showed the highest betacyanin content of all plant parts assayed. These results indicate that betacyanin accumulation and GPOD activity are defense responses to copper stress in actively growing organs.

Keywords: oxidative stress, betacyanin, abiotic stress, antioxidant compound, copper

Introduction

High levels of copper in the environment may originate from natural sources, industrial activity, or the use of fungicides in agriculture [1]. Copper is an essential plant micronutrient, involved in the photosynthetic electron transport, where plastocyanin is a copper-binding protein. It is also cofactor of several enzymes such as cytochrome oxidase, superoxide dismutase and ascorbate oxidase [2]. The exposure of plants to copper concentrations higher than those required for optimal growth results in phytotoxicity, causing growth inhibition and cell death. The inhibitory effects of copper on leaf growth in *Arabidopsis thaliana* arise from inhibition of cell expansion [3]. Inhibition of primary root growth by copper could result from the arrest of cellular division and/or nuclear membrane damage in the meristematic zone [4,5], whereas the stimulation of cell division at sites distant from root tips is associated with changes in the distributions of auxin and cytokinin,

which increase lateral root formation [6]. These observations, together with the demonstration that morphogenic growth responses to excess copper resemble those induced by sources of oxidative stress, such as the herbicide paraquat or tert-butylhydroperoxide, an analogue of hydrogen peroxide (H$_2$O$_2$) [3], suggest that the reorientation of plant growth in response to copper toxicity involves changes in phytohormone distribution and the metabolism of reactive oxygen species (ROS).

Elevated levels of intracellular copper generate oxidative stress through increasing the levels of ROS, which in turn increase rates of lipids and proteins oxidation [7–9]. The quantification of lipid peroxidation has been used as an indicator of oxidative damage in leaves and roots of copper-exposed plants [7,10–14]. Peroxidases such as guaiacol peroxidases (GPOD) are also considered to be reliable markers of metal stress. These enzymes catalyze the oxidation of several substrates, such as phenolics, lignin precursors, auxin, and secondary metabolites [15]. Given that copper stimulates the activities of coniferyl alcohol peroxidase and indole-3-acetic acid oxidase in *Pisum sativum* roots [16], it has been proposed that GPOD could contribute to cell wall lignification, growth regulation and antioxidant defense responses [11,17,18].

Several studies have demonstrated a differential toxicity of copper in leaves. Copper exposure inhibited the growth of *Phaseolus coccineus* young leaves, but the growth of older leaves remained unchanged even though they showed evidence of chlorosis [19]. Copper inhibited the growth of *Cucumis sativus* young leaves, but had no effect on the area of mature

* Corresponding author. Email: gsepulvedaj@ipn.mx

leaves, even though photosynthesis was inhibited [20]. At the cellular level, chlorophyll breakdown and lipid peroxidation in copper-exposed segments of old leaves were higher than in young leaves segments of *Avena sativa* [21]. However, studies concerning the relationship between toxicity of copper and the antioxidant responses in plant organs at different developmental stages are scarce. For example, Luna et al. [21] reported that young leaves segments of *Avena sativa* exposed to copper exhibited less oxidative damage and a higher increase in the activity of superoxide dismutase than segments of old leaves. Although it was reported that exposure of *Morus alba* plants to copper caused uniform yellowing in old leaves, the effects of copper toxicity on the activities of antioxidant enzymes were only evaluated in younger leaves [22].

The accumulation of secondary metabolites with antioxidant activity is also a defense response to copper exposure. Oxidative stress caused by exposure to high levels of copper was implicated in the accumulation of phenolic compounds in *Withania somnifera* plants and the accumulation of polyamines in *Triticum aestivum* plants [11,13]. Furthermore, incubation of wheat leaf segments in the presence of spermine prior to copper treatment prevented subsequent lipid peroxidation [13]. The copper accumulation in roots and leaves of *Matricaria chamomilla* increased the content of phenolic acids that may contribute to the avoidance of lipid peroxidation in leaves [12,23]. In *Raphanus sativus* shoots, the increases in cinnamic and benzoic acid derivatives and reduction in levels of ascorbate are dependent of the copper concentration. In this study a H_2O_2 detoxification system was suggested, that could be formed by phenolic acids and reduced ascorbate and a peroxidase [14]. The accumulation of secondary metabolites with antioxidant activity in response to copper stress may depend of the developmental stage of the organ. Skórzyńska-Polit et al. [24] reported that the increase in the flavonol content was higher in young leaves than in old leaves of *P. coccineus* after copper treatment.

Betacyanins are tyrosine-derived chromoalkaloids synthesized by members of nine families of the order Caryophyllales [25]. In vitro assays demonstrated the antioxidant activity of betacyanin present in *Beta vulgaris* L. roots [26,27]. Betacyanin accumulation in plants has been linked with oxidative stress factors such as exposure to UV radiation in *Mesembryanthemum crystallinum* [28], wounding and bacterial infiltration in leaves of *Beta vulgaris* L. [29], as well as low temperatures, high salinity and exposure to exogenous H_2O_2 in *Suaeda salsa* [30,31]. However, the effect of copper on betacyanin accumulation in leaves and roots at different developmental stages and its relationship with the toxicity symptoms has not been evaluated. Here we report for first time that the betacyanin accumulation and GPOD activation are defense responses in actively growing organs in plants of *Beta vulgaris* L. exposed to copper.

Material and methods

Seed germination and plant material
Seeds of red beet (*Beta vulgaris* L. var. Crosby Egyptian) were germinated between paper towels moistened with distilled water in darkness at room temperature (25 ±2°C). Five-day-old seedlings were transplanted into plastic pots filled with a sterile mixture of peat moss and metromix (1:3, w/w). Plants were grown at 21 ±2°C, with 60% relative humidity and a 16 h light/8 h dark photoperiod, with a photosynthetic photon flux density of 68 $\mu mol \times m^{-2} \times s^{-1}$ provided by fluorescent white light lamps (Osram L140W/20, Danvers, MA). Plants with three pairs of true leaves were treated with copper. Leaves between 7 and 17 days of age were classified as young, and those 34–36 days of age were classified as old leaves. Roots present at the time that treatment was initiated were referred to as old roots, and roots that emerged within 6 days of copper treatment were referred to as young roots.

Copper treatment
Roots of 7-week-old plants were washed twice with deionised water in order to eliminate excess growth substrate. Plants were then transferred to boxes (10 plants/box) containing Hoagland nutritive solution (pH 5.2 ±0.02), modified to contain 0.33 mM $NH_4H_2PO_4$, 1.3 mM $CaCl_2$, 0.66 mM $MgSO_4$, 2 mM KNO_3, 46 μM H_3BO_3, 0.8 μM $ZnSO_4$, 0.88 μM $MnCl_2$, 0.012 μM Na_2Mo $4H_2O$, 1.44 μM Fe-EDTA. The copper concentration in the control condition was 0.3 μM $CuSO_4$. After a 4-days adaptation period, plants were treated with copper for 6 days by adding additional $CuSO_4$ to the media. The copper concentrations evaluated were 50 μM, 100 μM, and 250 μM $CuSO_4$, and the nutritive solution was renewed every 3 days. For all determinations, tissue samples were collected from both old and young leaves and roots.

Levels of betacyanin, copper, and lipid peroxidation were determined at the time that treatments began, and at 3 and 6 days after exposure to copper. Chlorophyll content, leaf area and root length were measured after 6 days of treatment.

Copper content
Metal ions adhered to all roots were removed by incubating the roots in deionised water for 30 min, followed by incubation 10 mM $CaCl_2$ for 10 min. Leaves and roots were dried in an oven (Lab-Line Instruments, Inc., Melrose Park, IL) at 70°C for 24 h prior to being ground using a pestle and mortar. The powder was digested with $HNO_3{:}H_2O_2$ (1:1, v/v) mixture in a MARSXP15000 Plus microwave (CEM Corporation, Matthews, NC) at 150°C for 15 min. Total copper concentration was determined by atomic absorption spectroscopy using a Varian SpectrAA 55B flame spectrometer (Varian, Inc., Palo Alto, CA) at a wavelength of 324.8 nm. Experiments were performed at the Chemical Laboratory of the Science Institute, Benemérita Universidad Autónoma de Puebla, Mexico. Data were expressed as μg copper $\times g^{-1}$ dry weight (*DW*).

Determination of total chlorophylls
Two leaf discs (0.5 cm) were cut, frozen in liquid nitrogen and then immediately homogenized in 1 mL of 80% acetone. After centrifugation at 15000 $\times g$ for 15 min, the absorbance of the supernatant at 646.8 and 663.2 nm was measured using a Genesys 2 spectrophotometer (Thermo Spectronic, Madison, WI). Total chlorophyll content was calculated according to the equations proposed by Lichtenthaler [32].

Lipid peroxide determination
Lipid peroxidation was determined by measuring thiobarbituric acid reactive substances (TBARS) as described previously Alia et al. [33]. Six 0.5 cm discs of fresh leaves were homogenized in 1 mL of trichloroacetic acid (5%, w/v) and extracts were centrifuged at 15000 $\times g$ for 10 min. The supernatant was transferred to vials containing 1 mL of reaction buffer (0.5% thiobarbituric acid in 20% trichloroacetic acid

solution), incubated for 30 min in a water bath at 95°C, and cooled in an ice bath for 10 min. Samples were centrifuged at 9500 × g for 10 min at 15°C. The absorbance of the supernatant was measured at 532 nm, and the non-specific absorbance at 600 nm was subtracted. Results were expressed as malondialdehyde (MDA) equivalents in nmol × g⁻¹ fresh weigh (*FW*) and calculated using a standard curve prepared using known concentrations of MDA.

Quantification of betacyanins

Root segments (ca. 25 mg) or four 0.5 cm disc of fresh leaves were ground with a pestle and motor-driven grinder (Sigma-Aldrich Co., St. Louis, MO) in Eppendorf tubes. Samples were homogenized with 1 mL distilled water, and incubated at 85°C in agitation (300 rpm) for 5 min. Extracts were centrifuged at 9500 × g for 10 min at 25°C. The absorbance of the supernatant was measured at 540 nm, and the betacyanin content was calculated according to standard curve with known concentrations of betanin purified by high-performance liquid chromatography as described previously Schwartz and von Elbe [34]. Data were expressed in mg betanin equivalents × g⁻¹ *FW*.

GPOD activity

For each sample, three 0.5 cm discs from leaves were frozen in liquid nitrogen and ground in Eppendorf tubes with 0.005 g of poly(vinylpolypyrrolidone) and activated charcoal mixture (1:1, w/w) to eliminate pigments. Samples were homogenized in 300 µL of cold extraction buffer [50 mM sodium phosphate buffer (pH 6.0) containing 1 mM ethylenediamine tetraacetic acid, 1 mM DL-dithiothreitol, 1 mM phenylmethanesulfonyl fluoride]. The homogenate was centrifuged at 15000 × g for 30 min at 4°C. The GPOD activity was estimated by monitoring the formation of tetraguaiacol at 470 nm for 3 min, as described previously by Stasolla and Yeung [35]. The reaction mixture consisted of 0.799 mL of 50 mM sodium phosphate buffer (pH 6.0), 0.001 mL of 3 M guaiacol, 0.1 mL of enzyme extract, and 0.1 mL of 0.3% (v/v) H_2O_2. The enzyme activity was calculated using the molar extinction coefficient for tetra-guaiacol (26.6 mM⁻¹ × cm⁻¹).

Foliar area and root length

Young and old leaves were scanned in black/white 200 dpi format using a HP F4280 scanner (Hewlett-Packard Company, Palo Alto, CA). The areas were calculated using Image J 1.43 software and the results were expressed in cm². Lengths of the longest roots were measured using a ruler.

Statistical analysis

The results reported in tables and figures represent mean values ±*SE* of at least two independent experiments. For data related to copper content, leaf area, root length, chlorophyll content and betacyanin content in young leaves and roots, one-way analysis of variance (ANOVA) was performed and data transformation was used when was necessary (ln*x*). Pairwise comparisons were done using the Tukey test at $p < 0.05$. Kruskal-Wallis analysis followed by Dunn's test for comparisons of means was used only for analyses of betacyanin contents in old leaves. Student's t-test was used to evaluate the statistical significance of any differences in the GPOD activity data. All statistical analyses were done using the statistical software package SigmaStat ver. 3.5 (Systat Software Inc., San Jose, CA).

Results

Copper content in roots and leaves

At the beginning of treatment (day 0), young and old leaves of control plants (0.3 µM) showed a copper content of 13.30 ±3.16 and 15.09 ±1.34 µg × g⁻¹ *DW*, respectively. The Tab. 1 shows that these copper contents in leaves of control plants did not change during the 6 days of incubation. After treatment of copper for 3 days, only 50 µM of copper caused in young leaves an increase of 34% in the copper content. Whereas, the copper content in old leaves was not modified after treatment with copper for 3 and 6 days.

The copper content of old roots in control plants (day 0) was 71.30 ±15.67 µg × g⁻¹ *DW*. This copper content was ca. 4 times higher than those found in young and old leaves of control plants (day 0). Nonetheless, accumulation of the metal was evident in old roots from day 3 of treatment, and the extent of the accumulation was proportional to the increase of copper concentration in the nutrient solution. The maximum copper accumulation was found in old roots of plants treated with 250 µM of copper for 6 days. This copper content is 208-fold higher than that found in control roots (Tab. 1). In general, old roots had copper contents 1.8, 85.6, 221.5, and 480.2-fold higher than those found in leaves of plants treated with 0.3 µM, 50 µM, 100 µM, and 250 µM of copper for 6 days, respectively. Due young roots were developed after 3 days of treatment, the copper content only was determined at day 6 and the highest copper content was found in plants treated with 250 µM of copper (Tab. 1).

Tab. 1 Copper content in leaves and roots of *Beta vulgaris* L. plants treated with copper for 3 and 6 days.

CuSO₄ (µM)	Copper content (µg × g⁻¹ *DW*)			
	Leaves		Roots	
	Old	Young	Old	Young
Day 3				
0.3	17.05 ±3.26ᵃ	14.09 ±0.20ᵃ	35.12 ±3.88ᵃ	nd
50	18.40 ±2.32ᵃ	18.84 ±0.92ᵇ	730.12 ±251.75ᵇ	nd
100	23.19 ±2.67ᵃ	16.39 ±1.51ᵃᵇ	1860.47 ±22.23ᵇᶜ	nd
250	13.94 ±1.91ᵃ	12.93 ± 0.89ᵃ	5333.69 ±1372.26ᶜ	nd
ANOVA: F	2.30 ns	9.29***	90.22***	
Day 6				
0.3	15.19 ±3.15ᵃ	14.72 ±2.59ᵃ	42.11 ±15.21ᵃ	nd
50	19.41 ±3.07ᵃ	18.41 ±2.68ᵃ	1637.27 ±214.08ᵇ	922.80 ±127.67ᵃ
100	16.85 ±3.08ᵃ	14.56 ±2.08ᵃ	3495.67 ±980.79ᵇᶜ	1815.07 ±207.68ᵃ
250	20.22 ±2.59ᵃ	16.42 ±2.81ᵃ	8815.56 ±1456.79ᶜ	1980.85 ±328.44ᵃ
ANOVA: F	0.56 ns	0.50 ns	76.88***	3.03 ns

Values are means ±*SE*. Mean values in the same column followed by different letters are significantly different according to Tukey's test ($p < 0.05$). One-way ANOVA shows statistical significant difference at *** $p < 0.001$. nd – not determined; ns – statistically not different.

Effect of copper on chlorophyll content

At the beginning of treatment (day 0) in control plants, the chlorophyll content in young leaves (7.85 µg × disco⁻¹) was 44 % higher than that found in older ones (5.45 µg × disco⁻¹). These chlorophyll contents in young and old leaves of control plants were not modified after 6 days of incubation (Tab. 2).

After copper treatment, young leaves were characterized by a turgid appearance, whereas old leaves appeared chlorotic and dry (Fig. 1a,b). The chlorophyll content in young leaves did not change with exposure to increasing concentrations of copper and was always higher than that found in the old leaves. In old leaves treated with 50 µM, 100 µM, and 250 µM of copper, the respective chlorophyll contents were 32%, 39%, and 57% lower of those in control plants (Tab. 2).

Tab. 2 Chlorophyll content in leaves and growth of leaves and roots in *Beta vulgaris* plants treated with copper for 6 days.

CuSO$_4$ (µM)	Chlorophyll (µg × disc⁻¹)	Leaf area (cm²)	Root length (cm)
		Young	
0.3	8.75 ±0.30ª	41.30 ±1.92ª	nd
50	9.59 ±0.55ª	36.26 ±1.48ᵃᵇ	nd
100	8.73 ±0.41ª	32.66 ±1.40ᵇᶜ	nd
250	9.26 ±0.36ª	29.28 ±0.94ᶜ	nd
ANOVA: F	0.81 ns	13.99***	
		Old	
0.3	5.00 ±0.36ª	5.10 ±0.14ª	21.85 ±0.61ª
50	3.39 ±0.72ᵇ	5.31 ±0.30ª	17.06 ±0.72ᵇ
100	3.04 ±0.26ᵇᶜ	5.40 ±0.29ª	15.47 ±0.78ᵇ
250	2.14 ±0.26ᶜ	4.59 ±0.32ª	15.76 ±0.74ᵇ
ANOVA: F	16.72***	1.63 ns	16.77***

Values are means ±*SE*; chlorophyll content: n = 6–9; leaf area: n = 13–21; roots growth: n = 14–21. Mean values in the same column followed by different superscript letters are significantly different according to Tukey's test ($p < 0.05$). One-way ANOVA shows statistical significant difference at *** $p < 0.001$. nd – not determined; ns – statistically not different.

Differential effects of copper on leaf and root growth

Leaf growth was differentially affected by copper, depending on the age of the leaves. Young leaves of plants treated with 100 and 250 µM of copper showed reductions in foliar surface areas of 21% and 29%, respectively, in comparison to young leaves from control plants. In contrast, foliar surface areas of old leaves did not change in plants treated with copper (Tab. 2).

The root growth was reduced by 22%, 29%, and 28% following exposure to 50 µM, 100 µM, and 250 µM of copper, respectively (Tab. 2). Another difference was the formation of young roots at the top of the principal roots of plants exposed to all concentrations of copper tested.

Effect of copper on lipid peroxidation

In control plants incubated for 3 and 6 days, the level of lipid peroxidation was 1.4-fold higher in old leaves than in young leaves (Fig. 2a,b). The lipid peroxidation in young leaves was increased 3.9, 3.4 and 5.8 times with exposure to 50 µM, 100 µM, and 250 µM of copper for 3 days. Even though the average value of lipid peroxidation in young leaves treated for 3 days with 250 µM of copper is higher than that found with 100 µM of copper, data of treatment with 250 µM have high dispersion and did not show statistically significant difference.

The lipid peroxidation in young leaves of plants treated with 50 and 100 µM of copper for 3 days remained at the same level until day 6 (Fig. 2a). However, the level lipid peroxidation was reduced in young leaves of plants exposed to 250 µM of copper for 6 days. Unlike young leaves, old leaves did not show any change in the lipid peroxidation with copper treatment (Fig. 2b).

Differential accumulation of betacyanin

Young and old leaves of control plants (0.3 µM) showed a low level of betacyanin after 3 and 6 days of incubation (Fig. 3). Copper treatment induced in young leaves the development of red pigmentation associated with the betacyanin content (Fig. 1a). Young leaves accumulated betacyanin after 3 days of copper treatment. Maximum accumulation was observed after 6 days of copper treatment, with betacyanin contents 5.2-, 5.7-, and 6.4-fold higher than that found in the control (Fig. 3a). In contrast, the level of betacyanin in old leaves increased until day 6. Betacyanin contents were 3.6, 4.8, and 3 times lower than those found in young leaves at the same copper concentrations (Fig. 3b).

At the beginning of treatment in control plants, old roots showed a yellow-red pigmentation and betacyanin contents did not change during the incubation period. After copper treatment, old roots developed a dark-brown pigmentation and this change in pigmentation correlated with a decrease in betacyanin content (Fig. 4). In contrast, young roots developed in plants treated with copper for 6 days showed a red-violet pigmentation and accumulated betacyanin at levels close to 30 mg × g⁻¹ *FW* with all copper concentrations (Fig. 1d, Fig. 4). This mean betacyanin content was 2.6- and 5.9-fold higher than the highest betacyanin contents found in leaves and old roots, respectively (Fig. 3, Fig. 4).

Activity of GPOD

At the beginning of treatment (day 0), young and old leaves of the control plants showed GPOD activities of 0.15 and 0.07 µmoles × min⁻¹ × g⁻¹ *FW*, respectively. This GPOD activity in old leaves increased 2.8-fold after 6 days of incubation, whereas in young leaves did not change (Fig. 5).

The treatment with 50 µM of copper causes a negative effect on root growth, chlorophyll content and lipid peroxidation. Likewise, this copper concentration induced betacyanin accumulation. Thus, we decided that this concentration is adequate to evaluate defense responses to copper stress, such as GPOD activity. GPOD activity in young leaves increased 3.3-fold with the copper exposure for 3 days, and was sustained until day 6 (Fig. 5a). GPOD activity was also enhanced 1.4-fold after 3 days of copper treatment and did not change at day 6 (Fig. 5b).

Discussion

Plants have developed mechanisms to maintain the appropriate concentrations of metals ions in different plant organs and cellular compartments. The exposure to high levels

Fig. 1 Chlorosis, red pigmentation and growth of plants treated with copper (Cu) for 6 days. **a** Young leaves. **b** Old leaves. **c** Control roots. **d** Roots treated with copper.

essential metals ions, such as copper activates in plants the homeostatic mechanism to minimize the toxic effects of metal. The retention of metals in roots to prevent their translocation to shoots is a strategy used by many plants to tolerate heavy metal stress [36]. The accumulation of copper was evident in *B. vulgaris* roots with exposure to copper for 6 days, but the metal was not translocated to leaves (Tab. 1). Similar results have been reported for *Brassica napus*, where copper content increased in the roots, but not in the leaves throughout the course of exposure to elevated levels of copper [37]. Lequeux et al. [6] suggested that the lignin accumulation in cell walls of endodermis or xylem limits copper efflux from the vascular cylinder to the shoot. Accordingly, exposure of *M. chamomilla* plants to copper was associated with increased lignin deposition in the roots [23]. Forty percent of the copper retained in the roots of rice plants results from immobilization of the metal in the cell wall [38]. Cysteine-rich proteins that bind copper to form stable complexes have been detected in the cell walls, vesicles, and cytoplasm of meristematic cells in the root of *Allium sativum* after copper treatment [5].

Old leaves of control plants had consistently less chlorophyll than young leaves. The observation that the level of lipid peroxidation in old leaves was higher than that found in young leaves at either day 3 or day 6 (Tab. 2, Fig. 2) further suggests that the old leaves used in this study may already have been in the early stages of senescence when treatment began. Given that senescence is a catabolic process that leads to nutrient recycling, chloroplast degeneration is one of the earliest events in the process. It is accompanied by a decrease in photosynthetic activity, chlorophyll degradation, and protein loss [39].

Chlorophyll content in old leaves of *B. vulgaris* plants decreased progressively with the increase of copper concentration, with the leaves turning completely yellow and sometimes dying at the highest copper concentration tested. In young leaves treated with copper, chlorophyll content did not change and the leaves remained turgid (Fig. 1, Tab. 2). Therefore, consistent with previous studies that demonstrated a connection between copper-induced senescence and the developmental stage of the leaves [9,22]; we observed that copper accelerated the senescence of old leaves, but not had effect on young leaves.

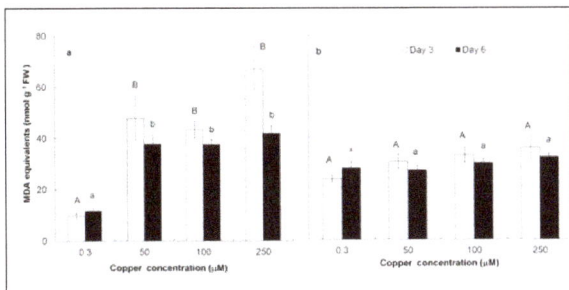

Fig. 2 **a** Lipid peroxidation in young leaves of *Beta vulgaris* L. plants treated with copper for 3 days (white bars) and 6 days (black bars); **b** in old leaves. Values are means ±*SE*; young leaves: *n* = 5–11, old leaves: *n* = 6–12. Different letters above the bars indicate statistical significant differences between treatments ($p < 0.05$) according to Tukey's test.

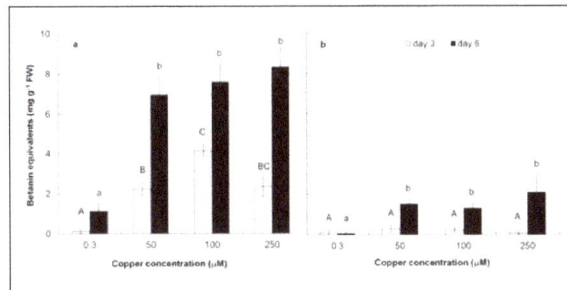

Fig. 3 **a** Betacyanin content in young leaves of *Beta vulgaris* L. plants treated with copper for 3 days (white bars) and 6 days (black bars); **b** in old leaves. Values are means ±*SE* (*n* = 4–10). Different letters above bars indicate statistical significant differences between treatments ($p < 0.05$) according to Tukey's test.

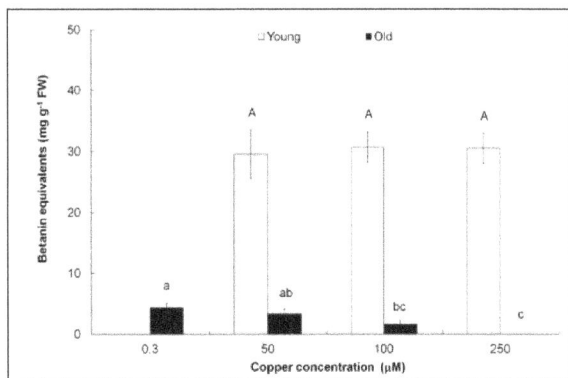

Fig. 4 Betacyanin content in young (white bars) and old (black bars) roots of *Beta vulgaris* L. plants treated with copper for 6 days. Values are mean $\pm SE$ (n = 5–9). Different letters above bars indicate statistical significant differences between treatments ($p < 0.05$) according to Tukey's test.

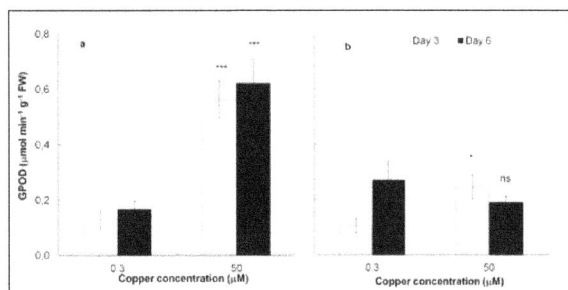

Fig. 5 **a** GPOD activity in young leaves of *Beta vulgaris* L. plants treated with copper for 3 days (white bars) and 6 days (black bars); **b** in old leaves. Values are mean $\pm SE$; young leaves: n = 6; old leaves: n = 3–7. * and *** – statistically different from the control at $p < 0.05$ and $p < 0.001$; ns – statistically not different, according to Student's t-test ($p < 0.05$).

Although copper was not translocated to leaves, the differential betacyanin accumulation was evident in young and old leaves from plants treated with copper, whereas betacyanin levels remained unchanged in young and old leaves of control plants (Tab. 1, Fig. 3). The accumulation of betacyanin in young leaves, which was evident at day 3 and day 6 after the beginning of treatment, was accompanied by increases in lipid peroxidation (Fig. 2a, Fig. 3a). In old leaves, betacyanins accumulated until day 6, although their levels were lower than those found in young leaves from plants exposed to the same copper concentrations. Moreover, copper had no effect on rates of lipid peroxidation in old leaves (Fig. 2b, Fig. 3b). These results suggest that betacyanin accumulation in young leaves of *B. vulgaris* may be a defense response to copper-induced oxidative stress, and that a signal molecule may be transferred from roots to leaves. Accordingly, Russo et al. [37] reported an increase in the level of reduced glutathione and the induction of glutathione reductase activity in *Brassica napus* leaves, where no translocation of copper occurred. In *Phaseolus vulgaris* leaves, increases in the amount of reduced ascorbate and the activities of the enzymes monodehydroascorbate reductase and dehydroascorbate reductase were observed before the

accumulation of copper in leaves [40]. Studies involving *A. thaliana* plants exposed to copper suggest that oxylipins or jasmonates could be involved in this interorgan signaling through the lipoxygenase genes expression [41,42].

Results of this study shown that organ age also affected the accumulation of betacyanin in *B. vulgaris* roots exposed to copper. Betacyanin content decreased in old roots with the increase in the copper content. However, new roots that were formed as result of copper treatment showed the highest betacyanin content of any of the plant organs we tested (Fig. 4). Similar to the copper effect observed in leaves, results in roots suggest that betacyanin are accumulated in actively growing organs. Accordingly, Skórzyńska-Polit et al. [24] found that copper induced higher rates of flavonol accumulation in young leaves than in old leaves of *P. coccineus* plants.

Age-dependent differences in the effects of copper stress on leaves have been reported for other factors capable of causing oxidative stress. In *Pisum sativum* plants exposed to the herbicide paraquat, the youngest and more photosynthetically active leaves were less susceptible to injury than older leaves, possibly as a result of higher basal levels of transcripts and activities of the antioxidant enzymes superoxide dismutase and glutathione reductase [43]. Relative to younger leaves, old tobacco leaves showed constitutively lower levels of antioxidant compounds and higher levels of H_2O_2 that resulted in photooxidative stress following exposure to paraquat [44]. Non-enzymatic antioxidants were also activated differentially in young and mature leaves of *A. thaliana* subjected to drought stress. Young leaves showed an increase in anthocyanin content and less photooxidative stress [45]. All of these studies showed that young leaves have an active defense system that ensures plant survival under stressful conditions.

The results obtained in this work show for first time that betacyanin accumulation is induced by copper stress in planta. Secondary metabolites may contribute to antioxidant defense responses to copper stress in plants. Copper-induced accumulation of chlorogenic acid in *M. chamomilla* leaves could prevent lipid peroxidation [12]. Pre-treatment of leaf segments with the polyamine spermine prevented the loss of glutathione reductase enzyme activity and reduced increases in levels of lipid peroxidation and H_2O_2 content induced by copper [13].

It has been suggested that betacyanin accumulation in *B. vulgaris* plants may contribute to counteracting the oxidative stress caused by wounding and bacterial infiltration [29]. Similarly, watering the roots of *Suaeda salsa* with H_2O_2 induces betacyanin synthesis in leaves [31]. However, other components of the antioxidant system of *B. vulgaris* L. plants may also be involved in the higher tolerance to copper stress of young organs compared with older ones. Our results show that induction of GPOD activity and lipid peroxidation in copper-treated plants was higher in young leaves than in old leaves (Fig. 2a, Fig. 5a). Consistent with these findings, Mediouni et al. [10] reported an increase in both levels of MDA and GPOD activity in leaves of *Solanum lycopersicon* plants exposed to elevated levels of copper. Some authors suggested a role for GPOD enzymes in scavenging ROS or mediating the lignification of cell walls in copper exposed plants [11,17,18]. Sgherri et al. [14] have proposed a hypothetical H_2O_2 scavenging system that comprises phenolics and peroxidase enzymes. The antioxidant activity of betacyanin has been demonstrated in vitro, and betacyanin oxidation by peroxidases of *B. vulgaris* roots has been described [26,27,46]. The betacyanin-peroxidase system may thus be part of an antioxidant response in *B. vulgaris* plants to copper stress. However, studies about the activation of enzymes involved on betacyanin synthesis and the levels of ROS are needed to analyze the role of these metabolites in tolerance of young leaves to metal stress.

The inhibition of leaf growth of *B. vulgaris* plants also varied with age. Whereas copper inhibited the growth of young leaves, the sizes of old leaves did not change upon exposure to copper (Tab. 2). These results indicated a differential effect with the leaf age. The most likely explanation for this observation is that whereas old leaves were totally expanded when the copper treatment was applied, young leaves were in the process of expanding. Pasternak et al. [3] attributed the decrease in growth of *A. thaliana* leaves to the effect of copper on cell expansion. Accordingly, leaves of *Phaseolus coccineus* in a stage of intensive growth showed reduced foliar surface areas and fresh weights after copper treatment. The metal caused chlorosis in older leaves but the foliar surface areas were unaffected [19]. The results in *B. vulgaris* leaves suggest the participation of a signal molecule, which is transported from roots to leaves and triggers a series of events leading to inhibition of leaf expansion in young leaves and accelerated senescence in old leaves.

Inhibition of root growth in *B. vulgaris* L. correlated with the increasing copper accumulation in this organ (Tab. 1, Tab. 2). Previous studies showed that copper accumulation in *Zea mays* roots led to the inhibition of cell division and root growth [4]. Moreover, growth inhibition of primary and lateral roots in copper-exposed *A. thaliana* plants was related to a decrease in viability of the apex cells [6]. Copper exposure also inhibited cell elongation, thereby reducing root diameter in *Sorghum bicolor* [47]. A decrease in the mitotic indices of meristematic cells of *A. sativum* roots was accompanied by degenerative changes that finally led to disruption of nuclear membranes, disintegration of organelles and cell death [5].

Elevated levels of copper in *B. vulgaris* L. induced the emergence of new short roots with red pigmentation (Fig. 1d). In addition to phytotoxic effects on roots, copper excess triggers reorganization of the radical system. According to Lequeux et al. [6], changes in the distributions of auxin and cytokinin are involved in remodeling of the *A. thaliana* root system. Furthermore, treatment with tert-butyl hydroperoxide, an analogue of H_2O_2, resulted in similar root architecture to that obtained in *A. thaliana* plants treated with copper, suggesting the participation of ROS in the control of these morphological changes induced by stress [3].

We conclude that the copper accumulated in *B. vulgaris* L. roots may activate a signaling process to the leaf and triggers a series of events that promote oxidative damage in young leaves and accelerate the senescence of older leaves. The betacyanin accumulation and GPOD activity in actively growing organs are induced as a defense response to oxidative damage caused by copper.

Acknowledgements

This research project was financially supported by SIP (Grant 20111038) and CONACYT (Grant 49950-Z). We thank Dr. A. René Arzuffi Barrera for assisting in statistical analyses and Dr. Kalina Bermúdez Torres for helpful discussions. Janet M. León Morales thanks CONACYT (171078) and PIFI-IPN for fellowships awarded. G. Sepúlveda-Jiménez and M. Rodríguez-Monroy thank EDI and SIBE-IPN.

References

1. Nagajyoti PC, Lee KD, Sreekanth TVM. Heavy metals, occurrence and toxicity for plants: a review. Environ Chem Lett. 2010;8(3):199–216.

2. Salisbury FB, Ross CW. Plant physiology. Belmont CA: Wadsworth Pub. Co.; 1992. (vol 1).

3. Pasternak T, Rudas V, Potters G, Jansen M. Morphogenic effects of abiotic stress: reorientation of growth in seedlings. Environ Exp Bot. 2005;53(3):299–314.

4. Jiang W, Liu D, Liu X. Effects of copper on root growth, cell division, and nucleolus of *Zea mays*. Biol Plant. 2001;44(1):105–109.

5. Liu D, Jiang W, Meng Q, Zou J, Gu J, Zeng M. Cytogenetical and ultrastructural effects of copper on root meristem cells of *Allium sativum* L. Biocell. 2009;33:25–32.

6. Lequeux H, Hermans C, Lutts S, Verbruggen N. Response to copper excess in *Arabidopsis thaliana*: impact on the root system architecture, hormone distribution, lignin accumulation and mineral profile. Plant Physiol Biochem. 2010;48(8):673–682.

7. Zhang H, Xia Y, Wang G, Shen Z. Excess copper induces accumulation of hydrogen peroxide and increases lipid peroxidation and total activity of copper-zinc superoxide dismutase in roots of *Elsholtzia haichowensis*. Planta. 2007;227(2):465–475.

8. Maksymiec W. Signaling responses in plants to heavy metal stress. Acta Physiol Plant. 2007;29(3):177–187.

9. Drażkiewicz M, Skórzyńska-Polit E, Krupa Z. Copper-induced oxidative stress and antioxidant defence in *Arabidopsis thaliana*. BioMetals. 2004;17(4):379–387.

10. Mediouni C, Ammar WB, Houlné G, Chabouté ME, Jemal F. Cadmium and copper induction of oxidative stress and antioxidative response in tomato (*Solanum lycopersicon*) leaves. Plant Growth Regul. 2008;57(1):89–99.

11. Khatun S, Ali MB, Hahn EJ, Paek KY. Copper toxicity in *Withania somnifera*: Growth and antioxidant enzymes responses of in vitro grown plants. Environ Exp Bot. 2008;64(3):279–285.

12. Kováčik J, Grúz J, Bačkor M, Tomko J, Strnad M, Repčák M. Phenolic compounds composition and physiological attributes of *Matricaria chamomilla* grown in copper excess. Environ Exp Bot. 2008;62(2):145–152.

13. Groppa MD, Tomaro ML, Benavides MP. Polyamines and heavy metal stress: the antioxidant behavior of spermine in cadmium- and copper-treated wheat leaves. BioMetals. 2006;20(2):185–195.

14. Sgherri C, Cosi E, Navari-Izzo F. Phenols and antioxidative status of *Raphanus sativus* grown in copper excess. Physiol Plant. 2003;118(1):21–28.

15. Jouili H, Bouazizi H, El Ferjani E. Plant peroxidases: biomarkers of metallic stress. Acta Physiol Plant.

2011;33(6):2075–2082.

16. Chaoui A, Jarrar B, EL Ferjani E. Effects of cadmium and copper on peroxidase, NADH oxidase and IAA oxidase activities in cell wall, soluble and microsomal membrane fractions of pea roots. J Plant Physiol. 2004;161(11):1225–1234.

17. Zhang H, Zhang F, Xia Y, Wang G, Shen Z. Excess copper induces production of hydrogen peroxide in the leaf of *Elsholtzia haichowensis* through apoplastic and symplastic CuZn-superoxide dismutase. J Hazard Mater. 2010;178(1–3):834–843.

18. Lin CC, Chen LM, Liu ZH. Rapid effect of copper on lignin biosynthesis in soybean roots. Plant Sci. 2005;168(3):855–861.

19. Maksymiec W, Russa R, Urbanik-Sypniewska T, Baszynski T. Effect of excess Cu on the photosynthetic apparatus of runner bean leaves treated at two different growth stages. Physiol Plant. 1994;91(4):715–721.

20. Vinit-Dunand F, Epron D, Alaoui-Sossé B, Badot PM. Effects of copper on growth and on photosynthesis of mature and expanding leaves in cucumber plants. Plant Sci. 2002;163(1):53–58.

21. Luna CM, González CA. Oxidative damage caused by an excess of copper in oat leaves. Plant Cell Physiol. 1993;35:11–15.

22. Tewari RK, Kumar P, Sharma PN. Antioxidant responses to enhanced generation of superoxide anion radical and hydrogen peroxide in the copper-stressed mulberry plants. Planta. 2005;223(6):1145–1153.

23. Kováčik J, Bačkor M, Kaduková J. Physiological responses of *Matricaria chamomilla* to cadmium and copper excess. Environ Toxicol. 2008;23(1):123–130.

24. Skórzyńska-Polit E, Drażkiewicz M, Wianowska D, Maksymiec W, Dawidowicz AL, Tukiendorf A. The influence of heavy metal stress on the level of some flavonols in the primary leaves of *Phaseolus coccineus*. Acta Physiol Plant. 2004;26(3):247–254.

25. Strack D, Steglich W, Wray V. Betalains. In: Waterman PG, editor. Methods in plant biochemistry. New York: Academic Press; 1993. p. 421–450. (vol 8).

26. Escribano J, Pedreño MA, García-Carmona F, Muñoz R. Characterization of the antiradical activity of betalains from *Beta vulgaris* L. roots. Phytochem Anal. 1998;9(3):124–127.

27. Kanner J, Harel S, Granit R. Betalains: a new class of dietary cationized antioxidants. J Agric Food Chem. 2001;49(11):5178–5185.

28. Ibdah M, Krins A, Seidlitz HK, Heller W, Strack D, Vogt T. Spectral dependence of flavonol and betacyanin accumulation in *Mesembryanthemum crystallinum* under enhanced ultraviolet radiation. Plant Cell Environ. 2002;25(9):1145–1154.

29. Sepúlveda-Jiménez G, Rueda-Benítez P, Porta H, Rocha-Sosa M. Betacyanin synthesis in red beet (*Beta vulgaris*)

leaves induced by wounding and bacterial infiltration is preceded by an oxidative burst. Physiol Mol Plant Pathol. 2004;64(3):125–133.

30. Wang CQ, Zhao JQ, Chen M, Wang BS. Identification of betacyanin and effects of environmental factors on its accumulation in halophyte *Suaeda salsa*. J Plant Physiol Mol Biol. 2006;32:195–201.

31. Wang CQ, Chen M, Wang BS. Betacyanin accumulation in the leaves of C3 halophyte *Suaeda salsa* L. is induced by watering roots with H_2O_2. Plant Sci. 2007;172(1):1–7.

32. Lichtenthaler HK. Chlorophylls and carotenoids: pigments of photosynthetic biomembranes. Method Enzymol. 1987;148:350–382.

33. Alia, Prasad KVSK, Pardha Saradhi P. Effect of zinc on free radicals and proline in *Brassica* and *Cajanus*. Phytochemistry. 1995;39(1):45–47.

34. Schwartz SJ, Von Elbe JH. Quantitative determination of individual betacyanin pigments by high-performance liquid chromatography. J Agric Food Chem. 1980;28(3):540–543.

35. Stasolla C, Yeung EC. Cellular ascorbic acid regulates the activity of major peroxidases in the apical poles of germinating white spruce (*Picea glauca*) somatic embryos. Plant Physiol Biochem. 2007;45(3–4):188–198.

36. Fernandes JC, Henriques FS. Biochemical, physiological, and structural effects of excess copper in plants. Bot Rev. 1991;57(3):246–273.

37. Russo M, Sgherri C, Izzo R, Navari-Izzo F. *Brassica napus* subjected to copper excess: phospholipases C and D and glutathione system in signalling. Environ Exp Bot. 2008;62(3):238–246.

38. Chen CT, Chen TH, Lo KF, Chiu CY. Effects of proline on copper transport in rice seedlings under excess copper stress. Plant Sci. 2004;166(1):103–111.

39. Lim PO, Nam HG. Aging and senescence of the leaf organ. J Plant Biol. 2007;50(3):291–300.

40. Cuypers A, Vangronsveld J, Clijsters H. Biphasic effect of copper on the ascorbate-glutathione pathway in primary leaves of *Phaseolus vulgaris* seedlings during the early stages of metal assimilation. Physiol Plant. 2000;110(4):512–517.

41. Remans T, Opdenakker K, Smeets K, Mathijsen D, Vangronsveld J, Cuypers A. Metal-specific and NADPH oxidase dependent changes in lipoxygenase and NADPH oxidase gene expression in *Arabidopsis thaliana* exposed to cadmium or excess copper. Funct Plant Biol. 2010;37(6):532.

42. Cuypers A, Smeets K, Ruytinx J, Opdenakker K, Keunen E, Remans T, et al. The cellular redox state as a modulator in cadmium and copper responses in *Arabidopsis thaliana* seedlings. J Plant Physiol. 2011;168(4):309–316.

43. Donahue JL, Okpodu CM, Cramer CL, Grabau EA, Alscher RG. Responses of antioxidants to paraquat in pea leaves (relationships to resistance). Plant Physiol.

1997;113(1):249–257.

44. Ohe M, Rapolu M, Mieda T, Miyagawa Y, Yabuta Y, Yoshimura K, et al. Decline in leaf photooxidative-stress tolerance with age in tobacco. Plant Sci. 2005;168(6):1487–1493.

45. Jung S. Variation in antioxidant metabolism of young and mature leaves of *Arabidopsis thaliana* subjected to drought. Plant Sci. 2004;166(2):459–466.

46. Martínez-Parra J, Muñoz R. Characterization of betacyanin oxidation catalyzed by a peroxidase from *Beta vulgaris* L. roots. J Agric Food Chem. 2001;49(8):4064–4068.

47. Kasim WA. Changes induced by copper and cadmium stress in the anatomy and grain yield of *Sorghum bicolor* (L.) Moench. Int J Agri Biol. 2006;8:123–128.

2

Glutathione-dependent responses of plants to drought: a review

Mateusz Labudda[1]*, Fardous Mohammad Safiul Azam[2]

[1] Department of Biochemistry, Warsaw University of Life Sciences – SGGW, Nowoursynowska 159, 02-776 Warsaw, Poland

[2] Department of Biotechnology and Genetic Engineering, University of Development Alternative, 80 Satmasjid Road, Dhanmondi R/A, Dhaka-1209, Bangladesh

Abstract

Water is a renewable resource. However, with the human population growth, economic development and improved living standards, the world's supply of fresh water is steadily decreasing and consequently water resources for agricultural production are limited and diminishing. Water deficiency is a significant problem in agriculture and increasing efforts are currently being made to understand plant tolerance mechanisms and to develop new tools (especially molecular) that could underpin plant breeding and cultivation. However, the biochemical and molecular mechanisms of plant water deficit tolerance are not fully understood, and the data available is incomplete. Here, we review the significance of glutathione and its related enzymes in plant responses to drought. Firstly, the roles of reduced glutathione and reduced/oxidized glutathione ratio, are discussed, followed by an extensive discussion of glutathione related enzymes, which play an important role in plant responses to drought. Special attention is given to the S-glutathionylation of proteins, which is involved in cell metabolism regulation and redox signaling in photosynthetic organisms subjected to abiotic stress. The review concludes with a brief overview of future perspectives for the involvement of glutathione and related enzymes in drought stress responses.

Keywords: abiotic stress; glutathione peroxidase; glutathione reductase; glutathione S-transferase; GSH; reactive oxygen species; S-glutathionylation; water deficit

Introduction

Drought is considered to be one of the major components of abiotic stress. Water deficit inhibits photosynthesis, induces changes in chlorophyll content and composition, and damages the photosynthetic apparatus [1]. Moreover, dehydration of tissue causes a reduction in the activity of Calvin–Benson–Bassham cycle enzymes and inhibits photochemical activities [2].

It is well established that chloroplast, mitochondria and peroxisomes are a major source of reactive oxygen species (ROS) such as superoxide radicals($O^{\cdot-}_2$), hydroxyl radicals (OH$^{\cdot}$), singlet oxygen ($^1O_2$), hydrogen peroxide (H_2O_2) and peroxide radicals ($O_2^{\cdot 2-}$). ROS play a dual role in plant biochemistry and physiology. They are important secondary signaling molecules, but equally, they are toxic products of aerobic metabolism that accumulate within cells during oxidative stress [3].

The equilibrium between the generation and the enzymatic and non-enzymatic elimination of ROS may be disturbed by drought. During water deficit, these disturbances in equilibrium result in a sudden increase in cellular redox

potential, which can damage many cell components, including lipids, proteins, and nucleic acids [4,5].

The polyunsaturated fatty acid (PUFA) components of membrane phospholipids are especially susceptible to ROS activity. When ROS levels exceed the capacity of the plant to scavenge, lipid peroxidation (LP) in biological membranes increases. This is supported by data collected over a number of years for a range of plant species under water deficit conditions (Tab. 1). The final products of oxidative modification of lipids are responsible for cell membrane damage including changes to the intrinsic properties of the membrane, such as fluidity, ion transport, loss of enzyme activity and protein cross-linking. These changes eventually result in cell death [6].

The ascorbate-glutathione (AsA-GSH) pathway, also known as the Foyer–Halliwell–Asada cycle, is a central antioxidant defense system for the efficient scavenging of ROS and is thus important for the maintenance of redox homeostasis in plants tissues under stress conditions. Indeed, the AsA-GSH pathway is a key element in the network of biochemical reactions involving antioxidant enzymes and low molecular weight antioxidants with redox properties for the efficient elimination of ROS, and thereby prevents the ROS-mediated oxidative damage of plant tissues [7,8].

Of the low molecular weight antioxidants, tripeptide glutathione (GSH, γ-L-Glutamyl-L-cysteinylglycine) is

* Corresponding author. Email: mateusz_labudda@sggw.pl

Handling Editor: Grażyna Kłobus

Tab. 1 The lipid peroxidation processes induced by water deficit in various plants.

Species	Reference No.
Oats (*Avena* spp. L.)	[56]
Jutes (*Corchorus* spp. L.)	[57]
Apple trees (*Malus* spp. Mill.)	[58]
Maize (*Zea mays* L.)	[59–62]
Soya bean [*Glycine max* (L.) Merr.]	[5]
Rapeseed (*Brassica napus* L.)	[29]
Upland cotton (*Gossypium hirsutum* L.)	[63]
Wheat (*Triticum aestivum* L.)	[44,64,65]
Common bean (*Phaseolus vulgaris* L.)	[66]
Sesame (*Sesamum indicum* L.)	[67]
Chickpea (*Cicer arietinum* L.)	[68]
Ramie [*Boehmeria nivea* (L.) Gaudich.]	[69]
White poplar (*Populus alba* L.)	[70]
Black locust (*Robinia pseudoacacia* L.)	[70]
Japanese pagoda tree [*Styphnolobium japonicum* (L.) Schott]	[70]
Bee bee tree [*Tetradium daniellii* (Benn.) T.G. Hartley]	[70]
Lime (*Tilia* sp. L.)	[70]
Black ash (*Fraxinus* sp. L.)	[70]
Pistachio (*Pistacia vera* L.)	[71]
Cowpea [*Vigna unguiculata* (L.) Walp.]	[72]
Moroccan alfalfa (*Medicago sativa* L.)	[73]
Spikemoss [*Selaginella lepidophylla* (Hook. & Grev.) Spring]	[74]
Moss [*Dicranella palustris* (Dickson) E.F. Warburg]	[75]
Bermudagrass [*Cynodon dactylon* (L.) Pers.]	[76]
Strawberry (*Fragaria orientalis* Losinsk.)	[77]
Moss [*Tortula ruralis* (Hedw.) Gaertn.]	[30]

considered the most important defense thiol in the prevention of oxidative damage in plants. GSH acts as a disulphide reductant and protects protein thiol (–SH) groups, regenerates ascorbate and acts as the substrate for important GSH-metabolism enzymes such as glutathione peroxidases (GPXs, EC 1.11.1.9) and glutathione S-transferases (GSTs, EC 2.5.1.18; Tab. 2).

Plants maintain a high cellular ratio of GSH to its oxidized form GSSG (about 20:1 in unstressed conditions), but GSH reacts with oxidants during environmental stress and becomes converted into GSSG. The intracellular homoeostasis between GSH and GSSG ensures the signaling of a stress response and modulates plant tolerance to abiotic stress. Glutathione reductase (GR, EC 1.6.4.2) catalyzes the NADPH-dependent conversion of GSSG to its GSH form (Tab. 2). This reaction provides the molecules of GSH necessary for active protein function under non-stress and stress conditions [9–11].

Consequently, GPXs, GSTs and GR, in association with superoxide dismutases (SODs), catalase (CAT) and peroxidases, provide an effective way of defending plants against the potential effects of oxidative stress [12]. The components of cellular "glutathione machinery" for the control of plant responses to different abiotic stresses, including drought, are summarized in Fig. 1.

The following review describes recent studies of changes in total reduced GSH, glutathione redox state, the key GSH-related enzymes and their significance in plant responses to water deficit.

Glutathione in the drought response of plants

GSH and GSH/GSSG ratio

GSH or GSH homologues are present in all plant species, where the C-terminal glycine is replaced by other amino acid, for example, glutamate, β-alanine or serine. GSH is produced in two steps. Firstly, γ-glutamyl-cysteine is synthesised in an ATP-dependent reaction catalyzed by glutamate-cysteine ligase (γ-GCL, EC 6.3.2.2). Then, glutathione synthetase (GSS, EC 6.3.2.3) catalyzes the addition of glycine to γ-glutamyl-cysteine [13]. Biosynthesis of GSH takes place in the chloroplasts, mitochondria and cytosol [14], and both enzymatic proteins are encoded by single genes with alternate transcription start points related to their subcellular localization [15]. Results collected over a number of years confirm that glutathione related parameters change in various plant species subjected to water deficit conditions (Tab. 3).

Tab. 2 Summary of the glutathione-dependent enzymes, reactions catalyzed, function and their tissue localization.

Enzyme	Reactions catalyzed	Function	Localization[z]
Glutathione peroxidase (GPXs)	$H_2O_2 + 2GSH \rightarrow 2H_2O + GSSG$	Detoxifies H_2O_2 and lipid hydroperoxides with GSH as reductor.	cyt, chl, mit, er
Glutathione S-transferase (GSTs)	$RX + GSH \rightarrow HX + R\text{-}S\text{-}S\text{-}GSH$[y]	Detoxifies lipid hydroperoxides and exhibit DHAR activity. Acts as non-catalythic cariers that facilitate the distribution and transport of various molecules. Degluthathionylation.	apo, cyt, chl, mit, nuc
Glutathione reductase (GR)	$GSSG + NAD(P)H \rightarrow 2GSH + NAD(P)^+$	Reduces GSSG with NADPH as the reductor.	cyt, chl, mit, per

[y] R may be an aromatic, heterocyclic or aliphatic group; X may be a halide, nitrite or sulphate group. [z] Gechev et al. [78] and Anjum et al. [25] are used as references for localization of enzymes. apo – apoplast; ch – chloroplasts; cyt – cytosol; DHAR – dehydroascorbate reductase; er – endoplasmic reticulum; mit – mitochondria; nuc – nucleus; per – peroxisomes.

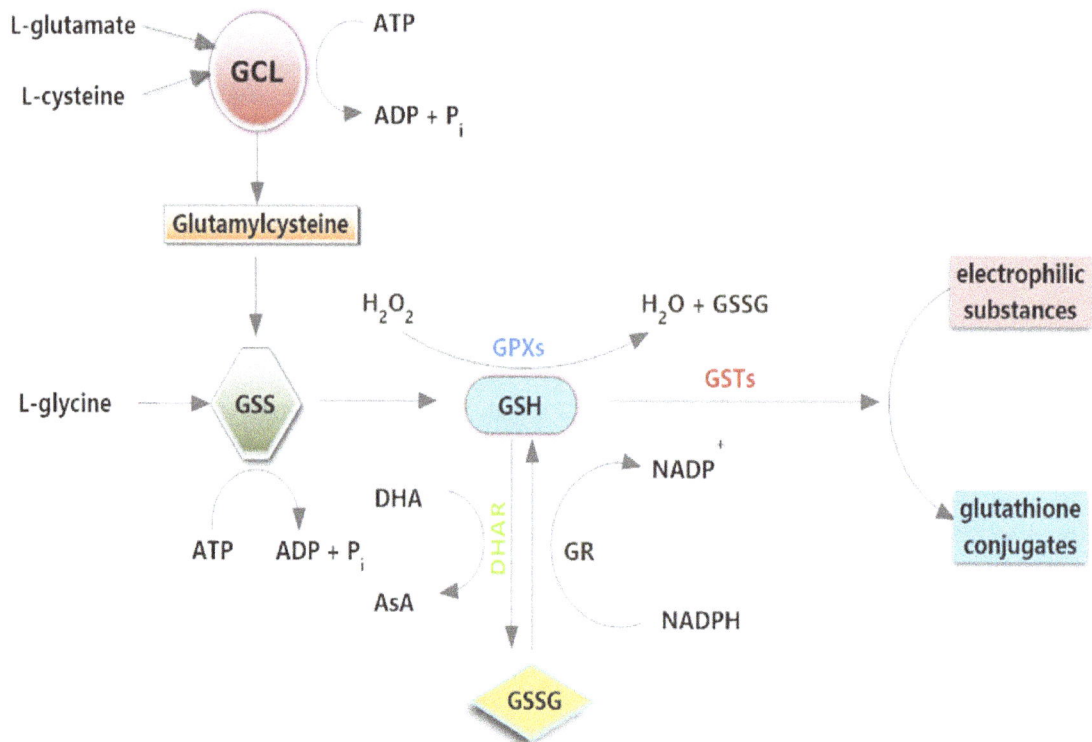

Fig. 1 The cellular "glutathione machinery" in plant responses to abiotic stress.

Sengupta et al. [16] demonstrated a decline in γ-glutamyl-cysteine synthetase activity and its transcript levels in the roots of mung bean [*Vigna radiata* (L.) Wilczek] during long-term water deficit. This is incompatible with the hypothesis that abiotic stress tolerance is associated with an increase in γ-glutamyl-cysteine synthetase level and activity, together with increases in GSH and Cys concentrations, as demonstrated for salt stress [17]. It should be noted that loss of function of γ-glutamyl-cysteine synthetase proved lethal during early developmental stages, and GSH deficiency resulted in increased sensitivity to cadmium in *Arabidopsis thaliana* [18].

Homoglutathione (hGSH), which is characteristic of members of the family Fabaceae, is a homologue of GSH in which the C-terminal glycine is replaced by β-alanine and it has the same functions as GSH [13]. This compound is an important regulator of nodulation, nitrogen fixation and symbiotic interactions, has antioxidant potential and is involved in the transport of reduced sulphur [11,19]. Researchers have reported increased hGSH synthetase (hGSS) mRNA levels in the leaves of a drought-tolerant cultivar (EPACE-1) of cowpea [*Vigna unguiculata* (L.) Walp.] during drought stress and desiccation [20]. By contrast, however, water deficit was shown to have no significant effect on the concentrations of GSH and hGSH in the nodules of alfalfa (*Medicago sativa* L.) [21].

Experimental studies indicate that GSH concentration increases in response to water deficit in sunflower (*Helianthus annuus* L. cv. Licia Stella) [22]. Furthermore, Herbinger et al. [23] showed that the concentration of GSH increased in flag leaf tissues of drought-sensitive wheat (*Triticum aestivum* Desf. cv. Nandu) and drought-resistant durum wheat (*Triticum durum* L. cv. Extradur) cultivars grown in open-top chambers using a water regime equivalent to 40% soil water capacity.

Pyngrope et al. [24] reported a consistent decline in GSH in a drought-sensitive cultivar (Malviya-36) of indica rice (*Oryza sativa* L.) in response to an increase in the intensity and duration of water deficiency. However, such changes in GSH level were not detected in the roots of a drought-tolerant cultivar (Brown Gora) subjected to water deficit, even though a statistically significant reduction in GSH levels was observed when its shoots were subjected to an osmotic potential of −2.1 MPa for 72 h. Conversely, drought-sensitive seedlings treated with 30% (v/v) polyethylene glycol (PEG-6000) in order to achieve an osmotic potential of −2.1 MPa for 72 h showed a 41% reduction in root GSH and a 61% reduction in shoot GSH, whereas similarly stressed tolerant plants showed a 22% reduction in shoot GSH concentration compared with the controls.

Many reports indicate that the GSH/GSSG ratio is an effective marker of cellular redox homeostasis and may be involved in ROS activity perception by plants. In this way, GSH/GSSG may have a direct or indirect key role in regulating and signaling at the transcriptional and/or post-translational level due to the interaction of these molecules with other cellular redox systems such as glutaredoxin, thioredoxin, peroxiredoxin and, mitogen-activated protein kinases (MAP kinases) [25].

Tab. 3 Summary of modulation of glutathione and its dependent enzymes in plant responses to water deficit.

Parameter	Species	Response	Reference No.
GSH pool	*Helianthus annuus*	+	[22]
	Triticum durum and *Triticum aestivum*	+	[23]
	Oryza sativa	−/×	[24]
	Medicago sativa	×	[21]
	Brassica campestris	+	[27]
	Brassica juncea	+	[28]
	Brassica napus	−	[29]
hGSH pool	*Medicago sativa*	×	[21]
GSSG pool	*Brassica campestris*	+	[27]
	Brassica juncea	+	[28]
	Tortula ruralis	+	[30]
total glutathione pool	*Brassica napus*	−	[29]
GSH/GSSG ratio	*Pinus canariensis*	−	[26]
	Brassica campestris	−	[27]
	Brassica napus	+	[29]
GCL (activity and mRNA level)	*Vigna radiata*	−	[16]
hGSS mRNA level	*Vigna unguiculata*	+	[20]
GPXs activity	*Helianthus annuus*	+	[31]
	Glycine max	+	[79]
	Beta vulgaris	+	[33]
	Cicer arietinum	+	[68]
GSTs activity	*Zea mays*	+	[36]
GSTs (activity and mRNA level)	*Triticum aestivum*	+	[37]
GR activity	*Gossypium hirsutum* and *Gossypium barbadense*	+	[41]
	Anoda cristata	+	[41]
	Populus przewalskii	+	[42]
	Robusta coffee	+/−	[43]
	Brassica napus	+	[29]
	Lycopersicon esculentum	+	[45]
	Pisum sativum	−	[47]
	Triticum aestivum	+	[46]

"−", "+" and "×" signs indicate decrease, increase, or unaltered/unaffected, respectively (see main text for more details).

Tausz el al. [26] demonstrated a slight reduction in the GSH/GSSG ratio of the needles of a species of pine tree (*Pinus canariensis* Chr. Sm. ex DC) exposed to short-term, moderate drought. The authors concluded that the glutathione redox cycling and the equilibrium between GSH and GSSG are sensitive elements of the antioxidative response in pine tree needles and suggested that they possibly have a role in longer-term adaptation processes.

Hossain et al. [27] showed that the concentration of GSSG increased in mustard (*Brassica campestris* L.) seedlings treated with 20% (v/v) polyethylene glycol (PEG-6000) in order to achieve drought stress. The authors speculated that the formation of GSSG under drought stress might be due to the reaction of GSH with oxyradicals generated by oxidative stress, antioxidative enzyme activity that decomposes H_2O_2 and organic hydroperoxide or an insufficient increase in glutathione reductase activity.

Furthermore, in seedlings of a second species of mustard (*Brassica juncea* L. cv. BARI Sharisha 11) subjected to short-term drought stress conditions, GSH levels increased by 32%

and 25% with 10% (v/v) and 20% (v/v) PEG, respectively. Conversely, the concentration of GSSG increased significantly in response to increased levels of water deficit and it was demonstrated that, compared with control plants, the GSSG pools were 48% and 101% greater at 10% (v/v) and 20% (v/v) PEG, respectively [28].

Liu et al. [29] assessed the effect of 5-aminolevulinic acid (ALA) on the growth of oilseed rape (*Brassica napus* L. cv. ZS758) seedlings under water deficit (−0.3 MPa) conditions induced by PEG-6000 treatment and demonstrated that dehydration of the tissues significantly reduced GSH and total glutathione levels, while simultaneously increasing the GSH/GSSG ratio. The compound 5-aminolevulinic acid is an important precursor of tetrapyrrols, such as porphyrins for the synthesis of chlorophylls and heme groups. Recently, it has been suggested that a low concentration of exogenous ALA has a beneficial effect on abiotic stress tolerance/resistance, regulates plant growth and increases the yields of crops. In the afore-mentioned publication, treatment with 0.1–10 mg l^{-1}ALA remarkably improved GSH levels, the total glutathione pool and, in particular, the GSH/GSSG ratio, which increased by at least 70% in relation to control oilseed rape seedlings under water deficit conditions.

Other studies demonstrated that GSSG levels increased in gametophytes of the drought-tolerant moss *Tortula ruralis* (Hedw.) Gaertn. subjected to water deficit. Moreover, it was observed that GSSG content was correlated negatively with protein synthesis and positively with lipid peroxidation. The author claims that the GSSG level is a good biochemical indicator of oxidative stress induced by drought and suggests that the oxidized glutathione mediates, at least in part, the water deficiency-induced inhibition of protein synthesis [30].

Glutathione peroxidases

Glutathione peroxidases (GPXs, EC 1.11.1.9) are a diverse group of isozymes having generous substrate spectrum and serve as antioxidant enzymes. They occur in plant cells in the cytosol, chloroplasts, mitochondria, and the endoplasmic reticulum, and catalyze the detoxification of H_2O_2 and lipid hydroperoxides with GSH as reductor, and thus protect biomolecules from oxidative damage (Fig. 1, Tab. 2, Tab. 3). It is now known that plant glutathione peroxidases exhibit substrate specificity and can use both GSH and thioredoxins (Trxs) as reductants. However, Trxs are more efficient reducing factors, and thus, the enzymes can functionally be considered to be peroxiredoxins rather than GPXs [25].

Pourtaghi et al. [31] demonstrated that water deficit significantly increased the activity of GPXs in sunflower plants compared with fully-irrigated control plants. Moreover, the relationship between seed yield and GPXs activity in fully irrigated (0.78) and moderately water stressed (0.91) plants was both positive and significant. The authors proposed that GPXs can be used as a marker of drought tolerance in selecting tolerant genotypes under moderate and extreme water deficiency conditions.

Cultivars of soybean (*Glycine max* L.) seedlings subjected to water deficit stress exhibited a significant increase in GPXs activity [31]. Similarly, Masoumi et al. [32] reported a positive and significant correlation between GPX activity and seed yield under optimal irrigation (0.99) conditions, mild water deficit stress (0.74) and, high water deficit stress (0.95).

Other research suggests that drought increases enzymatic GPXs activity in leaves of sugar beet (*Beta vulgaris* L.) genotypes. Thus, sugar beet plants might both tolerate and be protected from oxidative damage such as lipid peroxidation by increasing GPXs activity [33].

It has also been found that the over-expression of GPXs enhances plant tolerance to drought. In transgenic *Arabidopsis* seedlings, over-expressing *Synechocystis* PCC 6803 GPX-2 in the chloroplasts (ApGPX2) and cytosol (AcGPX2) showed that lipid peroxidation levels were elevated in both the transgenic and wild-type plants, however, the lipid hydroperoxide content in transgenic plants was significantly lower than that in the wild-type. On the basis of the results described in this work, it is clear that the lines of transgenic plants (ApGPX2 and AcGPX2) expressing *S.* PCC 6803 GPX-2 had enhanced tolerance to oxidative stress caused by drought [34].

Miao et al. [35] isolated two T-DNA insertion mutants of *Arabidopsis thaliana glutathione peroxidase3* (*ATGPX3*) and reported that the *ATGPX3* has a dual role in plant biochemistry, the first being the general control of H_2O_2 equilibrium, and the second specifically linking abscisic acid (ABA) and H_2O_2 signaling during stomatal closure and thus regulating water transpiration. The authors emphasized that the deficiency and over-expression of *ATGPX3* reduced and enhanced drought stress tolerance, respectively.

Glutathione S-transferases

Glutathione S-transferases (GSTs, EC 2.5.1.18) are important phase II, GSH-dependent ROS-scavenging enzymes found in the plant apoplast, cytosol, chloroplasts, mitochondria, and nucleus. This group of enzymes catalyzes the conjugation of GSH to electrophilic sites on a wide range of phytotoxic substrates (Fig. 1, Tab. 2). Currently, very few reports are available on the involvement of GSTs in response to drought (Tab. 3).

Kojić et al. [36] showed that the activity of GSTs increased in the roots of maize (*Zea mays* L.) at 20% soil (sand) humidity (drought conditions). In this study, GST activity was detected only in roots. More specifically, GSTs activity of the control group [70% soil (sand) humidity] increased from 255.5 to 711.6 U/mg protein under drought stress conditions. The authors claimed that the significant increase in GSTs activity under drought conditions agrees with the induction of oxidative stress in plant tissues evoked by drought.

Gallé el al. [37] analyzed GSTs activity and expression patterns in flag leaves of wheat genotypes differing in their tolerance to dehydration during the grain-filling period. GSTs activity and expression were measured for *Triticum aestivum* cv. MV Emese, cv. Plainsman (drought tolerant), cv. GK Élet and, cv. Cappelle Desprez (drought sensitive). *TaGSTU1B* and *TaGSTF6* sequences for *Triticum aestivum* mRNA glutathione transferases, investigated by real-time PCR, showed high-expression levels induced by drought in all of the four analyzed cultivars, but extremely high transcript contents were detected in drought tolerant cv. Plainsman. These data also indicate that expression levels and early induction of two senescence-associated GSTs under drought conditions are correlated with high yield stability. Further, induction of GSTs activity following water deficit was detected earlier in tolerant cultivars than in sensitive ones.

More recently, Chen et al. [38] reported the role of *Arabidopsis thaliana glutathione S-transferases U17* (*AtGSTU17*) in adaptive responses to drought stress by functioning as a negative component of stress-mediated signal transduction pathways. They showed that, when *AtGSTU17* was mutated, plants were more tolerant to drought than wild-type *Arabidopsis* ecotype Columbian plants. Moreover, two knockout T-DNA insertion mutants *atgstu17-1* and *atgstu17-2* accumulated higher levels of GSH and ABA and exhibited hyposensitivity to ABA during seed germination, smaller stomatal apertures, a lower transpiration rate, better development of primary and lateral root systems, and longer period of vegetative growth compared with wild-type *Arabidopsis*.

Some experimental studies suggest that the over-expression of GSTs increases drought tolerance in plants. A chloroplastic GST from *Prosopis juliflora* [39] and a τ class of the GST gene, *GsGST* from *Glycine soja* [40], improved drought stress tolerance in transgenic tobacco.

Glutathione reductase

Of the many components of the plant antioxidant system, glutathione reductase (EC 1.6.4.2) is the last enzyme of the ascorbate/glutathione cycle and plays a principal role in the protection of cells from damage induced by oxidative stress. In drought conditions, GR favors maintenance of the GSH pool, thereby intensifying the antioxidative response of the plant (Fig. 1, Tab. 2, Tab. 3).

Ratnayaka et al. [41] examined the effect of mild drought on glutathione reductase activity in two species of cotton (*Gossypium hirsutum* L. cv. Delta Pine 5415, and *Gossypium barbadense* L. cv. Pima S-7), together with spurred anoda (*Anoda cristata* L. Schlecht.). In this study, GR activity was greater in drought-stressed plants of all three species during recovery, but not during drought. Therefore, the authors proposed that elevated GR activity in drought-stressed plants during recovery strongly indicates that drought may result in acclimation to greater water deficit and/or cross-tolerance to other stresses later.

In poplar (*Populus przewalskii* Maximowicz) cuttings grown under three different watering regimes (100, 50, and 25% of the field capacity), GR activity significantly increased under progressive drought. Moreover, two contrasting populations of *P. przewalskii* were used in this study. They were originally obtained from wet and dry climate regions and it was demonstrated that the plants from the dry climate population presented greater GR activity than those from the wet climate population grown under the same watering regime. The researchers concluded that the combination of drought avoidance and tolerance mechanisms (including induction of GR activity) conferred on the poplar a high degree of plasticity in response to drought [42].

That induction of GR activity in response to oxidative stress triggered by drought is greater in drought sensitive than in drought tolerant individuals is noteworthy and there is some evidence of intensification of oxidative damage to tissues of sensitive plants. Pinheiro et al. [43], working on clones of robusta coffee (*Coffea canephora* Pierre ex Froehner) representing both drought tolerant and drought sensitive genotypes subjected to gradual water deficit until the water potential of their leaves was approximately −3.0 MPa, showed that GR activity either increased significantly (67%, drought sensitive clone) or became reduced (48%, drought tolerant clone) in drought-stressed plants compared with well irrigated control plants.

As mentioned above, treatment with 0.1–10 mg l^{-1} ALA remarkably improved GSH, the total glutathione pool and, in particular, the GSH/GSSG ratio in oilseed rape seedlings under water deficit stress. Similarly, GR expression and activity were significantly greater following treatment with 1 mg l^{-1} ALA during drought. It is probable that greater GR activity triggered by ALA in plants resulted in a large pool of GSH capable of increasing the efficiency of the AsA-GSH cycle [29]. Furthermore, Singh et al. [44] observed a strong and positive correlation between GR activity and ascorbate content in wheat roots (0.90) and leaves (0.87) under water deficit.

Recently, the role of the phytohormone ABA in the induction of antioxidant enzymes has been the subject of advanced research. It is worth stating that young and mature leaves of tomato seedlings (*Lycopersicon esculentum* Mill.) cv. Ailsa Craig (AC) and an ABA-deficient mutant (notabilis) exhibited differences in GR activity. However, the activity of GR remained unchanged following ABA treatment of young and mature leaves of AC, but enzyme activity declined in ABA-treated young leaves of notabilis. Furthermore, the exogenous ABA treatment increased GR activity in young leaves of notabilis under drought, compared with the control plants and those exposed to drought. On the basis of these studies, it can be concluded that enzymatic antioxidants (including GR) synthesized by plants are regulated not only by stress and ABA but also by the developmental stage of individual plants [45].

Another plant hormone, salicylic acid (SA) is a phenolic compound that is able to modulate plant responses to abiotic stresses. Greater drought tolerance was observed in *Triticum aestivum* L. cv. Yumai 34 seedlings following treatment with exogenous 0.5 mM salicylic acid under drought conditions compared with the stressed plants. This enhanced tolerance is related to the increased transcription of GR and other AsA-GSH cycle-related genes, as well as the increased content and biosynthesis of AsA and GSH [46].

In contrast to the results presented above, other researchers have found that GR activity diminishes under drought conditions. For example, Iturbe-Ormaetxe et al. [47] analyzed pea (*Pisum sativum* L. cv. Lincoln) plants grown both under optimal water (leaf Ψ_w values of −0.50 ±0.02 MPa) and water deficit conditions (leaf Ψ_w values of −1.30 ±0.04 MPa (S1) and −1.93 ±0.05 MPa (S2). In S1 and S2 plants, the activity of GR decreased in both regimes as compared with unstressed plants.

S-Glutathionylation

S-Glutathionylation is a redox post-translational modification of protein cysteine residues by the addition of glutathione. Protein S-glutathionylation is promoted by reactive oxygen and nitrogen species activity, but also occurs in unstressed cells. This biochemical process may serve to regulate a variety of cellular processes by modulating protein function and preventing irreversible oxidation of protein thiols [48]. Recent studies have identified S-glutathionylation as a significant mechanism of cell regulation and redox signaling in photosynthetic organisms. For example, it regulates Calvin cycle enzymes such as phosphoribulokinase, glyceraldehyde-3-phosphate dehydrogenase, ribose-5-phosphate isomerase, and phosphoglycerate kinase in the green alga (*Chlamydomonas reinhardtii* P.A. Dang.) growing under oxidative stress conditions [49].

Desiccation is not synonymous with drought, and desiccation tolerance is defined as the ability of a living plant structure to survive drying with low relative humidity and maintain low intracellular water concentrations. Whereas drought tolerance is survival of low environmental water availability while maintaining high internal water contents. In desiccation tolerant plants, the enormous changes in the water content of tissues during wetting and drying cycles are accompanied by equally extreme fluctuations in their cellular redox state. S-glutathionylation of proteins is a biochemical factor that is likely to contribute towards protection mechanisms that confer desiccation tolerance [50].

Some of the best described stress-related proteins that may be subject to S-glutathionylation belong to the annexin group. The annexin protein family comprises multigene, multifunctional membrane and Ca^{2+}-binding proteins with expected enzymatic activity involved in the signal transduction pathway. The characteristic attribute of annexins is that they can bind membrane phospholipids in a reversible, Ca^{2+}-dependent manner [51].

In vitro studies demonstrated that *Arabidopsis* annexin 1 (AnnAt1) can be S-glutathionylated on two Cys residues providing important data that show these residues to be chemically reactive [52]. It has been suggested that owing to the reactivity of these Cys residues, AnnAt1 may be one of the plant cellular proteins involved in H_2O_2 perception. Furthermore, Konopka-Postupolska et al. [52] found that the Cys residues in AnnAt1 are S-glutathionylated in vivo in response to ABA treatment, which provides evidence that this post-translational modification of AnnAt1 is physiologically relevant during drought responses.

Drought tolerance and adaptation processes are also regulated at the molecular level. DREBs (dehydration responsive element binding) are important plant transcription factors that regulate the expression of many stress-inducible genes in the ABA-independent pathways and play an important role in increasing the abiotic stresses tolerance of plants by interacting with a cis-element present in the promoter region in abiotic stress-responsive genes. DRE (dehydration responsive element) with a 9 bp conserved DNA sequence (5'-TACCGACAT-3') was first described in the promoter of the drought-responsive gene *rd29A* [53].

On perceiving a water deficit, the plant cell produces a biochemical signal, which is transduced via activation of DNA-binding proteins called CBF, which then bind to DREs on the *rd29A* promoter. This precipitates intensified transcription of the gene and finally the accumulation of rd29A proteins, which probably participate in the response to drought. Furthermore, the *rd29A* promoter also contains elements, which respond to ABA (ABREs). The DRE and ABRE elements probably function together to increase the rate of transcription [54].

Transgenic tomato homozygous T_2 (cv. Kashi Vishesh) plants over-expressing *Arabidopsis thaliana AtDREB1A/*

CBF3 driven by stress-inducible *rd29A* promoter showed significantly greater activity of GR when exposed to water deficit for 7, 14, and 21 days compared with non-transgenic plants under the same water deficit conditions. The contents of total ascorbate, total glutathione and GSH were greater in transgenic plants and increased with ROS levels. The authors demonstrated that AtDREB1A transgenic tomato lines were better adapted to water deficit, since they showed lower drought induced oxidative stress due to activation of the antioxidant response. In summary, the up-regulation of genes responsible for antioxidant defense might be a consequence of the over-expression of *AtDREB1A* in all the five transgenic tomato lines tested under drought conditions [55].

Concluding remarks and future challenges

This article gives a clear overview of the biochemical aspects of GSH and its related enzymes in a variety of plant species subjected to drought. The reviewed studies confirm that GSH plays a central role in the metabolism of plant cells during abiotic stress. Also, by acting as a key component of the Foyer–Halliwell–Asada pathway, the reduced glutathione and its related enzymes play a very important role in the protection of plants against oxidative stress induced by water deficit in tissues. GSH, its redox couple (GSH/GSSG) and related enzymes (GPXs, GSTs, GR) have been shown to be closely correlated in terms of their metabolic functions in plants during drought. Thus, the GSH system is often regarded as a useful marker in plant ecophysiological studies. However, many questions have yet to be answered, in particular regarding the regulation of S-glutathionylation and the molecular characterization of GSH-dependent enzymes in model plant organisms, wild species, and economically important crops growing under drought conditions. The present authors wish to highlight that transgenic plants over-expressing or expressing antisense constructs resulting in inhibition of specific GSH related enzymes, or mutants with impaired reactive oxygen species generation may be extremely useful in basic researches, and are likely to be valuable in subsequent analyzes of plant antioxidative mechanisms, and the role of glutathione in response to drought.

Acknowledgments
This work was financially supported by the Department of Biochemistry, Warsaw University of Life Sciences – SGGW.

Authors' contributions
The following declarations about authors' contributions to the research have been made: compiled the literature: ML; prepared the figure: FMSA; wrote the manuscript: ML, FMSA.

References
1. Nayyar H, Gupta D. Differential sensitivity of C3 and C4 plants to water deficit stress: association with oxidative stress and antioxidants. Env Exp Bot. 2006;58(1–3):106–113.

2. Monakhova OF, Chernyad'ev II. Protective role of kartolin-4 in wheat plants exposed to soil draught. Appl Biochem Microbiol. 2002;38(4):373–380.

3. Huang GT, Ma SL, Bai LP, Zhang L, Ma H, Jia P, et al. Signal

transduction during cold, salt, and drought stresses in plants. Mol Biol Rep. 2012;39(2):969–987.

4. Mittler R. Oxidative stress, antioxidants and stress tolerance. Trends Plant Sci. 2002;7(9):405–410.

5. Anjum SA, Wang L, Farooq M, Khan I, Xue L. Methyl jasmonate-induced alteration in lipid peroxidation, antioxidative defence system and yield in soybean under drought. J Agron Crop Sci. 2011;197(4):296–301.

6. Sharma P, Jha AB, Dubey RS, Pessarakli M. Reactive oxygen species, oxidative damage, and antioxidative defense mechanism in plants under stressful conditions. J Bot. 2012;2012:1–26.

7. Noctor G, Foyer CH. Ascorbate and glutathione: keeping active oxygen under control. Annu Rev Plant Physiol Plant Mol Biol. 1998;49:249–279.

8. Foyer CH, Noctor G. Redox homeostasis and antioxidant signaling: a metabolic interface between stress perception and physiological

responses. Plant Cell. 2005;17(7):1866–1875.

9. Gill SS, Tuteja N. Reactive oxygen species and antioxidant machinery in abiotic stress tolerance in crop plants. Plant Physiol Biochem. 2010;48(12):909–930.

10. Boguszewska D, Zagdańska B. ROS as signaling molecules and enzymes of plant response to unfavorable environmental conditions. In: Lushchak V, Semchyshyn HM, editors. Oxidative stress – molecular mechanisms and biological effects. Rijeka, Croatia: InTech; 2012. p. 341–362.

11. Zagorchev L, Seal C, Kranner I, Odjakova M. A central role for thiols in plant tolerance to abiotic stress. Int J Mol Sci. 2013;14(4):7405–7432.

12. Boguszewska D, Grudkowska M, Zagdańska B. Drought-responsive antioxidant enzymes in potato (Solanum tuberosum L.). Potato Res. 2010;53(4):373–382.

13. Noctor G, Mhamdi A, Chaouch S, Han Y, Neukermans J, Marquez-Garcia B, et al. Glutathione in plants: an integrated overview. Plant Cell Env. 2012;35(2):454–484.

14. Zechmann B, Müller M. Subcellular compartmentation of glutathione in dicotyledonous plants. Protoplasma. 2010;246(1–4):15–24.

15. Wachter A, Wolf S, Steininger H, Bogs J, Rausch T. Differential targeting of GSH1 and GSH2 is achieved by multiple transcription initiation: implications for the compartmentation of glutathione biosynthesis in the Brassicaceae. Plant J. 2005;41(1):15–30.

16. Sengupta D, Ramesh G, Mudalkar S, Kumar KRR, Kirti PB, Reddy AR. Molecular cloning and characterization of γ-glutamyl cysteine synthetase (VrγECS) from roots of Vigna radiata (L.) Wilczek under progressive drought stress and recovery. Plant Mol Biol Rep. 2012;30(4):894–903.

17. Nazar R, Iqbal N, Masood A, Syeed S, Khan NA. Understanding the significance of sulfur in improving salinity tolerance in plants. Env. Exp. Bot. 2011;70(2–3):80–87.

18. Lim B, Meyer AJ, Cobbett CS. Development of glutathione-deficient embryos in Arabidopsis is influenced by the maternal level of glutathione. Plant Biol Stuttg. 2011;13(4):693–697.

19. El Msehli S, Lambert A, Baldacci-Cresp F, Hopkins J, Boncompagni E, Smiti SA, et al. Crucial role of (homo)glutathione in nitrogen fixation in Medicago truncatula nodules. New Phytol. 2011;192(2):496–506.

20. Cruz de Carvalho MH, Brunet J, Bazin J, Kranner I, d'Arcy-Lameta A, Zuily-Fodil Y, et al. Homoglutathione synthetase and glutathione synthetase in drought-stressed cowpea leaves: expression patterns and accumulation of low-molecular-weight thiols. J Plant Physiol. 2010;167(6):480–487.

21. Naya L, Ladrera R, Ramos J, González EM, Arrese-Igor C, Minchin FR, et al. The response of carbon metabolism and antioxidant defenses of alfalfa nodules to drought stress and to the subsequent recovery of plants. Plant Physiol. 2007;144(2):1104–1114.

22. Sgherri CLM, Navari-Izzo F. Sunflower seedlings subjected to increasing water deficit stress: oxidative stress and defence mechanisms. Physiol Plant. 1995;93(1):25–30.

23. Herbinger K, Tausz M, Wonisch A, Soja G, Sorger A, Grill D. Complex interactive effects of drought and ozone stress on the antioxidant defence systems of two wheat cultivars. Plant Physiol Biochem. 2002;40(6–8):691–696.

24. Pyngrope S, Bhoomika K, Dubey RS. Reactive oxygen species, ascorbate-glutathione pool, and enzymes of their metabolism in drought-sensitive and tolerant indica rice (Oryza sativa L.) seedlings subjected to progressing levels of water deficit. Protoplasma. 2013;250(2):585–600.

25. Anjum NA, Ahmad I, Mohmood I, Pacheco M, Duarte AC, Pereira E, et al. Modulation of glutathione and its related enzymes in plants' responses to toxic metals and metalloids – a review. Env Exp Bot. 2012;75:307–324.

26. Tausz M, Wonisch A, Peters J, Jiménez MS, Morales D, Grill D. Short-term changes in free radical scavengers and chloroplast pigments in Pinus canariensis needles as affected by mild drought stress. J Plant Physiol. 2001;158(2):213–219.

27. Anwar Hossain M, Golam Mostofa M, Fujita M. Heat-shock positively modulates oxidative protection of salt and drought-stressed mustard (Brassica campestris L.) seedlings. J Plant Sci Mol Breed. 2013;2(1):1–14.

28. Alam MM, Hasanuzzaman M, Nahar K, Fujita M. Exogenous salicylic acid ameliorates short-term drought stress in mustard (Brassica juncea L.) seedlings by up-regulating the antioxidant defense and glyoxalase system. Aust J Crop Sci. 2013;7(7):1053–1063.

29. Liu D, Pei ZF, Naeem MS, Ming DF, Liu HB, Khan F, et al. 5-aminolevulinic acid activates antioxidative defence system and seedling growth in Brassica napus L. under water-deficit stress. J Agron Crop Sci. 2011;197(4):284–295.

30. Dhindsa RS. Drought stress, enzymes of glutathione metabolism, oxidation injury, and protein synthesis in Tortula ruralis. Plant Physiol. 1991;95(2):648–651.

31. Pourtaghi A, Darvish F, Habibi D, Nourmohammadi G, Daneshian J. Effect of irrigation water deficit on antioxidant activity and yield of some sunflower hybrids. Aust J Crop Sci. 2011;5(2):197–204.

32. Masoumi H, Masoumi M, Darvish F, Daneshian J, Nourmohammadi G, Habibi D. Change in several antioxidant enzymes activity and seed yield by water deficit stress in soybean (Glycine max L.) cultivars. Bot Hort Agrobot Cluj. 2010;38(3):86–94.

33. Sayfzadeh S, Rashidi M. Response of antioxidant enzymes activities of sugar beet to drought stress. J Agric Biol Sci. 2011;6(4):27–33.

34. Gaber A, Yoshimura K, Yamamoto T, Yabuta Y, Takeda T, Miyasaka H, et al. Glutathione peroxidase-like protein of Synechocystis PCC 6803 confers tolerance to oxidative and environmental stresses in transgenic Arabidopsis. Physiol Plant. 2006;128(2):251–262.

35. Miao Y, Lv D, Wang P, Wang XC, Chen J, Miao C, et al. An Arabidopsis glutathione peroxidase functions as both a redox transducer and a scavenger in abscisic acid and drought stress responses. Plant Cell. 2006;18(10):2749–2766.

36. Kojić D, Pajević S, Jovanović-Galović A, Purać J, Pamer E, Škondrić S, et al. Efficacy of natural aluminosilicates in moderating drought effects on the morphological and physiological parameters of maize plants (Zea mays L.). J Soil Sci Plant Nutr. 2012;12(1):113–123.

37. Gallé A, Csiszár J, Secenji M, Guóth A, Cseuz L, Tari I, et al. Glutathione transferase activity and expression patterns during grain filling in flag leaves of wheat genotypes differing in drought tolerance: response to water deficit. J Plant Physiol. 2009;166(17):1878–1891.

38. Chen JH, Jiang HW, Hsieh EJ, Chen HY, Chien CT, Hsieh HL, et al. Drought and salt stress tolerance of an Arabidopsis glutathione S-transferase U17 knockout mutant are attributed to the combined effect of glutathione and abscisic acid. Plant Physiol. 2012;158(1):340–351.

39. George S, Venkataraman G, Parida A. A chloroplast-localized and auxin-induced glutathione S-transferase from phreatophyte Prosopis juliflora confer drought tolerance on tobacco. J Plant Physiol. 2010;167(4):311–318.

40. Ji W, Zhu Y, Li Y, Yang L, Zhao X, Cai H, et al. Over-expression of a glutathione S-transferase gene, GsGST, from wild soybean (Glycine soja) enhances drought and salt tolerance in transgenic tobacco. Biotechnol Lett. 2010;32(8):1173–1179.

41. Ratnayaka HH, Molin WT, Sterling TM. Physiological and antioxidant responses of cotton and spurred anoda under interference and

mild drought. J Exp Bot. 2003;54(391):2293–2305.

42. Lei Y, Yin C, Li C. Differences in some morphological, physiological, and biochemical responses to drought stress in two contrasting populations of Populus przewalskii. Physiol Plant. 2006;127(2):182–191.

43. Pinheiro HA, DaMatta FM, Chaves ARM, Fontes EPB, Loureiro ME. Drought tolerance in relation to protection against oxidative stress in clones of Coffea canephora subjected to long-term drought. Plant Sci. 2004;167(6):1307–1314.

44. Singh S, Gupta AK, Kaur N. Differential responses of antioxidative defence system to long-term field drought in wheat (Triticum aestivum L.) genotypes differing in drought tolerance. J Agron Crop Sci. 2012;198(3):185–195.

45. Ünyayar S, Çekiç FÖ. Changes in antioxidative enzymes of young and mature leaves of tomato seedlings under drought stress. Turk J Biol. 2006;29(4):211–216.

46. Kang GZ, Li GZ, Liu GQ, Xu W, Peng XQ, Wang CY, et al. Exogenous salicylic acid enhances wheat drought tolerance by influence on the expression of genes related to ascorbate-glutathione cycle. Biol Plant. 2013;57(4):718–724.

47. Iturbe-Ormaetxe I, Escuredo PR, Arrese-Igor C, Becana M. Oxidative damage in pea plants xxposed to water deficit or paraquat. Plant Physiol. 1998;116(1):173–181.

48. Dalle-Donne I, Rossi R, Colombo G, Giustarini D, Milzani A. Protein S-glutathionylation: a regulatory device from bacteria to humans. Trends Biochem Sci. 2009;34(2):85–96.

49. Zaffagnini M, Bedhomme M, Groni H, Marchand CH, Puppo C, Gontero B, et al. Glutathionylation in the photosynthetic model organism Chlamydomonas reinhardtii: a proteomic survey. Mol Cell Proteomics. 2012;11(2):M111.014142.

50. Colville L, Kranner I. Desiccation tolerant plants as model systems to study redox regulation of protein thiols. Plant Growth Regul. 2010;62(3):241–255.

51. Talukdar T, Gorecka KM, de Carvalho-Niebel F, Downie JA, Cullimore J, Pikula S. Annexins - calcium- and membrane-binding proteins in the plant kingdom: potential role in nodulation and mycorrhization in Medicago truncatula. Acta Biochim Pol. 2009;56(2):199–210.

52. Konopka-Postupolska D, Clark G, Goch G, Debski J, Floras K, Cantero A, et al. The role of annexin 1 in drought stress in Arabidopsis. Plant Physiol. 2009;150(3):1394–1410.

53. Lata C, Prasad M. Role of DREBs in regulation of abiotic stress responses in plants. J Exp Bot. 2011;62(14):4731–4748.

54. Hamill JD. Gene expression modified by external factors. In: Turnbull CGN, Atwell BJ, Kriedemann PE, editors. Plants in action. Adaptation in nature, performance in cultivation. Melbourne, Australia: Macmillan Education Australia Pty Ltd; 1999. p. 10.3.4.

55. Rai GK, Rai NP, Rathaur S, Kumar S, Singh M. Expression of rd29A::AtDREB1A/CBF3 in tomato alleviates drought-induced oxidative stress by regulating key enzymatic and non-enzymatic antioxidants. Plant Physiol Biochem. 2013;69:90–100.

56. Pandey HC, Baig MJ, Chandra A, Bhatt RK. Drought stress induced changes in lipid peroxidation and antioxidant system in genus Avena. J Env. Biol. 2010;31(4):435–440.

57. Chowdhury SR, Choudhuri MA. Hydrogen peroxide metabolism as an index of water stress tolerance in jute. Physiol Plant. 1985;65(4):476–480.

58. Wang S, Liang D, Li C, Hao Y, Ma F, Shu H. Influence of drought stress on the cellular ultrastructure and antioxidant system in leaves of drought-tolerant and drought-sensitive apple rootstocks. Plant Physiol Biochem. 2012;51:81–89.

59. Bai LP, Sui FG, Ge TD, Sun ZH, Lu YY, Zhou GS. Effect of soil drought stress on leaf water status, membrane permeability and enzymatic antioxidant system of maize. Pedosphere. 2006;16(3):326–332.

60. Ali Q, Ashraf M. Induction of drought tolerance in maize (Zea mays L.) due to exogenous application of trehalose: growth, photosynthesis, water relations and oxidative defence mechanism. J Agron Crop Sci. 2011;197(4):258–271.

61. Anjum SA, Wang LC, Farooq M, Hussain M, Xue LL, Zou CM. Brassinolide application improves the drought tolerance in maize through modulation of enzymatic antioxidants and leaf gas exchange. J Agron Crop Sci. 2011;197(3):177–185.

62. Anjum SA, Wang L, Farooq M, Xue L, Ali S. Fulvic acid application improves the maize performance under well-watered and drought conditions. J Agron Crop Sci. 2011;197(6):409–417.

63. Yildiz-Aktas L, Dagnon S, Gurel A, Gesheva E, Edreva A. Drought tolerance in cotton: involvement of non-enzymatic ROS-scavenging compounds. J Agron Crop Sci. 2009;195(4):247–253.

64. Sairam RK, Shukla DS, Saxena DC. Stress induced injury and antioxidant enzymes in relation to drought tolerance in wheat genotypes. Biol Plant. 1997;40(3):357–364.

65. Marcińska I, Czyczyło-Mysza I, Skrzypek E, Grzesiak MT, Janowiak F, Filek M, et al. Alleviation of osmotic stress effects by exogenous application of salicylic or abscisic acid on wheat seedlings. Int J Mol Sci. 2013;14(7):13171–13193.

66. Zlatev ZS, Lidon FC, Ramalho JC, Yordanov IT. Comparison of resistance to drought of three bean cultivars. Biol Plant. 2006;50(3):389–394.

67. Fazeli F, Ghorbanli M, Niknam V. Effect of drought on biomass, protein content, lipid peroxidation and antioxidant enzymes in two sesame cultivars. Biol Plant. 2007;51(1):98–103.

68. Mohammadi A, Habibi D, Rohami M, Mafakheri S. Effect of drought stress on antioxidant enzymes activity of some chickpea cultivars. Am-Eurasian J Agric Env Sci. 2011;11(6):782–785.

69. Huang C, Zhao S, Wang L, Anjum SA, Chen M, Zhou H, et al. Alteration in chlorophyll fluorescence, lipid peroxidation and antioxidant enzymes activities in hybrid ramie (Boehmeria nivea L.) under drought stress. Aust J Crop Sci. 2013;7(5):594.

70. Štajner D, Orlović S, Popović BM, Kebert M, Galić Z. Screening of drought oxidative stress tolerance in Serbian melliferous plant species. Afr J Biotechnol. 2011;10(9):1609–1614.

71. Habibi G, Hajiboland R. Alleviation of drought stress by silicon supplementation in pistachio (Pistacia vera L.) plants. Folia Hort. 2013;25(1):21–29.

72. El-Enany AE, AL-Anazi AD, Dief N, Al-Taisan WA. Role of antioxidant enzymes in amelioration of water deficit and waterlogging stresses on Vigna sinensis plants. J Biol Earth Sci. 2013;3(1):B144–B153.

73. Farissi M, Bouizgaren A, Faghire M, Bargaz A, Ghoulam C. Agrophysiological and biochemical properties associated with adaptation of Medicago sativa populations to water deficit. Turk J Bot. 2013;37:1166–1175.

74. Yobi A, Wone BWM, Xu W, Alexander DC, Guo L, Ryals JA, et al. Metabolomic profiling in Selaginella lepidophylla at various hydration states provides new insights into the mechanistic basis of desiccation tolerance. Mol Plant. 2013;6(2):369–385.

75. Bednarski W, Hendry G, Atherton N, Lee J. Radical formation and accumulation in vivo, in desiccation tolerant and intolerant mosses. Free Radic Res Commun. 1991;15(3):133–141.

76. Shi H, Wang Y, Cheng Z, Ye T, Chan Z. Analysis of natural variation in bermudagrass (Cynodon dactylon) reveals physiological responses underlying drought tolerance. PLoS ONE. 2012;7(12):e53422.

77. Zhang Y, Zhang Q, Sammul M. Physiological integration ameliorates negative effects of drought stress in the clonal herb *Fragaria orientalis* PLoS ONE. 2012;7(9):e44221.

78. Gechev TS, Van Breusegem F, Stone JM, Denev I, Laloi C. Reactive oxygen species as signals that modulate plant stress responses and programmed cell death. Bioessays. 2006;28(11):1091–1101.

79. Masoumi H, Darvish F, Daneshian J, Normohammadi G, Habibi D. Effects of water deficit stress on seed yield and antioxidants content in soybean (*Glycine max* L.) cultivars. Afr J Agr Res. 2011;6(5):1209–1218.

3

Isolation of biosynthesis related transcripts of 2,3,5,4′-tetrahydroxy stilbene-2-O-β-D-glucoside from *Fallopia multiflora* by suppression subtractive hybridization

Wei Zhao[1,2], Shujing Sheng[2], Zhongyu Liu[2], Di Lu[2], Kuanpeng Zhu[1,2], Xiaoze Li[1,2], Shujin Zhao[2]*, Yan Yao[3]

[1] School of Bioscience and Bioengineering, South China University of Technology, Guangzhou 510006, People's Republic of China

[2] Department of Pharmacy, General Hospital of Guangzhou Military Command, Guangzhou 510010, People's Republic of China

[3] School of Life Sciences, Guangzhou University, Guangzhou 510006, People's Republic of China

Abstract

2,3,5,4′-tetrahydroxy stilbene-2-O-ß-D-glucoside (THSG) exerts multiple pharmacodynamic actions, found in *Fallopia multiflora*, but the biosynthesis pathway of THSG is still unclear. To clear this ambiguity, we constructed suppression subtractive hybridization (SSH) libraries to screen the genes involved in THSG biosynthesis from two *F. multiflora* varieties, which vary significantly in THSG content. Twelve non-redundant differentially expressed sequence tags were obtained and the full lengths of 4 unreported fragments were amplified by rapid amplification of cDNA ends. We totally got 7 full-length transcripts, and all of them were aligned to the transcriptome and digital gene expression tag profiling database of four *F. multiflora* tissues (root, stem and leaf from Deqing *F. multiflora* and another root from Chongqing *F. multiflora*; data unpublished) using local BLAST. The results showed that there was a significant, organ specific difference in the expression of fragments and full-length sequences. All the sequences were annotated by aligning to nucleotide and protein databases. Kyoto Encyclopedia of Genes and Genomes pathway analysis indicated that THSG biosynthesis was correlated with multiple life activities.

Keywords: *Fallopia multiflora*; gene expression difference; transcriptome; 2,3,5,4′-tetrahydroxy stilbene-2-O-β-D-glucooside

Introduction

Fallopia multiflora is a traditional Chinese medicinal herb, which has been widely used for thousands of years. Research showed that *F. multiflora* has potent antioxidative and cytoprotective properties [1], enhanced purgative effects, promoted diuresis and choleretic effects [2], and exerted a neuroprotective effect against glutamate-induced neurotoxicity [3]. 2,3,5,4′-tetrahydroxy stilbene-2-O-ß-D-glucoside (THSG) is an active component in *F. multiflora*, which possesses anti-hyperlipidemic [4], anti-oxidative, anti-inflammatory, endothelial-protective activities [5]. Furthermore, THSG can protect osteoblastic MC3T3-E1 cells via inhibiting the release of bone-resorbing mediators and oxidative damage of the cells [6], and suppress atherosclerosis by altering the expression of key proteins that may be novel molecular targets responsible for atherogenesis [7]. Sun et al. who reported that THSG might provide a potentially new strategy for preventing and treating neurodegenerative

disorders such as Parkinson's disease, also showed that THSG may protect neurons against MPP+- induced cell death through improving mitochondrial function, decreasing oxidative stress and inhibiting apoptosis [8]. The pharmacological research revealed that THSG has a potential impact on human health. As a stilbene, THSG and resveratrol (3,5,4′-trihydroxy-trans-stilbene) belong to phenylpropanoids characterized by a 1,2-diphenylethylene backbone. Plant stilbenes are derived from phenylalanine biosynthesis via the general phenylpropanoid pathway. One p-coumaroyl-CoA or cinnamoyl-CoA derived from the phenylpropanoid pathway is ligated with three malonyl-CoA under the catalysis of stilbene synthase (STS) and stilbene is the product [9]. STS is a polyketide synthase (PKS) belonging to type III PKSs. Other PKSs III include resveratrol synthase (RS), chalcone synthases (CHS), bibenzyl synthase (BBS), stilbene carboxylate synthase (STCS), 4-coumaroyltriacetate lactone synthase (CTAS) and more [10]. Every PKSs III catalyzes the synthesis of one unique stilbene. Up to now, no enzyme was found to catalyze the biosynthesis of THSG. Sheng et al. isolated a stilbene synthase gene *FmPKS* from the rhizomes of *F. multiflora*. The gene expression pattern in the plant correlated with the THSG content in different tissues, but

---

* Corresponding author. Email: gzzsjzhs@163.com

Handling Editor: Przemysław Wojtaszek

THSG was still not detectable in transgenic *Arabidopsis thaliana* in which *FmPKS* was inserted and expressed [11]. Shao et al. showed precursor feeding of methyl jasmonate and salicylic acid in suspension cultures of *F. multiflora* that could increase THSG production [12]. It is still not clear if THSG is synthesized via phenylalanine pathway or some other way. Suppression subtractive hybridization (SSH) is helpful in identifying differentially expressed genes. Extensive studies showed that SSH is a powerful tool in the analysis of stress resistance [13], pathological mechanism [14] and developmental physiology [15]. Moreover, Wang et al. found that light could be effective for activation of the biosynthesis of phenylpropanoids by establishing SSH cDNA libraries of tea calli [16]. In this study, we attempted to identify the genes that are related to THSG biosynthesis by SSH. Full-length sequences were obtained using 3'5'RACE, and then the gene functions were annotated by blastn to nucleotide databases nt, and blastx to protein databases nr, Swiss-Prot, Kyoto Encyclopedia of Genes and Genomes (KEGG) and clusters of orthologous groups (COG). Meanwhile, comparative analysis with the transcriptome and digital expression profile revealed the expression differences for each gene in various tissues of *F. multiflora*. This may indicate that these genes are associated with THSG synthesis.

Material and methods

Plant material collection

F. multiflora plants were gathered from 13 cities (counties) in March 2010, which includes Guangxi Province (Nanning City, Guilin City, Jingxi County, Tianlin County), Guangdong Province (Guangzhou City, Shenzhen City, Dongguan City, Deqing City, Zhaoqing City, Gaozhou City, Zhanjiang City), Jiangxi Province (Xinyu City), Chongqing City. As some samples were gathered in the field, the age was unclear. All plant materials were maintained in the medicinal plant garden of the Department of Pharmacy, Guangzhou Liuhuaqiao Hospital, Guangzhou, China.

HPLC analysis of THSG

To quantify the THSG in *F. multiflora* roots, fresh plant materials were frozen in liquid nitrogen and ground to fine powder in a mortar. After vacuum freeze-drying, 0.2 g of each sample was taken up in 25 ml 50% (v/v) methanol and then refluxed at room temperature for 16 h. After being filtered through a 0.22 μm film, 10 μl of filtrate was analyzed by HPLC using a Dikma Diamonsil C18 column (250 × 4.6 mm, tablets path 5 μm). Chromatographic separation was performed using a solvent system of H_2O and CH_3CN with the ratio of 3:1 (v/v) over 10 min. The flow rate was 1 ml/min, with detection at 320 nm. Each data point represents the average of three independent experiments.

Total RNA extraction and mRNA purification

Total RNA was extracted from leaves and roots using the Plant Total RNA Isolation kit (Bioteke, China) and treated with DNase I (TaKaRa, Dalian, China). Ethidium bromide (EtBr) staining, agarose gel electrophoresis and spectrophotometric (NanoDrop 2000, USA) analysis were performed to examine the quality and concentration of total RNA. mRNA was purified using Oligo tes™-dT₃₀<SUPER> mRNA Purification Kit (from Total RNA) according to the manufacturer's instructions (TaKaRa, Dalian, China).

Suppression subtractive hybridization (SSH)

Driver and tester cDNA was synthesized from two samples at equal amounts of purified mRNA, using SMARTer PCR cDNA Synthesis K it (Clontech, USA) according to the manufacturer's instructions. After repeated twice hybridization, target genes were amplified using nested PCR by Advantage™ cDNA PCR Kit (Clontech, USA). To raise the PCR efficiency, an adapter was added to the nested PCR primers (Tab. 1; homo nested primer). The following target genes were amplified using Fermentas Dream Taq (Fermentas, USA) with homo PCR primer (Tab. 1; homo PCR primer).

The PCR products were inserted into the pMD19-T vector (TaKaRa, Dalian, China), and the ligated products were transformed into *Escherichia coli* DH5α competent cells by heat shock. Then plated onto LB medium containing 100 μg/ml ampicillin, 24 μg/ml IPTG and 20 μg/ml X-gal, and incubated overnight at 37°C. Recombinant white colonies were randomly selected for colony PCR with universal primers (Tab. 1; universal primers). Colonies containing cloned fragments were sent for sequencing. Besides control sample (Poly A⁺ RNA from human skeletal muscle) provided with the kit, which was used as a control driver cDNA, the experiments were performed with 3 different testers: I – cDNA from Guilin's root sample as tester and Chongqing's root sample as driver; II – Chongqing's root sample as tester and Guilin's root sample as driver; III – Guilin's leaf sample as tester and Guilin's root sample as driver.

Cloning genes with full length by RACE

The 3'5'RACE of the candidate genes was performed in purified mRNA using PCR-select™ cDNA subtraction kit (Clontech, USA) according to the manufacturer's instruction. To improve the specificity, nested primers were designed for Nested-PCR besides gene specific primers. Fragments of 4 genes were selected to do 3'5'RACE from SSH library after BLAST in NCBI (Tab. 1). The Nested-PCR reaction was performed by Advantage cDNA Polymerase Mix (Clontech, USA) with the following thermal cycling parameters: 94°C for 4 min, followed by 35 cycles of 94°C for 30 s, 68°C for 30 s and 72°C for 1.5 min, and the final extension was performed at 72°C for 10 min. The PCR product was cloned into the pMD19-T vector (TaKaRa, Dalian, China) and sequenced.

Sequence analysis

Gene fragments obtained from SSH and RACE were compared with the sequences in nucleotide collection database and expressed sequence tag (EST) database at NCBI using the BLASTN algorithm. Further annotation was carried out using nr, Swiss-Prot, gene ontology database (GO), COG and KEGG. Clustalx1.83 and Contig Express software were used for multiple sequence alignment, sequence identity and linkage. Moreover the obtained sequences were compared with the transcriptome and digital gene expression tag profiling (DGE) database of four different tissue samples (Cr: roots of Chongqing *F. multiflora*; Dr: roots of Deqing

Tab. 1 SSH primers and RACE primers of 4 selected gene fragments.

Gene	Primer	Sequence
	Homo nested primer	5'-GCGACCTACAACATGGCTACCGTCGAGCGGCCGCCCGGGCAGGT
	homo PCR primer	5'-GCGACCTACAACATGGCTACCG
	universal primer	5'-GAGCGGATAACAATTTCACACAGG
GE	3' GSP	5'-CTGGCAACTGTTGAAGGACGCGAAGA
	5' GSP	5'-GCCCAAGCACCAAATGCCTCTGC
	3' Nested primer	5'-AGAGAGGTAGCAACAGATGGAGG
	5' Nested primer	5'-CCTCCATCTGTTGCTACCTCTCT
GW	3' GSP	5'-GTCTCATGCGCTCCTCCTCCAGCC
	5' GSP	5'-GGCGAGCAGAGAAGCAAAGGCTGG
	3' Nested primer	5'-AGCCTTTGCTTCTCTGCTCGCCT
	5' Nested primer	5'-GGAGGAGCGCATGAGACAGAACA
CF	3' GSP	5'-AGGCAATGGAGAAAAGCCGCTCGC
	5' GSP	5'-TAGGCAACGGCAATCCCAGCACGG
	3' Nested primer	5'-GGATTGCCGTTGCCTAAAGTGTG
	5' Nested primer	5'-TTGCTCATTCCTACACTCCTCGC
CA	3' GSP	5'-CTTGGGCTTTTTCAGGGACAGACG
	5' GSP	5'-CCGGTAGAGGCCATCAGGATGCAG
	3' Nested primer	5'-CAGACGAAAAGACGAGAATGAGG
	5' Nested primer	5'-TTCCTCTTTGGCTTTCCCACTTC

"G" represents the origin of the sample, Guilin city; "C" represents Chongqing city.

F. multiflora; Dl: leaves of Deqing *F. multiflora*; Ds: stems of Deqing *F. multiflora*) of *F. multiflora* (unpublished data) using local BLAST.

Results

Quantification of THSG
To select the specific samples with robust difference in THSG concentration, HPLC was used to analyze the THSG contents. Results indicated that there was a significant difference (P-value < 0.001) in THSG content in *F. multiflora* root from different origins (Fig. 1, Tab. 2). Root samples from Guilin City and Zhaoqin City have the highest accumulation of THSG, with a content of 6.37% and 5.25% respectively. Chongqing samples have the lowest THSG level, with a content of 0.001%. Significant differences were found even within Guilin samples. For example, Guilin samples, the lowest is 1.46%, while the highest is 6.37%. According to this data, we choose Guilin and Chongqing samples as the following materials.

SSH fragments comparison analysis
SSH fragments were amplified and cloned into pMD19-T vector. A total of 136 colonies were picked up, and the insertions were confirmed by PCR using universal primers. 75 clones among the 136 colonies were positive (55.1%) and sequenced, with insertion length ranging from 0.2 to 0.5 kb (data not shown). Removing redundant sequences, 12 fragments were obtained, and they were compared with the sequences in nucleotide and EST database at NCBI using the BLASTN algorithm. Results are shown in Tab. 3. There are

4 sequences with no significant similarity in the databases. The 4 sequences were all obtained from the roots and named GE, GW, CF, CA for subsequent research ("G" represent the origin of the sample, Guilin city; "C" represent Chongqing city.). All sequences were submitted to the NCBI GenBank (accession Nos. JZ469200 to JZ469209) for public domain use, except YCTS4 and GTS2 (ribosomal RNA).

Then the 12 sequences were compared with the transcriptome and DEG database of *F. multiflora* (unpublished data) using local BLAST. Significantly similar genes identified in the transcriptome database were initially aligned by blastx to protein databases nr, Swiss-Prot, KEGG and COG (e-value <0.00001), and then aligned by blastn to nucleotide database nt (e-value <0.00001). Proteins with the highest sequence similarity were retrieved with the given Unigenes and their functional annotations (Tab. 4).

3'5'RACE and full-length sequences comparison analysis
The full-length sequences of GE, GW, CF and CA were cloned by 3'RACE and 5'RACE. More than one full-length sequence was obtained for all genes but one, CF. All sequences have been submitted to the NCBI GenBank (accession Nos. KF054163 to KF054169) for public domain. Meanwhile the sequences were compared with the transcriptome and DEG database using local BLAST. Significantly similar Unigenes in transcriptome were aligned by blastx to protein databases, and aligned by blastn to nucleotide databases nt to annotate gene function (Tab. 5).

Pathway annotation
Different genes generally interact with each other to sustain their biological functions. Pathway-based analysis

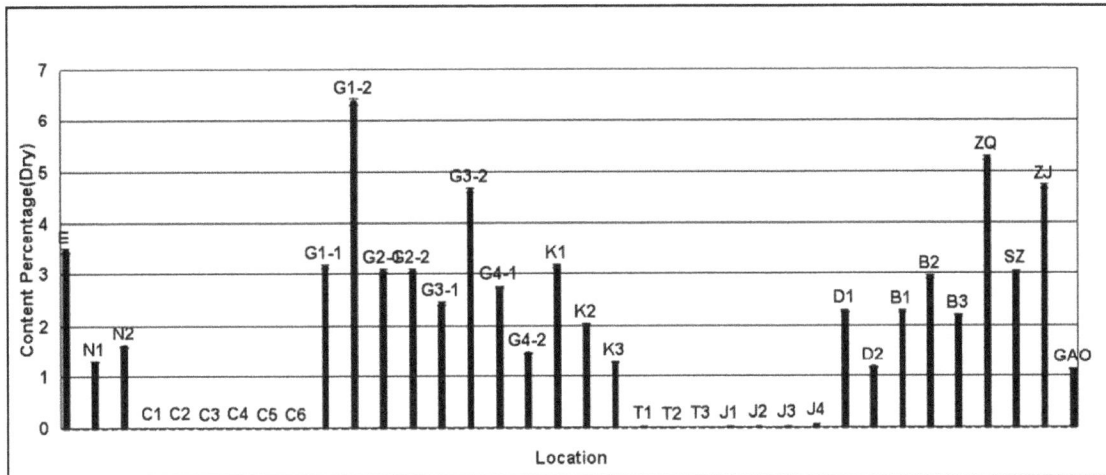

Fig. 1 THSG content in *F. multiflora* plants from different areas. Content percentage (dry): (THSG quality/quality of *F. multiflora* powder) × 100. All samples are the roots of *F. multiflora*, except Gl, Gs, Dl and Ds. B – Guangzhou City; C – Chongqing City; D – Deqing City; E – Dongguan City; G – Guilin City; GAO – Gaozhou City; J – Xinyu City; K – Jingxi County; N – Nanning City; SZ – Shenzhen City; T – Tianlin County; ZJ – Zhanjiang City; ZQ – Zhaoqing City. Dl – leaves of Deqing *F. multiflora*; Ds: stems of Deqing *F. multiflora*; Gl – leaves of Guilin *F. multiflora*; Gs: stems of Guilin *F. multiflora*.

Tab. 2 Quantification of THSG in *F. multiflora* plants from different areas ($n = 3$).

Sample	Average concentration (ng/μl)	Average concentration (%)	Standard deviation	Sample	Average concentration (ng/μl)	Average concentration (%)	Standard deviation
E	272.68	3.35	0.4487	K1	252.88	3.15	0.0125
N1	103.36	1.29	0.0105	K2	162.80	2.01	0.0091
N2	128.56	1.61	0.0004	K3	101.32	1.26	0.0009
C1	0.79	0.01	0.0004	T1	2.08	0.03	0.0001
C2	1.08	0.01	0.0002	T2	0.28	0.00	0.0002
C3	0.08	0.00	0.0000	T3	1.32	0.02	0.0069
C4	0.88	0.01	0.0001	J1	2.24	0.03	0.0005
C5	0.60	0.01	0.0002	J2	3.08	0.04	0.0001
C6	0.36	0.00	0.0006	J3	1.36	0.03	0.0007
G1-1	253.52	3.15	0.0116	J4	6.40	0.08	0.0013
G1-2	509.96	6.37	0.0582	D1	185.68	2.27	0.0244
G2-1	244.96	3.05	0.0127	D2	96.72	1.16	0.0104
G2-2	254.00	3.06	0.0086	B1	182.36	2.26	0.0758
G3-1	194.08	2.43	0.0006	B2	245.04	2.94	0.0995
G3-2	372.20	4.64	0.0179	B3	148.08	2.18	0.0534
G4-1	221.28	2.73	0.0135	ZQ	230.04	5.25	0.0411
G4-2	117.64	1.46	0.0039	ZJ	390.50	4.70	0.0572
SZ	248.25	3.03	0.0092	GAO	93.00	1.12	0.0121
Gl	3.22	0.04	0.0007	Dl	1.56	0.02	0.0105
Gs	84.36	1.01	0.0120	Ds	45.91	0.58	0.0057

Concentration %: (THSG quality/quality of *F. multiflora* powder) × 100. All samples are the roots of *F. multiflora*, except Gl, Gs, Dl and Ds. B – Guangzhou City; C – Chongqing City; D – Deqing City; E – Dongguan City; G – Guilin City; GAO – Gaozhou City; J – Xinyu City; K – Jingxi County; N – Nanning City; SZ – Shenzhen City; T – Tianlin County; ZJ – Zhanjiang City; ZQ – Zhaoqing City. Dl – leaves of Deqing *F. multiflora*; Ds: stems of Deqing *F. multiflora*; Gl – leaves of Guilin *F. multiflora*; Gs: stems of Guilin *F. multiflora*.

Tab. 3 Parameters of high similarity SSH sequence alignments, identified by BLASTN in nucleotide and EST NCBI databases.

Gene	Description	Query cover	e–value	Max ident.	Accession
YCTS1	*Rheum australe* catalase mRNA	27%	4e-35	94%	EU931220.1
YCTS2	*Salicornia europaea* TUB mRNA for alpha tubulin, partial cds	95%	2e-71	90%	AB437373.1
	Panicum virgatum clone PV_ABa073-K05, complete sequence	96%	2e-70	90%	AC243247.1
	Phyllostachys edulis cDNA clone: bphylf024i19, full insert sequence	96%	2e-70	90%	FP093842.1
YCTS3	*Mesembryanthemum crystallinum* major latex protein homolog mRNA, complete cds	44%	4e-29	85%	AF054445.1
YCTS4	*Sesbania drummondii* clone SSH-47_02_F03_T7, mRNA sequence	80%	5e-56	95%	DQ465800.1
GTS1	LEAF460 *Polygonum sibiricum* leaf *Knorringia sibirica* cDNA clone LEAF460, mRNA sequence	48%	8e-40	93%	FE903382.1
	XHBC-B22 *Anas platyrhynchos* muscle-related, library constructed by SSH *Anas platyrhynchos*	28%	2e-26	99%	HS410797.1
	cDNA, mRNA sequence, leaf_cn90 cDNA library from leaf of *Polygonum minus*	49%	2e-30	87%	JG745134.1
	Persicaria minor cDNA clone leaf_cn90, mRNA sequence				
GTS2	*Fallopia multiflora* voucher DB1 18S ribosomal RNA, gene, complete sequence	85%	9e-58	98%	EF153706.1
YTS1	*Brassica rapa* subsp. *pekinensis* thionin mRNA, complete cds	36%	4e-09	79%	AF090836.1
CTS1	*Rheum australe* catalase mRNA, complete cds	46%	2e-35	94%	EU931220.1

The character before "TS" represent the tester; "YCTS" means the fragment was found from both "Y" and "C" testers. C – roots of Chongqing *F. multiflora*; G – roots of Guilin *F. multiflora*; Y – leaves of Guilin *F. multiflora*.

Tab. 4 Parameters and annotation of high similarity SSH sequence alignments, identified by local BLAST in *F. multiflora* transcriptome databases.

Gene	Unigene	Score	e-value	Gene length	Nt-annotation	Nr-annotation	Accession
GE	CL3180.Contig2_TrCD	244	2e-64	1125	*Vitis vinifera* clone SS0AEB22YD16	PREDICTED: vacuolar protein-sorting-associated protein 37 homolog 2 (*Vitis vinifera*)	FQ388706
	CL3180.Contig1_TrCD	244	2e-64	1234	*Vitis vinifera* clone SS0AEB22YD16	PREDICTED: vacuolar protein-sorting-associated protein 37 homolog 2 (*Vitis vinifera*)	FQ388706
GW	CL5835.Contig2_TrCD	254	2e-67	1520	PREDICTED: *Vitis vinifera* GPN-loop GTPase 1 homolog-like (LOC100249973), mRNA	hypothetical protein VITISV_029075 (*Vitis vinifera*)	XM_002279556
	CL5835.Contig1_TrCD	254	2e-67	1523	PREDICTED: *Vitis vinifera* GPN-loop GTPase 1 homolog-like (LOC100249973), mRNA	hypothetical protein VITISV_029075 (*Vitis vinifera*)	XM_002279556
CF	CL9534.Contig2_TrCD	480	e-135	1155	PREDICTED: *Glycine max* uncharacterized protein LOC100797178 (LOC100797178), mRNA	predicted protein (*Populus trichocarpa*)	XM_003517340
	CL9534.Contig1_TrCD	480	e-135	1168	PREDICTED: *Glycine max* uncharacterized protein LOC100797178 (LOC100797178), mRNA	predicted protein (*Populus trichocarpa*)	XM_003517340
	CL8316.Contig2_TrCD	480	e-135	1137	PREDICTED: *Glycine max* uncharacterized protein LOC100797178 (LOC100797178), mRNA	predicted protein (*Populus trichocarpa*)	XM_003517340

Tab. 4 *(continued)*

Gene	Unigene	Score	*e*-value	Gene length	Nt-annotation	Nr-annotation	Accession
	CL8316.Contig1_TrCD	480	e-135	1150	PREDICTED: *Glycine max* uncharacterized protein LOC100797178 (LOC100797178), mRNA	predicted protein (*Populus trichocarpa*)	XM_003517340
CA	*Unigene22710_TrCD*	418	e-117	1174	PREDICTED: *Vitis vinifera* uncharacterized LOC100248615 (LOC100248615), mRNA	PREDICTED: uncharacterized protein LOC100248615 (*Vitis vinifera*)	XM_002274928
YCTS1	*CL3542.Contig3_TrCD*	234	5e-61	2031	*Rheum australe* catalase mRNA, complete cds	catalase (*Rheum australe*)	EU931220
	CL3542.Contig1_TrCD	111	5e-24	487	*Rheum australe* catalase mRNA, complete cds	catalase (*Rheum australe*)	EU931220
YCTS2	*CL2077.Contig2_TrCD*	434	e-121	1663	*P. amygdalus* mRNA for alpha-tubulin	tubulin alpha chain, putative (*Ricinus communis*)	X67162
	CL2077.Contig1_TrCD	111	3e-24	564	*Populus trichocarpa* tubulin alpha-8 chain (TUA8), mRNA	alpha-tubulin (*Gossypium hirsutum*)	EF151304
	CL2077.Contig7_TrCD	96	2e-19	394	PREDICTED: *Vitis vinifera* tubulin alpha-3 chain-like (LOC100258974), mRNA	alpha-tubulin (*Miscanthus sinensis*)	XM_002281631
YCTS3	*CL5036.Contig1_TrCD*	521	e-147	756	*Mesembryanthemum crystallinum* major latex protein homolog mRNA, complete cds	major latex protein homolog (*Mesembryanthemum crystallinum*)	AF054445
	CL5036.Contig2_TrCD	210	5e-54	455	-	major latex protein homolog (*Mesembryanthemum crystallinum*)	AF054445
YCTS4	*CL8135.Contig1_TrCD*	236	5e-62	1672	*Polygonum sachalinense* 26S ribosomal RNA gene, complete sequence	hypothetical protein MTR_5g050970 (*Medicago truncatula*)	AF479085
GTS1	*CL8017.Contig1_TrCD*	212	6e-55	834	*Phyllostachys edulis* cDNA clone: bphyst022a10, full insert sequence	Chain A, structure of Psbp protein from *Spinacia oleracea* at 1.98 A resolution	FP094014
GTS2	*CL10591.Contig1_TrCD*	248	e-65	2806	*Fallopia multiflora* voucher DX 18S ribosomal RNA gene, complete sequence	Cytochrome P450 likeTBP (*Medicago truncatula*)	EF153701
CTS1	*CL3542.Contig3_TrCD*	230	e-60	2031	*Rheum australe* catalase mRNA, complete cds	catalase (*Rheum australe*)	EU931220
	CL3542.Contig1_TrCD	111	e-24	487	*Rheum australe* catalase mRNA, complete cds	catalase (*Rheum australe*)	EU931220
YTS1	*CL5290.Contig2_TrCD*	486	e-137	780	*Brassica napus* clone Bn 3872 thionin precursor, mRNA, complete cds	crambin precursor = thionin variant Thi2Ca10 (*Crambe abyssinica*, seeds, Peptide Partial, 134 aa)	EU887266
	CL5290.Contig1_TrCD	486	e-137	1094	*Boechera divaricarpa* isolate SLW-C-C05 mRNA sequence	crambin precursor = thionin variant Thi2Ca10 (*Crambe abyssinica*, seeds, Peptide Partial, 134 aa)	DQ226844
	Unigene31799_TrCD	113	9e-25	712	*Boechera divaricarpa* isolate SLW-C-C05 mRNA sequence	crambin precursor = thionin variant Thi2Ca10 (*Crambe abyssinica*, seeds, Peptide Partial, 134 aa)	DQ226844
	CL5290.Contig3_TrCD	113	9e-25	427	*Brassica napus* clone Bn 3872 thionin precursor, mRNA, complete cds	thionin precursor (*Brassica napus*)	EU887266

Tab. 5 Parameters and annotation of high similarity full-length sequences alignments, identified by local BLAST in *F. multiflora* transcriptome databases.

Gene	Unigene	Gene length	Similarity bp	e-value	Nt-annotation	Nr-annotation	Accession
GE*fl*	*CL3180.Contig2_TrCD*	1125	805	0	*Vitis vinifera* clone SS0AEB22YD16	PREDICTED: vacuolar protein-sorting-associated protein 37 homolog 2 (*Vitis vinifera*)	FQ388706
	CL3180.Contig1_TrCD	1234	805	0	*Vitis vinifera* clone SS0AEB22YD16	PREDICTED: vacuolar protein-sorting-associated protein 37 homolog 2 (*Vitis vinifera*)	FQ388706
GW*fl1*	*Unigene6178_TrCD*	532	451	0	PREDICTED: *Vitis vinifera* 40S ribosomal protein S30-like (LOC100241142), mRNA	unnamed protein product (*Vitis vinifera*)	XM_002281340
	Unigene20297_TrCD	467	190	e-46	PREDICTED: *Vitis vinifera* 40S ribosomal protein S30-like (LOC100252467), mRNA	unnamed protein product (*Vitis vinifera*)	XM_002282997
GW*fl2*	*CL3625.Contig1_TrCD*	1188	436	0	PREDICTED: *Vitis vinifera* katanin p80 WD40 repeat-containing subunit B1 homolog 1-like (LOC100247509), mRNA	PREDICTED: katanin p80 WD40 repeat-containing subunit B1 homolog 1-like (*Glycine max*)	XM_002264675
GW*fl3*	*CL3504.Contig1_TrCD*	868	664	0	*Vigna radiata* ubiquitin-conjugating enzyme E2 mRNA, complete cds	PREDICTED: ubiquitin-conjugating enzyme E2 10-like (*Glycine max*)	FJ436357
	Unigene6181_TrCD	979	445	e-62	*Camellia sinensis* ubiquitin-conjugating enzyme E2 mRNA, complete cds	PREDICTED: ubiquitin-conjugating enzyme E2 10-like (*Glycine max*)	JN400596
	CL3504.Contig3_TrCD	973	443	4e-47	PREDICTED: *Vitis vinifera* ubiquitin-conjugating enzyme (LOC100232965), mRNA	predicted protein (*Populus trichocarpa*)	XM_003633920
	CL3504.Contig4_TrCD	915	405	7e-46	PREDICTED: *Vitis vinifera* ubiquitin-conjugating enzyme (LOC100232965), mRNA	ubiquitin-conjugating enzyme E2-17 kDa (*Solanum lycopersicum*)	XM_003633920
	Unigene27674_TrCD	818	306	2e-39	PREDICTED: *Vitis vinifera* ubiquitin-conjugating enzyme (LOC100232965), mRNA	E2 ubiquitin-conjugating enzyme UBC10 (*Brassica napus*)	XM_003633920
	Unigene7679_TrCD	520	243	5e-16	*Solanum tuberosum* clone 116D11 ubiquitin-conjugating protein-like mRNA, complete cds	ubiquitin-conjugating enzyme E2 28 (*Arabidopsis thaliana*)	DQ284472
CF*fl*	*CL9534.Contig2_TrCD*	1155	144	3e-72	PREDICTED: *Glycine max* uncharacterized protein LOC100797178 (LOC100797178), mRNA	predicted protein (*Populus trichocarpa*)	XM_003517340
	CL9534.Contig1_TrCD	1168	144	3e-72	PREDICTED: *Glycine max* uncharacterized protein LOC100797178 (LOC100797178), mRNA	predicted protein (*Populus trichocarpa*)	XM_003517340
	CL8316.Contig2_TrCD	1137	144	3e-72	PREDICTED: *Glycine max* uncharacterized protein LOC100797178 (LOC100797178), mRNA	predicted protein (*Populus trichocarpa*)	XM_003517340

Tab. 5 *(continued)*

Gene	Unigene	Gene length	Similarity bp	e-value	Nt-annotation	Nr-annotation	Accession
	CL8316.Contig1_TrCD	1150	144	3e-72	PREDICTED: *Glycine max* uncharacterized protein LOC100797178 (LOC100797178), mRNA	predicted protein (*Populus trichocarpa*)	XM_003517340
CWfl1	Unigene22710_TrCD	1174	316	3e-158	PREDICTED: *Vitis vinifera* uncharacterized LOC100248615 (LOC100248615), mRNA	PREDICTED: uncharacterized protein LOC100248615 (*Vitis vinifera*)	XM_002274928
	CL4034.Contig3_TrCD	2016	209	7e-104	PREDICTED: *Vitis vinifera* pumilio homolog 1-like (LOC100243338), mRNA	PREDICTED: pumilio homolog 1-like (*Vitis vinifera*)	XM_002283155
CWfl2	CL2293.Contig2_TrCD	2481	165	9e-82	PREDICTED: *Vitis vinifera* uncharacterized LOC100246727 (LOC100246727), mRNA	expressed protein, putative (*Ricinus communis*)	XM_002273173
	CL2293.Contig1_TrCD	2572	165	9e-82	PREDICTED: *Vitis vinifera* uncharacterized LOC100246727 (LOC100246727), mRNA	expressed protein, putative (*Ricinus communis*)	XM_002273173

helps to further understand genes' biological functions. According to KEGG, the pathways of isolated genes are diverse in function. These genes were mainly involved in metabolic pathways; ubiquitin mediated proteolysis, pathogenic *E. coli* infection, amyotrophic lateral sclerosis (ALS), tryptophan metabolism, photosynthesis, gap junction, RIG-I-like receptor signaling pathway, mRNA surveillance pathway, peroxisome, endocytosis, ribosome, biosynthesis of secondary metabolites, phagosome, protein processing in endoplasmic reticulum, methane metabolism, glyoxylate and dicarboxylate metabolism, microbial metabolism in diverse environments, and RNA transport (Tab. 6).

Comparison of genes expression and THSG content in different tissues

Genes expression of SSH fragments in different plant tissues according to DGE database was shown in Fig. 2. Expression of *GE*, *GW*, *CF* and *YTS1* in different plant tissues has no obvious difference. Gene expression of full- length sequences in different plant tissues has significant differences (Fig. 3), except *GEfl* and *CFfl*. Comparison between the expression of isolated genes and THSG content in different tissues revealed that the expression of *GWfl1* (*Unigene20297_TrCD*), *GWfl3* (*Unigene27674*) and *CWfl1* (*CL4034.Contig3_TrCD*) is in parallel with THSG content in different tissues. Expression of *CA*, *YCTS4* and *CWfl1* (*Unigene22710_TrCD*) is contrary to THSG content in different tissues. Expression of *CA*, *CTS1*, *GWfl2* (*CL3625.Contig1_TrCD*), *GWfl3* (*CL3504. Contig1_TrCD*) , *YCTS4* and *CWfl1* (*Unigene22710_TrCD*) in Cr are higher than in all D sample tissues.

Discussion

As of any secondary metabolite in plant, the production of THSG in *F. multiflora* is influenced by multiple factors, such as temperature, climate, altitude, precipitation etc. Gene

screening between samples of distinct product content by SSH could display the mRNA information associated with the production of THSG. DGE profile of different plant tissues contains genes expression information in these tissues. Comparison with DGE profile showed the high level of activity of targeted genes in each tissue. The genes whose expression was significantly different were selected. Similar pattern of difference in gene abundance and THSG content in different tissues confirm that the gene might be associated with THSG biosynthesis.

In our study, the expression of *GWfl1* (*Unigene20297_TrCD*), *GWfl3* (*Unigene27674_TrCD*) and *CWfl1* (*CL4034. Contig3_TrCD*) is in parallel with THSG content in different tissues. Blast in NCBI database showed they might code 40S ribosomal protein, ubiquitin-conjugating enzyme and RNA-binding protein of the Puf family. The predicted protein function indicated there should be active protein degradation and expression behavior in root sample. Expression of *CA*, *YCTS4* and *CWfl1* (*Unigene22710_TrCD*) is contrary to THSG content in different tissues. And *CTS1* is higher expressed in Cr than Dr, Ds and Dl. They might be inhibitors of THSG biosynthesis. As roots always contain the highest amount of THSG, we consider the root is most likely the site of THSG synthesis. The fragments, which were obtained from Guilin root tester should be focused on. *GW* (*CL5835.Contig1_TrCD/CL5835.Contig2_TrCD*) and *GTS1* (*CL8017.Contig1_TrCD*) might be involved in providing energy for secondary metabolism via GTP metabolism and photosynthesis. *GWfl2* (*CL3625.Contig1_TrCD*) is annotated as Katanin p80 WD40 repeat-containing subunit B1, *GWfl3* (*CL3504.Contig1_TrCD*) as Ubiquitin-protein ligase. As these fragments were identified in SSH when Guilin root tester was applied, it means the intermediate process of THSG biosynthesis should be fast and efficient, and related protein is likely to be subjected to rapid degradation. *GTS2* (*CL10591. Contig1_TrCD*) has high homology with Cytochrome P450

Tab. 6 Pathway annotation of SSH sequences and full-length sequences participated.

Gene	KOID	Definition	Pathway
GE	K12185	VPS37; ESCRT-I complex subunit VPS37	ko04144 endocytosis
GW	K06883		
CF	-		
CA	-		
YCTS1	K03781	katE, CAT, catB, srpA; catalase (EC:1.11.1.6)	ko01120 microbial metabolism in diverse environments; ko01100 metabolic pathways; ko01110 biosynthesis of secondary metabolites; ko00630 glyoxylate and dicarboxylate metabolism; ko00680 methane metabolism; ko00380 tryptophan metabolism; ko04146 peroxisome; ko05014 amyotrophic lateral sclerosis (ALS)
YCTS2	K07374	TUBA; tubulin alpha	ko04145 phagosome; ko04540 gap junction; ko05130 pathogenic *Escherichia coli* infection
YCTS3	-		
YCTS4	-		
GTS1	K02717	psbP; photosystem II oxygen-evolving enhancer protein 2	ko01100 metabolic pathways; ko00195 photosynthesis
GTS2	K14325	RNPS1; RNA-binding protein with serine-rich domain 1	ko03013 RNA transport; ko03015 mRNA surveillance pathway
CTS1	K03781	katE, CAT, catB, srpA; catalase (EC:1.11.1.6)	ko01120 microbial metabolism in diverse environments; ko01100 metabolic pathways; ko01110 biosynthesis of secondary metabolites; ko00630 glyoxylate and dicarboxylate metabolism; ko00680 methane metabolism; ko00380 tryptophan metabolism; ko04146 peroxisome; ko05014 amyotrophic lateral sclerosis (ALS)
YTS1	-		
GEfl	K12185	VPS37; ESCRT-I complex subunit VPS37	ko04144 endocytosis
GWfl1	K02983	RP-S30e, RPS30; small subunit ribosomal protein S30e	ko03010 ribosome
GWfl2	-		
GWfl3	K06689	UBE2D_E, UBC4, UBC5; ubiquitin-conjugating enzyme E2 D/E (EC:6.3.2.19)	ko04141 protein processing in endoplasmic reticulum; ko04120 ubiquitin mediated proteolysis
CFfl	-		
CWfl1	K14844	PUF6; pumilio homology domain family member 6	ko03009 ribosome biogenesis
CWfl2	K12655	OTUD5, DUBA; OTU domain-containing protein 5 (EC:3.1.2.15)	ko04622 RIG-I-like receptor signaling pathway

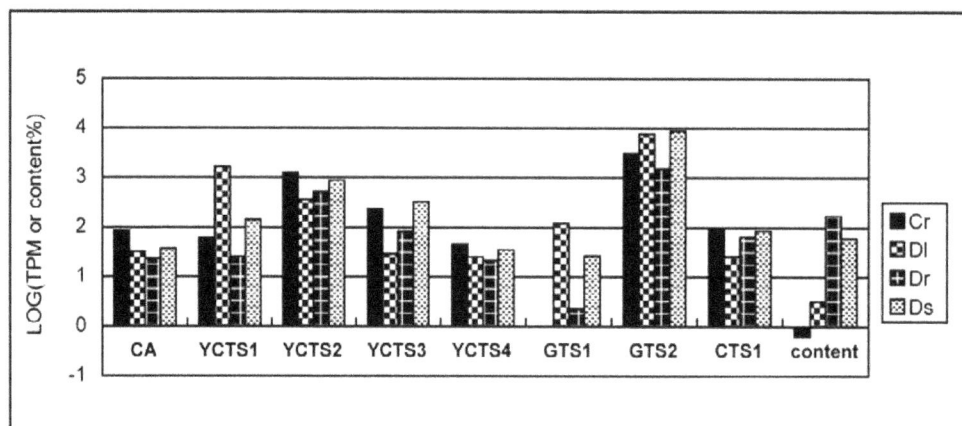

Fig. 2 Comparison of SSH fragments expression and THSG content in different tissues. Cr – roots of Chongqing *F. multiflora*; Dl – leaves of Deqing *F. multiflora*; Dr – roots of Deqing *F. multiflora*; Ds – stems of Deqing *F. multiflora*; TPM – transcripts per million clean tags (all data was calculated as the base of log10).

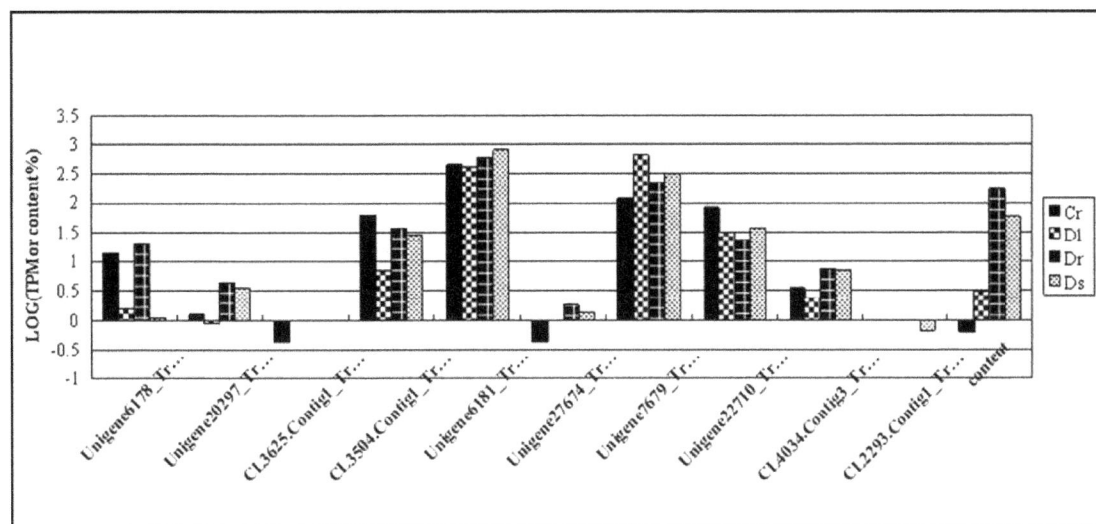

Fig. 3 Comparison of full-length sequences expression and THSG content in different tissues . Cr – roots of Chongqing *F. multiflora*; Dl – leaves of Deqing *F. multiflora*; Dr – roots of Deqing *F. multiflora*; Ds – stems of Deqing *F. multiflora*; TPM – transcripts per million clean tags (all data was calculated as the base of log10).

in *Medicago truncatula*. Cytochrome P450 was found be involved as multifunctional oxidases in the biosynthesis of many secondary metabolites like triterpenoid [17], glycyrrhizin [18], hemolytic saponins [19] and many more. Does it take part in THSG biosynthesis?

If THSG is derived from phenylalanine pathway with other stilbenes, then the *CTS1* gene obtained from Chongqing root tester may be related with the inhibition of THSG biosynthesis. *CTS1* gene is annotated as catalase and participates in "biosynthesis of secondary metabolites (ko01110)"; "glyoxylate and dicarboxylate metabolism (ko00630)" and

"tryptophan metabolism (ko00380)". It may regulate the metabolism of tryptophan to influence phenylalanine synthesis. Whether or not, THSG biosynthesis must be complicated and correlate with multiple pathways. More than just a phenylalanine pathway also must have common effect of ATP energy metabolism active and other related reactions to strengthen or weaken the process. Our research isolated several candidate genes that might participate in the THSG biosynthesis. Further research will focus on the characterization of candidate genes and gene screening to transcriptome in more diverse tissues.

Acknowledgments
This work was funded by grants from Natural Science Foundation of Guangdong Province (10151001002000012), Science and Technology Department of Guangdong Province (2010B060200009).

Authors' contributions
The following declarations about authors' contributions to the research have been made: conceived and designed the experiments: WZ, SS, ZL; performed the experiments: WZ, DL, KZ, XL, YY; performed the analysis: WZ, SZ; wrote the paper: WZ; revised the paper: SZ, SS.

References
1. Steele ML, Truong J, Govindaraghavan S, Ooi L, Sucher NJ, Münch G. Cytoprotective properties of traditional Chinese medicinal herbal extracts in hydrogen peroxide challenged human U373 astroglia cells. Neurochem Int. 2013;62(5):522–529.

2. Xie W, Zhao Y, Du L. Emerging approaches of traditional Chinese medicine formulas for the treatment of hyperlipidemia. J Ethnopharmacol. 2012;140(2):345–367.

3. Jang JY, Kim HN, Kim YR, Choi YW, Choi YH, Lee JH, et al. Hexane extract from *Polygonum multiflorum* attenuates glutamate-induced apoptosis in primary cultured cortical neurons. J Ethnopharmacol. 2013;145(1):261–268.

4. Wang M, Zhao R, Wang W, Mao X, Yu J. Lipid regulation effects of Polygoni Multiflori Radix, its processed products and its major substances on steatosis human liver cell line L02. J Ethnopharmacol. 2012;139(1):287–293.

5. Lv G, Lou Z, Chen S, Gu H, Shan L. Pharmacokinetics and tissue distribution of 2,3,5,4'-tetrahydroxystilbene-2-O-β-D-glucoside from traditional Chinese medicine *Polygonum multiflorum* following oral administration to rats. J Ethnopharmacol. 2011;137(1):449–456.

6. Zhang JK, Yang L, Meng GL, Fan J, Chen JZ, He QZ, et al. Protective effect of tetrahydroxystilbene glucoside against hydrogen peroxide-induced dysfunction and oxidative stress in osteoblastic MC3T3-E1 cells. Eur J Pharmacol. 2012;689(1–3):31–37.

7. Yao WJ, Fan WJ, Huang C, Zhong H, Chen XF, Zhang W. Proteomic analysis for anti-atherosclerotic effect of tetrahydroxystilbene glucoside in rats. Biomed Pharmacother. 2013;67(2):140–145.

8. Sun FL, Zhang L, Zhang RY, Li L. Tetrahydroxystilbene glucoside protects human neuroblastoma SH-SY5Y cells against MPP+-induced cytotoxicity. Eur J Pharmacol. 2011;660(2/3):283–290.

9. Chong J, Poutaraud A, Hugueney P. Metabolism and roles of stilbenes

in plants. Plant Sci. 2009;177(3):143–155.

10. Hartmann T. From waste products to ecochemicals: fifty years research of plant secondary metabolism. Phytochemistry. 2007;68(22–24):2831–2846.

11. Sheng SJ, Liu ZY, Zhao W, Shao L, Zhao SJ. Molecular analysis of a type III polyketide synthase gene in *Fallopia multiflora*. Biologia (Bratisl.). 2010;65(6):939–946.

12. Shao L, Zhao SJ, Cui TB, Liu ZY, Zhao W. 2,3,5,4′-tetrahydroxystilbene-2-O-β-D-glycoside biosynthesis by suspension cells cultures of *Polygonum multiflorum* Thunb and production enhancement by methyl jasmonate and salicylic acid. Molecules. 2012;17(12):2240–2247.

13. Peng T, Zhu XF, Fan QJ, Sun PP, Liu JH. Identification and characterization of low temperature stress responsive genes in *Poncirus trifoliata* by suppression subtractive hybridization. Gene. 2012;492(1):220–228.

14. Gao M, Wang Q, Wan R, Fei Z, Wang X. Identification of genes differentially expressed in grapevine associated with resistance to *Elsinoe ampelina* through suppressive subtraction hybridization. Plant Physiol Biochem. 2012;58:253–268.

15. Pimentel P, Salvatierra A, Moya-León MA, Herrera R. Isolation of genes differentially expressed during development and ripening of *Fragaria chiloensis* fruit by suppression subtractive hybridization. J Plant Physiol. 2010;167(14):1179–1187.

16. Wang YS, Gao LP, Wang ZR, Liu YJ, Sun ML, Yang DQ, et al. Light-induced expression of genes involved in phenylpropanoid biosynthetic pathways in callus of tea [*Camellia sinensis* (L.) O. Kuntze]. Sci Hortic. 2012;133:72–83.

17. Fukushima EO, Seki H, Ohyama K, Ono E, Umemoto N, Mizutani M, et al. CYP716A subfamily members are multifunctional oxidases in triterpenoid biosynthesis. Plant Cell Physiol. 2011;52(12):2050–2061.

18. Seki H, Sawai S, Ohyama K, Mizutani M, Ohnishi T, Sudo H, et al. Triterpene functional genomics in licorice for identification of CYP72A154 involved in the biosynthesis of glycyrrhizin. Plant Cell. 2011;23(11):4112–4123.

19. Carelli M, Biazzi E, Panara F, Tava A, Scaramelli L, Porceddu A, et al. *Medicago truncatula* CYP716A12 is a multifunctional oxidase involved in the biosynthesis of hemolytic saponins. Plant Cell. 2011;23(8):3070–3081.

4

Production of intergeneric allotetraploid between autotetraploid non-heading Chinese cabbage (*Brassica campestris* ssp. *chinensis* Makino) and autotetraploid radish (*Raphanus sativus* L.)

Sun Cheng-Zhen, Li Ying, Zhang Shu-Ning*, Zheng Jin-Shuang

State Key Laboratory of Crop Genetics & Germplasm Enhancement, Horticultural College, Nanjing Agricultural University, Nanjing, 210095, China

Abstract

Intergeneric hybrids between non-heading Chinese cabbage (*Brassica campestris* ssp. *chinensis* Makino; 2n = 4x = 40) and radish (*Raphanus sativus* L.; 2n = 4x = 36) were obtained through ovary culture and embryo rescue. Some hybrid embryos (0.11 per ovary) were produced, but only 4 of them germinated. As most hybrid embryos failed to develop into plantlets directly, plants were regenerated by inducing shoots on the cultured cotyledon and inducing roots on the root induction medium. All hybrid plants were morphologically uniform. They resembled the non-heading Chinese cabbage in the long-lived habit, the plant status, the vernalization requirement and the petiole color, while the petiole shape, leaf venation pattern and flowers were more similar to those of radish. Upon examination of the flowers, these were found to have normal pistil, but rudimentary anthers with non-functional pollen grains. The somatic chromosome number of F1 plants was 38. Analysis of SSR banding patterns provided additional confirmation of hybridity.

Keywords: intergeneric allotetraploid; embryo culture; simple sequence repeat

Introduction

It is estimated that 47% of all flowering plants and 95% of all pteridophytes are polyploids and that the majority of these are allopolyploids [1,2], which is a widespread and major force of evolution in plants [3]. Allopolyploidy is often accompanied by major structural, cytogenetic, epigenetic and functional changes to the genome, leading to new phenotypes and to reproductive isolation [4,5]. In addition, the permanent heterozygosity fixation of the allopolyploid [6] has the potential to offer a substantial heterozygote advantage. Despite these potential benefits, allopolyploid is an enormous challenge with the orchestration of gene expression, DNA replication, and chromosome pairing. For these reasons, investigation of allopolyploids is very important. The newly synthesized allopolyploid is an ideal model system since it can offer an opportunity to study the response to this genomic change from defined parents.

Allopolyploid formation can occur by two main pathways, the so-called "one-step" and "two-step" models [5]. In the one-step model, the allopolyploid arises directly from an interspecific cross by the fusion of either two unreduced (2n) gametes from diploid parents or two normal (n = 2x) gametes from tetraploid parents. By contrast, in the two-step model,

an interspecific F1 hybrid is first formed and the polyploidy is derived from it either from a fertile shoot generated by meristematic tissues having experienced a somatic doubling or by fusion of two 2n gametes produced by the F1 hybrid itself [6]. The production of 2n gametes appears to occur at a surprisingly low rate. Ramsey and Schemske estimated its frequency at 0.56% [7]. Although with the use of colchicine the frequency of the somatic doubling has increased a lot, the effective diploidization rate is still very low (10.5%) [8]. In addition, problems with chimeras, abnormal phenotypes and sterility also occur [9]. Little attention, however, has been focused on the use of this method, although a "synthetic" allotetraploid had been obtained by crossing a tetraploid *Arabidopsis thaliana* (2n = 4x = 20) and *A. arenosa* (2n = 4x = 32) [10].

In the crop *Brassica*, breeders have resorted to varying degrees of hybridization involving close relatives of it in their search for novel traits in developing new and improved varieties [11]. The non-heading Chinese cabbage (*Brassica campestris* ssp. *chinensis* Makino) is a main vegetable, which grows in south of China, and it has a long history of cultivation in our country. Radish (*Raphanus sativus* L.) is cultivated worldwide. It possesses desirable agronomic characters, such as resistance to white rust (*Albugo candida* [12], BCN (*Heterodera schachtii*) [13,14] and culbroot (*Plasmodiophora brassicae*) [15], as well as resistance to pod shattering [16]. Besides, various related wild species have attracted research attention as potential germplasms

* Corresponding author. Email: snzhang@njau.edu.cn

Handling Editor: Beata Zagórska-Marek

for improvement of *Brassica* crops [17]. Therefore, it can be used as a gene donor for the modification of the non-heading Chinese cabbage. Here, we report the successful production of allotetraploid by intergeneric hybridization between autotetraploid non-heading Chinese cabbage (*Brassica campestris* ssp. *chinensis* Makino) and autotetraploid radish (*Raphanus sativus* L.) aimed at enriching the gene pool of the non-heading Chinese cabbage (*Brassica campestris* ssp. *chinensis* Makino) and creating useful research material for further understanding of the relationship and the genomic structure between the two genera.

Material and methods

Plant material and intergeneric hybridization

The plant material consisted of the two autotetraploid cultivars, non-heading Chinese cabbage (*Brassica campestris* ssp. *chinensis* Makino; maternal parent, 2n = 4x = 40) and radish (*Raphanus sativus* L.; paternal parent, 2n = 4x = 36). Seeds of two cultivars were grown on experimental fields of Jiangpu Farm, Nanjing Agricultural University. Flowers were protected from foreign pollen two days before anthesis and intergeneric crosses were made by hand.

Embryo culture

The ovaries were excised 5–10 days after pollination and sterilized with 70% ethanol for 30 s followed by a sodium hypochlorite solution containing 1% active chlorite for 10 min. After washing in sterile distilled water three times, the ovaries were cultured on MS [18] hormone free solid medium containing 500 mg/l casein hydrolysate. Fifteen days later, the embryos were isolated from the ovaries and transferred to hormone-free MS medium supplemented with 5% coconut milk and 500 mg/l casein hydrolysate. As most embryos failed to develop directly into plantlets [19], cotyledons were cut off and cultured on MS medium containing 2 mg/l 6-benzyladenine (BA) and 0.1 mg/l NAA and 5% coconut milk in order to induce shoot regeneration. After the shoots regenerated, they were transplanted to hormone-free ½ MS medium supplemented with 0.2 mg/ml NAA for root induction. All cultures were incubated at 25°C in a 16-hour photoperiod. The embryo-rescued plants with well-developed root system were hardened for 4–8

days at 10–12°C, 14-hour photoperiod and then transferred into pots with soil for normal growth under glasshouse conditions [20].

Characterization of hybrids

Hybrid identity of F1 plants was confirmed by morphological examination, chromosome analysis and further characterize by simple sequence repeat (SSR) analysis.

Chromosome counts were carried out on root tips from hybrid plants and pretreated with 0.002 M 8-hydroxyquinoline for 4 hours. Material was fixed in 3:1 alcohol-glacial acetic acid, hydrolyzed in 1 N HCl 60°C for 6 min and stained with leuco-basic fuchsine 30 min, and then squash preparations were made using 45% acetic acid [21]. About 0.3 g fresh leaves were used to extract genomic DNA using the cetyl-trimethyl-ammonium bromide (CTAB) method [22]. An Eppendorf protein and nucleic determine instrument was used for determining DNA concentration. Primer sequences for SSR markers obtained from various sources [23–25] were used (Tab. 1). The SSR reactions were performed in a 20 µl volume containing 60 ng DNA, 0.5 µmol/l forward and reverse primer, 0.2 mmol/l dNTPs, 1.0 mmol/l MgCl$_2$, and 0.5 U Taq DNA polymerase. The PCR procedure was programmed at 95°C for 2 min, 94°C for 30 s, 55°C for 30 s, 72°C for 30 s for 35 cycles, and then 72°C for 10 min. The products were separated on 5% vertical polyacrylamide gel. The gel was run at a 150 V constant voltage for 1 to 1.5 h before silver staining.

For cytological studies young anthers were fixed in Carnoy solution and squashed in 1% acetocarmine [26].

Results

Embryo culture

After 1–2 weeks of culture, the ovaries were observed turgid. Of all the 100 ovaries only 11 embryos developed to the mature cotyledonary stage after 3 weeks in culture (Tab. 2). Of 11 cultured embryos, 4 embryos germinated (germination rate = 36.3%) with the help of embryo culture in vitro. The 4 embryos germinated halted their development when their shoots were 1 cm in length, and did not lead to whole plants. Regeneration plantlets were induced from the cotyledon sections of the 4 germinated embryos on MS

Tab. 1 Primers for SSR marker assays.

Primer name	Forward primers	Reverse primers
FITO 137	ATGGGTAAGTCTCGTAAATG	AAACCGAATAAACCGAAA
Na10F06	CTCTTCGGTTCGATCCTCG	TTTTTAACAGGAACGGTGGC
Na12H09	AGGCGTCTATCTCGAAATGC	CGTTTTTCAGAATCTCGTTGC
Ni4A03	ACACAGAAACATCAAACATACC	GGACCGGTTTTATTTGTTCG
Ol09A06	TGTGTGAAAGCTTGAAACAG	TAGGATTTTTTTGTTCACCG
Ol10F11	TTTGGAACGTCCGTAGAAGG	CAGCTGACTTCGAAAGGTCC
Ol12F11	AAGGACTCATCGTGCAATCC	GTGTCAGTGGCTACAGAGAC
Na14D09	GATCAACGTAAGGTCGCCTC	GAATCCAACGGATCAGAAGC

Tab. 2 Embryo production from ovary culture in hybridizations between *B. campestris* ssp. *chinensis* Makino and *R. sativus*.

Crosses	No. of cultured ovaries	No. of embryos produced	No. of embryos survived	Rate of germination (%)
B. campestris ssp. *chinensis* Makino × *R. sativus*	100	11	4	36.3

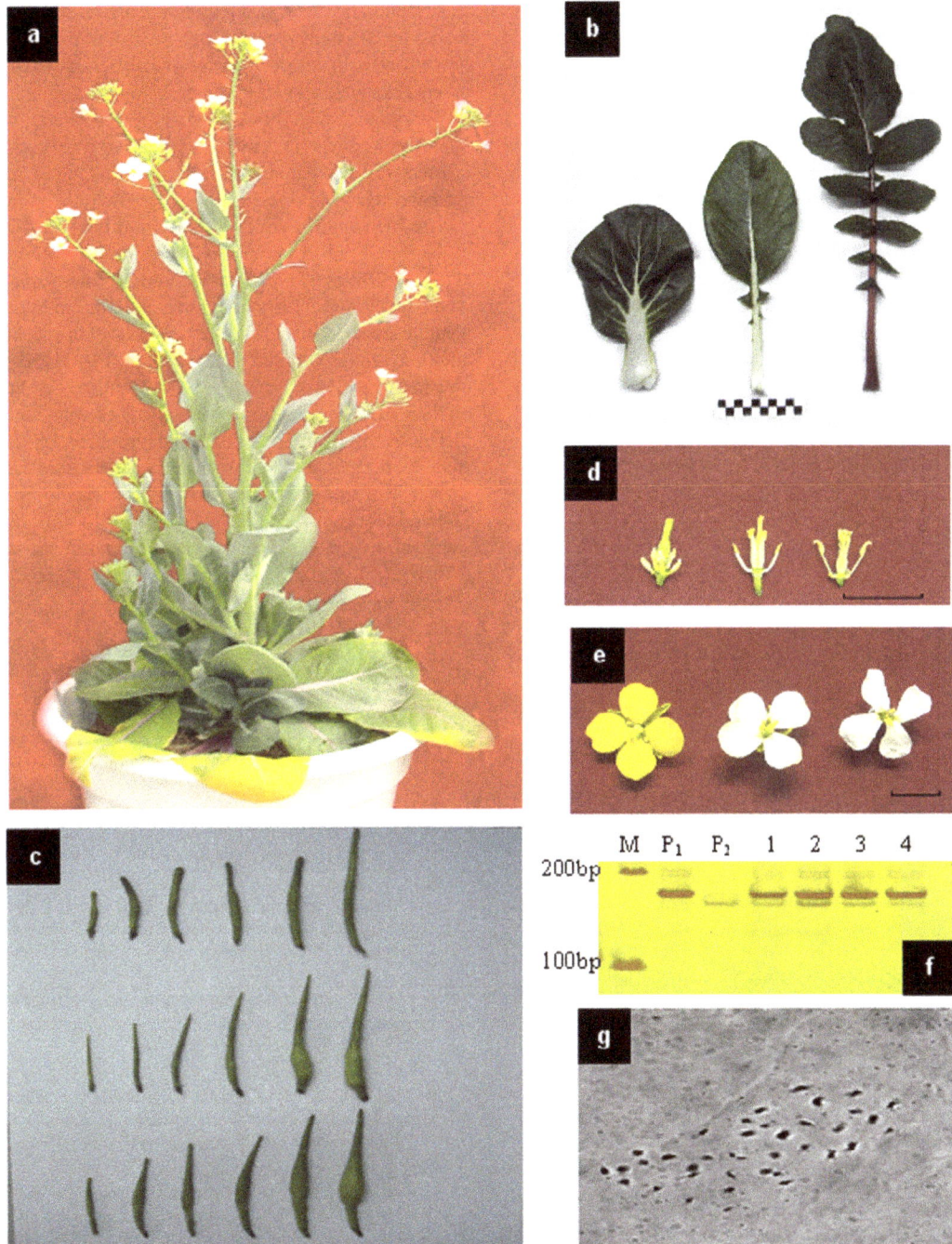

Fig. 1 Hybrid identity of F1 plants. **a** Flowering stage of F1. **b** Leaves of *B. campestris* (left), *R. sativus* (right) and F1 (middle). **c** Siliques of *B. campestris* (up), *R. sativus* (middle) and F1 (bottom). **d** Stamens and pistil of *B. campestris* (left), *R. sativus* (right) and F1 (middle). **e** Flower of *B. campestris* (left), *R. sativus* (right) and F1 (middle). **f** Amplification results of primer Na14D09 on *B. campestris* (P1), *R. sativus* (P2) and F1 (1, 2, 3 and 4). **g** Chromosome number of F1 (2n = 4x = 38; 1000×). Scale bars: **b–e** 1 cm.

medium supplemented with 2.0 mg/l 6-benzyladenine and 0.1 mg/l NAA, and then they were transplanted to the root induction medium. A total of 4 lines were obtained with several numbers of plants.

Characterization of hybrids

All hybrid plants were morphologically uniform and grew vigorously.

The hybrids were intermediate in size and shape (Fig. 1a). They resembled the non-heading Chinese cabbage in the long-lived habit, the plant status, the vernalization requirement, the petiole color (Fig. 1b). Petiole and silique shape, leaf venation pattern and flower color were more similar to those of radish (Fig. 1c–e). Upon examination of the flowers, these were found to have normal pistil, but rudimentary anthers with non-functional pollen grains (Fig. 1d).

The somatic chromosome number of the regenerated plants was counted at the middle stage of cell division. The results showed that the chromosome number of all plants tested was 38 (Fig. 1g), indicating that these regenerated plants were all true hybrids of $B. campestris$ (2n = 4x = 40) $\times R. sativus$ (2n = 4x = 36).

Out of 8 SSR primer pairs, only 1 pair (12.5%) SSR primers (Tab. 1) had polymorphic between parents gnomic DNA. All of the hybrids were tested by SSR analysis with 8 primer pairs. SSR analysis indicated that the hybrids had hybrid patterns containing characteristic bands from $R. sativus$ in addition to the $B. campestris$, which exhibits codominant (Fig. 1f).

Discussion

The present paper describes the production of a new allotetraploid by intergeneric hybridization between autotetraploid non-heading Chinese cabbage ($Brassica campestris$ ssp. $chinensis$ Makino) and autotetraploid radish ($Raphanus sativus$ L.) through ovary culture and embryo rescue. We found that the production of allotetraploid was about 4%, which compared favorably with percentages reported for $B. napus \times D. siifolia$ (11.5%) [27] and $B. campestris \times M. arvensis$ (5%) [19]. The interspecific hybridizations we obtained confirm that crossing between tetraploid parents is a useful method in producing synthetic allotetraploid. In addition, the use of two tetraploid parents offers a good model for the study of a one-step process of polyploidization whereby, in nature, unreduced gametes of two diploid parents can yield allotetraploid progeny.

For cytological studies, no functional pollen grains have been found and no seeds were obtained by selfing and crossed with two parents, which shows that the F1 maybe male sterile.

The intergeneric hybrids between non-heading Chinese cabbage and radish were successfully produced and characterized. These will be the base material for developing the whole set of $R. sativus$-$B. campestris$ additions in future for genome analysis and chromosomal localization of genes. In addition, the agronomical potential of the hybrid progenies obtained by selfing or backcross are under current evaluation on their advancement, improvement and exploitation. Furthermore, this hybrid plant offers an ideal model system to study the response to genomic changes from defined parents, such as structural rearrangements on the chromosome level [28] and sequence level [29,30], regulation of gene expression [31], activation of transposons [32], and amplification, reassortment, or elimination of highly repetitive sequences [33] and low-copy sequences [34].

Acknowledgments

Authors are grateful to Jiangsu province science and technology support program (BE201130172) for financial support.

Authors' contributions

The following declarations about authors' contributions to the research have been made: designing research: ZSN; performing experiments: ZSN, SCZ, ZJS; writing the manuscript: ZSN, SCZ, LY.

References

1. Grant V. Plant speciation. New York, NY: Columbia University Press; 1981.

2. Haufler CH, Soltis DE. Genetic evidence suggests that homosporous ferns with high chromosome numbers are diploid. Proc Natl Acad Sci USA. 1986;83(12):4389–4393.

3. Shaked H, Kashkush K, Ozkan H, Feldman M, Levy AA. Sequence elimination and cytosine methylation are rapid and reproducible responses of the genome to wide hybridization and allopolyploidy in wheat. Plant Cell. 2001;13(8):1749–1759.

4. Adams KL, Wendel JF. Novel patterns of gene expression in polyploid plants. Trends Genet. 2005;21(10):539–543.

5. Chen ZJ, Ni Z. Mechanisms of genomic rearrangements and gene expression changes in plant polyploids. BioEssays. 2006;28(3):240–252.

6. Beaulieu J, Jean M, Belzile F. The allotetraploid $Arabidopsis thaliana$-$Arabidopsis lyrata$ subsp. $petraea$ as an alternative model system for the study of polyploidy in plants. Mol Genet Genomics. 2009;281(4):421–435.

7. Ramsey J, Schemske DW. Pathways, mechanisms, and rates of polyploid formation in flowering plants. Ann Rev Ecol Syst. 1998;29(1):467–501.

8. Thao NTP, Ureshino K, Miyajima I, Ozaki Y, Okubo H. Induction of tetraploids in ornamental $Alocasia$ through colchicine and oryzalin treatments. Plant Cell Tissue Organ Cult. 2003;72(1):19–25.

9. Möllers C, Iqbal MCM, Röbbelen G. Efficient production of doubled haploid $Brassica napus$ plants by colchicine treatment of microspores. Euphytica. 1994;75(1–2):95–104.

10. Chen ZJ, Wang J, Tian L, Lee HS, Wang JJ, Chen M, et al. The development of an $Arabidopsis$ model system for genome-wide analysis of polyploidy effects. Biol J Linn Soc. 2004;82(4):689–700.

11. Lelivelt CLC, Lange W, Dolstra O. Intergeneric crosses for the transfer of resistance to the beet cyst nematode from $Raphanus sativus$ to $Brassica napus$. Euphytica. 1993;68(1–2):111–120.

12. Kolte SJ, Bordoloi DK, Awasthi RP. The search for resistance to major diseases of rapeseed and mustard in India. In: Proceedings of the 8th international rapeseed congress. Saskatoon: Organizing Committee of the Eighth International Rapeseed Congress; 1991. p. 219–225.

13. Thierfelder A, Hackenberg E, Nichterlein K, Friedt W. Development of nematode-resistant rapeseed genotypes via interspecific hybridization. In: Proceedings of the 8th international rapeseed congress. Saskatoon: Organizing Committee of the Eighth International Rapeseed Congress; 1991. p. 269–273.

14. Peterka H, Budahn H, Schrader O, Ahne R, Schütze W. Transfer of resistance against the beet cyst nematode from radish (*Raphanus sativus*) to rape (*Brassica napus*) by monosomic chromosome addition. Theor Appl Genet. 2004;109(1):30–41.

15. Long MH, Xing GM, Okubo H, Fujieda K. Cross compatibility between Brassicoraphanus (*Brassica oleracea* × *Raphanus sativus*) and cruciferous crops, and rescuing the hybrid embryos through ovary and embryo culture. J Fac Agric Kyushu Univ. 1992;37(1):29–39.

16. Agnihotri A, Shivanni KR, Lakshmikumaran MS. Micropropagation and DNA analysis of wide hybrids of cultivated *Brassica*. In: Proceedings of the 8th international rapeseed congress. Saskatoon: Organizing Committee of the Eighth International Rapeseed Congress; 1991. p. 151.

17. Warwick SI, Black LD. Molecular systematics of *Brassica* and allied genera (subtribe Brassicinae, Brassiceae) – chloroplast genome and cytodeme congruence. Theor Appl Genet. 1991;82(1):81–92.

18. Murashige T, Skoog F. A revised medium for rapid growth and bio assays with tobacco tissue cultures. Physiol Plant. 1962;15(3):473–497.

19. Takahata Y. Production of intergeneric hybrids between a C3-C4 intermediate species *Moricandia arvensis* and a C3 species *Brassica oleracea* through ovary culture. Euphytica. 1990;46(3):259–264.

20. Zhang GQ, Tang GX, Song WJ, Zhou WJ. Resynthesizing *Brassica napus* from interspecific hybridization between *Brassica rapa* and *B. oleracea* through ovary culture. Euphytica. 2004;140(3):181–187.

21. Tashiro Y. Cytogenetic studies on the origin of *Allium wakegi* Araki. Agr Saga Univ. 1984;56:1–63.

22. Saghai-Maroof MA, Soliman KM, Jorgensen RA, Allard RW. Ribosomal DNA spacer-length polymorphisms in barley: mendelian inheritance, chromosomal location, and population dynamics. Proc Natl Acad Sci USA. 1984;81(24):8014–8018.

23. Kim JS, Chung TY, King GJ, Jin M, Yang TJ, Jin YM, et al. A sequence-tagged linkage map of *Brassica rapa*. Genetics. 2006;174(1):29–39.

24. Choi SR, Teakle GR, Plaha P, Kim JH, Allender CJ, Beynon E, et al. The reference genetic linkage map for the multinational *Brassica rapa* genome sequencing project. Theor Appl Genet. 2007;115(6):777–792.

25. Cui XM, Dong YX, Hou XL, Cheng Y, Zhang JY, Jin MF. Development and characterization of microsatellite markers in *Brassica rapa* ssp. *chinensis* and transferability among related species. Agr Sci Chin. 2008;7(1):19–31.

26. Nanda Kumar PB, Shivanna KR. Intergeneric hybridization between *Diplotaxis siettiana* and crop brassicas for the production of alloplasmic lines. Theor Appl Genet. 1993;85(6–7):770–776.

27. Batra V, Prakash S, Shivanna KR. Intergeneric hybridization between *Diplotaxis siifolia*, a wild species and crop brassicas. Theor Appl Genet. 1990;80(4):537–541.

28. Leitch IJ, Bennett MD. Polyploidy in angiosperms. Trends Plant Sci. 1997;2(12):470–476.

29. Song K, Lu P, Tang K, Osborn TC. Rapid genome change in synthetic polyploids of *Brassica* and its implications for polyploid evolution. Proc Natl Acad Sci USA. 1995;92(17):7719–7723.

30. Wendel JF, Schnabel A, Seelanan T. Bidirectional interlocus concerted evolution following allopolyploid speciation in cotton (*Gossypium*). Proc Natl Acad Sci USA. 1995;92(1):280–284.

31. Comai L, Tyagi AP, Winter K, Holmes-Davis R, Reynolds SH, Stevens Y, et al. Phenotypic instability and rapid gene silencing in newly formed Arabidopsis allotetraploids. Plant Cell. 2000;12(9):1551–1567.

32. Hanson RE, Islam-Faridi MN, Crane CF, Zwick MS, Czeschin DG, Wendel JF, et al. Ty1-copia-retrotransposon behavior in a polyploid cotton. Chromosome Res. 2000;8(1):73–76.

33. Salina EA, Ozkan H, Feldman M. Subtelomeric repeat reorganization in synthesized amphiploids of wheat. In: Proceedings of the international conference on biodiversity and dynamics of systems, North Eurasia. Novosibirsk: ICG; 2000. p. 102–105.

34. Liu B, Vega JM, Segal G, Abbo S, Rodova M, Feldman M. Rapid genomic changes in newly synthesized amphiploids of *Triticum* and *Aegilops*. I. Changes in low-copy noncoding DNA sequences. Genome. 1998;41(2):272–277.

AtDeg2 – a chloroplast protein with dual protease/chaperone activity

Przemysław Jagodzik*, Małgorzata Adamiec, Grzegorz Jackowski

Institute of Experimental Biology, Adam Mickiewicz University, Umultowska 89, 61-614 Poznań; Poland

Abstract

Chloroplast protease AtDeg2 (an ATP-independent serine endopeptidase) is cytosolically synthesized as a precursor, which is imported into the chloroplast stroma and deprived of its transit peptide. Then the mature protein undergoes routing to its functional location at the stromal side of thylakoid membrane. In its linear structure AtDeg2 molecule contains the protease domain with catalytic triad (HDS) and two PDZ domains (PDZ1 and PDZ2). In vivo AtDeg2 most probably exists as a supposedly inactive haxamer, which may change its oligomeric stage to form active 12-mer, or 24-mer. AtDeg2 has recently been demonstrated to exhibit dual protease/chaperone function. This review is focused on the current awareness with regard to AtDeg2 structure and functional significance.

Keywords: AtDeg2; chaperone; chloroplast; hexamer; protease; PDZ domain

Introduction

The results of studies performed in recent years show that regulatory protein hydrolysis, catalyzed by proteolytic enzymes occurs ubiquitously and is directly or indirectly involved in a majority (if not all) cellular processes of living organisms. It is widely believed that proteases are involved in protein quality control and protein turnover processes. The protein quality control comprises a hydrolysis of proteins, which have been damaged due to mutations or exposition of plants to stressing environmental conditions, synthesized at redundant quantity or sorted to incorrect cell compartment. Then protein turnover involves a hydrolysis of proteins, which become unnecessary in a defined ontogenetical context under comfortable environmental conditions. Aminoacids released during the course of protein turnover are reused for protein synthesis as a part of a continuous breakdown/resynthesis of cellular components. Proteolysis is considered to regulate extensively whole-organism maintenance, structure and functions according to changes of environmental conditions and to progression of ontogenetic stages. For some proteases ATP binding and hydrolysis is necessary for their catalytic activity. All of them possess a conservative AAA+ domain which is responsible for ATP binding and hydrolysis, necessary to unfold protein substrates so that they can enter a catalytical chamber of the proteases molecule through a narrow entrance [1]. Nevertheless, some other proteases, including the Deg group, function in an ATP-independent manner, most probably because the availability of catalytic

center is less restricted for unfolded substrates than in the case of ATP-dependent enzymes [2].

Deg proteases comprise a very important group of proteolytic enzymes occurring in all domains of life, including Archaea, Bacteria and Eukarya [3]. Deg proteases were first discovered by studying *Escherichia coli* mutants unable to hydrolyze periplasmic proteins damaged under heat shock conditions (>37°C) [4,5]. Later it was found that the quality of *E. coli* periplasmic proteins is controlled by three Deg proteins namely DegP, DegQ and DegS whose structure and function have been precisely solved [6–8]. A unique feature of DegP protein is that it switches from protease to chaperone activity in a temperature-dependent manner. Namely at low temperatures (28°C) protease activity was found to be hardly detectable and chaperone activity was high; on the other hand at elevated temperatures (>28°C) the proteolytic activity rises abruptly whereas the chaperone activity is strongly diminished [9,10]. The chaperone activity of DegP consists in an ability to refold denatured, MalS (a native substrate) [10] and to prevent aggregation of a lysozyme (an artificial substrate) [9].

According to MEROPS database (9.10 release) count of known proteases in the model plant species *Arabidopsis thaliana* is 783, representing about 3% of all proteins identified in this taxon. Sixteen genes coding for proteins orthologous to DegP, Q and S have been identified in the *A. thaliana* nuclear genome and designated *AtDEG1–16* [11,12]. The genes code for proteins, which are targeted either to chloroplasts (*AtDEG1, 2, 5, 7* and *8*) or mitochondria (*AtDEG10*) or peroxisomes (*AtDEG15*). The localization of nine proteins encoded by remaining *AtDEG* genes is unclear; moreover some of those nine *AtDEGs* may be pseudogenes [13] and two potential AtDeg proteases may

---

* Corresponding author. Email: przemyslaw.jagodzik@amu.edu.pl

Handling Editor: Beata Zagórska-Marek

be proteolytically inactive [12]. A majority of *AtDEG* genes have their orthologues in other model plant species, as *Populus trichocarpa*, *Oryza sativa*, *Physcomitrella patens* and *Chlamydomonas reinhardtii* [12].

AtDeg2 protease is an ATP-independent serine endopeptidase containing a trypsin type catalytic domain with a catalytic triad (HDS). It belongs to the DegP (*E. coli*) type subfamily (S1C) of the clan PA according to the MEROPS nomenclature and may be found in this database as Deg2 chloroplast peptidase (*Arabidopsis thaliana*). However, MEROPS database gives also the other name – AtDegP2. All AtDeg proteases were originally named as AtDegPX ones and then renamed just AtDegX [11]. It has been shown recently that AtDeg2 exhibits chaperone-like activity in vitro [14] yet no data exist with regard to how protease and chaperone activities are interlocked within a single AtDeg2 molecule.

This review focuses on current awareness with respect to structure and functions of AtDeg2 – a chloroplast protein with dual protease/chaperone activity.

AtDEG2 gene

AtDeg2 protease is encoded by nuclear *AtDEG2* gene (*DEGRADATION OF PERIPLASMIC PROTEINS 2*) with a locus tag AT2G47940 and DDBJ/EMBL/GeneBank accession No. NC_00307, and exists at a single locus in the *Arabidopsis* genome. *AtDEG2* gene comprises 4266 bp and is composed of 19 exons. The gene has been identified between 19617986 bp and 19622251 bp on the edge of 2 chromosome of *Arabidopsis* genome (DDBJ/EMBL/GeneBank accession No. AC005309) and is "read" in opposite direction (19622251→19617986) *AtDEG2* gene is

transcribed onto mRNA (NM_130361) that contains 2278 bp and encodes a 607 aminoacids long pre-AtDeg2 protein (NP_566115.1). NCBI database indicates one more mRNA of *ATDEG2* gene (NM_001125072.1) composed of 2294 bp, which is regarded as a splice variant 2. This mRNA encodes a protein (NP_001118544.1) composed of 606 amino acids (Fig. 1). The lacking aminoacid residue (K62) is localized within an amino terminal stroma-targeting transit peptide, which is cleaved off after pre-AtDeg2 import into the plastid stoma therefore a single mature AtDeg2 isoform is found in the chloroplast.

AtDeg2 protein

Localization and structure

Hydropathy plots demonstrated mature AtDeg2 to be mostly a hydrophilic protein with no predicted transmembrane α-helices [15]. In fact studies based on immunoblot analysis of individual chloroplast subcompartments demonstrated that AtDeg2 is associated peripherally to the stromal side of stroma thylakoid membranes (80–90%) and non-appressed regions of grana stacks (10–20%) [15], the observation supported by the results of LC-MS/MS analyses of proteome of subchloroplast fractions [16]. AtDeg2 is cytosolically synthesized as a precursor, which is imported into the plastid stroma and deprived of its transit peptide (positions 1–69). The mature protein (positions 70–607, i.e. 538 aminoacid residues in total) is routed to its functional location at the stromal side of thylakoid membrane [15]. Its molecular weight was assessed to be 60 kDa [17]. According to HHPred platform (http://toolkit.tuebingen. mpg.de/hhpred/) [18] mature AtDeg2 molecule contains

```
Splice Variant 1:  MAASVANCCFSVLNASVKIQSSSISSPWCFVSASSLTPRASSNIKRKSSRSDSPSPILNP  60
Splice Variant 2:  MAASVANCCFSVLNASVKIQSSSTSSPWCFVSASSLTPRASSNIKRKSSRSDSPSPILNP  60

Splice Variant 1:  EKNYPGRVRDESSNPPQKMAFKAFGSPKKEKKESLSDFSRDQQTDPAKIHDASFLNAVVK  120
Splice Variant 2:  E-NYPGRVRDESSNPPQKMAFKAFGSPKKEKKESLSDFSRDQQTDPAKIHDASFLNAVVK  119

Splice Variant 1:  VYCTHTAPDYSLPWQKQRQFTSTGSAFMIGDGKLLTNAHCVEHDTQVKVKRRGDDRKYVA  180
Splice Variant 2:  VYCTHTAPDYSLPWQKQRQFTSTGSAFMIGDGKLLTNAHCVEHDTQVKVKRRGDDRKYVA  179

Splice Variant 1:  KVLVRGVDCDIALLSVESEDFWKGAEPLRLGHLPRLQDSVTVVGYPLGGDTISVTKGVVS  240
Splice Variant 2:  KVLVRGVDCDIALLSVESEDFWKGAEPLRLGHLPRLQDSVTVVGYPLGGDTISVTKGVVS  239

Splice Variant 1:  RIEVTSYAHGSSDLLGIQIDAAINPGNSGGPAFNDQGECIGVAFQVYRSEETENIGYVIP  300
Splice Variant 2:  RIEVTSYAHGSSDLLGIQIDAAINPGNSGGPAFNDQGECIGVAFQVYRSEETENIGYVIP  299

Splice Variant 1:  TTVVSHFLTDYERNGKYTGYPCLGVLLQKLENPALRECLKVPTNEGVLVRRVEPTSDASK  360
Splice Variant 2:  TTVVSHFLTDYERNGKYTGYPCLGVLLQKLENPALRECLKVPTNEGVLVRRVEPTSDASK  359

Splice Variant 1:  VLKEGDVIVSFDDLHVGCEGTVPFRSSERIAFRYLISQKFAGDIAEIGIIRAGEHKKVQV  420
Splice Variant 2:  VLKEGDVIVSFDDLHVGCEGTVPFRSSERIAFRYLISQKFAGDIAEIGIIRAGEHKKVQV  419

Splice Variant 1:  VLRPRVHLVPYHIDGGQPSYIIVAGLVFTPLSEPLIEEECEDTIGLKLLTKARYSVARFR  480
Splice Variant 2:  VLRPRVHLVPYHIDGGQPSYIIVAGLVFTPLSEPLIEEECEDTIGLKLLTKARYSVARFR  479

Splice Variant 1:  GEQIVILSQVLANEVNIGYEDMNNQQVLKFNGIPIRNIHHLAHLIDMCKDKYLVFEFEDN  540
Splice Variant 2:  GEQIVILSQVLANEVNIGYEDMNNQQVLKFNGIPIRNIHHLAHLIDMCKDKYLVFEFEDN  539

Splice Variant 1:  YVAVLEREASNSASLCILKDYGIPSERSADLLEPYVDPIDDTQALDQGIGDSPVSNLEIG  600
Splice Variant 2:  YVAVLEREASNSASLCILKDYGIPSERSADLLEPYVDPIDDTQALDQGIGDSPVSNLEIG  599

Splice Variant 1:  FDGLVWA  607
Splice Variant 2:  FDCLVWA  606
```

Fig. 1 Alignment of two splice variants of AtDeg2 protein. Sequences were obtained by BLAST search.

protease domain as well PDZ1 and PDZ2 domains (positions 110–313, 314–422 and 423–577, respectively [14]). The protease domain is preceded by and PDZ2 is followed by short regions, which are lacking any conserved domains (positions 70–109 and 578–607, respectively; Fig. 2). The protease domain involves a catalytic H159 D190 S268 triad.

A great progress has been made recently with regard to understanding structural organization of AtDeg2 since recombinant version of this protein (spanning 110–577 positions) has been crystallized and the crystal structure solved based on X-ray diffraction analysis [14]. When analyzed by size exclusion chromatography AtDeg2 in solution forms supposedly inactive hexamer (a dimer of trimers) of "sealed cage" type, consisting of two trimeric rings stacked upon each other in such a way that they form an inner, catalytic chamber the entrance to which is restricted to six pores.

The formation of trimer is enabled by interactions between protease and PDZ1 domains of individual monomers (Fig. 3). In the AtDeg2 hexamer the assembly is stabilized through multiple interactions involving two interfaces formed for each monomer, consisting of the protease domain and PDZ2 (interface 1) and PDZ1 and PDZ2 (interface 2) [14]. While PDZ1 domain resembles conventional PDZ domains of bacterial Deg proteases, PDZ2 domain has a few unique features with respect to canonical PDZ domains of

other Deg proteases. The most interesting feature is presence of an unusual β-strand (β21) comprising aminoacids 440–444, which acts as an intramolecular internal PDZ ligand. The internal PDZ2 ligand was shown to interact with the protease domain of the same monomer by binding its LA loop (Fig. 4) thereby fixing the protease domain and making AtDeg2 monomer and hexamer rigid molecules. Apart from making AtDeg2 monomer and hexamer rigid, internal PDZ2 ligand may play a crucial role in conversion of the supposedly resting hexameric state into the enzymatically active 12-mers and 24-mers by dissociating from PDZ2 so that trimeric units may be rearranged into higher oligomeric states. The oligomerization of AtDeg2 in solution was demonstrated to be pH-independent and to increase with the incubation time with an artificial substrate (but the hexameric state existed predominantly all the time [14]).

Dual protease/chaperone activity (in vitro and in vivo)
Recombinant AtDeg2 catalyzed in vitro hydrolysis of various artificial protein substrates as gelatin [15], fluorescence-labelled casein [19] and β-casein [14] thereby demonstrating to be bona fide proteolytic enzyme. AtDeg2 proteolityc activity in vitro was found to be regulated in redox- and pH-dependent manner [14,19] and it was suggested that cysteine residues found in AtDeg2 molecule may trigger

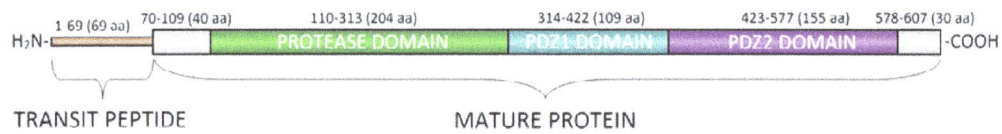

Fig. 2 The distribution of conservative domains in a linear structure of the precursor of AtDeg2 protein. The distribution of the domains is shown after [14] and the determination of processing site was performed using TARGETP 1.1 Server (http://www.cbs.dtu.dk/services/TargetP/). aa – aminoacids.

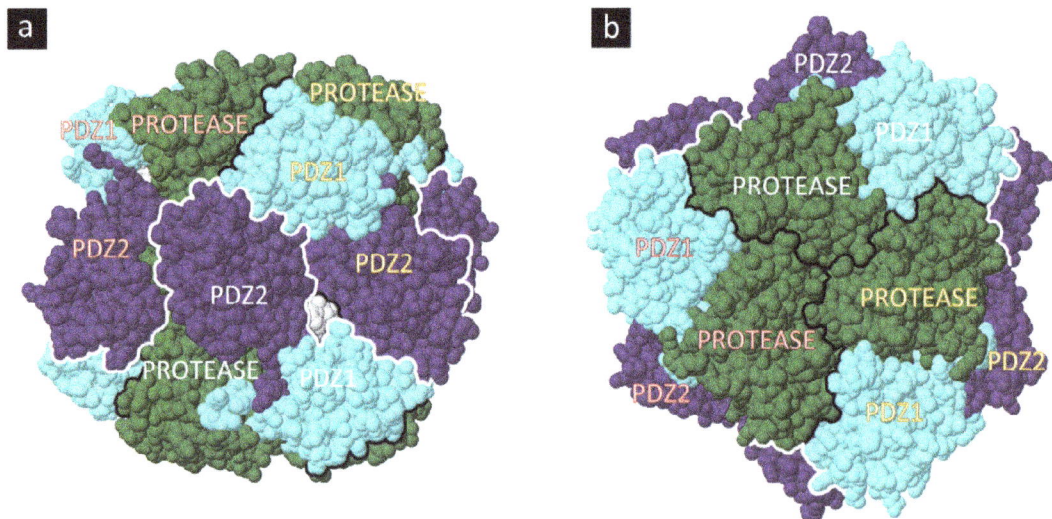

Fig. 3 Space fill representation of side (**a**) and top (**b**) views of structure of AtDeg2 hexamer (MMDB: 103069; PDB ID: 4FLN). Appropriate domains are distinguished by different colours (green – protease domain; cyan – PDZ1 domain; indigo – PDZ2 domain; grey – co-crystalized peptide). Monomers are divided by black line and trimers by white line. View in Jmol (Jmol: an open-source Java viewer for chemical structures in 3D – http://www.jmol.org/).

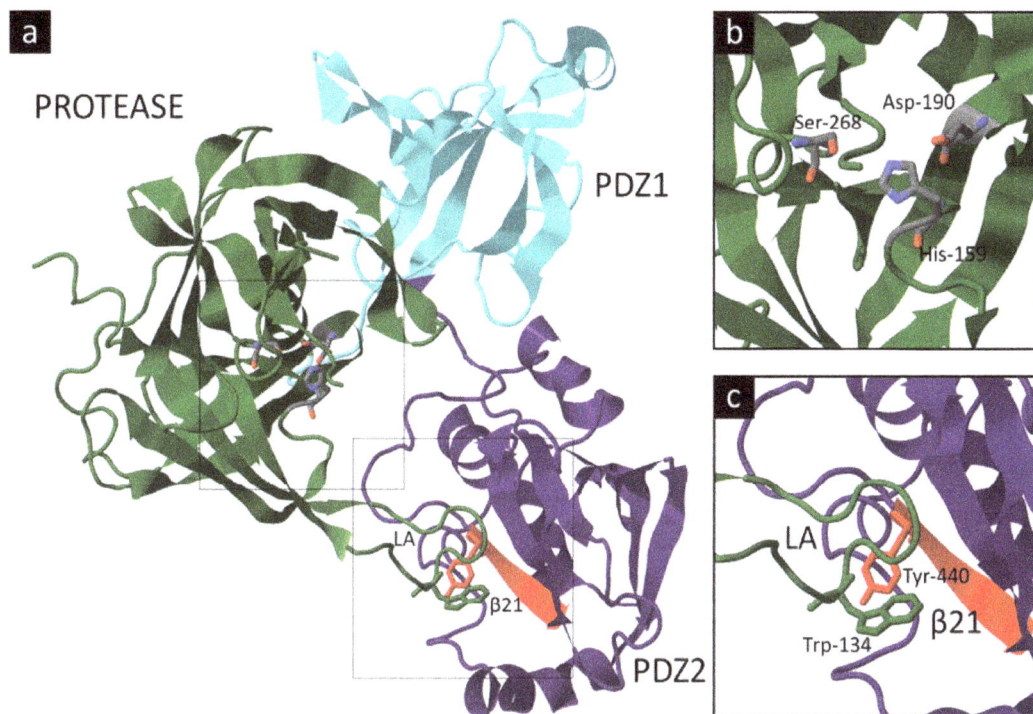

Fig. 4 Cartoon representation of structure of AtDeg2 monomer (**a**), catalytic triad (**b**) and PDZ2 ligand red colored (**c**). Appropriate domains are distinguished by different colors (green – protease domain; cyan – PDZ1 domain; indigo – PDZ2 domain). The catalytic triad and amino acids participating in LA loop binding are shown as a stick model (MMDB ID: 103069; PDB ID: 4FLN). View in Jmol (Jmol: an open-source Java viewer for chemical structures in 3D – http://www.jmol.org/).

redox-dependent conformational changes of the molecule leading to modifications of proteolytic activity observed in various redox surroundings [19]. However, extremely little is known about the identity of native substrates for AtDeg2 under stressing or non-stressing conditions in vivo. It was revealed in this respect that AtDeg2 might be involved in vivo in degradation of photodamaged PsbA protein by catalyzing the primary cleavage of stroma-exposed DE loop [15]. Yet, the role of this protease in degradation of PsbA was questioned since later it was shown that photodamaged PsbA protein had been effectively degraded in *A. thaliana* mutant lacking AtDeg2 [20]. The protease has recently been demonstrated to take part in protection against photoinhibition [17] again indicating possible involvement of this protease in removing photodamaged D1 protein in vivo. The only more straightforward data concerning identity of native substrates for AtDeg2 come from our laboratory – an apoprotein Lhcb6 has recently been found to be a target for short stress (3 h)-related degradation catalyzed by AtDeg2, as judged by an inability of *A. thaliana* mutants devoid of AtDeg2 to cleave Lhcb6 in leaves exposed to various stresses (elevated irradiance, heat, high salt and wounding). Most probably Lhcb6 apoprotein experiences an oxidative damage as result of short stresses in a manner that marks this protein for AtDeg2-dependent recognition and hydrolysis [17].

It was thought for several years that in contrast to what is observed with respect to DegP, chloroplast-targeted AtDegs function uniquely as proteases, i.e. are not able to switch reversibly from protease to chaperone and it was only in last four years that intriguing observations have been published demonstrating that AtDeg2 (and AtDeg1) may function as chaperones as well. Namely recombinant AtDeg1 was demonstrated to stimulate refolding of unfolded version of MalS protein [21] and recombinant AtDeg2 inhibited aggregation of denatured lysozyme in vitro [14]. However, both AtDeg1 and AtDeg2 exhibited dual activity in a wide range of temperatures in contrast to temperature-dependent shifting between protease and chaperone functions demonstrated for DegP. It was found that chaperone activity of AtDeg1 is confined to catalytic triad of protease domain [21] yet no data exist concerning a localization of chaperone activity in a linear structure of AtDeg2 molecule [14].

Functional importance

AtDEG2 is expressed almost ubiquitously, practically through all ontogenetic stages of the model plant species [22,23]. When AthaMap (http://www.athamap.de/) [24] was used to search for transcription binding elements within *AtDEG2* promoter predicted by AGRIS (http://arabidopsis.med.ohio-state.edu/) [25] eight well known and highly conserved cis-regulatory elements have been identified, specific for ARR-B, GATA, CAMTA, LFY, NAC and AP2/EREBP families of transcription factors. Namely, *AtDEG2* promoter contains two individual ARR-B recognition sequences, known as ARR1 and ARR2 and identified as key transcription activators in cytokinin downstream signaling pathway

[26]. Two potential targets for GATA transcription factors were also detected. One of the motifs designated GATA-1 was found to be involved in light regulation of nuclear genes expression [27] whereas the second one (described as ZML2) was identified as playing a crucial role in the cryptochrome1-dependent response to excess light [28]. In addition, three identical cis-regulatory elements recognizing LFY and single cis-regulatory elements specific for CAMTA3 (CAMTA family), NAP (NAC family) and RAP2.2 (AP2/EREBP family) transcription factors have been detected. LFY transcription factor is thought to play significant role in photoperiodic-dependent gene expression regulatory pathways [29], CAMTA3 in turn is considered to be involved in responses to biotic stresses [30], cold and freezing tolerance [31] as well as in regulation of ethylene-induced senescence [32]. NAP plays a key role in leaf senescence by participation in regulatory pathway which controls stomatal movement and water loss during leaf senescence [33] and the RAP2.2 was found to be involved in transactivation of two genes coding for enzymes engaged in carotenogenesis [34]. It is not easy to imagine how *AtDEG2* transactivation, exerted by interactions between the above mentioned transcription factors and cis-regulatory motifs within *AtDEG2* promoter may influence growth and development events since reliable data regarding the role of AtDeg2 in controlling the course of growth and developmental processes at the whole plant level are very scarce. It was demonstrated by us that under non-stressing conditions AtDeg2 is involved in regulation of both morphology and chloroplast life cycle of juvenile rosette leaves – at the moment when first flower was opened the area of juvenile leaves of mutants devoid of AtDeg2 was reduced significantly with respect to those of wild type plants (the alteration was found to be photoperiod independent). At the same moment chloroplasts of the wild plants' juvenile leaves showed signs of entering a senescence phase (e.g. the presence of numerous, large plastoglobules and periodic undulations of thylakoids parallel to those of the envelope – Fig. 5) which chloroplasts of mutants devoid of AtDeg2 appear not to do [17]. *AtDEG2* transactivation phenomena may mediate regulatory events triggered in response to exposure of *A. thaliana* plants to abiotic stress conditions as well, as judged by the fact that AtDeg2 mRNA accumulation is remarkably downregulated in response to short-term (2 h) exposure of detached mature leaves to a variety of abiotic stress conditions including heat, high salt

or desiccation [15] and upregulated in response to short term exposure (2.5 h) to elevated irradiance [35]. On the contrary long-term exposure (5 h) to elevated irradiance was found to be accompanied by a reduction in AtDeg2 transcript level [36]. Usually only a weak correlation exists between the accumulation of numerous chloroplast protease transcript and their proteins including AtDeg2 [15,37] thereby complex AtDeg2 regulatory mechanisms including transcriptional/translational as well as posttranscriptional phenomena [38] have to be triggered in response to the action of exogenous stressing factors as well as in response to progression of ontogenetic phases.

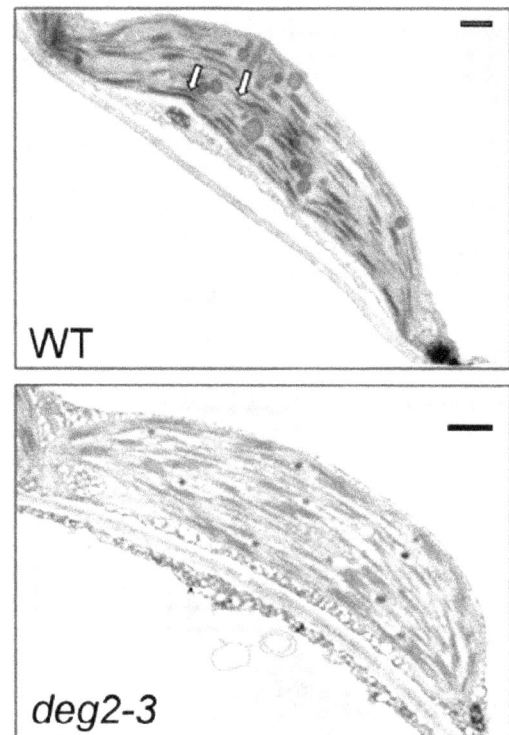

Fig. 5 Transmission electron microscopy of chloroplast of juvenile leaves of *Arabidopsis thaliana* wild type plants (WT) and mutants devoid of AtDeg2 protease (*deg2–3*) [17]. Micrographs show chloroplasts of mesophyll cells of plants which reached the moment when the first flower opened. [17]. Arrows point to the undulations of thylakoids of chloroplast of WT plants; their appearance marks the onset of early senescence. Scale bar: 500 nm.

Acknowledgments
The project was supported by grant from Polish National Science Center based on decision DEC-2013/09/B/NZ3/00449.

Authors' contributions
The following declarations about authors' contributions to the research have been made: worked out the idea of the review: GJ, PJ; wrote the text: GJ, PJ, MA; performed the analyses with AGRIS and AthaMap tools: MA.

References
1. Narberhaus F, Obrist M, Führer F, Langklotz S. Degradation of cytoplasmic substrates by FtsH, a membrane-anchored protease with many talents. Res Microbiol. 2009;160(9):652–659.

2. Kley J, Schmidt B, Boyanov B, Stolt-Bergner PC, Kirk R, Ehrmann M, et al. Structural adaptation of the plant protease Deg1 to repair photosystem II during light exposure. Nat Struct Mol Biol. 2011;18(6):728–731.

3. Clausen T, Southan C, Ehrmann M. The HtrA family of proteases: implications for protein composition and cell fate. Mol Cell. 2002;10(3):443–455.

4. Lipinska B, Sharma S, Georgopoulos C. Sequence analysis and regulation of the *htrA* gene of *Escherichia coli*: a σ^{32}-independent mechanism of heat-inducible transcription. Nucleic Acids Res. 1988;16(21):10053–10067.

5. Strauch KL, Beckwith J. An *Escherichia coli* mutation preventing degradation of abnormal periplasmic proteins. Proc Natl Acad Sci USA. 1988;85(5):1576–1580.

6. Wilken C, Kitzing K, Kurzbauer R, Ehrmann M, Clausen T. Crystal structure of the DegS stress sensor: how a PDZ domain recognizes misfolded protein and activates a protease. Cell. 2004;117(4):483–494.

7. Jiang J, Zhang X, Chen Y, Wu Y, Zhou ZH, Chang Z, et al. Activation of DegP chaperone-protease via formation of large cage-like oligomers upon binding to substrate proteins. Proc Natl Acad Sci USA. 2008;105(33):11939–11944.

8. Bai XC, Pan XJ, Wang XJ, Ye YY, Chang LF, Leng D, et al. Characterization of the structure and function of *Escherichia coli* DegQ as a representative of the DegQ-like proteases of bacterial HtrA family proteins. Structure. 2011;19(9):1328–1337.

9. Skórko-Glonek J, Krzewski K, Lipińska B, Bertoli E, Tanfani F. Comparison of the structure of wild-type HtrA heat shock protease and mutant HtrA proteins. A Fourier transform infrared spectroscopic study. J Biol Chem. 1995;270(19):11140–11146.

10. Spiess C, Beil A, Ehrmann M. A temperature-dependent switch from chaperone to protease in a widely conserved heat shock protein. Cell. 1999;97(3):339–347.

11. Huesgen PF, Schuhmann H, Adamska I. The family of Deg proteases in cyanobacteria and chloroplasts of higher plants. Physiol Plant. 2005;123(4):413–420.

12. Schuhmann H, Huesgen PF, Adamska I. The family of Deg/HtrA proteases in plants. BMC Plant Biol. 2012;12(1):52.

13. Schuhmann H, Adamska I. Deg proteases and their role in protein quality control and processing in different subcellular compartments of the plant cell. Physiol Plant. 2012;145(1):224–234.

14. Sun R, Fan H, Gao F, Lin Y, Zhang L, Gong W, et al. Crystal structure of *Arabidopsis* Deg2 protein reveals an internal PDZ ligand locking the hexameric resting state. J Biol Chem. 2012;287(44):37564–37569.

15. Haussuhl K, Andersson B, Adamska I. A chloroplast DegP2 protease performs the primary cleavage of the photodamaged D1 protein in plant photosystem II. EMBO J. 2001;20(4):713–722.

16. Ferro M, Brugiere S, Salvi D, Seigneurin-Berny D, Court M, Moyet L, et al. AT_CHLORO, a comprehensive chloroplast proteome database with subplastidial localization and curated information on envelope proteins. Mol Cell Proteomics. 2010;9(6):1063–1084.

17. Luciński R, Misztal L, Samardakiewicz S, Jackowski G. The thylakoid protease Deg2 is involved in stress-related degradation of the photosystem II light-harvesting protein Lhcb6 in *Arabidopsis thaliana*. New Phytol. 2011;192(1):74–86.

18. Soding J, Biegert A, Lupas AN. The HHpred interactive server for protein homology detection and structure prediction. Nucleic Acids Res. 2005;33:W244–W248.

19. Ströher E, Dietz KJ. The dynamic thiol–disulphide redox proteome of the *Arabidopsis thaliana* chloroplast as revealed by differential electrophoretic mobility. Physiol Plant. 2008;133(3):566–583.

20. Huesgen PF, Schuhmann H, Adamska I. Photodamaged D1 protein is degraded in *Arabidopsis* mutants lacking the Deg2 protease. FEBS Lett. 2006;580(30):6929–6932.

21. Sun X, Ouyang M, Guo J, Ma J, Lu C, Adam Z, et al. The thylakoid protease Deg1 is involved in photosystem-II assembly in *Arabidopsis thaliana*: chaperone function of Deg1. Plant J. 2010;62(2):240–249.

22. Schmid M, Davison TS, Henz SR, Pape UJ, Demar M, Vingron M, et al. A gene expression map of *Arabidopsis thaliana* development. Nat Genet. 2005;37(5):501–506.

23. Nakabayashi K, Okamoto M, Koshiba T, Kamiya Y, Nambara E. Genome-wide profiling of stored mRNA in *Arabidopsis thaliana* seed germination: epigenetic and genetic regulation of transcription in seed: molecular profiling in *Arabidopsis* seed. Plant J. 2005;41(5):697–709.

24. Yilmaz A, Mejia-Guerra MK, Kurz K, Liang X, Welch L, Grotewold E. AGRIS: the *Arabidopsis* gene regulatory information server, an update. Nucleic Acids Res. 2011;39:D1118–D1122.

25. Hehl R, Bülow L. AthaMap web tools for the analysis of transcriptional and posttranscriptional regulation of gene expression in Arabidopsis thaliana. In: Staiger D, editor. Plant circadian networks. New York, NY: Springer New York; 2014. p. 139–156. (Methods in molecular biology; vol 1158).

26. Hwang I, Sheen J. Two-component circuitry in *Arabidopsis* cytokinin signal transduction. Nature. 2001;413(6854):383–389.

27. Jeong MJ, Shih MC. Interaction of a GATA factor with cis-acting elements involved in light regulation of nuclear genes encoding chloroplast glyceraldehyde-3-phosphate dehydrogenase in *Arabidopsis*. Biochem Biophys Res Commun. 2003;300(2):555–562.

28. Shaikhali J, de Dios Barajas-Lopez J, Otvos K, Kremnev D, Garcia AS, Srivastava V, et al. The CRYPTOCHROME1-dependent response to excess light is mediated through the transcriptional activators ZINC FINGER PROTEIN EXPRESSED IN INFLORESCENCE MERISTEM LIKE1 and ZML2 in *Arabidopsis*. Plant Cell. 2012;24(7):3009–3025.

29. Benlloch R, Kim MC, Sayou C, Thévenon E, Parcy F, Nilsson O. Integrating long-day flowering signals: a LEAFY binding site is essential for proper photoperiodic activation of *APETALA1*. Plant J. 2011;67(6):1094–1102.

30. Galon Y, Nave R, Boyce JM, Nachmias D, Knight MR, Fromm H. Calmodulin-binding transcription activator (CAMTA) 3 mediates biotic defense responses in *Arabidopsis*. FEBS Lett. 2008;582(6):943–948.

31. Doherty CJ, van Buskirk HA, Myers SJ, Thomashow MF. Roles for *Arabidopsis* CAMTA transcription factors in cold-regulated gene expression and freezing tolerance. Plant Cell. 2009;21(3):972–984.

32. Nie H, Zhao C, Wu G, Wu Y, Chen Y, Tang D. SR1, a calmodulin-binding transcription factor, modulates plant defense and ethylene-induced senescence by directly regulating *NDR1* and *EIN3*. Plant Physiol. 2012;158(4):1847–1859.

33. Zhang K, Gan SS. An abscisic acid-AtNAP transcription factor-SAG113 protein phosphatase 2C regulatory chain for controlling dehydration in senescing *Arabidopsis* leaves. Plant Physiol. 2012;158(2):961–969.

34. Welsch R, Maass D, Voegel T, DellaPenna D, Beyer P. Transcription factor RAP2.2 and its interacting partner SINAT2: stable elements in the carotenogenesis of *Arabidopsis* leaves. Plant Physiol. 2007;145(3):1073–1085.

35. Sinvany-Villalobo G, Davydov O, Ben-Ari G, Zaltsman A, Raskind A, Adam Z. Expression in multigene families. Analysis of chloroplast and mitochondrial proteases. Plant Physiol. 2004;135(3):1336–1345.

36. Adamiec M, Luciński R, Jackowski G. The irradiance dependent transcriptional regulation of *AtCLPB3* expression. Plant Sci. 2011;181(4):449–456.

37. Zheng B, Halperin T, Hruskova-Heidingsfeldova O, Adam Z, Clarke AK. Characterization of chloroplast Clp proteins in *Arabidopsis*: localization, tissue specificity and stress responses. Physiol Plant. 2002;114(1):92–101.

38. Żelisko A, Jackowski G. Senescence-dependent degradation of Lhcb3 is mediated by a thylakoid membrane-bound protease. J Plant Physiol. 2004;161(10):1157–1170.

6

Effect of cobalt chloride on soybean seedlings subjected to cadmium stress

Jagna Chmielowska-Bąk[1]*, Isabelle Lefèvre[2], Stanley Lutts[2], Agata Kulik[1], Joanna Deckert[1]

[1] Department of Plant Ecophysiology, Institute of Experimental Biology, Faculty of Biology, Adam Mickiewicz University, Umultowska 89, 61-614 Poznań, Poland

[2] Groupe de Recherche en Physiologie végétale (GRPV), Earth and Life Institute, Université catholique de Louvain, Croix du Sud, 4-5, bte L7.07.13, 1348 Louvain-la-Neuve, Belgium

Abstract

Contamination of the environment with heavy metals such as Cd is a serious problem of modern world. Exposure of plants to Cd leads to oxidative stress, inhibition of respiration and photosynthesis, increased rate of mutation and, as a consequence, stunted growth and yield decrease. One of the common reactions of plants to cadmium stress is over-production of ethylene, however the exact role of this hormone in plants response to Cd is still unrecognized. The aim of the present study is evaluation of the impact of an ethylene synthesis inhibitor, Co, on the response of soybean seedlings to cadmium stress. The experiments included measurements of growth, cell viability, ethylene production and expression of genes associated with cellular signaling in soybean seedlings exposed to $CdCl_2$ (with Cd in a concentration of 223 μM) and/or $CoCl_2$ (with Co in concentration of 4.6 μM). Surprisingly, the results show that Co has no effect on ethylene biosynthesis, however, it affects cell viability and expression of Cd-induced genes associated with plant signaling pathways. The affected genes encode mitogen-activated protein kinase kinase2 (MAPKK2), nitrate reductase and DOF1 and bZIP2 transcription factors. The role of Co in plants response to cadmium stress and its potential use as an ethylene inhibitor is discussed.

Keywords: cobalt; heavy metal; gene expression; signaling; *Glycine max*

Introduction

Contamination of the environment with heavy metals, including Cd, is a serious problem of the modern world. Cadmium toxicity in plants leads to the generation of oxidative stress, chlorosis, inhibition of photosynthesis, disturbances in mineral homeostasis, increased rate of mutations and initiation of apoptotic and necrotic processes [1–3]. The described toxic symptoms lead to inhibition of plants growth and decrease in the obtained yield. Moreover, Cd might accumulate in crop plants and enter human organisms through the food chain [4]. One of the common responses of plants to cadmium stress is enhanced production of ethylene [5–8]. However, the exact role of the observed Cd-dependent induction of this hormone's production is still unrecognized. It is known that ethylene constitutes an important stress-related signaling molecule [9]. There are individual reports stating that it mediates Cd-dependent growth inhibition, hydrogen peroxide accumulation and programmed cell death (PCD) [7,10–12]. In mustard plants ethylene has been shown to participate in sulfur dependent alleviation of Cd toxicity through stimulation of antioxidant system [7,13].

Due to the important role of ethylene in flowering, fruit ripening and response of plants to stress factors several inhibitors of its synthesis and perception were developed. One of the commonly used inhibitor is Co, which affects the activity of a key enzyme in ethylene's biosynthesis pathway – 1-aminocyclopropane-1-carboxylic acid oxidase (ACO) [14]. However, although relatively low concentration of Co might be beneficial for plants, higher concentrations exhibit toxic effect [15,16]. This metal has been shown to inhibit plants growth, cause oxidative stress, DNA damage and disturbances in photosynthesis [16–20]. Excess of Co also leads to alterations in germination, sex ratio, photoperiodism and uptake of other elements [16].

In the previous study we have shown that cadmium stress causes induction of ethylene biosynthesis and elevated expression of several genes associated with signaling pathways. Interestingly the promoters of Cd-induced genes contained cis-acting elements connected with ethylene signaling [6]. The aim of present study was to investigate the effect of Co, as a potential ethylene inhibitor, on soybean seedlings subjected to short term cadmium stress. The conducted research includes evaluation of the impact of Co on soybean growth parameters, viability, ethylene production and expression of six Cd-induced genes.

* Corresponding author. Email: jagna_c20@wp.pl

Handling Editor: Grażyna Kłobus

Material and methods

Plant material, growth conditions and treatment procedures

If not stated differently the reagents were purchased at Sigma-Aldrich company (St. Louis, Missouri, USA). The applied growth conditions, Cd and Co concentrations and treatment periods were based on previous studies [6,21].

Soybean (*Glycine max* L cv. Nawiko) seeds were surface sterilized with 75% ethanol for 5 min and for 10 min with 1% sodium hyperchlorite. Seeds were washed for 30 min, soaked in distilled water for 2 h and germinated on Petri dishes with moistened blotting paper for 48 h. Germinated seedlings were transferred to new Petri dishes and treated with 5 ml of either: distilled water (control), 4.6 µM Co in CoCl$_2$ solution (corresponding 10 µM CoCl$_2$), 223 µM Cd in CdCl$_2$ solution (corresponding to 25 mg l^{-1}) or combined Cd and Co (corresponding to 4.6 µM and 223 µM respectively). After 3 and 6 hours of treatment seedlings root tips (100 mg) were cut off and frozen in −80°C for RNA isolation. Due to the lack of significant effect in shorter treatment periods (data not shown) the measurements of roots growth and cell viability were carried out after 24 hours.

If not stated differently the measurements were performed on samples from 3 independent experimental repetitions, each sample consisted of a pool of 20 seedlings.

Measurements of growth parameters

After 24 hours of treatment the roots of soybean seedlings were straightened and their length was measured with the use of ruler. The fresh and dry weight of roots was measured on the WPS60/C scale (RadWag, Radom, Poland). The fresh weight was evaluated immediately after cutting off, while the dry weight was measured after 72 h of incubation at 55°C.

Estimation of cell viability

The measurements of cell death were carried out according to the modified method described by Lehotai et al. [22]. After 24 hours of treatment with appropriate solutions root tips (200 mg) of soybean seedlings were cut off and incubated 20 min in 0.25% Blue Evans solution. Root tips were washed 3 times with distilled water and homogenized with 1.2 ml of 1% of SDS dissolved in 50% ethanol. The samples were incubated for 15 min at 50°C and centrifuged 15 min at 12 000 *g*. The absorbance of supernatant was measured with Biomate 3S spectrophotometer (Thermo Scientific, Waltham, USA) at $\lambda = 600$ nm.

Measurements of ethylene biosynthesis

The ethylene production was measured with the use of ethylene detector ETD-300 (Sensor Sense, Nijmegen, The Netherlands). Soybean seedlings were placed in Petri dishes on two layers of filter paper moistened with 5 ml of either distilled water (control), 4.6 µM Co in CoCl$_2$ solution, 223 µM Cd in CdCl$_2$ solution or combined Cd and Co (corresponding to 4.6 µM and 223 µM, respectively). The bottom part of the dish was covered with a Plexiglas plate with an inlet and outlet for gas flow, and tightly closed. The flow from each cuvette was directed into a photoacoustic cell where ethylene was quantified. The measurements were conducted in the dark during 24 hours, in stop-and-flow mode, with each cuvette being alternatively flushed with a flow of 3 l h^{-1}. The amount of produced ethylene was detected every 12 minutes. As a control from the obtained emission rates the levels of ethylene in a cuvette containing moistened filter papers without seedlings was also measured. A detailed description of the system has been given previously [23]. The obtained results were analyzed with the use of Valve Controller software and expressed as nl per hour per 1 g of roots fresh weight. Measurements were performed on samples from 3 independent experimental repetitions; each sample consisted of a pool of 10 seedlings.

Measurements of genes expression

The RNA was isolated from 100 mg of frozen root tips with the use of TriReagent according to the manufacture's instructions. The concentration of the obtained RNA was evaluated on NanoCell Accessory coupled with spectrophotometer Biomate 3S (Thermo Scientific, Waltham, USA).

For the reverse transcription 1 µg of RNA was purified with Deoxyribonuclease Kit and processed with the use of Reverse Transcription Kit (Thermo Scientific Fermentas, Waltham, USA): incubated with 1 µl oligo dT (100 µM, 0.5 µg/µl) at 65°C for 5 min followed by the incubation with 4 µl of 5× Reaction Buffer, 1 µl of RiboLock™ RNase Inhibitor (20 u/µl), 2 µl of 10 mM dNTP Mix and 1 µl of RevertAid™ Reverse Transcriptase at 42°C for 10 min. The reaction was stopped by incubation at 70°C for 10 min. The obtained cDNA was diluted 5×.

The measurements of genes expression were carried out with the use of real-time PCR reaction performed on Rotor-Gene 6000 Thermocycler (Qiagen, Venlo, The Netherlands). The primers (listed in Tab. 1) were designed on the basis of sequences accessible in Soybase.org with the use of Primer3 software (http://bioinfo.ut.ee/primer3-0.4.0/). The reaction mixture contained 0.1 µM of each primer, 1 µl of diluted cDNA, 10 µl of Power SYBR Green PCR Master Mix (Applied Biosystems, Foster City, California, USA) and DEPC treated water (BioShop, Burlington, Canada) to the total volume of 20 µl. The real-time PCR reaction started with initial denaturation at 95°C for 5 min, followed by 13 cycles of touchdown PCR (15 s at 95°C, 20 s at 68°C decreasing by 1°C each cycle and 30 s at 72°C) and 45 cycles of 10 s at 95°C, 20 s at 55°C and 30 s at 72°C. The reaction was finalized by denaturation at a temperature rising from 72°C to 95°C by one degree every 5 s.

The relative gene expression was calculated with the use of Pfaffl equation [24] based on the efficiency and Ct values determined by Real-time PCR Miner [25]. Ubiquitin was chosen as reference gene. Measurements were performed on samples from 2–3 independent experimental repetitions; each sample consisted of a pool of 20 seedlings.

Statistical analysis

For evaluation of statistically significant differences the obtained data was analyzed with the use of ANOVA ($\alpha = 0.05$). In the case of the measurements of genes expression, due to the non-normal distribution of data, Mann–Whitney *U* post-hoc test has been used. In all the other cases Scheffe's test has been applied. Results, which showed no statistically significant differences, are marked with the same letter.

Tab. 1 Sequences of primers used for the real-time PCR reaction.

Gene number in Soybase.org	Primers	Encoded protein
Glyma05g37410	Left: TGTGCTATGCCAACATGGAT Right: GAGGTATGGGGGAGTGAGGT	1-aminocyclopropane-1-carboxylate synthase (ACS)
Glyma17g06020	Left: AGCAGGTGCTGAAGGGTCTA Right: TTCCTGGCTTCCATTGATTC	mitogen-activated protein kinase kinase 2 (MAPKK2)
Glyma13g02510	Left: AAATCCCATGCAAGCTCATC Right: GGTGCACCCCTTTGAAGTAA	nitrate reductase (NR)
Glyma13g42820	Left: AAGCCAAAACTTGGAGCAGA Right: CCTTGTCGACGGAGGAATTA	DOF1 transcription factor
Glyma11g11450	Left: GAATCGACCCTGCAACTCAT Right: ACCCAAACTGCAAACGAAAC	MYBZ2 transcription factor
Glyma06g08390	Left: GCCCCATTGCTGTTCCTCATGT Right: GCTGAGACTGGGCTCCCAACA	bZIP62 transcription factor
Glyma20g27950	Left: GAAGTCGAAAGCTCCGACAC Right: TGTT TTGGGAACACATCCAA	ubiquitin – reference gene

Results

Treatment with Cd for 24 hours caused inhibition of roots length and dry weight (Tab. 2). In turn Co had no effect on any of the growth parameters. At this time point there were also no differences in the viability of cells in the roots of control seedlings and seedlings treated with Co (Fig. 1). However, treatment with Cd caused increase in the Evans Blue uptake providing a significant increase in cells mortality. The cells mortality increased even stronger in the case of the roots of seedlings treated simultaneously with Cd and Co.

Tab. 2 Growth parameters of the roots of soybean seedlings.

Experimental variant	Roots length (mm)	Roots fresh weight (mg)	Roots dry weight (mg)
Control	43 ±1[a]	54 ±4[a]	2.9 ±0.2[a]
CoCl$_2$	47 ±5[a]	62 ±8[a]	3.3 ±0.3[a]
Cadmium	30 ±2[b]	53 ±3[a]	2.7 ±0.1[b]
Cadmium+CoCl$_2$	28 ±1[b]	50 ±3[a]	2.7 ±0.2[b]

The results are means of 3 independent experiments ±SE. Results which showed no statistically significant differences are marked with the same letter (a or b).

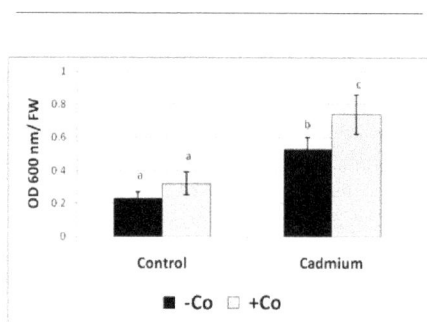

Fig. 1 Mortality of cell of soybean seedling roots represented as uptake of Evans Blue dye. The results are means of 4 independent experiments ±SE. Results, which showed no statistically significant differences, are marked with the same letter (a, b or c).

Application of Cd lead to augmented ethylene biosynthesis starting from the 5th hour of treatment (Fig. 2). After 24 h the levels of ethylene were four times higher in the roots of seedlings exposed to Cd than in the roots of control seedlings and reached 8 nl/h × fresh weight. Co had no significant effect on ethylene production neither in control nor Cd-treated seedlings.

Accordingly to our previous study [6] treatment with CdCl$_2$ for 3 h caused increase in the expression levels of genes encoding aminocyclopropane-1-carboxylic acid synthase (ACS), mitogen-activated kinase kinase2 (MAPKK2) and DOF1 and MYBZ2 transcription factors (Fig. 3). Present research shows that Co diminished the Cd-dependent induction of *MAPKK2* and *DOF1* expression (Fig. 3b,c).

Longer (6 h) exposure to Cd resulted in elevated levels of mRNA encoding nitrate reductase (NR), ACS and MYBZ2 and bZIP62 transcription factors (Fig. 4). Co caused augmentation of Cd-dependent induction of NR gene (Fig. 4b) and at the same time decrease in Cd-dependent stimulation of gene encoding bZIP62 transcription factor (Fig. 4d).

Discussion

Although Cd caused significant reduction of roots length and dry weight after 24 hours of application, Co did not affect any of the measured growth parameters (Tab. 2). Treatment with Cd also led to the increase in the amount of dead cells in the roots soybean seedlings (Fig. 1). Application of Co additionally increased Cd toxicity as the amount of dead cells was significantly higher in seedlings treated with both Co and Cd than in the seedlings treated with only Cd (Fig. 1). Co is known to have beneficial effects on plant growth at moderate levels [15], however, it is possible that the combined effect of Co and Cd leads to the aggravation of the metals' toxicity. Indeed, combination of Cu and Zn was shown to be more toxic to black lentil than the separate effect of both heavy metals, while combined Cd and Pb stress was found to be more harmful to mustard plants than application of Cd or Pb alone [26,27]. In higher concentrations Co is toxic. It was shown to inhibit plants growth, germination rate, cause leaf fall, hamper photosynthesis and respiration and lead

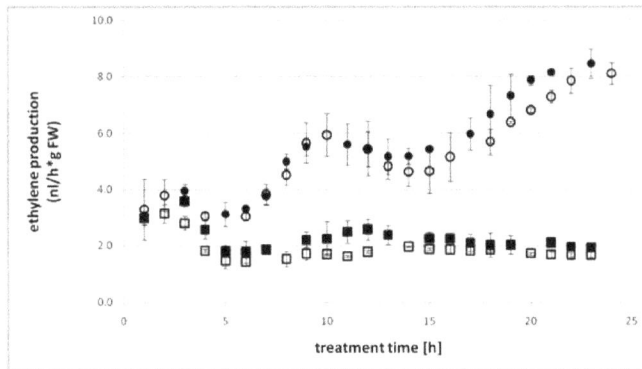

Fig. 2 Production of ethylene in roots of control soybean seedlings (dark square), seedlings treated with Co (light square), seedlings treated with Cd (dark circle) and seedlings treated with Cd and Co (light circle). The ethylene production is presented in nanoliters per hour in one gram of roots fresh weight. The results are means of 3 independent experiments ±SE.

to DNA damage as well as decrease in RNA levels [16–20]. One admitted function of Co is to impair ethylene synthesis at various concentrations, treatment durations and plant species, by inhibiting ACC oxidase [28–32]. Surprisingly in the applied experimental conditions Co had no effect on hormones production (Fig. 2). It is possible that the inhibitory action of Co requires longer treatment periods or higher concentrations. Indeed, in the majority of studies performed by other researchers Co was applied for several days [28,30–32]. In the research conducted by Koehl et al. [29] tobacco suspension culture was also treated with Co for short time periods (3, 9 and 24 h), however, the applied CoCl$_2$ concentration was much higher (100 μM) and the inhibitory effect on ethylene biosynthesis was noticed only after 9 h of treatment [29]. It is also possible that ethylene is synthesized without the participation of ACC oxdiase – a target gene for Co inhibitory action. It has been shown

that ACC might be oxidized nonenzymatically through superoxide anion generated in response to wounding [33].

In our previous study we have demonstrated that Cd causes induction of several genes associated with plant signaling pathways [6]. Application of Cd for 3 h induced expression of genes encoding 1-aminocyclopropane-1-carboxylic acid synthase (ACS), mitogen-activated kinase kinase2 (MAPKK2) and DOF1 and MYBZ2 transcription factors, while 6 h long treatment led to the increase in the expression of *nitrate reductase* (NR), *ACS*, *MYBZ2* and *bZIP62* genes. In the present study the influence of Co on the expression of mentioned, signaling associated genes after the same treatment has been evaluated.

Interestingly the results of present research show that Co influences expression of a key enzyme engaged in ethylene synthesis: 1-aminocyclopropane-1-carboxylic acid synthase. The genes expression was slightly repressed after 3 h (Fig. 3a)

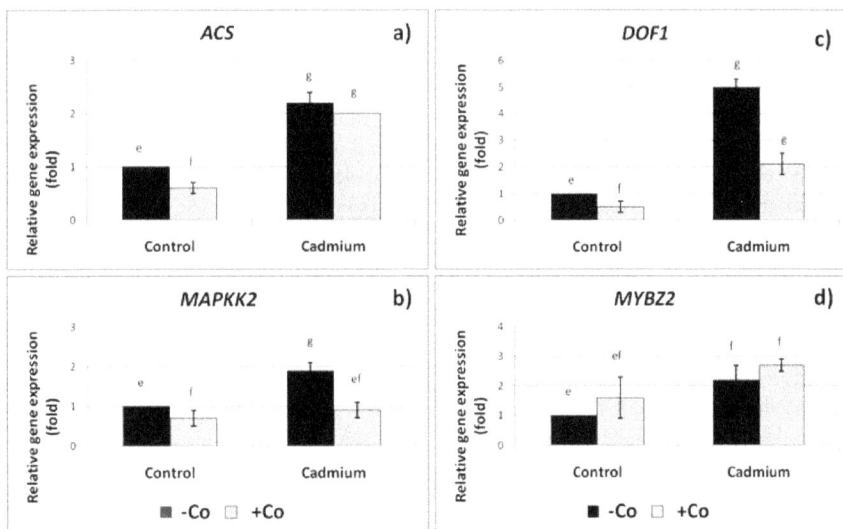

Fig. 3 Relative expression of gene encoding 1-aminocyclopropane-1-carboxylic acid synthase (**a**), mitogen-activated kinase kinase2 (**b**), DOF1 transcription factor (**c**) and MYBZ2 transcription factor (**d**) in soybean seedlings treated with appropriate solutions for 3 h. The results are the means of 2–3 independent experiments ±SE. Results, which showed no statistically significant differences, are marked with the same letter (e, f or g).

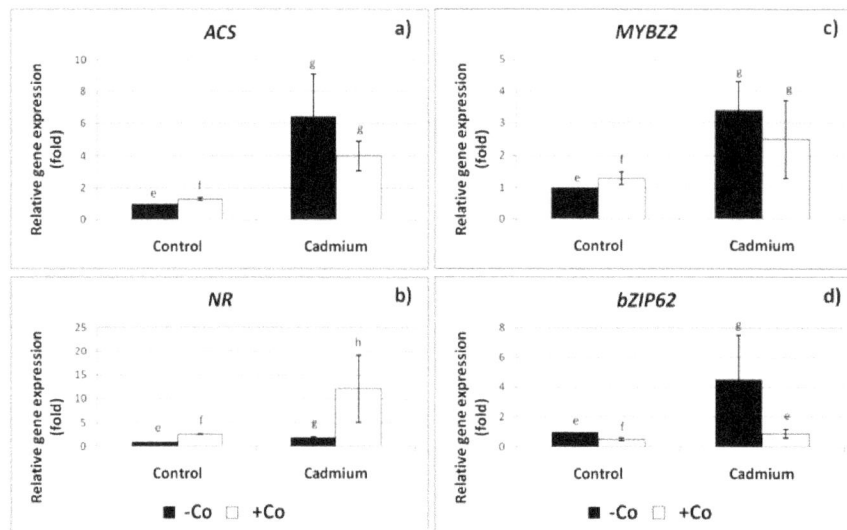

Fig. 4 Relative expression of gene encoding 1-aminocyclopropane-1-carboxylic acid synthase (**a**), nitrate reductase (**b**), MYBZ2 transcription factor (**c**) and bZIP62 transcription factor (**d**) in soybean seedlings treated with appropriate solutions for 6 h. The results are the means of 2–4 independent experiments ±*SE*. Results, which showed no statistically significant differences, are marked with the same letter (e, f, g or h).

and up-regulated after 6 h of treatment (Fig. 4a). Therefore it is possible that Co affects ethylene biosynthesis not solely by inhibiting ACC oxidase.

The obtained results also show that Co reversed the Cd-dependent induction of genes encoding MAPKK2 (Fig. 3b) and DOF1 transcription factor (Fig. 3c) after 3 h as well as gene encoding bZIP62 transcription factor (Fig. 4d) after 6 h of treatment. A possible explanation of the described phenomenon is the inhibitory effect of Co on Cd uptake. Indeed, Co was shown to reduce Cd uptake in bush beans and green alga *Chlamydomonas reinhardatii* [16,34]. However, in the referenced experiments the treatment times were much longer (21 and 60 days respectively). In the present study Co was applied for short time (3, 6 or 24 h) and even after 24 h did not reverse the Cd-dependent inhibition of roots growth (Tab. 2). Moreover, application of Co for 24 h led to the intensification of Cd toxicity expressed by higher cells mortality (Fig. 1). Therefore, it is unlikely that Co caused a significant reduction of Cd uptake. The measurements of genes expression also showed that, after 6 h of treatment Co led to strong augmentation of Cd-dependent induction of *NR* gene (Fig. 4b).

The impact of Co on the expression of analyzed genes might lead to altered Cd sensing. There are several reports implying that MAPK cascades are involved in the transduction of Cd signals. Stimulation of various MAPKs by Cd has been observed in rice, alfalfa and *Arabidopsis* plants. Moreover, MPK6 identified in *Arabidopsis* plants, was shown to participate in Cd-dependent initiation of programmed cell death (PCD) [10,35–38]. Therefore, observed in the present research alterations in *MAPKK2* expression in response to Co might lead to disorders in the transduction of Cd signal and contribute to exacerbation of Cd toxicity.

There are reports stating that both applied metals, Cd and Co, cause inhibition of nitrate reductase activity after long treatment times [17,39–41]. Interestingly, the present study shows that short-term cadmium stress causes induction of *NR* gene, which is strongly augmented by Co (Fig. 4b). Nitrate reductase is an important enzymes engaged in nitrogen metabolism – it catalyses reduction of nitrates to nitrites [42]. Therefore, the observed stimulation of *NR* expression might constitute a defense mechanism, which aims to sustain nitrogen homeostasis. Nitrate reductase is also one of the enzymes engaged in NO formation [42]. Accumulation of NO in response to Cd has been observed in various plant species [43], however, its role is still debatable. The treatment with nitric oxide donor, SNP, caused attenuation of chlorophyll degradation and oxidative stress in plant exposed to Cd [44,45]. Nitric oxide also mediates induction of several signaling associated genes in response to short-term cadmium stress [46]. On the other hand NO contributes to Cd toxicity through increase in Cd uptake [1, 47]. The observed Co-dependent induction of *NR* gene might lead to over-production of NO, however, the exact role of such induction would need further investigation.

Increase in *DOF1* and *bZIP62* mRNA levels in response to Cd [6] and the fact that DOF1 was induced by drought stress in amaranth roots and bZIP62 conferred tolerance to low temperatures and salt stress in transgenic *Arabidopsis* plants, suggests that both transcription factors are involved in the regulation of genes expression under various stress conditions [48,49]. Therefore, observed in the present study Co-dependent decrease in *DOF1* and *bZIP62* expression might lead to alterations in the expression pattern of Cd-responsive genes.

It can be concluded that Co may increase Cd toxicity, at least partly, through alterations in MAPK and NO signaling as well as disorders in regulation of genes expression mediated by DOF1 and bZIP62 transcription factors. The observed effects are independent from ethylene action.

Acknowledgments
The project was financed by National Science Center granted on the basis of decision number DEC-2011/03/N/NZ9/00214. The first author is a scholarship holder within the project "Scholarship support for Ph.D. students specializing in majors strategic for Wielkopolska's development", Sub-measure 8.2.2 Human Capital Operational Programme, co-financed by European Union under the European Social Fund. This work was also supported by the Fonds National de la Recherche Scientifique (FNRS, Belgium; conventions Nos. 1.5117.11 and 1.5114.11).

Authors' contributions
The following declarations about authors' contributions to the research have been made: designed the experiments: JD, JCB; supervised the research: JD, SL; conducted the experiments: JCB, AK; analyzed the results: JCB, IL; wrote the manuscript: JCB; revised the manuscript: JD, SL, IL.

References

1. Arasimowicz-Jelonek M, Floryszak-Wieczorek J, Deckert J, Rucińska-Sobkowiak R, Gzyl J, Pawlak-Sprada S, et al. Nitric oxide implication in cadmium-induced programmed cell death in roots and signaling response of yellow lupine plants. Plant Physiol Biochem. 2012;58:124–134.

2. Nedjimi B, Daoud Y. Cadmium accumulation in *Atriplex halimus* subsp. schweinfurthii and its influence on growth, proline, root hydraulic conductivity and nutrient uptake. Flora. 2009;204(4):316–324.

3. Sun Z, Wang L, Chen M, Wang L, Liang C, Zhou Q, et al. Interactive effects of cadmium and acid rain on photosynthetic light reaction in soybean seedlings. Ecotoxicol Env Saf. 2012;79:62–68.

4. Yang Y, Li F, Bi X, Sun L, Liu T, Jin Z, et al. Lead, zinc, and cadmium in vegetable/crops in a zinc smelting region and its potential human toxicity. Bull Env Contam Toxicol. 2011;87(5):586–590.

5. Arteca RN, Arteca JM. Heavy-metal-induced ethylene production in Arabidopsis thaliana. J Plant Physiol. 2007;164(11):1480–1488.

6. Chmielowska-Bąk J, Lefèvre I, Lutts S, Deckert J. Short term signaling responses in roots of young soybean seedlings exposed to cadmium stress. J Plant Physiol. 2013;170(18):1585–1594.

7. Masood A, Iqbal N, Khan NA. Role of ethylene in alleviation of cadmium-induced photosynthetic capacity inhibition by sulphur in mustard: Ethylene in S-mediated alleviation of Cd stress. Plant Cell Env. 2012;35(3):524–533.

8. Rodriguez-Serrano M, Romero-Puertas MC, Zabalza A, Corpas FJ, Gomez M, Del Rio LA, et al. Cadmium effect on oxidative metabolism of pea (*Pisum sativum* L.) roots. Imaging of reactive oxygen species and nitric oxide accumulation in vivo. Plant Cell Env. 2006;29(8):1532–1544.

9. Wu L, Zhang Z, Zhang H, Wang XC, Huang R. Transcriptional modulation of ethylene response factor protein JERF3 in the oxidative stress response enhances tolerance of tobacco seedlings to salt, drought, and freezing. Plant Physiol. 2008;148(4):1953–1963.

10. Liu XM, Kim KE, Kim KC, Nguyen XC, Han HJ, Jung MS, et al. Cadmium activates *Arabidopsis* MPK3 and MPK6 via accumulation of reactive oxygen species. Phytochemistry. 2010;71(5–6):614–618.

11. Maksymiec W. Effects of jasmonate and some other signalling factors on bean and onion growth during the initial phase of cadmium action. Biol Plant. 2011;55(1):112–118.

12. Yakimova ET, Kapchina-Toteva VM, Laarhoven LJ, Harren FM, Woltering EJ. Involvement of ethylene and lipid signalling in cadmium-induced programmed cell death in tomato suspension cells. Plant Physiol Biochem. 2006;44(10):581–589.

13. Asgher M, Khan NA, Khan MIR, Fatma M, Masood A. Ethylene production is associated with alleviation of cadmium-induced oxidative stress by sulfur in mustard types differing in ethylene sensitivity. Ecotoxicol Env Saf. 2014;106:54–61.

14. Serek M, Woltering EJ, Sisler EC, Frello S, Sriskandarajah S. Controlling ethylene responses in flowers at the receptor level. Biotech Adv. 2006;24(4):368–381.

15. Marschner H. Mineral nutrition of higher plants. 2nd ed. London: Academic Press; 1995.

16. Palit S, Sharma A, Talukder G. Effects of cobalt on plants. Bot Rev. 1994;60(2):149–181.

17. Hasen SA, Hayat S, Wani AS, Ahmed A. Establishment of sensitive and resistant variety of tomato on the basis of photosynthesis and antioxidative enzymes in the presence of cobalt applied as shotgun approach. Braz J Plant Physiol. 2011;23:175–185.

18. Jaleel CA, Jayakumar K, Chang-Xing Z, Azooz MM. Antioxidant potentials protect *Vigna radiata* (L.) Wilczek plants from soil cobalt stress and improve growth and pigment composition. Plant Omics. 2009;2(3):120–126.

19. Rastgoo L, Alemzadeh A. Biochemical responses of Gouan (*Aeluropus littoralis*) to heavy metal stress. Aust J Crop Sci. 2011;5:375–383.

20. Yıldız M, Ciğerci İH, Konuk M, Fatih Fidan A, Terzi H. Determination of genotoxic effects of copper sulphate and cobalt chloride in *Allium cepa* root cells by chromosome aberration and comet assays. Chemosphere. 2009;75(7):934–938.

21. Chmielowska J, Deckert J. Activity of peroxidases and phenylalanine ammonia-lyase in lupine and soybean seedlings treated with copper and an ethylene inhibitor. Biol Lett. 2008;45:59–67.

22. Lehotai N, Pető A, Bajkán S, Erdei L, Tari I, Kolbert Z. In vivo and in situ visualization of early physiological events induced by heavy metals in pea root meristem. Acta Physiol Plant. 2011;33(6):2199–2207.

23. Cristescu SM, De Martinis D, te Lintel Hekkert S, Parker DH, Harren FJM. Ethylene production by *Botrytis cinerea* in vitro and in tomatoes. Appl Env Microbiol. 2002;68(11):5342–5350.

24. Pfaffl MW. A new mathematical model for relative quantification in real-time RT-PCR. Nucl Acids Res. 2001;29(9):45e–45.

25. Zhao S, Fernald RD. Comprehensive algorithm for quantitative real-time polymerase chain reaction. J Comput Biol. 2005;12(8):1047–1064.

26. Ahmad P, Ozturk M, Gucel S. Oxidative damage and antioxidants induced by heavy metal stress in two cultivars of mustard (*Brassica juncea* L.). Fresen Env Bull. 2012;12:2953–2961.

27. Dhankhar R, Solanki R. Effect of copper and zinc toxicity on physiological and biochemical parameters in *Vigna mungo* (L.) Hepper. Int J Pharma Bio Sci. 2011;2:553–565.

28. Chang C, Wang B, Shi L, Li Y, Duo L, Zhang W. Alleviation of salt stress-induced inhibition of seed germination in cucumber

Therefore, another important conclusion is that $CoCl_2$ should be used as an ethylene inhibitor with cautions, as it might be difficult to distinguish if the observed effects of $CoCl_2$ are dependent on changes in ethylene production or on the action of Co itself.

(*Cucumis sativus* L.) by ethylene and glutamate. J Plant Physiol. 2010;167(14):1152–1156.

29. Koehl J, Djulic A, Kirner V, Nguyen TT, Heiser I. Ethylene is required for elicitin-induced oxidative burst but not for cell death induction in tobacco cell suspension cultures. J Plant Physiol. 2007;164(12):1555–1563.

30. Locke JM. Contrasting effects of ethylene perception and biosynthesis inhibitors on germination and seedling growth of barley (*Hordeum vulgare* L.). J Exp Bot. 2000;51(352):1843–1849.

31. Santana-Buzzy N, Canto-Flick A, Iglesias-Andreu LG, Montalvo-Peniche MC, López-Puc G, Barahona-Pérez F. Improvement of in vitro culturing of habanero pepper by inhibition of ethylene effects. HortScience. 2006;41(2):405–409.

32. Tamimi SM, Timko MP. Effects of ethylene and inhibitors of ethylene synthesis and action on nodulation in common bean (*Phaseolus vulgaris* L.). Plant Soil. 2003;257(1):125–131.

33. Kumar GNM, Knowles NR. Wound-induced superoxide production and PAL activity decline with potato tuber age and wound healing ability. Physiol Plant. 2003;117(1):108–117.

34. Lavoie M, Fortin C, Campbell PGC. Influence of essential elements on cadmium uptake and toxicity in a unicellular green alga: the protective effect of trace zinc and cobalt concentrations. Env Toxicol Chem. 2012;31(7):1445–1452.

35. Agrawal GK, Rakwal R, Iwahashi H. Isolation of novel rice (*Oryza sativa* L.) multiple stress responsive MAP kinase gene, *OsMSRMK2*, whose mRNA accumulates rapidly in response to environmental cues. Biochem Biophys Res Commun. 2002;294(5):1009–1016.

36. Agrawal GK, Tamogami S, Iwahashi H, Agrawal VP, Rakwal R. Transient regulation of jasmonic acid-inducible rice MAP kinase gene (*OsBWMK1*) by diverse biotic and abiotic stresses. Plant Physiol Biochem. 2003;41(4):355–361.

37. Jonak C, Nakagami H, Hirt H. Heavy metal stress. Activation of distinct mitogen-activated protein kinase pathways by copper and cadmium. Plant Physiol. 2004;136(2):3276–3283.

38. Ye Y, Li Z, Xing D. Nitric oxide promotes MPK6-mediated caspase-3-like activation in cadmium-induced *Arabidopsis thaliana* programmed cell death: NO and MPK6 regulate Cd2+-induced PCD. Plant Cell Env. 2013;36(1):1–15.

39. Dguimi HM, Debouba M, Ghorbel MH, Gouia H. Tissue-specific cadmium accumulation and its effects on nitrogen metabolism in tobacco (*Nicotiana tabaccum*, *Bureley* v. Fb9). CR Biol. 2009;332(1):58–68.

40. Gill SS, Khan NA, Tuteja N. Cadmium at high dose perturbs growth, photosynthesis and nitrogen metabolism while at low dose it up regulates sulfur assimilation and antioxidant machinery in garden cress (*Lepidium sativum* L.). Plant Sci. 2012;182:112–120.

41. Huang H, Xiong ZT. Toxic effects of cadmium, acetochlor and bensulfuron-methyl on nitrogen metabolism and plant growth in rice seedlings. Pestic Biochem Physiol. 2009;94(2–3):64–67.

42. Gupta KJ, Fernie AR, Kaiser WM, van Dongen JT. On the origins of nitric oxide. Trends Plant Sci. 2011;16(3):160–168.

43. Chmielowska-Bąk J, Deckert J. A common response to common danger? Comparison of animal and plant signaling pathways involved in cadmium sensing. J Cell Commun Signal. 2012;6(4):191–204.

44. Chen F, Wang F, Sun H, Cai Y, Mao W, Zhang G, et al. Genotype-dependent effect of exogenous nitric oxide on Cd-induced changes in antioxidative metabolism, ultrastructure, and photosynthetic performance in barley seedlings (*Hordeum vulgare*). J Plant Growth Regul. 2010;29(4):394–408.

45. Kopyra M, Stachoń-Wilk M, Gwóźdź EA. Effects of exogenous nitric oxide on the antioxidant capacity of cadmium-treated soybean cell suspension. Acta Physiol Plant. 2006;28(6):525–536.

46. Chmielowska-Bąk J, Deckert J. Nitric oxide mediates Cd-dependent induction of signaling- associated genes. Plant Signal Behav. 2013;8(12):e26664.

47. Besson-Bard A, Gravot A, Richaud P, Auroy P, Duc C, Gaymard F, et al. Nitric oxide contributes to cadmium toxicity in *Arabidopsis* by promoting cadmium accumulation in roots and by up-regulating genes related to iron uptake. Plant Physiol. 2009;149(3):1302–1315.

48. Huerta-Ocampo JA, León-Galván MF, Ortega-Cruz LB, Barrera-Pacheco A, De León-Rodríguez A, Mendoza-Hernández G, et al. Water stress induces up-regulation of DOF1 and MIF1 transcription factors and down-regulation of proteins involved in secondary metabolism in amaranth roots (*Amaranthus hypochondriacus* L.): proteomic and transcriptomic analysis of amaranth roots under drought stress. Plant Biol. 2011;13(3):472–482.

49. Liao Y, Zou HF, Wei W, Hao YJ, Tian AG, Huang J, et al. Soybean *GmbZIP44*, *GmbZIP62* and *GmbZIP78* genes function as negative regulator of ABA signaling and confer salt and freezing tolerance in transgenic *Arabidopsis*. Planta. 2008;228(2):225–240.

7

Identification and quantitative determination of pinoresinol in *Taxus* ×*media* Rehder needles, cell suspension and shoot cultures

Paulina Mistrzak[1,2]*, Hanna Celejewska-Marciniak[1], Wojciech J. Szypuła[1], Olga Olszowska[1], Anna K. Kiss[2]

[1] Department of Pharmaceutical Biology and Medicinal Plant Biotechnology, Medical University of Warsaw, Banacha 1, 02-097 Warsaw, Poland

[2] Department of Pharmacognosy and Molecular Basis of Phytotherapy, Medical University of Warsaw, Banacha 1, 02-097 Warsaw, Poland

Abstract

The aim of our study was to investigate the presence and quantitative contents of lignans in the tissues of *Taxus* ×*media*. The presence of the lignans: pinoresinol, matairesinol and secoisolariciresinol was assessed in needles, shoots cultures and suspension culture. Pinoresinol was the only lignan found in the tissue of *T.* ×*media*. The total pinoresinol content in the needles and in the shoots was 1.24 mg/g dry weight (dw) and 0.69 mg/g dw, respectively. Most of the pinoresinol identified was appeared glycosidically bound. In needles, the amount of glycosidically bound pinoresinol (0.81 mg/g dw) was about twice as high as that of free pinoresinol (0.43 mg/g dw). The content of free and glycosidically bound pinoresinol showed the level of 0.18 mg/g dw and 0.51 mg/g dw, respectively in the in vitro shoot cultures. In the cell culture, no pinoresinol was found.

Keywords: pinoresinol; *Taxus* ×*media*; shoot culture; suspension culture

Introduction

The species of *Taxus* genus are noted due to their anticancer properties of taxane diterpenoids, mainly paclitaxel [1,2]. Nowadays, there is a growing interest in other groups of secondary metabolites found in *Taxus* species, including lignans with a considerable pharmacological and hence clinical potential. Lignans are a class of phenylpropanoids that present a large structural and biological variety as they have antitumor, antiviral, hepatprotective, antioxidant, antiallergic and antiosteoporotic properties [3]. They are generally detected in vascular plants from different families [4]. Their primary physiological role in plants is defensive, particularly in heartwood and seed-forming tissues [5]. Important developments have been reached in elucidating the biosynthesis and chemistry of lignans [4–6]. Lignans are formed of two units of hydroxycinnamoyl alcohol, mostly coniferyl alcohol, via C8-C8′ linkage. The connection of coniferyl alcohol forms pinoresinol. The twice reduction of the product leads to the formation of secoisolariciresinol. Next, the lactone ring is closed to provide matairesinol, which can be the initial point for all lactone ring lignan paths [7–11] (Fig. 1). Topcu and Demirkiran [12] published a review of isolation and structural clarification studies of *Taxus* lignans, with their biological properties. About 50 lignans have been found in eight *Taxus* species. The most common lignans of *Taxus* species are α-coniferin, taxiresinol, secoisolariciresinol, isolariciresinol and lariciresinol.

The highest lignan diversity was observed in *T. mairei*, with over 35 lignans described in this species [13–15], followed by *T. baccata*, with 18 lignans identified [16–22]. The most studies of the lignans, inclusive *Taxus* species, have concentrated on the bioactivities, to ensure the future lead drugs. The results of the studies imply the use of isotaxiresinol in postmenopausal osteoporosis treatment, mainly in the prevention of estrogen deficiency induced bone fracture [23].

Cytotoxic activity analyzes conducted on α-conidendrin, secoisolariciresinol, isotaxiresinol and taxiresinol, proved efficient cytotoxicity against KB-16, A-549 and HT-29 tumor cell lines [24].

Secoisolariciresinol, taxiresinol and isotaxiresinol, the main compounds obtained from the wood of *T. yunnanensis* tree, were assessed for the antiproliferative properties against human fibrosarcoma cell lines and murine colon carcinoma [25]. Taxiresinol, isolated from *T. wallichiana*, showed noteworthy in vitro anticancer action against colon, ovarian, liver and breast cancer [26]. Matairesinol and pinoresinol were proved to have antileukemia activity and the ability to inhibit cAMP [27], as well as they were found to be an anti-HIV agents [28].

The aim of our study was to verify the presence and quantitative contents of lignans in the tissues of the *T.* ×*media* tree with the help of already developed by Schmitt and Petersen [29] method.

* Corresponding author. Email: mistrzak.paulina@gmail.com

Handling Editor: Jan Rybczyński

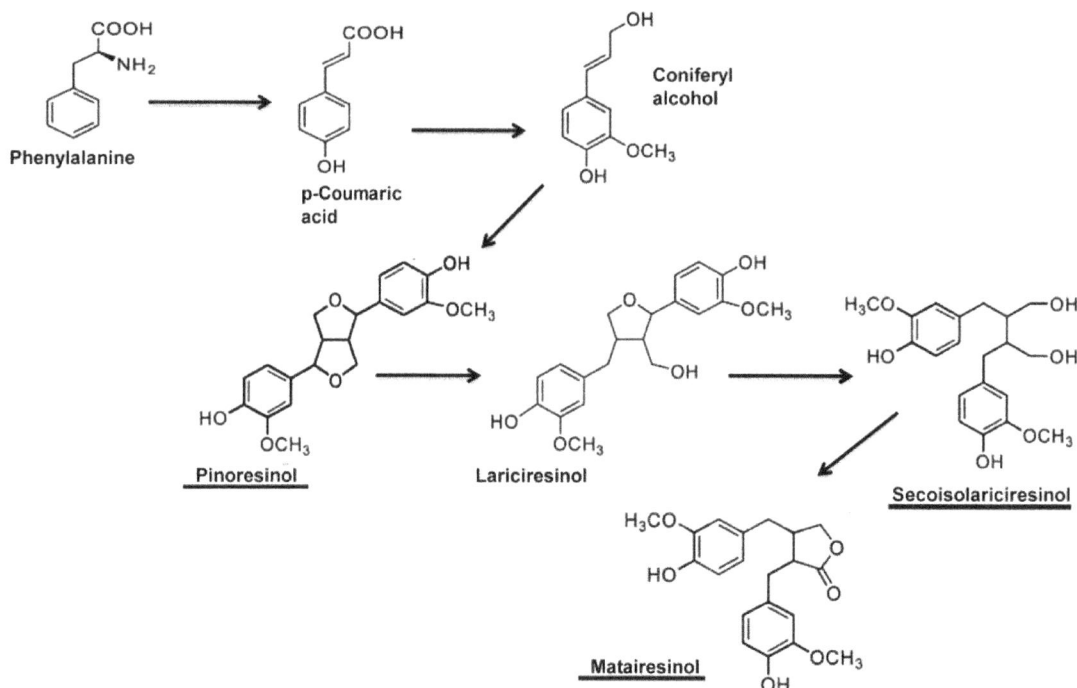

Fig. 1 Biosynthetic pathway of pinoresinol, secoisolariciresinol and matairesinol.

Material and methods

Plant material

The experimental plant material originated from *T. ×media* plant growing in The Botanical Garden of Warsaw University, Warsaw, Poland. Needles for lignans determination and establishment of cell suspension culture were excited from the plant. Shoot cultures were established from shoot of this plant. The initial plant material both needles and shoots were harvested in May 2009.

Establishment of shoot culture

Shoots of *T. ×media* were soaked for 15 min in tap water with a detergent, then rinsed several times with distilled water, incubated for 15 min in 70° EtOH, followed by 30 min in 2.5% chlorine disinfectant solution. After repeated rinsing of the plant material in sterile water, the shoots were cut into 2 cm long pieces and placed onto 30 ml of solid Gupta and Durzan (DCR) medium [30] and woody plant medium/Gamborg medium (WPM/B5) medium in 100 ml conical flasks. WPM/B5 medium is the combination of mineral salts of woody plant medium (WPM) [31] and vitamins of B5 medium [32]. WPM/B5 medium was supplemented with 1 mg/l NAA (1-naphthaleneacetic acid) and 0.1 mg/l BA (6-benzyladenine) and 1.0 mg/l active charcoal. DCR medium was supplemented with 0.8 mg/l NAA and 1.0 mg/l BA. Both media contained 30 g/l sucrose. The shoot culture was maintained at 25°C in the day-light (50 µM m^{-2}s^{-1} and 14 h photoperiod) and subcultured every four weeks on the same fresh media (Tab. 1).

Establishment of cell suspension culture

The needles were cut along the main nerve and placed onto solid Rangaswany medium (WR) [33] in 100 ml conical flasks for callus initiation. WR medium was supplemented with 80 mg/l KNO$_3$, 500 mg/l casamino acids, 30 g/l sucrose and 5 µM picloram. The callus culture was maintained at 25°C in the day-light light intensity. After three months of culture 2 g of developed callus excised and transferred into 250 ml conical flasks containing 50 ml WPM /B5 liquid medium supplemented with 500 mg/l casamino acids, 10 mg/l glycine-betaine (GB–Green Steem®, Finnsugar

Tab. 1 Growth parameters of *Taxus ×media* shoot cultures on DCR (0.8 mg/l NAA and 1.0 mg/l BA) and WPM/B5 (1 mg/l NAA, 0.1 mg/l BA and 1.0 mg/l active charcoal) solid media*.

Medium	Contamination (%)		Necrosis (%)		Survival (%)	Shooting (%)**	
DCR	1 month	15	1 month	7	73	2 months	32
	3 months	5	3 months	-		3 months	58
	Total	20	Total	7		4 months	79
WPM/B5	1 month	13	1 month	9	74	2 months	25
	3 months	5	3 months	-		3 months	55
	Total	18	Total	9		4 months	69

* The experiment was performed for 120 shoot explants in 120 flasks on DCR solid medium as well as on WPM/B5 solid medium.
** The shooting was calculated as the percentage ratio of the number of developed side shoots to the total number of survivors shoot explants.

Bioproducts), 10 mg/l adenine sulphate, 5 µM picloram, 20 g/l sucrose and phosphoric buffer according to Tóth et al. [34]. The culture was maintained at 25°C in the continuous light (50 µM m^{-2}s^{-1}) on a Gyrotory Shaker (New Brunswick Scientific Co.) at 110 rpm. The established cell suspension culture consisted of 30 flasks, was subcultured every three weeks by transferring about 2 g fresh weight of 21 day old cell culture into 50 ml fresh medium in 250-ml conical flasks. Growth parameters and lignans contents of T. ×media cell suspension culture were determined over a 3 passages. The culture of the three conical flasks was harvested every 7 days. Fresh and dry weight were determined separately for each sample and then presented as average value (Tab. 2).

Tab. 2 Development of fresh weight (fw), dry weight (dw) of *Taxus ×media* suspension culture in WPM/B5 liquid medium supplemented with 500 mg/l casamino acids, 10 mg/l glycine-betaine, 10 mg/l adenine sulphate, 5 µM picloram.

Growth (days)	Passage 1		Passage 2		Passage 3	
	Fresh weight (g)*	Dry weight (g)	Fresh weight (g)	Dry weight (g)	Fresh weight (g)	Dry weight (g)
7	2.75 ±0.30	0.42 ±0.40	2.54 ±0.24	0.45 ±0.61	2.32 ±0.17	0.47 ±0.18
14	4.28 ±0.22	0.85 ±0.42	5.00 ±0.28	0.76 ±0.24	4.99 ±0.78	0.67 ±0.15
21	4.00 ±0.56	0.55 ±0.20	4.04 ±0.23	0.58 ±0.19	3.90 ±0.52	0.53 ±0.11

* Values are means of three samples ±SD.

Lignan extraction and determination

For free and glycosidically bound lignans Schmitt and Petersen [29] method was used. Needles harvested in May 2009, side shoots from 4 months old in vitro shoot culture on both solid media (Tab. 1) and biomass from cell suspension culture (Tab. 2) were lyophilized and ground in a mortar. A portion of 0.1 g plant material was suspended in 1 ml MeOH and twice extracted in a ultrasonic bath for 30 s with indirect cooling on ice. After adding of 4 ml H$_2$O (adjusted to pH 5 with 1M H$_3$PO$_4$), the obtained extract was mixed carefully for 30 min with 5 ml CH$_2$Cl$_2$. The organic phase, containing lignan aglyca, was removed, evaporated to dryness and resuspended 3 times in 250 µl MeOH. After evaporation of the solvent, the residue was redissolved in 150 µl 40% MeOH for HPLC analysis. The remaining H$_2$O phase, containing lignan glucosides, was incubated with 0.5 mg β-glucosidase (1000 U mg^{-1}, Roth, Karlsruhe, Germany) per 5 ml for 3.5 h at 35°C and then again extracted with 5 ml CH$_2$Cl$_2$ as described above. By this procedure, lignan glucosides are hydrolyzed and then extracted as aglyca. Samples of all H$_2$O and CH$_2$Cl$_2$ phases were analyzed by HPLC [29]. We also extended the duration of plant material MeOH-extraction up to 24 h according to Theodoridis et al. [35] method used for paclitaxel determination. A portion of 0.1 g freeze-dried, powdered material was suspended in 1 ml MeOH, mixed, sonicated for 15 min, and next centrifuged on the Gyrotory Shaker at 110 rpm for 24 h. The plant sample was centrifuged at 5°C, 15 500 rpm for 15 min, then the methanolic extract was collected and the residue was again extracted with 1 ml MeOH, mixed, sonicated for 15 min and centrifuged

as above. A second methanolic extract was collected and combined with the first one. After evaporation of the solvent, the residue was redissolved in 1 ml MeOH and treated according to Schmitt and Petersen [29] method.

HPLC-DAD analysis of pinoresinol

HPLC analysis was performed on the Shimadzu system consisting of LC-10A pump, UV ASD-10A detector, CBM-20A integrator and LC Solution program. The reversed phase, a Nova-Pak Phenyl 3.9 × 150 mm C18 column (Waters) was used with the following gradient program: 0 min 0% B, 5 min 35% B, 25 min 70% B, 40 min 100% B at a flow rate of 1 ml min^{-1} (solvent A: H$_2$O plus 0.01% H$_3$PO$_4$ and solvent B: 50% acetonitrile plus water). Eluted substances were monitored at 280 nm. The peak areas and retention times were compared to authentic pinoresinol, matairesinol (PhytoLab), and secoisolariciresinol (ChromaDex) standards for identification and quantification. To check the identity and purity of lignans from T. ×media tissues, the samples were subjected to HPLC analysis with diode array detection on the DIONEX system consisting of P580 pump, UVD 340S detector, automated sample injector ASI-100 and Dionex Data System – Chromeleon Version 6.1. For this purpose, the same column, analytical parameters and solvents were used as above. The content of pinoresinol aglyca and pinoresinol glucosides were calculated from three samples of the plant material with every sample measurement performed in triplicate and presented as mean values.

Results

Shoot and suspension culture

To determine and compare lignans content in shoot cultures WPM/B5 and DCR solid media were chosen. Both WPM/B5 and DCR media affected shoot survival and growth (data not shown; Fig. 2a). We analyzed the shooting percentage of explants derived from the yew tree on WPM/B5 and DCR media over a culture of period of 4 months. After three months of culture up to 80% of explants on both solid media did not show any signs of contamination. The necrosis of explants during the first month of culture was mainly caused by damage during surface disinfection (Tab. 1). The suspension culture, obtained from callus on WR medium (Fig. 2b) and maintained in WPM/B5 liquid medium, which consisted of green cells aggregates, was growing fast and required the subculture every 3 weeks (Fig. 2c). The development of the biomass of the suspension culture is shown in Tab. 2 for three consecutive passages.

Phytochemical analysis

Pinoresinol was the only lignan present in the extracts of T. ×media needles as well as in shoot cultures on WPM/B5 and DCR medium. No lignans were detected in the suspension culture. The retention time (Rt) for the pinoresinol standard was 10.7 min (Fig. 3a). The UV spectrum of the pinoresinol standard is presented in Fig. 3b. Sample chromatogram of needles is shown in Fig. 4a before hydrolysis and in Fig. 4b after β-glucosidase hydrolysis. Based on the chromatogram of secoisolariciresinol and matairesinol

Fig. 2 **a** *Taxus ×media* shoot culture on WPM/B5 solid medium supplemented with 1 mg/l NAA, 0.1 mg/l BA and 1.0 mg/l active charcoal after 3 months of cultivation. **b** The callus culture of *T. ×media* on WR solid medium supplemented with 80 mg/l KNO₃, 500 mg/l casamino acids, 30 g/l sucrose and 5 μM picloram after 2 months of cultivation. **c** Cell suspension culture of *T. ×media* in WPM/B5 liquid medium supplemented with 500 mg/l casamino acids, 10 mg/l glycine-betaine, 10 mg/l adenine sulphate, 5 μM picloram after 3 weeks of cultivation.

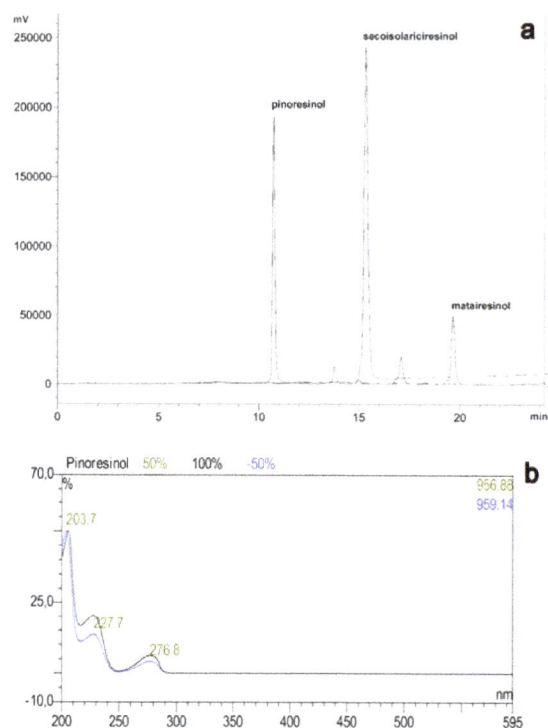

Fig. 3 **a** Chromatogram of the pinoresinol, secoisolariciresinol and matairesinol standards. **b** UV spectrum of the pinoresinol standard.

(dw) and was about twice as high as that that found in the shoot culture on WPM/B5 and DCR media. Most of the pinoresinol found was glycosidically bound and recovered after enzymatic hydrolysis with β-glucosidase. The amount of glycosidically bound pinoresinol was about twice as high as that of free pinoresinol in needles and about four times higher than in shoots (Tab. 3).

Discussion

The most studies of the lignans, along with *Taxus* species, have concentrated on the bioactivities, to provide the future lead drugs. Based on the effects of numerous biological research, it was established that pinoresinol is the potential protecting agent of human health. The carried out investigations indicated its analgesic and local anesthetic properties [36]. Pinoresinol is an effective inhibitor of cAMP [37] and also showed selective inhibitory activity against NF-kB mediated transcription of HIV-1 [38]. The inhibitory ability of the inflammatory responses in lipopolysaccharide (LPS)-activated microglia, attenuation mRNA and protein levels of inducible nitric oxide synthase (iNOS), cyclooxygenase-2 (COX-2) as well as proinflammatory cytokines in LPS-activation suggest that pinoresinol could be potentially useful in modulation of inflammatory status in brain disorders [39]. Pinoresinol was also recognized as a

standards (Fig. 3a), we did not find any of these lignans in the analyzed samples. The highest total content of pinoresinol was obtained from the needles 1.24 mg/g dry weight

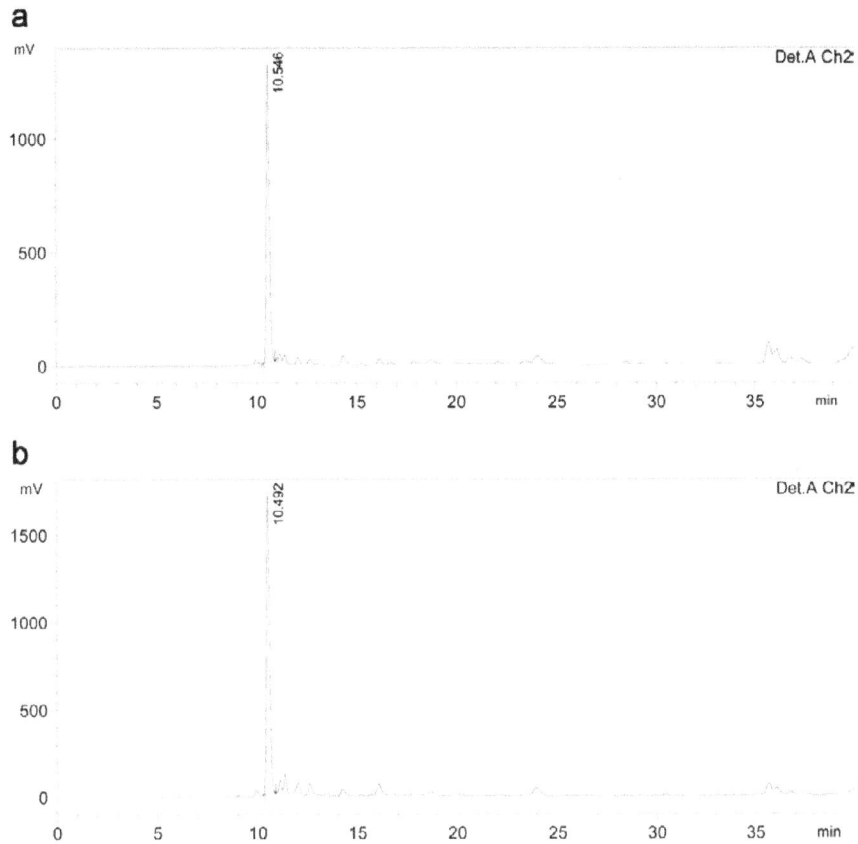

Fig. 4 Chromatogram of sample of *Taxus ×media* needles extract before (**a**) and after (**b**) enzymatic hydrolysis.

Tab. 3 Pinoresinol content in *Taxus ×media* needles and side shoots after 4 months of the in vitro cultivation.

Plant material	Schmitt and Petersen [29] method of extraction	Free pinoresinol (mg/g dw)	Glycosidically bound Pinoresinol (mg/g dw)	Total pinoresinol content (mg/g dw)
Needles	Original	0.34 ±0.03*	0.65 ±0.01	0.99 ±0.01
	With extended MeOH extraction	0.43 ±0.06	0.81 ±0.05	1.24 ±0.01
Shoots on WPM/B5	Original	0.14 ±0.03	0.43 ±0.07	0.57 ±0.05
	With extended MeOH extraction	0.18 ±0.05	0.51 ±0.01	0.69 ±0.01
Shoots on DCR	Original	0.11 ±0.05	0.40 ±0.04	0.51 ±0.05
	With extended MeOH extraction	0.15 ±0.05	0.53 ±0.02	0.67 ±0.06

* Values are means of three samples ±*SD*. Variations among accumulation of free and glycosidically bound pinoresinol between original and extended MeOH extractions and between media used for shoot culture (Tukey–Kramer post-hoc test.) do not differ statistically at $P < 0.05$.

putative hypoglycemic agent that inhibited intestinal maltase. The investigations proved that an assorted-type inhibition mechanism showed pinoresinol as an agent responsible for delaying enzyme work by straight binding to maltase and interfering maltase-moltose intermediate [40]. Due to the maltase inhibitory activity of pinoresinol by noncompetitive way, it would demonstrate synergistic impact with antidi-betic drugs such as acarbose in extinguishing level of blood glucose, consequently intensely reducing the applying dose of acarbose [41]. Furthermore pinoresinol is metabolized in human digestive tract to enterolactone and enterodiol, the principal lignans related with decreased risk of peculiar cancers and cardiovascular diseases [42]. The studies carried out on pinoresinol-rich extra virgin olive oil exhibited its ability to decrease the cell viability, induces cell cycle arrest and apoptosis by specifically upregulation the ATM-p53 cascade. It indicates that pinoresinol-rich extra virgin olive oil might be an effective agent in the chemoprevention colorectal cancer cells [43].

Glycosidically bound pinoresinol: pinoresinol 4′-O-β-D-glucopyranoside and pinoresinol di-O-β-D-glucopyranoside, has been shown to possess anti-hypertensive effects, could increase luciferase activity in both estrogen receptor (ER) ERβ and ERα, they were equally potent in eliciting trans-activation through the two ER subtypes [44]. Blood vessels express ERs, and ERβ plays an essential role in the regulation of vascular function and blood pressure [45]. Activating ERs reinforces their anti-hypertensive effect.

The method developed by Schmitt and Petersen [29] for the quantitative estimation of free and glycosidically bound pinoresinol, matairesinol and secoisolariciresinol in *Forsythia ×intermedia* cell suspension culture appeared effective in our experiments concerning preparation of extracts of *T. ×media* tissues. Using the Schmitt and Petersen [29] method, pinoresinol was the only lignan found and quantified in needles of *T. ×media* tree as well as in shoots growing in vitro. *Taxus ×media* suspension culture did not contain any of the lignans in question. After extended duration of methanolic extraction according to Theodoridis et al. [35] method used for paclitaxel determination, the pinoresinol content was about 20 percent higher comparing Schmitt and Petersen [29] method employed (Tab. 1).

So far, the presence of pinoresinol has been determined solely in *T. mairei* [15] and *T. cuspidata* [46]. Pinoresinol in wood, twigs (0.02 mg/g dw) and roots (0.01 mg/g dw) was presented. The needles did not contain the lignan in question [46].

Our studies indicated that examined *T. ×media* is a rich sourse of pinoresinol, the amount determined in needles (1.24 mg/g dw) and shoot culture (0.69 mg/g dw) is comparable to that of pinoresinol isolated from knots of *Abies alba* (Pinaceae) which ranged from 0.36 mg/g dw to 1.0 mg/g dw [47,48].

In the flaxseed (*Linum usitatissimum* L.), the richest known source of precursors of phytoestrogens (enterolactone and enterodiol), the content of pinoresinol was 0.03 mg/g dw [49]. Only a few of a large variety of plant lignans are converted into the enterodiol and enterolactone by the intestinal microflora. It was initially considered that only secoisolariciresinol and matairesinol were enterolignan precursors, but later identified precursors lariciresinol and pinoresinol have a high degree of conversion [50].

The presence of lignans has been confirmed in almost all parts of the plant: bark, heartwood, needles and roots. Taking into account the large amount of lignans isolated from *Taxus* species, the explanation of great structural variety of Taxaceae family lignans is problematic.

To date, two lignans 7-hydroxymatairesinol and epinortrachelogenin have been found in *T. ×media* [51]. In *T. cuspidata* [46] with the considerable amount of pinoresinol, the presence of taxiresinol (0.37 mg/g dw) and secoisolariciresinol (9.5 mg/g dw) in the needles was also determined.

Examinations of gymnosperms and angiosperms exhibited the presence of a general lignan biosynthetic pathway where it performs as enzymes, especially pinoresinol/lariciresinol reductase [52] and secoisolariciresinol dehydrogenase [53].

The absence of secoisolariciresinol and matairesinol in the needles and shoot cultures of the *T. ×media* tree might indicate that pinoresinol/lariciresinol reductase and secoisolariciresinol dehydrogenase involved in the next steps of lignan biosynthesis, are not active in these organs. In the suspension culture, no products of biosynthetic pathway of the lignans in questions were found. The issue of lignan biosynthesis of the *T. ×media* tree requires further study.

Acknowledgments
We would like to thank Agnieszka Pietrosiuk, Phar D, PhD for her help in performing the phytochemical analysis and Ireneusz Rudnicki for photographs of the in vitro cultures.

Authors' contributions
The following declarations about authors' contributions to the research have been made: concept of the study: PM, OO; laboratory research and data analyses: PM, HCM, AKK, WJS; writing of the manuscript: PM, AKK, OO, WJS.

Competing interests
No competing interests have been declared.

References
1. Kingston DGI, Molinero AA, Rimoldi JM. The taxane diterpenoids. In: Herz W, Kirby GW, Moore RE, Steglich W, Tamm C, editors. Progress in the chemistry of organic natural products. New York, NY: Springer; 1993. p. 1–206.

2. Parmar VS, Jha A, Bisht KS, Taneja P, Singh SK, Kumar A, et al. Constituents of the yew trees. Phytochemistry. 1999;50:1267–1304.

3. Lee KH, Xiao Z. Lignans in treatment of cancer and other diseases. Phytochem Rev. 2003;2:341–362.

4. Umezawa T. Diversity in lignan biosynthesis. Phytochem Rev. 2003;2:371–390.

5. Lewis NG, Davin LB. Lignans: biosynthesis and function. In: Sankawa U, editor. Comprehensive natural products chemistry. Amsterdam: Elsevier; 1999. p. 639–712. (vol 1).

6. Umezawa T. Biosynthesis of lignans and related phenylpropanoid compounds. Reg Plant Growth Dev. 2001;36:57–67.

7. Umezawa T, Davin LB, Lewis NG, 1990. Formation of the lignan, (−)secoisolariciresinol, by cell free extracts of *Forsythia intermedia*. Biochem Biophys Res Commun. 1990;171:1008–1014.

8. Katayama T, Davin LB, Lewis NG. An extraordinary accumulation of (−)-pinoresinol in cell-free extracts of *Forsythia intermedia*: evidence

for enantiospecific reduction of (+)-pinoresinol. Phytochemistry. 1992;31:3875–3881.

9. Katayama T, Davin LB, Alex CA, Norman G. Lewis NG. Novel benzylic ether reductions in lignan biogenesis in *Forsythia intermedia*. Phytochemistry. 1993;33:581–591.

10. Umezawa T, Kuroda H, Isohata T, Higuchi T, Shimada M. Enantioselective lignan synthesis by cell-free extracts of *Forsythia koreana*. Biosci Biotechnol Biochem 1994;58:230–234.

11. Davin LB, Wang HB, Crowell AL, Bedgar DL, Martin DM, Sarkanen S, et al. Stereoselective bimolecular phenoxy radical coupling by an auxiliary (Dirigent) protein without an active center. Science. 1997;275:362–367.

12. Topcu G, Demirkiran O. Lignans from Taxus species. Top Heterocycl Chem. 2007;11:103–144.

13. Shen YC, Chen CY, Lin YM, Kuo YH. A lignan from roots of *Taxus mairei*. Phytochemistry. 1997;46:1111–1113.

14. Shi QW, Oritani T, Sugiyama T, Yamada T. Taxane diterpenoids from the seeds of Chinese yew, *Taxus mairei*. Nat Prod Lett. 1999;13:179–186.

15. Yang SJ, Fang JM, Cheng YS. Lignans, flavonoids and phenolic derivatives from *Taxus mairei*. J Chin Chem Soc. 1999;46:811–818.

16. Das B, Takhi M, Srinivas KVNS, Yadav JS. Phenolics from needles of himalayan *Taxus baccata*. Phytochemistry. 1993;33:1489–1491.

17. Das B, Takhi M, Srinivas KVNS, Yadav JS. A lignan from needles of himalayan *Taxus baccata*. Phytochemistry. 1994;36:1031–1033.

18. Das B, Padma Rao S, Srinivas KVNS, Yadav JS. Lignans, biflavones and taxoids from Himalayan *Taxus baccata*. Phytochemistry. 1995;38(3):715–717.

19. Erdemoglu N, Sener B, Ozcan Y, Ide S. Structural and spectroscopic characteristics of two new dibenzylbutane type lignans from *Taxus baccata* L. J Mol Struct. 2003;655:459–466.

20. Erdemoglu N, Sener B, Choudhary MI. Bioactivity of lignans from *Taxus baccata*. Z Naturforsch C. 2004;59c:494–498.

21. King FE, Jurd L, King TJ. Iso-taxiresinol (3′-demethylisolariciresinol) a new lignan extracted from the heartwood of the English yew, *Taxus baccata*. J Chem Soc. 1952;17–24.

22. Mujumdar RB, Srinivasan R, Venkataraman K. Taxiresinol, a new lignan in the heartwood of *Taxus baccata*. Indian J Chem. 1972;10:677–680.

23. Yin J, Tezuka Y, Subehan SL, Nobukawa M, Nobukawa T, Kadota S. In vivo anti-osteoporotic activity of isotaxiresinol, a lignan from wood of *Taxus yunnanensis*. Phytomedicine. 2006;13:37–42.

24. Shen YC, Chen CY, Chen YJ, Kuo YH, Chien CT, Lin YM. Bioactive lignans and taxoids from the roots of formosan *Taxus mairei*. Chin Pharm J. 1997;49:285–296.

25. Banskota AH, Usia T, Tezuka Y, Kouda K, Nguyen NT, Kadota S. Three new C-14 oxygenated taxanes from the wood of *Taxus yunnanensis*. J Nat Prod. 2002;65(11): 1700–1702.

26. Chattopadhyay SK, Kumar TRS, Maulik PR, Srivastava S, Garg A, Sharon A, et al. Absolute configuration and anticancer activity of taxiresinol and related lignans of *Taxus wallichiana*. Bioorg Med Chem. 2003;11:4945–4948.

27. Tsukamoto H, Hisada A, Nishibe S. Lignans from bark of the *Olea* plants. Chem Pharm Bull. 1984;32:2730–2735.

28. Ishida J, Wang HK, Oyama M, Cosentino ML, Hu CQ, Lee KH.

Anti-AIDS agents. 46.1 anti-HIV activity of harman, an anti-HIV principle from *Symplocos setchuensis*, and its derivatives. J Nat Prod. 2001;64:958–960.

29. Schmitt J, Petersen M. Pinoresinol and matairesinol accumulation in a *Forsythia ×intermedia* cell suspension culture. Plant Cell Tissue Organ Cult. 2002;68:91–98.

30. Gupta PK, Durzan DJ. Shoot multiplication from mature trees of Douglas-fir (*Pseudotsuga menziesii*) and sugar pine (*Pinus lambertiana*). Plant Cell Rep. 1985;4:177–179.

31. Lloyd G, McCown B. Commercially-feasible micropropagation of mountain laurel, *Kalmia latifolia*, by use of shoot-tip culture. Comb Proc Int Plant Prop Soc. 1980;30:420–427.

32. Gamborg OL, Miller RA, Ojima O. Nutrient requirements of suspension cultures of soybean root cell. Exp Cell Res. 1968;50:151–158.

33. Rangaswamy NS. Experimental studies on female reproductive structures of *Citrus microcarpa* Bunge. Phytomorphology. 1961;11:109–127.

34. Tóth S, Scott P, Sorvari S, Toldi O. Effective and reproducible protocols for in vitro culturing and plant regeneration of the physiological model plant *Ramonda myconi* (L.) Rchb. Plant Sci. 2004;166:1027–1034.

35. Theodoridis G, de Jong CF, Laskaris G, Verpoorte R. Application of SPE for the HPLC analysis of taxanes from *Taxus* cell cultures. Chromatographia. 1998;47:25–34.

36. Okuyama E, Suzumurak K, Kamazaki M. Pharmacologicaly active compounds of Todopon Puok (*Fragraea racemosa*), a medical plant from Borneo. Chem Pharm Bull. 1995;43:2200–2204.

37. Tsukamoto H, Hisada S, Nishibe S. Lignans from the bark of *Olea* plants. Chem Pharm Bull. 1984;32:2730–2735.

38. Mitsuhashi S, Kishimoto T, Uraki Y, Okamoto T, Ubukata M. Low molecular weight lignin suppresses activation of NF-κB and HIV-1 promoter. Bioorg Med Chem. 2008;16:2645–2650.

39. Hyo WJ, Ramalingam M, Jong GL, Seung HL, Young SK, Yong-Ki P. Pinoresinol from the fruits of *Forsythia koreana* inhibits inflammatory responses in LPS-activated microglia. Neurosci Lett. 2010;480:215–220.

40. Wikul A, Damsud T, Kataoka K, Phuwapraisirisan P. (+)-Pinoresinol is a putative hypoglycemic agent in defatted sesame (*Sesamum indicum*) seeds though inhibiting a-glucosidase. Bioorg Med Chem Lett. 1012;22:5215–5217.

41. Wang Y, Ma L, Pang C, Huang M, Huang, Z, Gu L. Synergetic inhibition of genistein and d-glucose on α-glucosidase. Bioorg Med Chem Lett. 2004;14:2947–2950.

42. Liu Z, Saarinen NM, Thompson LU. Sesamin is one of the major precursors of mammalian lignans in sesame seed (*Sesamum indicum*) as observed in vitro and in rats. J Nutr. 2006;136:906–912.

43. Fini L, Hotchkiss E, Fogliano V, Graziani G, Romano M, de Vol EB, et al. Chemopreventive properties of pinoresinol-rich olive oil involve a selective activation of the ATM-p53 cascade in colon cancer cell lines. Carcinogenesis. 2008;29(1):139–146.

44. Wang H, Li MC, Yang J, Yang D, Su YF, Fan GW, et al. Estrogenic properties of six compounds derived from *Eucommia ulmoides* Oliv. and their differing biological activity through estrogen receptors alpha and beta. Food Chem. 2011;129:408–416.

45. Zhu Y, Bian Z, Lu P, Karas RH, Bao L, Cox D, et al. Abnormal vascular function and hypertension in mice deficient in estrogen receptor beta. Science. 2002;295(5554):505–508.

46. Kawamura F, Kikuchi Y, Ohira T, Yastagai M. Phenolic constituents of *Taxus cuspidata*. I: lignans from the roots. J Wood Sci. 2000;46:167–171.

47. Willför S, Nisula L, Hemming J, Reunanen M, Holmbom B. Bioactive phenolic substances in industrially important tree species. Part
1: knots and stemwood of different spruce species. Holzforschung.
2005;58(4):335–344.

48. Willför S, Nisula L, Hemming J, Reunanen M, Holmbom B. Bioactive
phenolic substances in industrially important tree species. Part 2: knots
and stemwood of fir species. Holzforschung. 2005;58(6):650–659.

49. Milder IE, Arts IC, van de Putte B, Venema DP, Hollman PC. Lignan contents of Dutch plant foods: a database including lariciresinol, pinoresinol, secoisolariciresinol and matairesinol. Br J Nutr.
2005;93(3):393–402.

50. Heinonen S, Nurmi T, Liukkonen K, Poutanen K, Wähälä K, Deyama
T, et al. In vitro metabolism of plant lignans: new precursors of
mammalian lignans enterolactone and enterodiol. J Agric Food Chem.
2001;49:3178–3186.

51. Apendino G, Cravotto G, Enriu R, Gariboldi P, Barboni L, Torregiani
E, et al. Taxoids from the roots of *Taxus* ×*media* cv. Hicksii. J Nat
Prod. 1994;57:607–613.

52. Katayama T, Masaoka T, Yamada H. Biosynthesis and stereochemistry
of lignans in *Zanthoxylum ailanthoides*. I. (+)-Lariciresinol formation
by enzymatic reduction of (±)-pinoresinols. Mokuzai Gakkaishi.
1997;43:580–588.

53. Xia ZQ, Costa MA, Pelissier HC, Davin LB, Lewis NG. Secoisolariciresinol dehydrogenase purification, cloning, and functional
expression. Implications for human health protection. J Biol Chem.
2001;276:12614–12623.

8

The impacts of BSMV on vegetative growth and water status in hulless barley (*Hordeum vulgare* var. *nudum*) in VIGS study

Junjun Liang[1], Xin Chen[1,2], Huanhuan Zhao[1], Shuiyang Yu[1], Hai Long[1], Guangbing Deng[1], Zhifen Pan[1], Maoqun Yu[1]*

[1] Chengdu Institute of Biology, Chinese Academy of Sciences, No. 9 Section 4, Renmin South Road, Chengdu 610041, China

[2] College of Life Sciences, Sichuan University, No. 24 South Section 1, Yihuan Road, Chengdu 610065, China

Abstract

Barley stripe mosaic virus (BSMV) is an established and extensively used virus-induced gene silencing (VIGS) vector for gene function analysis in monocots. However, the phenotypes generated by targeted gene silencing may be affected or masked by symptoms of BSMV infection. To better understand the potential effects of BSMV-VIGS in hulless barley (*Hordeum vulgare* var. *nudum*), the accumulation pattern of BSMV and its impacts on vegetative growth and water status were investigated. The results indicated that the vegetative growth of infected plants was significantly and continuously impacted by BSMV from 10 to 40 days post inoculation (dpi). When the accumulation of BSMV was extremely high (7 to 11 dpi), infected plants displayed twisted leaf tips with an increased water lose rate (WLR) and decreased water content (WC). Virus accumulation declined and stabilized after 25 dpi, at this stage, the WLR and WC were unaffected in the infected plants. The efficiency of VIGS was tested by the silencing of Phytoene desaturase (PDS). RT-qPCR indicated that BSMV-VIGS can be sustained with good efficiency for up to 40 dpi under an altered condition with lower temperature (22 ±1°C) and higher relative humidity (70 ±10%). It was concluded that 25 to 40 dpi was the appropriate time zone for drought-related gene analysis by BSMV-VIGS under such condition.

Keywords: hulless barley (*Hordeum vulgare* var. *nudum*); VIGS; BSMV; virus reproduction; vegetative growth; water retention capacity

Introduction

Virus-induced gene silencing (VIGS) has been developed as an important tool for gene function analysis [1–3]. This method has several advantages over conventional transgenic technologies in that it is easy to manipulate, saves time and has a low cost [4]. Barley stripe mosaic virus (BSMV) is a *Hordeivirus* which can cause mild to moderate mosaic symptoms [4,5]. It has been genetically reconstructed to a VIGS vector for monocots [6] and has been widely utilized in gene silencing [6–13].

However, there are some drawbacks in utilizing BSMV-VIGS, in that the high virus level can alter plant development, particularly in relation to height and leaf morphology [4]. The infection of BSMV can cause non-specific stress responses, resulting in delayed biosynthesis processes and accelerated senescence [14]. Such results could mask, or be confused as, effects of gene silencing. Progeny testing of infected plants can be used to obtain a mild viral symptom in

order to elucidate the true effects of VIGS [15,16], however the extra time taken mitigates one of the primary advantages of the quick results that this method offers.

Tibetan hulless barley (*Hordeum vulgare* var. *nudum*) is an annual, self-pollinated species with adaptations to drought, salinity and low temperature conditions associated with its domestication on the Qinghai-Tibet plateau. These hulless barleys grow on a broad range of environments with large differences in water availability, temperature and soil type, giving rise to a range of adaptive diversity to abiotic stresses. Thus they are likely to contain good sources of drought resistance alleles suitable for breeding purposes [17]. Identifying genes involved in the drought tolerance is a challenge as this a quantitative trait that involves many metabolic pathways. VIGS is a potential tool to help characterize the functions of prospective stress-response genes [18]. However, to apply BSMV-VIGS in characterizing drought stress related candidate genes, it is important to understand how BSMV accumulation affects plant morphology, vegetative growth and water retention capacity of Tibetan hulless barley.

Phytoene desaturase (PDS) is normally used as a reporter gene to test if a VIGS vector works in a certain species and can be used as an indicator to show the best time period

* Corresponding author. Email: yumaoqun@cib.ac.cn

Handling Editor: Przemysław Wojtaszek

for gene silencing. The reduction or loss of this enzyme results in inhibition of the carotenoid biosynthesis pathway leading to a photo-bleaching phenotype due to chlorophyll photo-oxidation [1,19–22].

In this study, the relationship between BSMV virus accumulation and its impact on vegetative growth and water retention capacity of Tibetan hulless barley was analyzed from 10 to 40 days post inoculation (dpi). BSMV accumulation was checked by quantitative reverse transcription polymerase chain reaction (RT-qPCR) and its effect on plant height, total dry weight (TDW), water loss rate (WLR) and water content (WC) was investigated. Plant growth conditions were optimized with lower temperature ($22 \pm 1°C$) and higher relative humidity ($70 \pm 10\%$). The efficiency of gene silencing under thus condition was analyzed from 7 to 60 dpi using *PDS* as an indicator. The impacts of *PDS* silencing on vegetative growth and water retention capacity and the possibility of setting *PDS*-silenced plants as a control group to compare phenotype changes caused by targeted gene silencing were also investigated.

Material and methods

Plant material and growth conditions

The hulless barley line Z033 is drought tolerant and has good early vigor [23] and was employed for BSMV inoculation. All seeds were germinated in half-strength Murashige and Skoog (MS) solid medium for 3 d and uniformly germinated seeds were transplanted into plastic pots (5 cm in height and 5 cm in diameter; one plant per pot) that contained 100 g of potting mixture. The potting mix consisted of local soil, nutrient soil and vermiculite (7:2:1, v:v:v). The seedlings were cultivated in a greenhouse with a temperature range of $22 \pm 1°C$, a relative humidity of $70 \pm 10\%$ and a photoperiod of 16 h/8 h (light/dark). All plants were watered every second day and pots were free draining.

BSMV inoculation

The BSMV vectors BSMV-α (U35767.1), BSMV-β (U35770.1), BSMV:GFP and BSMV:PDS were kindly provided by Prof. Daowen Wang (Institute of Genetics and Developmental Biology, Chinese Academy of Sciences, Beijing, China). Genomic organization of BSMV (ND18 strain) RNAs α, β and γ were described by Zhou et al. [24]. RNA β do not have the coat protein beta A deletion. The vector, BSMV:GFP was previously described in Haupt et al. [25] and its sequence is listed in Tab S1. A *PDS* of 200 bp was inserted in an antisense orientation (Fig. S1) to form the construct BSMV:PDS. The sequence of BSMV:PDS is listed in Tab. S2. The plasmid components of BSMV-α and BSMV-γ (BSMV:GFP and BSMV:PDS) were linearized by *Mlu*I while the BSMV-β plasmid was linearized by *Spe*I, extracted with phenol/chloroform and used as templates for in vitro transcription by the RiboMAX™ Large Scale RNA Production System-T7 (Promega, United States) with Ribo m⁷G Cap Analog (Promega) for 3 h at 37°C. The three viral RNA components were first mixed using a 1:1:1 (v:v:v) ratio, then the combined components were mixed with nuclease-free water and GKP buffer [50 mM glycine, 30 mM dipotassium

hydrogen phosphate, pH 9.2, 1% (w:v) bentonite, and 1% (w:v) celite] with a ratio of 1:3:4 (v:v:v). Three leaf stage seedlings (3rd leaf less than 5 cm) were inoculated with 8 μl of this mixture, with the solution being applied with gentle strokes using rubber gloves to the bottom of the second and the third leaves. For mock inoculation on control seedlings, 8 μl of nuclease-free water and GKP buffer in a 1:1 (v:v) ratio was applied as described above. After inoculation, the seedlings were fog sprayed with nuclease-free water and covered with plastic film to maintain high humidity for 3 d. All the inoculated and control plants were maintained in a greenhouse with a temperature range of $22 \pm 1°C$, a relative humidity of $70 \pm 10\%$, and a photoperiod of 16 h/8 h (light/dark). All plants were watered every alternate day to soil water holding capacity.

Biomass accumulation tests

Plants were harvested at 10, 20, 30 and 40 days post inoculation (dpi). Entire plants were harvested (soils adhering to the roots was gently removed), and placed in a 70°C oven for 3 d to dry. Roots and the aerial parts of the plant were weighed separately. Biomass accumulation was determined by TDW.

Water loss rate and water content analysis

WC was used for the investigation of water status of BSMV-infected and mock-inoculated control plants under optimized condition with normal water supply. A 10 cm leaf length was collected from the newest leaves by excising the newest leaf to its base (if this was less than 10 cm) and making up the remaining length with the penultimate leaf, as measured from the leaf base. These were weighed to calculate fresh weight (FW) and oven dried at 70°C for 3 d to calculate dry weight (DW). WC, as a percentage of fresh mass, was calculated according to the following formula: WC = (FW – DW) / FW.

To investigate the water holding capacity of BSMV-infected and mock-inoculated controls under detached drought stress, WLR of excised newly grown leaves were measured as described by a detached drought stress as previously reported [23] at 10, 20, 30 and 40 dpi. WLR over 6 h was calculated by the following formula: WLR_6 (g h⁻¹ g⁻¹ DW) = (FW – W_6) / (DW × 6), where FW: fresh weight, W_6: weight of 6 h, DW: dry weight).

RT-qPCR analysis

Total RNA from the newly grown leaves (10 cm leaf length, as described above) of BSMV:PDS-inoculated plants and BSMV:GFP-inoculated plants were isolated every two days from 7 to 19 dpi and every five days from 25 to 60 dpi (each sample contained leaves from at least 5 individual plants). A total of 5 μg RNA were treated with DNaseI and used for first-strand cDNA synthesis using M-MLV reverse transcriptase (TaKaRa). The cDNA reaction mixture was diluted tenfold, and 1 μl was used as a template in a 20-μl PCR reaction. PCR was performed after pre-incubation at 95°C for 5 min followed by 40 cycles of denaturation at 95°C for 15 s, annealing at 60°C for 15 s, and extension at 72°C for 15 s. All the reactions were performed in the Chromo4 real-time PCR detector system (Bio-Rad) using iQ SYBR green supermix (Bio-Rad). To normalize the cDNA templates, the housekeeping gene *elongation factor 1α* (*EF*) was co-amplified. All primers were synthesized by Invitrogen (Shanghai, China). Their sequences and efficiency are shown in Tab. 1. The primers, PDS P1 and P2, which are designed for *PDS* expression analysis, targeted regions of *PDS* other than the section used for silencing. The amplification specificity was checked with a heat-dissociation protocol (melting curves in the 65 to 95°C range) as a final step of the PCR. Primer pairs showed a single peak on the melting curve, and a single band with the expected size was detected by agarose gel electrophoresis.

Tab. 1 PCR primers used to amplify reference and target genes.

Name	Primer sequences	Product size	Efficiency
EF	P₁: 5'-AAGGATCTCAAGCGTGGG-3' P₂: 5'-GTGGGATGTGTGGCAGTC-3'	154 bp	102.30%
PDS	P₁: 5'-CACTTACTGACGGGACTCA-3' P₂: 5'-AGAAGGTGGTCGTATGTGTT-3'	198 bp	99.96%
BSMV-α	P₁: 5'-AGAAGGTAACAGGGACGGTG-3' P₂: 5'-TGAGTGTGATGGAGTTGACCC-3'	178 bp	89.10%
BSMV-β	P₁: 5'-ACTTTAGCGTTGCTTCCTCTC-3' P₂: 5'-CGATTGTTGTGCGGCTGA-3'	219 bp	103.00%
BSMV-γ	P₁: 5'-GTCCTTACGCTTTCATCACCT-3' P₂: 5'-TTCAGACGGGAGAACAGGCG-3'	245 bp	94.73%
PDSs	P₁: 5'-TGGCTAAGCTTGAAAGTGAGG-3' P₂: 5'-GTGGACTTGCAAACACTCC-3'	396 bp	

EF: elongation factor 1α; PDS: Phytoene desaturase; BSMV-α, -β and -γ represent the three components of barley stripe mosaic virus; PDSs: situated at both sides of the *PDS* insert on BSMV-γ

The stability of BSMV-VIGS

The insertion site flanking primers for BSMV-γ, named PDSs P1 and P2, were designed for the amplification of the silencing inducing fragment (the *PDS* insert) and subsequent sequencing (Fig. S1). Their sequences are shown in Tab. 1. The PCR products from the cDNA templates of control plants and BSMV:PDS-inoculated plants (7 to 60 dpi) were checked by agarose gel electrophoresis. Products with expected sizes from BSMV:PDS-inoculated plants at 45, 50, 55 and 60 dpi were sequenced at GENEWIZ (Suizhou, China).

Data analysis

Means, standard deviation (*SD*), standard error (*SE*) of plant height, TDW, WLR, WC and expression data (RT-qPCR results) were performed using the SPSS package (version 17.0, SPSS Inc.). All data obtained were subjected to one way analysis of variance (ANOVA) and the mean differences were compared by the least significant difference (LSD) test. All analyses were conducted according to a completely randomized design.

Results

Dynamic accumulation of BSMV from 7 to 60 dpi

The accumulation of BSMV-α, BSMV-β and BSMV-γ from 7 to 60 dpi was analyzed by RT-qPCR and the results showed similar accumulation patterns (Fig. 1). BSMV

accumulation in BSMV:PDS-inoculated plants reached a peak from 7 to 11 dpi and declined from 11 to 17 dpi. After 25 dpi, BSMV:PDS was at a low and relatively stable level.

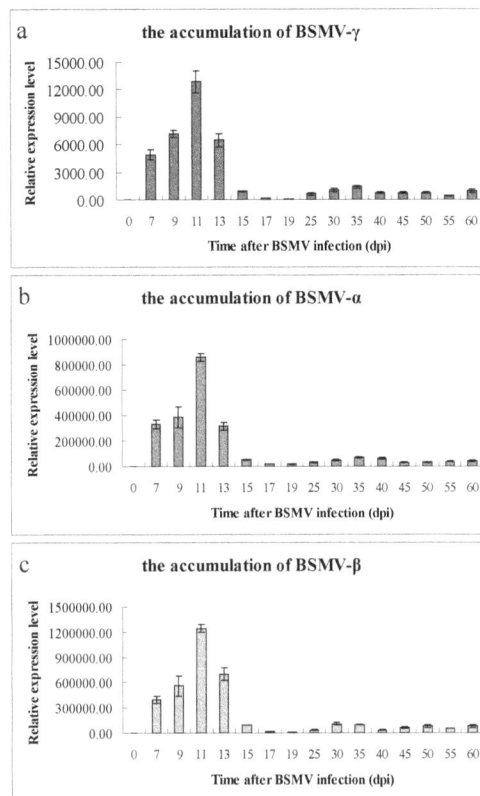

Fig. 1 The accumulation of BSMV-γ (**a**), BSMV-α (**b**) and BSMV-β (**c**) from 7 to 60 dpi. Barley stripe mosaic virus (BSMV) accumulation in BSMV:PDS-inoculated plants was analyzed by quantitative reverse transcriptase polymerase chain reaction (RT-qPCR) using *elongation factor 1α* (*EF*) as the reference from 7 to 60 days post inoculation (dpi). 0 dpi refers to mock-inoculated wild type (WT) control plants (three- to four-leaf stage). Each sample contained newly grown leaves from at least 5 individual plants. All data represent the average value of three technical repeats and are shown as the means ±*SD* (*n* = 3).

The impacts of BSMV accumulation on plant growth

Plant height and the TDW of BSMV-infected plants and mock-inoculated wild type (WT) controls were measured at 10, 20, 30 and 40 dpi (Fig. 2, Fig. 3a,b). Average height of BSMV-infected plants were significantly (*P* < 0.05) shorter than controls at all time points (Fig. 2, Fig. 3a). TDW of BSMV:GFP-inoculated and BSMV:PDS-inoculated plants also showed lower values than controls from 10 to 40 dpi (Fig. 3b). At 10 dpi, the mean TDW of BSMV:GFP-inoculated infected plants were 25 mg less than the control, and the difference increased to 53 mg and 98 mg at 20 and 30 dpi, and finally to 282 mg at 40 dpi. The morphology of the leaves were also monitored at 10, 20, 30 and 40 dpi (Fig. 4), with the blades in BSMV-infected plants being significantly smaller at 30 and 40 dpi when compared to controls. Furthermore,

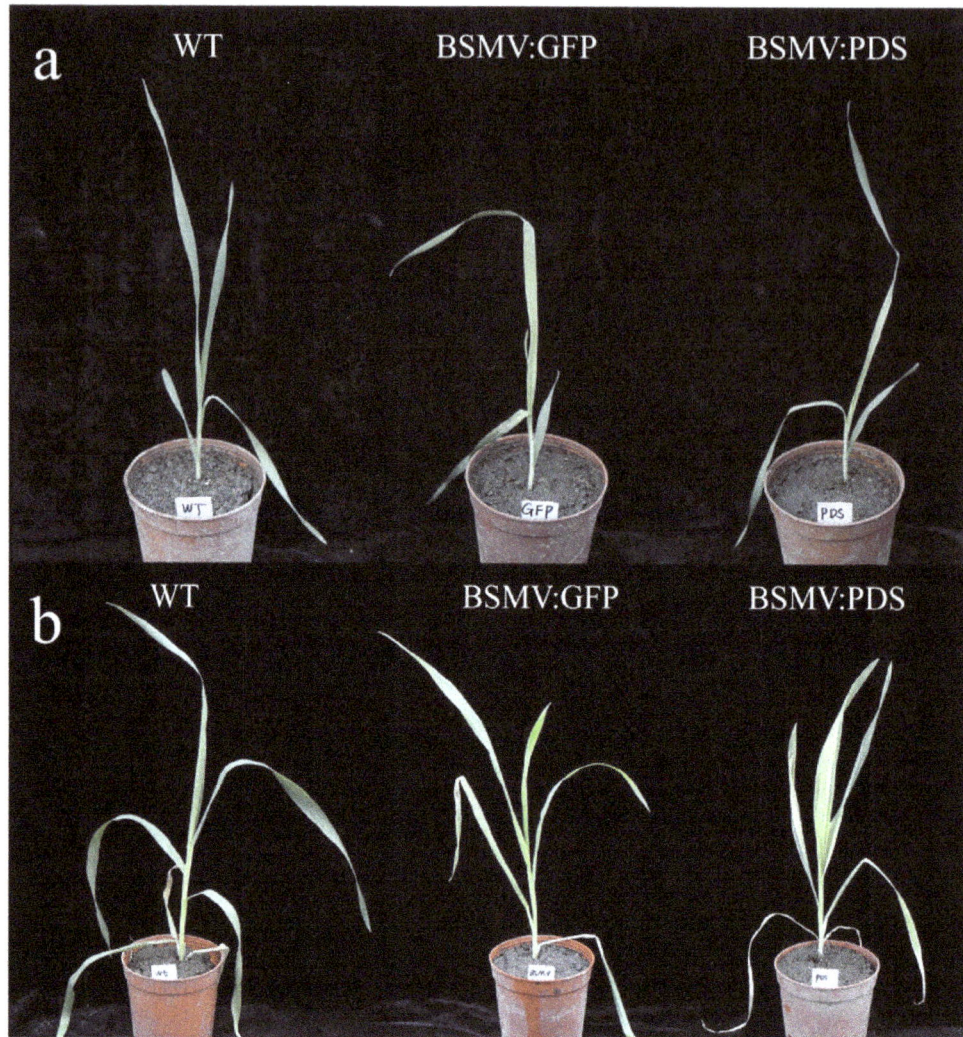

Fig. 2 Comparison of the vegetative growth state of mock-inoculated wild type (WT) control plants and BSMV-infected plants. (**a**) and (**b**) indicate the vegetative growth of mock-inoculated WT control plants, BSMV:GFP-infected and BSMV:PDS-infected plants at 10 and 30 days post inoculation (dpi) respectively. WT – mock-inoculated WT control plants; BSMV:GFP – plants inoculated by BSMV:GFP vector; BSMV:PDS – plants inoculated by BSMV:PDS vector.

lower values of TDW of the root system in all BSMV-infected plants indicated decreased biomass accumulation in roots after infection. (Fig. 3c).

The impacts of *PDS* silencing on vegetative growth was also measured. The height of BSMV:GFP-inoculated plants was significantly higher than BSMV:PDS-inoculated plants at 10 dpi, but showed no obvious differences from 20 to 40 dpi (Fig. 3a). The biomass accumulation of BSMV:GFP-inoculated plants was significantly higher than BSMV:PDS-inoculated plants at 20 and 30 dpi, but showed no significant ($P < 0.05$) differences at 40 dpi (Fig. 3b).

The impact of BSMV on the water retention capacity
The 6-h WLR of BSMV-infected plants were significantly ($P < 0.05$) higher than that of mock-inoculated WT controls at 10 and 20 dpi (Fig. 5a). The water status under normal growth conditions was also investigated using WC as an

indicator. The WC of all the BSMV-infected plants was significantly ($P < 0.05$) lower than that of controls at 10 dpi (Fig. 5b). Distinct from controls, the leaf tips of these plants became withered and twisted (Fig. 4).

The 6-h WLR of BSMV-infected plants showed no significant differences ($P < 0.05$) with the controls at 30 and 40 dpi (Fig. 5a). The WC of the BSMV:GFP-inoculated plants also showed no significant differences ($P < 0.05$) at 20 and 30 dpi, but was higher than the controls at 40 dpi (Fig. 5b). These results indicated that the impact of BSMV on WLR and WC reduced after 20 dpi.

The influence of PDS silencing on WLR and WC was also investigated. The 6-h WLR of BSMV:PDS-inoculated plants was higher than that of BSMV:GFP-inoculated plants at 10 dpi and significantly ($P < 0.05$) higher than BSMV:GFP-inoculated plants at 20 dpi (Fig. 5a). The WC of BSMV:PDS-inoculated plants was significantly ($P < 0.05$) lower than that

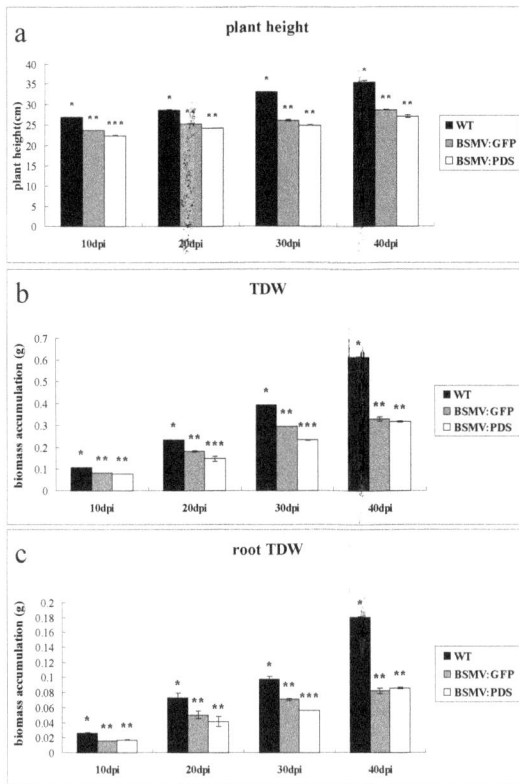

Fig. 3 Comparison of plant height and biomass accumulation of BSMV-infected plants and mock-inoculated WT control plants. a The height of BSMV-infected plants and mock-inoculated wild type (WT) control plants at 10, 20, 30 and 40 days post inoculation (dpi). b The biomass accumulated of BSMV-infected plants and mock-inoculated WT control plants at 10, 20, 30 and 40 dpi. c The biomass accumulated of the roots of BSMV-infected plants and mock-inoculated WT control plants at 10, 20, 30 and 40 dpi. All data represent the average of three experiments with at least 6 replicates. Data are shown as means $\pm SE$ ($n = 3$). The marker *, ** and *** indicate that the means are significantly different at $P = 0.05$ as determined by the least significant difference (LSD) test using Tukey's test (SPSS package, version 17.0). WT – mock-inoculated WT control plants; BSMV:GFP – plants inoculated by BSMV:GFP vector; BSMV:PDS – plants inoculated by BSMV:PDS vector.

---

of BSMV:GFP-inoculated plants at 10 and 20 dpi (Fig. 5b). These results indicated that the highly suppressed expression of *PDS* may impair the water retention capacity of hulless barley.

The silencing efficiency and the stability of BSMV-VIGS in Tibetan hulless barley

The silencing efficiency of BSMV-VIGS under our experimental condition was checked by RT-qPCR from 7 to 60 dpi using *PDS* as an indicator. From 7 to 17 dpi, the expression level of *PDS* in BSMV:PDS-inoculated plants was much lower than that of the BSMV:GFP-inoculated controls (4.46% to 10.20% of the latter). The *PDS* expression level in BSMV:PDS-inoculated plants increased gradually from 25 to 45 dpi and maintained at a relatively higher level after 45 dpi

(Fig. 6), although this was still lower than the expression observed for BSMV:GFP. The leaves of BSMV:PDS-inoculated plants showed the classic white photo-bleaching phenotype from 10 to 40 dpi, indicating that the BSMV-VIGS worked well and the silencing phenotype could be maintained at least 40 days (Fig. 4).

The stability of BSMV-VIGS was investigated by checking the PCR products of BSMV:PDS-inoculated plants from 7 to 60 dpi by agarose gel electrophoresis. All the products amplified prior to 45 dpi showed a single band with an expected size of around 400 bp (Fig. 7), while the products after 40 dpi had another obscure band at about 200 bp. The sequencing result showed that the PCR products were composed of the whole *PDS* insert which was designed for silencing and some of the flanking BSMV-γ sequences (Fig. S2). In all, these results indicated that the designed *PDS* insert existed in almost all BSMV:PDS-inoculated plants from 7 to 60 dpi, and suggested that the stability of BSMV:PDS under our condition is acceptable before 40 dpi.

Discussion

The pattern of BSMV accumulation in Tibetan hulless barley

Virus accumulation in plants is regulated by the balance of defense and counter-defense mechanisms [26]. At the very beginning of the infection, when the amount of BSMV is low, the virus may not be detected by the defense system [27] and thus accumulates quickly. When viral dsRNA is abundant, the RNA-induced silencing complex (RISC) begins to target viral RNA. The virus accumulation is then slowed down and even declined [27,28]. In this study, BSMV accumulation showed this similar pattern where the virus reproduced at a peak rate from 7 to 11 dpi and declined rapidly from 11 to 17 dpi. After 25 dpi, BSMV was rebalanced to a lower but relatively stable level.

The impact of BSMV on vegetative growth and leaf morphology

Previous researches indicated that virus infection could impair photosynthesis, reduce chlorophyll content, decrease activities of some photosynthetic enzymes, elevate sugar or starch contents [29–31] and reduce plant growth rate. In our study, the biomass accumulation was significantly retarded after BSMV infection. Although we suspected that after the virus level decreased, the biomass accumulation of BSMV infected plants would become comparable to the controls, the final results clearly indicated that the gap of TDW between BSMV infected plants and controls increased.

As BSMV affects plants in a systemic manner, we also checked the root system of the infected plants. The results showed that the root TDW of BSMV infected plants were also significantly less than that of WT controls. This may not be so important when scoring drought tolerance by a detached leaf assay, but may impact on other drought related assays.

Almási et al. [14] reported that BSMV infection can accelerated senescence. In our study, we noticed that the tip of new leaves lost water and withered at 10 dpi. Cells in new leaves senesced faster than those in mature leaves, suggesting that BSMV-VIGS analysis of genes related to the leaf morphology is probably not appropriate before 10 dpi.

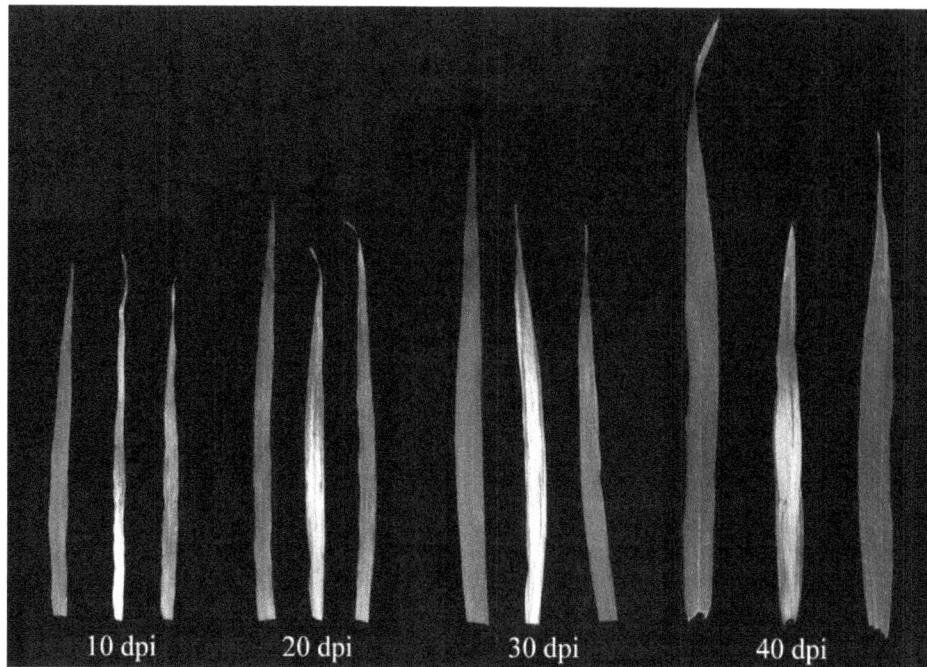

Fig. 4 Comparison of the leaf morphology of BSMV-infected plants and mock-inoculated wild type (WT) control plants. Leaves of mock-inoculated WT control plants, BSMV:PDS-inoculated and BSMV:GFP-inoculated plants (from left to right) at 10, 20, 30 and 40 days post inoculation (dpi), respectively.

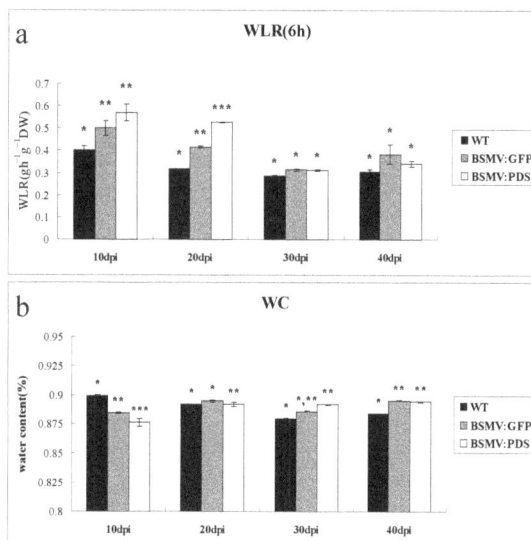

Fig. 5 Comparison of water loss rate (WLR) and water content (WC) of BSMV-infected plants and mock-inoculated wild type (WT) control plants. **a** The 6-h WLR of BSMV-infected plants and mock-inoculated WT control plants at 10, 20, 30 and 40 days post inoculation (dpi). **b** The WC of BSMV-infected plants and mock-inoculated WT control plants at 10, 20, 30 and 40 dpi. All data represent the average of three experiments with at least 6 replicates. Data are shown as means $\pm SE$ ($n = 3$). The marker *, ** and *** show that the means are significantly different at $P = 0.05$ as determined by the least significant difference (LSD) test using Tukey's test (SPSS package, version 17.0). WT – mock-inoculated WT control plants; BSMV:GFP – plants inoculated by BSMV:GFP vector; BSMV:PDS – plants inoculated by BSMV:PDS vector.

The relationship between BSMV accumulation and its impacts on water holding capacity

Our data indicated that the 6-h WLR of BSMV infected plants was significantly higher, while the WC of BSMV infected plants was significantly lower than the controls at 10 dpi, suggesting that the high accumulation of BSMV from 7 to 11 dpi greatly impacted the water holding capacity. When virus accumulation dropped from 11 to 17 dpi, its impact on water holding capacity also weakened. As our data indicated, the WC of BSMV:GFP infected plants had improves and had no significant difference compared to control plants at 20 dpi. After 25 dpi, the reproduction of BSMV virus declined gradually and became stable and its impact on WLR and WC also reduced to an undetectable level. These results suggested that the restricted reproduction of BSMV had only marginal impact on the water retention capacity of hulless barley seedlings. Therefore the best time to compare phenotypic changes related to water status is after 25 dpi.

As our results indicated that the silence of *PDS* impaired the water retention capacity of hulless barley at 10 and 20 dpi. Therefore, the *PDS*-silenced plants were not suitable for being a control when analyzing water status related genes before 20 dpi.

The prolonged VIGS obtained by lower temperature and high humidity

Previous studies showed that VIGS persists for longer if there is a continuous presence of the virus [5,32,33], however if the viral infection symptoms disappeared, the subsequent gene silencing phenotype would also disappear. It was reported that the most efficient time period for VIGS

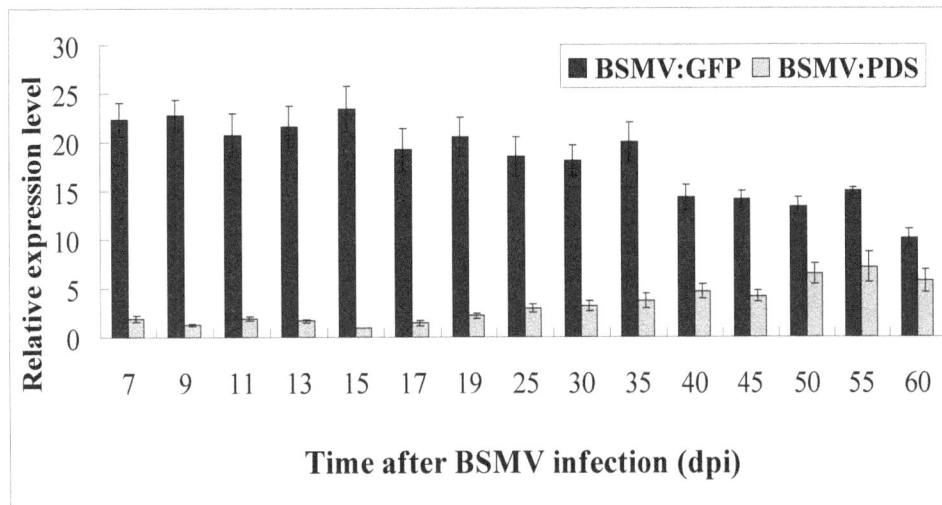

Fig. 6 Expression of *Phytoene desaturase* (*PDS*) in BSMV:GFP-inoculated plants and BSMV:PDS-inoculated control plants from 7 to 60 days post inoculation (dpi). The relative expression level of *PDS* in BSMV:GFP-inoculated plants and BSMV:PDS-inoculated control plants were analyzed by quantitative reverse transcription polymerase chain reaction (RT-qPCR) using *elongation factor 1α* (*EF*) as the reference from 7 to 60 dpi. Each sample contained newly grown leaves from at least 5 individual plants. All data represent the average value of three technical repeats and are shown as the means ±SD (*n* = 3). BSMV:GFP – plants inoculated by BSMV:GFP vector; BSMV:PDS – plants inoculated by BSMV:PDS vector.

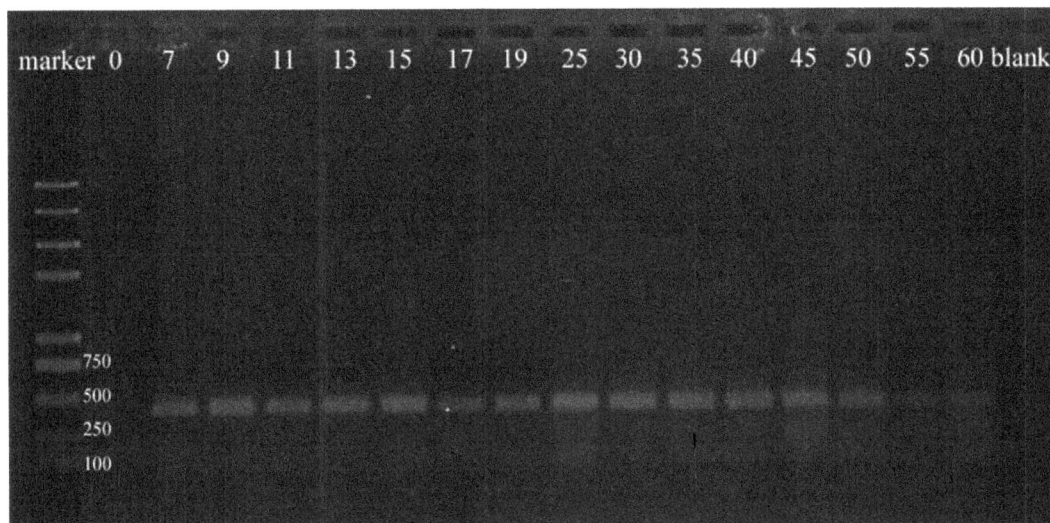

Fig. 7 The PCR products of PDSs of BSMV:PDS-inoculated plants from 7 to 60 days post inoculation (dpi). Agarose gel electrophoresis of the PCR products of using PDSs as the primer and the cDNA of BSMV:PDS-inoculated plants (from 7 to 60 dpi) as the templates. The predicted PCR product is 396 bp in length, including both the PDS insert and part of the BSMV-γ sequences as well. 0 – mock-inoculated wild type plant; 7, 9 ,11, 13, 15, 17, 19, 25, 30, 35, 40, 45, 50, 55, 60 – BSMV:PDS-inoculated plants from 7 to 60 dpi; blank – water template.

is within 3 weeks [34] and the efficiency of VIGS decreased after 1 month when plants start to recover from silencing [22]. The results of this study suggested that the best time for highly efficient silencing is from 7 to 17 dpi. The results also indicated that the most significant impacts of BSMV on the water status occurred prior to 25 dpi. This severely limited the potential for the conventional BSMV-VIGS assay to analyze drought related candidate genes. It was reported that lower temperatures can lead to better silencing phenotypes and targeted gene silencing [32,35–37]. Compared with the experimental condition of our previous BSMV-VIGS study [23], we found that by adjusting greenhouse temperatures from 24 ±1°C to 22 ±1°C, the time period for efficient BSMV-VIGS can be maintained to at least 40 days after inoculation.

Furthermore, Cakir and Tör [37] reported that the PDS bleaching can be extend further to the forth leaves and to the tillering leaves of plants treated at the lower temperatures. In our tests, we also noticed that the PDS-silencing phenotype on some plants can even persisted to 150 dpi (the initial shoot died, but the BSMV-VIGS still persisted in the tiller). Fu et al. [35] indicated that low humidity (30%) enhanced the silencing of PDS. But in our experiment, the rate of successful infection was extremely low under the low relative humidity of 30% (data not shown). On the contrary, we achieved much better infection rate and good silencing efficient with a higher relative humidity of 70 ±10%.

The unstable BSMV

The uneven distribution of BSMV in infected plants has been reported by many researchers [14,38]. The silencing induced by BSMV is not uniform from tissue to tissue, plant to plant or from experiment to experiment, and the phenotype of VIGS is not very stable [39,40]. Through our observation, the instability of VIGS is exacerbated after 40 dpi, a small part of PDS-silenced plants had completely lost the photo-bleaching phenotype, and some even lost the classic phenotype of BSMV infection (stripes and mosaics on leaves). While most of the plants retained their photobleaching phenotype until 60 dpi, the color of leaves turned from white into a green-white mix with only a few individuals retaining pure white leaves. The result of agarose gel electrophoresis also showed that the products after 40 dpi had

a second band, suggesting that some of plants had begun to lose the PDS insert on BSMV-γ vector. Considering that the efficiency of BSMV-VIGS was relative low and the instability of VIGS has also been exacerbated after 40 dpi, we suggest it is necessary to finish data sampling before 40 dpi and to use large numbers of infected plants for statistical analysis.

Summary

BSMV accumulated at high level from 7 to 13 dpi and significantly impacted the water holding capacity of Tibetan hulless barley. The virus reproduction level declined and became relatively stable after 25 dpi, correlating with reduced impacts on WLR and WC. For the purpose of maintaining efficient BSMV-VIGS for extended periods, we put our inoculated plants under a lower temperature and higher relative humidity condition. The efficiency and the stability of BSMV-VIGS were checked, and proved to be acceptable for VIGS prior to 40 dpi. Thus, we conclude that the optimal time period for testing the function of drought stress related genes is between 25 and 40 dpi. As the vegetative growth was affected by the BSMV virus throughout the experiments, the study on related genes should be carefully interpreted, regardless of whether the gene is expressed in the aboveground parts of the plant or in the roots. We suggest that it is more reliable if different control groups and large numbers of infected plants can be employed.

Acknowledgments
We thank Chengdu Institute of Biology, Chinese Academy of Sciences for the Senior Research Fellowship award. Prof. Daowen Wang of Institute of Genetics and Developmental Biology, Chinese Academy of Sciences, Beijing, are greatly acknowledged for providing BSMV vectors. We sincerely acknowledge Dr. Bin Li for technical expertise and advice on BSMV inoculation. Thanks to Dr. Garry Rosewarne for English corrections. This work was supported by National Science & Technology Pillar Program (2012BAD03B00), P.R. China and West Light Foundation of The Chinese Academy of Sciences.

Authors' contributions
The following declarations about authors' contributions to the research have been made: designed the research: JJL, HL; carried out most of the experiments: JJL; XC; analyzed the data and drafted the paper: JJL, SYY; supplied with hulless barley material and took the photos: GBD; helped to revise the manuscript: HHZ, GBD, ZFP; gave the final approval of the version to be published: MQY, HL.

Competing interests
No competing interests have been declared.

References
1. Kumagai M, Donson J, Della-Cioppa G, Harvey D, Hanley K, Grill L. Cytoplasmic inhibition of carotenoid biosynthesis with virus-derived RNA. Proc Natl Acad Sci USA. 1995;92(5):1679–1683.

2. Ratcliff F, Harrison BD, Baulcombe DC. A similarity between viral defense and gene silencing in plants. Science. 1997;276(5318):1558–1560.

3. Baulcombe DC. Fast forward genetics based on virus-induced gene silencing. Curr Opin Plant Biol. 1999;2(2):109–113.

4. Burch-Smith TM, Anderson JC, Martin GB, Dinesh-Kumar SP. Applications and advantages of virus-induced gene silencing for gene function studies in plants. Plant J. 2004;39(5):734–746.

5. Lee WS, Hammond-Kosack KE, Kanyuka K. Barley stripe mosaic virus-mediated tools for investigating gene function in cereal plants and their pathogens: VIGS, HIGS and VOX. Plant Physiol. 2012;160(2):582–590.

6. Holzberg S, Brosio P, Gross C, Pogue GP. Barley stripe mosaic virus-induced gene silencing in a monocot plant. Plant J. 2002;30(3):315–327.

7. Lacomme C, Hrubikova K, Hein I. Enhancement of virus-induced gene silencing through viral-based production of inverted-repeats. Plant J. 2003;34(4):543–553.

8. Scofield SR, Huang L, Brandt AS, Gill BS. Development of a virus-induced gene-silencing system for hexaploid wheat and its use in functional analysis of the Lr21-mediated leaf rust resistance pathway. Plant Physiol. 2005;138(4):2165–2173.

9. Hein I, Barciszewska-Pacak M, Hrubikova K, Williamson S, Dinesen M, Soenderby IE, et al. Virus-induced gene silencing-based functional

characterization of genes associated with powdery mildew resistance in barley. Plant Physiol. 2005;138(4):2155–2164.

10. Hu P, Meng Y, Wise RP. Functional contribution of chorismate synthase, anthranilate synthase, and chorismate mutase to penetration resistance in barley-powdery mildew interactions. Mol Plant Microbe Interact. 2009;22(3):311–320.

11. Meng Y, Moscou MJ, Wise RP. *Blufensin1* negatively impacts basal defense in response to barley powdery mildew. Plant Physiol. 2009;149(1):271–285.

12. Scofield SR, Nelson RS. Resources for virus-induced gene silencing in the grasses. Plant Physiol. 2009;149(1):152–157.

13. Cakir C, Gillespie ME, Scofield SR. Rapid determination of gene function by virus-induced gene silencing in wheat and barley. Crop Sci. 2010;50(S1):S-77–S-84.

14. Almási A, Apatini D, Bóka K, Böddi B, Gáborjányi R. BSMV infection inhibits chlorophyll biosynthesis in barley plants. Physiol Mol Plant Pathol. 2000;56(6):227–233.

15. Bennypaul HS. Genetic analysis and functional genomic tool development to characterize resistance gene candidates in wheat (*Triticum aestivum* L.) [PhD thesis]. Pullman, WA: Washington State University; 2008.

16. Senthil-Kumar M, Mysore KS. New dimensions for VIGS in plant functional genomics. Trends Plant Sci. 2011;16(12):656–665.

17. Qian G, Han Z, Zhao T, Deng G, Pan Z, Yu M. Genotypic variability in sequence and expression of *HVA1* gene in Tibetan hulless barley, *Hordeum vulgare* ssp. *vulgare*, associated with resistance to water deficit. Aust J Agric Res. 2007;58(5):425–431.

18. Senthil-Kumar M, Rame Gowda HV, Hema R, Mysore KS, Udayakumar M. Virus-induced gene silencing and its application in characterizing genes involved in water-deficit-stress tolerance. J Plant Physiol. 2008;165(13):1404–1421.

19. Demmig-Adams B, Adams Iii W. Photoprotection and other responses of plants to high light stress. Annu Rev Plant Physiol Plant Mol Biol. 1992;43:599–626.

20. Ruiz MT, Voinnet O, Baulcombe DC. Initiation and maintenance of virus-induced gene silencing. Plant Cell. 1998;10(6):937–946.

21. Angell SM, Baulcombe DC. Technical advance: potato virus X amplicon-mediated silencing of nuclear genes. Plant J. 1999;20:357–362.

22. Ratcliff F, Martin-Hernandez AM, Baulcombe DC. Technical advance: tobacco rattle virus as a vector for analysis of gene function by silencing. Plant J. 2001;25(2):237–245.

23. Liang J, Deng G, Long H, Pan Z, Wang C, Cai P, et al. Virus-induced silencing of genes encoding LEA protein in Tibetan hulless barley (*Hordeum vulgare* ssp. *vulgare*) and their relationship to drought tolerance. Mol Breed. 2012;30(1):441–451.

24. Zhou H, Li S, Deng Z, Wang X, Chen T, Zhang J, et al. Molecular analysis of three new receptor-like kinase genes from hexaploid wheat and evidence for their participation in wheat hypersensitive response to stripe rust fungus infection. Plant J. 2007;52(3):420–434.

25. Haupt S, Duncan GH, Holzberg S, Oparka KJ. Evidence for symplastic phloem unloading in sink leaves of barley. Plant Physiol. 2001;125(1):209–218.

26. Maule A, Leh V, Lederer C. The dialogue between viruses and hosts in compatible interactions. Curr Opin Plant Biol. 2002;5(4):279–284.

27. Lu R, Malcuit I, Moffett P, Ruiz MT, Peart J, Wu AJ, et al. High throughput virus-induced gene silencing implicates heat shock protein 90 in plant disease resistance. EMBO J. 2003;22(21):5690–5699.

28. Voinnet O. RNA silencing as a plant immune system against viruses. Trends Genet. 2001;17(8):449–459.

29. Montalbini P, Lupattelli M. Effect of localized and systemic tobacco mosaic virus infection on some photochemical and enzymatic activities of isolated tobacco chloroplasts. Physiol Mol Plant Pathol. 1989;34(2):147–162.

30. Funayama S, Sonoike K, Terashima I. Photosynthetic properties of leaves of *Eupatorium makinoi* infected by a geminivirus. Photosynth Res. 1997;53(2–3):253–261.

31. Funayama S, Terashima I. Effects of geminivirus infection and growth irradiance on the vegetative growth and photosynthetic production of *Eupatorium makinoi*. New Phytol. 1999;142(3):483–494.

32. Bruun-Rasmussen M, Madsen CT, Jessing S, Albrechtsen M. Stability of *Barley stripe mosaic virus*-induced gene silencing in barley. Mol Plant Microbe Interact. 2007;20(11):1323–1331.

33. Senthil-Kumar M, Mysore KS. Virus-induced gene silencing can persist for more than 2 years and also be transmitted to progeny seedlings in *Nicotiana benthamiana* and tomato. Plant Biotechnol J. 2011;9(7):797–806.

34. Ryu CM, Anand A, Kang L, Mysore KS. Agrodrench: a novel and effective agroinoculation method for virus-induced gene silencing in roots and diverse *Solanaceous* species. Plant J. 2004;40(2):322–331.

35. Fu DQ, Zhu BZ, Zhu HL, Zhang HX, Xie YH, Jiang WB, et al. Enhancement of virus-induced gene silencing in tomato by low temperature and low humidity. Mol Cells. 2006;21:153–160.

36. Tuttle JR, Idris AM, Brown JK, Haigler CH, Robertson D. Geminivirus-mediated gene silencing from *Cotton leaf crumple virus* is enhanced by low temperature in cotton. Plant Physiol. 2008;148(1):41–50.

37. Cakir C, Tör M. Factors influencing *Barley stripe mosaic virus*-mediated gene silencing in wheat. Physiol Mol Plant Pathol. 2010;74(3–4):246–253.

38. Lin NS, Langenberg W. Distribution of *Barley stripe mosaic virus* protein in infected wheat root and shoot tips. J Gen Virol. 1984;65(12):2217–2224.

39. Campbell J, Huang L. Silencing of multiple genes in wheat using *Barley stripe mosaic virus*. J Biotech Res. 2010;2:12–20.

40. Huang C, Qian Y, Li Z, Zhou X. Virus-induced gene silencing and its application in plant functional genomics. Sci China Life Sci. 2012;55(2):99–108.

9

Effects of selenium on the growth and photosynthetic characteristics of flue-cured tobacco (*Nicotiana tabacum* L.)

Chaoqiang Jiang, Chaolong Zu*, Jia Shen, Fuwen Shao, Tian Li

Tobacco Research Institute, Anhui Academy of Agricultural Sciences, Nongkenan Road 40, Hefei, Anhui, China

Abstract

The objective of this study was to investigate the effect of Selenium (Se) supply (0, 3, 6, 12, 24 mg kg^{-1}) on the growth, photosynthetic characteristics, Se accumulation and distribution of flue-cured tobacco (*Nicotiana tabacum* L.). Results showed that low-dose Se treatments (≤ 6 mg kg^{-1}) stimulated plant growth but high-dose Se treatments (≥ 12 mg kg^{-1}) hindered plant growth. Optimal Se dose (6 mg kg^{-1}) stimulated plant growth by reducing MDA content and improving photosynthetic capability. However, excess Se (24 mg kg^{-1}) increased MDA content by 28%, decreased net photosynthetic rate and carboxylation efficiency by 34% and 39%, respectively. The Se concentration in the roots, stems, and leaves of the tobacco plants significantly increased with increasing Se application. A linear correlation ($R = 0.95$, $P < 0.01$) was observed between Se level and tobacco plant tissue Se concentration. This correlation indicated that the tobacco plant tissues were not saturated within the concentration range tested. The pattern of total Se concentration in the tobacco plant tissues followed the order root > leaf > stem. The Se concentration in the roots was 3.17 and 7.57 times higher than that in the leaves and stems, respectively, after treatment with 24 mg kg^{-1} Se. In conclusion, the present study suggested that optimal Se dose (6 mg kg^{-1}) improved the plant growth mainly by enhancing photosynthesis, stomatal conductance, carboxylation efficiency and Rubisco content in the flue-cured tobacco leaves. However, the inhibition of excess Se on tobacco growth might be due to high accumulation of Se in roots and the damage of photosynthesis in leaves.

Keywords: flue-cured tobacco; selenium (Se); photosynthesis; carboxylation efficiency; Rubisco content

Introduction

Selenium (Se) is essential to animals and humans [1]. Recent research has shown that this trace element is also beneficial to plants [2]. However, high Se concentrations may elicit toxic effects on plants [3]. The difference between the deficiency and toxicity of Se, as in other essential trace elements, is narrow [4]. Plant species differ in Se uptake and accumulation in shoots and roots, as well as in tolerance to high Se concentrations in solution or soil [5]. For example, *Astragalus bisulcatus* and *Stanleya pinnata* exhibit high tolerance to Se in soil; these plants can hyperaccumulate Se up to 1% of their dry weights (DWs) [6]. By contrast, tobacco and soybean are sensitive to Se; these plants can be affected by low Se concentrations (e.g., 1 mg kg^{-1}) in culture media [7]. It's been well reported that the phytotoxicity of Se varies among agricultural crops [5].

Evidence to prove that nonaccumulator plants require Se remains lacking. However, numerous studies have reported that low Se concentrations benefit the growth of these plants.

Turakainen et al. [8] showed that appropriate Se concentrations has positive effects also on potato carbohydrate accumulation and possibly on yield formation. Similarly, other studies revealed that Se promotes the growth of ryegrass [3], tea [9], rice [10], and soybean [11]. However, excess Se accumulation (>0.1% plant DW) is generally toxic to plants, except for rare Se-hyperaccumulating plants [4].

Plants subjected to Se stress exhibit different physiological changes, including stunted root growth, reduced biomass, chlorosis, reduced photosynthetic efficiency, and ultimately plant death [4]. Previous studies reported that Se improves the antioxidant capacity of plants [3,12]. Feng et al. [2] have recently discovered that Se elicits protective effects on plants against abiotic stresses. Soil treatment with Se has been highly recommended to produce Se-enriched food for human consumption. Se-enriched products, such as tea [9], rice [10,13], and vegetables [14] have been developed. Furthermore, various studies have associated the consumption of Se-enriched vegetables with reduced risk of developing cancer [14,15]. Broccoli can accumulate Se and convert it into a form that is chemoprotective against cancer [15]. These findings suggest that Se-enriched vegetables benefit human nutrition and health [14]. Therefore, understanding

* Corresponding author. Email: lcz2468@sina.com

Handling Editor: Grażyna Kłobus

the effect and function of Se on the plant growth is important to the development of Se-enriched agricultural products.

Although Se is known to elicit detrimental effects on plants, the effects of Se stress on the photosynthetic characteristics of tobacco have yet to be elucidated. Tobacco (*Nicotiana tabacum* L.) is an economically important non-food crop worldwide; flue-cured tobacco accounts for approximately 80% of the world's tobacco production [16]. Therefore, the present study aims to determine the effects of treatments with different Se concentrations (0 mg kg^{-1} to 24 mg kg^{-1}) on the plant growth, gas exchange, chlorophyll concentration, Rubisco content, malondialdehyde (MDA) content, Se accumulation and distribution of flue-cured tobacco.

Material and methods

Experimental materials and growth conditions

A soil pot experiment was conducted in greenhouse conditions in major rice-growing areas of Anhui province, China. Tobacco (*N. tabacum* L., cv. Yunyan 87), a popular flue-cured tobacco cultivar in China, was used in this study. The seeds were provided by Chizhou Tobacco Corporation, Anhui province, China. Seeds were surfaced-sterilized with 2% (v/v) NaOCl for 10 min, rinsed with deionized water, and then sown in floating nursery. During the growing season, the plants were placed in a greenhouse with a daytime temperature of 25°C to 33°C, a nighttime temperature of 15°C to 23°C, and a relative humidity (RH) of 60% to 85%.

The soil was paddy soil, collected from the tobacco field in Chizhou, Anhui province, China. Main physical and chemical properties were as follows: pH$_{water\ 2.5:1}$ 5.4, organic matter 17.6 g kg^{-1}, available N 157.7 mg kg^{-1}, available P 16.6 mg kg^{-1}, available K 184.7 mg kg^{-1}, and total Se 0.16 mg kg^{-1}.

Experimental design

Five levels of Se (sodium selenite, Na$_2$SeO$_3$) treatment, i.e. 0 (CK), 3, 6, 12, and 24 mg kg^{-1} Se were performed in the experiment. Each pot (35 cm in diameter, 28 cm in height) was filled with 20 kg of air-dried and 2 mm-sieved soil. Therefore, each pot was added with 0, 60, 120, 240, or 480 mg of Na$_2$SeO$_3$ to produce the five treatments. The fertilizers for each pot were as follows: flue-cured tobacco special fertilizer (N:P:K = 9:13.5:22.5) 45.45 g, KNO$_3$ 10.10 g, K$_2$SO$_4$ 3.28 g, Ca(H$_2$PO$_4$)$_2$ 14.61 g. All the fertilizers and Na$_2$SeO$_3$ were mixed thoroughly and applied as basal dressing at 10 days before tobacco seedlings transplanting. Tobacco seedlings (approximately 12 cm in height) were transplanted into the pots on April 9, 2011, with one plant for each pot. The pots were arranged at 1.2 m inter-row spacing and 0.5 m intra-row spacing. Each treatment was replicated for four times, and each replicate included four plants.

Gas exchange measurements

At 40 d after the treatments, the gas exchange on the newly expanded leaves was measured from 8:30 to 12:00 using a Li-6400 portable photosynthesis system (Li-Cor, Inc., Lincoln, NB, USA) as previously described [17]. During the measurements, leaf temperature was maintained at 25 ±1°C

with a photosynthetic photon flux intensity of 1200 µmol m^{-2} s^{-1}. Ambient CO$_2$ concentration in the cuvette (C_{a-c}) was adjusted to Ca (380 µmol CO$_2$ mol^{-1}), and RH was maintained at 50% ±5%. After 10 min, C_{a-c} was controlled across the series of 1000, 800, 600, 400, 200, 100, and 50 µmol CO$_2$ mol^{-1}. Carboxylation efficiency (CE) was calculated as the initial slope of the A/Ci response curves. Data were recorded after equilibration to a steady state.

Leaf chlorophyll concentration determination

At 40 d after the treatments, the relative chlorophyll concentrations in the newly expanded leaves of the labeled leaf segments were determined using a SPAD-502 Chlorophyll Meter (Minolta, Mahwah, NJ, USA).

MDA content determination

At 40 d after the treatments, the MDA contents in the newly expanded leaves of the labeled leaf segments were measured as previously described [18]. The absorbance at 450, 532, and 600 nm was obtained with an ultraviolet spectrophotometer (UV-755B, Shanghai Precision and Scientific Instrument Co., Ltd., China). The concentration of MDA was calculated using the following formula: C = 6.45(D$_{532}$ − D$_{600}$) − 0.56D$_{450}$.

Rubisco measurements

After the gas exchange measurements, the Rubisco contents in the newly expanded leaves were measured according to the method of Li et al. [19]. Briefly, about 0.50 g newly expanded leaves were ground with a cooled extraction buffer containing 50 mmol l^{-1} Tris-HCl (pH 8.0), 5 mmol l^{-1} β-mercaptoethanol and 12.5% (v/v) glycerol at 0–4°C. The homogenate was centrifuged at 1500 g for 15 min at 4°C. The supernatant solution was mixed with a dissolving solution containing 2% (w/v) SDS, 4% (v/v) β-mercaptoethanol and 10% (v/v) glycerol. Then the mixture was boiled in water for 5 min for gel electrophoresis. An electrophoretic buffer system was used for SDS-PAGE with a 12.5% (w/v) stacking gel and a 4% (w/v) separating gel. Afterwards, the gels were washed with deionized water several times then dyed in 0.25% Coomassie Blue for 12 h and detained. Large subunits and relevant small subunits were transferred to a 10 ml cuvette with 2 ml of formamide and washed in a 50°C water bath for 8 h. The washed solutions were measured at 595 nm using background gel as a blank and bovine serum albumin (BSA) as the protein standard.

Determination of plant biomass

After the above measurements were completed, the plants were harvested and analyzed for root, stem, and leaf fresh weights. All samples were oven-dried at 105°C for 30 min and then at 60°C until constant weight were reached.

Total Se analysis

The roots, stems, and leaves of the plants were milled into powder using a mixer mill (Shanghai Bilon Instrument Co., China). The total Se concentrations in the samples were determined as previously described [20,21]. The dried plant powders (0.5 g) were digested overnight with 10.0 ml of HNO$_3$:HClO$_4$ (9:1) in a polypropylene sample tube at room temperature. The digested solution was heated on an electrical hot plate at 60°C for 2 h and then at 100°C for 1 h. The tube containing the solution was added with 10.0 ml of $HNO_3:HClO_4$ (9:1) and then stored at 170°C for 2 h until a white fume formed. After cooling to room temperature, the tube was added with 5 ml HCl (1:1) and then heated until the solution became colorless within at least 3 h. After cooling, the solution was filtered and diluted to 50 ml with deionized water. The total Se concentration in the solution was analyzed through inductively coupled plasma–mass spectrometry [X Series ICP–MS (Thermo Electron Corporation, United States)]. Four replicate determinations were performed for each material.

Statistical analysis

All measurements were carried out on replicate samples collected from four individual plants. Statistical analyses were performed using one-way ANOVA with SPSS statistical software. Data were presented as mean and *SE*. The results were verified via Duncan's multiple-range test.

Results

Effects of Se treatment on plant growth

Tobacco plants were grown in soil with fertilizers containing five Se concentrations (0, 3, 6, 12, and 24 mg kg^{-1})

under greenhouse conditions. As shown in Tab. 1, low-dose Se treatments (≤6 mg kg^{-1}) enhanced the growth of tobacco plants. Treatment with 6 mg kg^{-1} Se significantly enhanced root, stem, leaf, and whole-plant DWs by 11%, 29%, 18% and 19%, respectively, compared with CK. In contrast to low-dose Se treatments, high-dose Se treatments (≥12 mg kg^{-1}) reduced the growth of tobacco plants. For instance, treatment with 24 mg kg^{-1} Se decreased root, stem, leaf, and whole-plant DWs by 18%, 10%, 19%, and 16%, respectively, compared with CK. However, no significant difference in root/shoot ratio was detected between the CK- and Se-treated tobacco plants.

Effects of Se treatment on gas exchange

Tab. 2 presents the changes in the gas exchange parameters of newly expanded tobacco leaves after 40 d of Se treatment. Compared with CK, low-dose Se treatments (≤6 mg kg^{-1}) significantly increased the net photosynthetic rate (Pn) in the leaves, whereas high-dose Se treatments (≥12 mg kg^{-1}) significantly decreased this parameter. The Pn values of the tobacco leaves under 3, 6, 12, and 24 mg kg^{-1} Se treatments were 1.17-, 1.26-, 0.96-, and 0.66-fold higher than those of the tobacco leaves under CK treatment, respectively. Similar changes in stomatal conductance (gs) were observed under the different Se treatments. Compared with CK, low-dose Se treatments (≤6 mg kg^{-1}) increased the CE, whereas high-dose Se treatments (≥12 mg kg^{-1}) decreased

Tab. 1 Effects of different levels of Se treatment on biomass of tobacco plants.

Se treatments (mg kg^{-1})	Root DW (g plant^{-1})	Stem DW (g plant^{-1})	Leaf DW (g plant^{-1})	Whole-plant DW (g plant^{-1})	Root/shoot ration
CK	9.95 ±0.85 b	13.18 ±0.55 c	29.38 ±0.59 b	52.51 ±1.66 b	0.234 ±0.016 a
3	9.81 ±0.35 b	14.92 ±0.66 b	31.86 ±1.76 b	56.59 ±2.43 b	0.210 ±0.013 a
6	11.05 ±0.54 a	16.98 ±0.65 a	34.61 ±1.35 a	62.64 ±1.06 a	0.214 ±0.017 a
12	9.42 ±0.40 b	12.90 ±0.40 b	28.30 ±1.56 b	50.95 ±1.61 b	0.229 ±0.009 a
24	8.12 ±0.75 c	11.91 ±1.06 c	23.84 ±1.71 c	43.88 ±3.38 c	0.227 ±0.013 a

Tobacco plants were supplied with different concentrations of selenite (Na$_2$SeO$_3$) [0 (CK), 3, 6, 12, or 24 mg kg^{-1})] through soil application. After 40 d, the plants were harvested and analyzed for root, stem, and leaf dry weights, as well as root/shoot ratio. The values are presented as mean and *SE* (*n* = 4). Different letters indicate a significant difference in the same column at *P* < 0.05.

Tab. 2 Effects of different levels of Se treatment on net photosynthetic rate (Pn), stomatal conductance (gs), intercellular CO$_2$ concentration (Ci), and carboxylation efficiency (CE) measured from A/Ci curves of newly expanded tobacco leaves.

Se treatments (mg kg^{-1})	Pn (μmol CO$_2$ m^{-2} s^{-1})	gs (μmol H$_2$O m^{-2} s^{-1})	Ci (μmol CO$_2$ mol^{-1})	CE
CK	9.30 ±0.34 b	0.098 ±0.004 b	224.31 ±4.04 c	0.0558 ±0.0024 bc
3	10.89 ±0.61 a	0.118 ±0.010 ab	223.29 ±3.10 c	0.0593 ±0.0011 ab
6	11.69 ±1.11 a	0.129 ±0.020 a	227.12 ±11.99 bc	0.0629 ±0.0012 a
12	8.97 ±0.24 b	0.108 ±0.017 b	239.27 ±9.58 b	0.0521 ±0.0025 c
24	6.13 ±0.50 c	0.077 ±0.003 c	252.07 ±8.95 a	0.0339 ±0.0025 d

The values are presented as mean and *SE* (*n* = 4). Different letters indicate a significant difference in the same column at *P* < 0.05.

this parameter. The CEs of the plants under 3 and 6 mg kg^{-1} Se treatments increased by 3% and 13% compared with those of the plants under CK treatment, respectively. However, the CEs of the plants under 12 and 24 mg kg^{-1} Se treatments were only 93% and 61% those of the plants under CK treatment, respectively. No significant differences in intercellular CO_2 concentration (C_i) was observed between the plants treated with CK and low-dose Se concentrations ($\leq$6 mg kg^{-1}). By contrast, high-dose Se treatments ($\geq$12 mg kg^{-1}) significantly increased the C_i compared with CK ($P < 0.05$).

Effects of Se treatment on leaf chlorophyll concentration and Rubisco content

As shown in Fig. 1, the chlorophyll concentration (SPAD value) slightly increased in the newly expanded leaves under low-dose Se treatments ($\leq$6 mg kg^{-1}). However, the chlorophyll concentration declined with increasing Se concentration. Compared with CK treatment, 24 mg kg^{-1} Se treatment significantly decreased the chlorophyll concentration in the leaves. Similarly, compared with CK treatment, low-dose Se treatments ($\leq$6 mg kg^{-1}) increased the Rubisco content, whereas 24 mg kg^{-1} Se treatment decreased this parameter by 18% (Fig. 2).

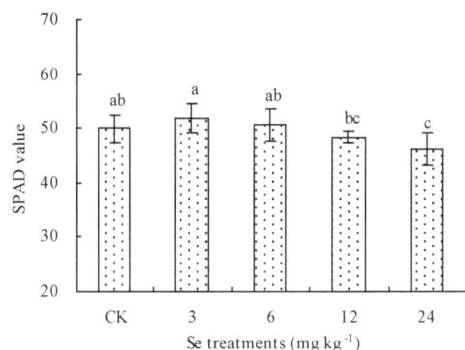

Fig. 1 Effects of different levels of Se treatment on the chlorophyll concentration of tobacco leaves. The results are presented as mean and *SE* ($n = 4$). Different letters indicate a significant difference at $P < 0.05$.

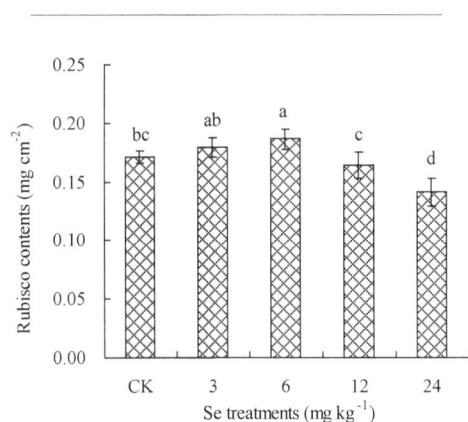

Fig. 2 Effects of different levels of Se treatment on Rubisco content of tobacco leaves The results are presented as mean and *SE* ($n = 4$). Different letters indicate a significant difference at $P < 0.05$.

Effects of Se treatment on MDA content

Compared with CK treatment, low-dose Se treatments ($\leq$6 mg kg^{-1}) significantly reduced the MDA contents in the tobacco leaves, whereas high-dose Se treatments ($\geq$12 mg kg^{-1}) increased this parameter (Fig. 3). The MDA contents in the leaves under 3 and 6 mg kg^{-1} Se treatments were only 86% and 82% those in the leaves under CK treatment, respectively. The MDA contents in the leaves under 12 and 24 mg kg^{-1} Se treatments were 1.05 and 1.28 times those in the leaves under CK treatment, respectively.

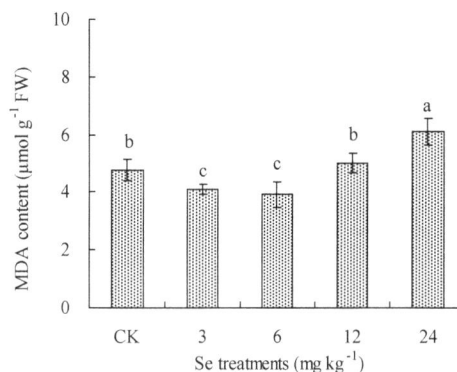

Fig. 3 Effects of different levels of Se treatment on MDA content of tobacco leaves. The results are presented as mean and *SE* ($n = 4$). Different letters indicate a significant difference at $P < 0.05$.

Se concentration and accumulation in tobacco plant tissues

The Se concentration in the different tobacco plant parts significantly increased ($P < 0.01$) with increasing Se application in soil (Tab. 3). For example, the Se concentrations in the roots, stems, and leaves under 24 mg kg^{-1} Se treatment were 6.59, 5.91 and 5.43 times higher than those in the same plant parts under 3 mg kg^{-1} Se treatment, respectively. Under the same Se treatment, the Se concentration in the roots was evidently higher than those in the stems and leaves. In general, the pattern of total Se concentration in tobacco plant tissues followed the order root > leaf > stem. Under the same treatment of 24 mg kg^{-1} Se, the Se concentration in the roots (31.36 mg kg^{-1}) was 3.17 and 7.57 times higher than those in leaves and stems, respectively (Tab. 3).

The accumulation of Se in the tobacco roots, stems, and leaves significantly increased with increasing Se application (Tab. 4). For example, the roots, stems, and leaves under 24 mg kg^{-1} Se treatment accumulated 254.64, 43.60, and 179.30 μg plant^{-1} of Se; these values were 6.28-, 5.51-, and 4.81-fold higher than those accumulated by the roots, stems, and leaves under 3 mg kg^{-1} Se treatment, respectively. Under high-dose Se treatments (24 mg kg^{-1}), the roots accumulated up to 254.64 μg plant^{-1} of Se, which was 1.42 and 5.84 times higher than those accumulated by the leaves and stems, respectively.

Relationship between Se concentration and tobacco plants

The relationship of the Se concentrations in the different treatments with those in the different tobacco plant parts

Tab. 3 Se concentration in the roots, stems, and leaves of tobacco plants under different levels of Se treatment.

Se treatments (mg kg^{-1})	Root (mg kg^{-1})	Stem (mg kg^{-1})	Leaf (mg kg^{-1})
CK	0.15 ±0.02 e	0.04 ±0.00 e	0.06 ±0.00 e
3	4.13 ±0.44 d	0.53 ±0.10 d	1.17 ±0.20 d
6	6.65 ±0.74 c	1.04 ±0.08 c	1.59 ±0.08 c
12	13.81 ±1.28 b	1.69 ±0.15 b	3.54 ±0.35 b
24	31.36 ±2.84 a	3.66 ±0.33 a	7.52 ±0.71 a

The values are presented as mean and *SE* (*n* = 4). Different letters indicate a significant difference in the same column at *P* < 0.05.

Tab. 4 Se accumulation in roots, stems, and leaves of tobacco plants under different levels of Se treatment.

Se treatments (mg kg^{-1})	Root (µg plant^{-1})	Stem (µg plant^{-1})	Leaf (µg plant^{-1})
CK	1.49 ±0.13 e	0.53 ±0.02 e	1.76 ±0.04 e
3	40.52 ±1.43 d	7.91 ±0.35 d	37.27 ±2.06 d
6	73.46 ±3.56 c	18.00 ±0.69 c	55.03 ±2.14 c
12	134.65 ±2.59 b	24.61 ±1.08 b	103.73 ±4.53 b
24	254.64 ±23.58 a	43.60 ±3.89 a	179.30 ±12.84 a

The values are presented as mean and *SE* (*n* = 4). Different letters indicate a significant difference in the same column at *P* < 0.05.

(root, stem and leaf) was analyzed and compared after 40 d of treatment (Fig. 4). The Se concentrations in the roots, stems, and leaves significantly correlated with those in the different treatments. The amounts of absorbed Se in the roots, stems, and leaves were closely related to the Se concentrations in the different treatments.

Discussion

Research has revealed that Se exerts a dual effect on the growth of different plant species [1,5,22]. At low doses, Se stimulates plant growth; at high doses, this element hinders plant growth [2]. In the present study, the exposure of plants to 6 mg kg^{-1} Se increased the yield of roots, shoots and leaves by 11%, 18% and 29%, respectively (Tab. 1). These findings indicate that the growth of flue-cured tobacco plant increased at low-dose Se treatments (≤6 mg kg^{-1}), which agreed with earlier observations of Yao et al. [22], who found that treatment with 1 mg kg^{-1} to 3 mg kg^{-1} Se promotes biomass accumulation in wheat seedlings. However, several reports have provided evidence that high Se addition levels decrease the biomass of non-accumulator plants like wheat [5] and rice [23]. Our results showed that selenite inhibited tobacco plant growth at concentrations up to 12 mg kg^{-1}. The decrease in the root, stem, and leaf DWs was much more apparent at concentrations up to 24 mg kg^{-1} (Tab. 1).

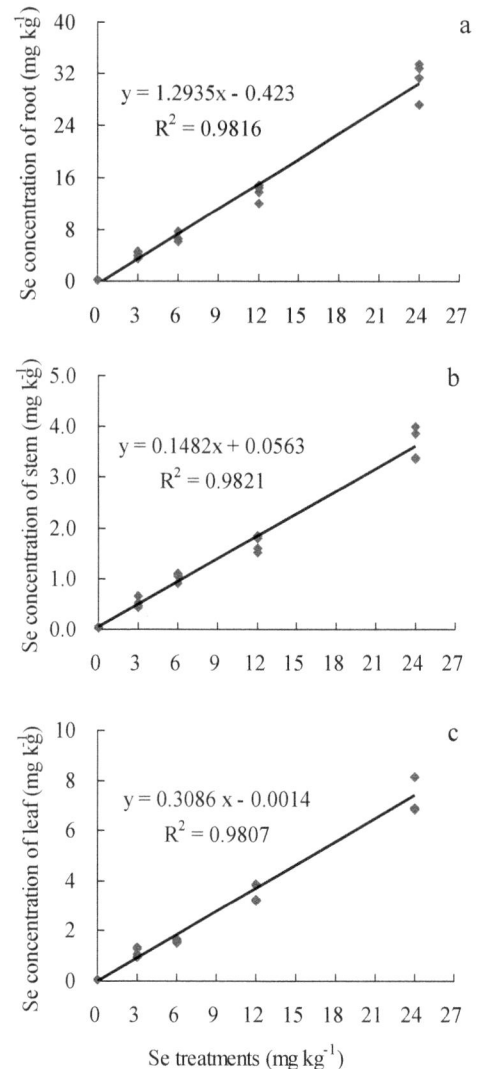

Fig. 4 Relationship between soil Se treatments and Se concentration in different parts of tobacco plants

The improvement of plant growth at low-dose Se treatments (Tab. 1) may be due to the significant increase in photosynthesis of tobacco plants (Tab. 2). In the present study, low-dose Se treatments (≤6 mg kg^{-1}) increased the photosynthetic rate, stomatal conductance and carboxylation efficiency in tobacco leaves (Tab. 2). A similar pattern was observed in rice [23] and sorghum leaves [11]. Wang et al. [23] revealed that in rice seedlings, low doses of Se enhanced photosynthesis. In addition, in sorghum, Se application significantly increased the photosynthetic rate and stomatal conductance [11]. In contrast, at high Se supply levels, the photosynthetic rate in tobacco leaves decreased significantly. High-dose Se treatments (24 mg kg^{-1}) reduced the Pn, CE and Rubisco content by 34%, 39%, and 18%, respectively (Tab. 2 and Fig. 2). This fact suggests that, growth inhibition of tobacco plants under high-dose Se treatments may result from impaired photosynthesis.

Excess Se was toxic to tobacco plants, leading to reduction of chlorophyll concentration (SPAD value; Fig. 1) that may cause photosynthesis suppression. In our experiment, the chlorophyll concentration in the leaves decreased significantly under 24 mg kg^{-1} Se treatments (Fig. 1). Similarly, Chen et al. [24] reported that *Chlorella vulgaris* has lower total chlorophyll content under high Se treatments than under low Se treatments, suggesting that Se affects chlorophyll synthesis or chlorophyllase activity. A decrease in pigment concentrations could also decrease photosynthetic functioning and elicit negative effects on the levels [24]. However, to acquire detailed regulatory mechanisms underlying these effects, further studies should focus on the nature of PSII photochemistry and photosynthetic apparatus under excess Se.

The inhibition of photosynthesis in tobacco plants under high Se application may be closely related to the increased MDA levels. As a product of lipid peroxidation, MDA is an indicator of oxidative damage [25]. Reports have shown that proper doses of Se can reduce MDA accumulation in various plants [2,26]. In the present study, Se application of 6 mg kg^{-1} significantly decreased the MDA content in tobacco leaves; however, 24 mg kg^{-1} Se remarkably increased this parameter (Fig. 3). Similarly, Cartes et al. [26] found that low-dose Se treatments (≤6.0 mg kg^{-1}) reduce the MDA content in ryegrass, and vice versa. The MDA content reflected the extent of lipid peroxidation and indirectly reflected the degree of cell damage. Therefore, our results suggested that the optimal Se dose (≤6 mg kg^{-1}) enhanced antioxidant capacity and reduced lipid peroxidation in flue-cured tobacco leaves, whereas excess Se accelerated lipid peroxidation.

Plant growth responses were closely related with the concentrations of Se in the plant tissues (Tab. 3). The Se concentrations in the different parts of tobacco plant increased as those in the different treatments increased (Tab. 3). A linear correlation ($R = 0.95$, $P < 0.01$) was found between soil and tobacco plant tissue Se concentrations (Fig. 4), indicating that the tobacco plant tissues were not saturated within the tested concentration range. Moreover, Se concentration was much more higher in the tobacco roots than in the leaves and stems (Tab. 3). Therefore, a significantly higher amount of Se was accumulated in the roots than in the leaves and stems (Tab. 4). The pattern of total Se concentration in tobacco plant tissues (root > leaf > stem) was similar to that previously observed in ryegrass [27], wherein Se was principally accumulated in the roots. Previous studies have demonstrated that hyperaccumulators were characterized by a high leaf Se concentration, and a higher shoot:root Se concentration ratio [28,29]. Valdez et al. [30] reported that as a Se hyperaccumulator, the pattern of total Se concentration in *Astragalus bisulcatus* follows the order root < leaf < stem. Under 24 mg kg^{-1} Se, the Se concentration in the roots (31.36 mg kg^{-1}) was 3.17 and 7.57 times higher than those in leaves and stems, respectively (Tab. 3). It suggested that high accumulation of Se in roots (31 mg kg^{-1}) caused tobacco plant toxicity. Generally, most cultivated plants contain less than 25 mg Se kg^{-1} DWs and are considered to be non-accumulators [31]. More recently, it was revealed that most plant species growing on seleniferous soils contain <10 mg Se kg^{-1} DWs, and experience toxicity at levels above ~100 mg Se kg^{-1} DWs [32]. Therefore, from our results it can be suggested that the flue-cured tobacco had a low tolerance to high Se levels, an should be classified as a Se non-accumulator.

Conclusions

The present results showed that Se stimulated tobacco plant growth at low-dose application (≤6 mg kg^{-1}) but inhibited the plant growth at high-dose (≥12 mg kg^{-1}), and the optimal dose of Se application was 6 mg kg^{-1}. Optimal Se dose (6 mg kg^{-1}) improved the plant growth mainly by enhancing photosynthesis, stomatal conductance, carboxylation efficiency and Rubisco content in the flue-cured tobacco leaves. However, the inhibition of excess Se on tobacco growth might be due to high accumulation of Se in roots and the damage of photosynthesis in leaves.

Acknowledgments
This work was jointly supported by 2012 Dean Youth Innovation Fund of Anhui Academy of Agricultural Sciences (12B0919) and Anhui Tobacco Corporation (20120551002).

Authors' contributions
The following declarations about authors' contributions to the research have been made: designed the research: JC, ZC, SF; performed the experiments: JC, SJ; analyzed the data: JC, LT; wrote the paper: JC.

Competing interests
No competing interests have been declared.

References
1. Terry N, Zayed AM, de Souza MP, Tarun AS. Selenium in higher plants. Annu Rev Plant Physiol Plant Mol Biol. 2000;51(1):401–432.

2. Feng R, Wei C, Tu S. The roles of selenium in protecting plants against abiotic stresses. Environ Exp Bot. 2013;87:58–68.

3. Hartikainen H, Xue T, Piironen V. Selenium as an anti-oxidant and pro-oxidant in ryegrass. Plant Soil. 2000;225(1–2):193–200.

4. van Hoewyk D. A tale of two toxicities: malformed selenoproteins and oxidative stress both contribute to selenium stress in plants. Ann Bot. 2013;112(6):965–972.

5. Lyons GH, Stangoulis JCR, Graham RD. Tolerance of wheat (*Triticum aestivum* L.) to high soil and solution selenium levels. Plant Soil. 2005;270(1):179–188.

6. Zhu YG, Pilon-Smits EAH, Zhao FJ, Williams PN, Meharg AA. Selenium in higher plants: understanding mechanisms for biofortification and phytoremediation. Trends Plant Sci. 2009;14(8):436–442.

7. Martin AL, Trelease SF. Absorption of selenium by tobacco and soy beans in sand cultures. Am J Bot. 1938;5(7):380–385.

8. Turakainen M, Hartikainen H, Seppänen MM. Effects of selenium treatments on potato (*Solanum tuberosum* L.) growth and concentrations of soluble sugars and starch. J Agr Food Chem. 2004;52(17):5378–5382.

9. Hu Q, Xu J, Pang G. Effect of selenium on the yield and quality

of green tea leaves harvested in early spring. J Agr Food Chem. 2003;51(11):3379–3381.

10. Liu Q, Wang DJ, Jiang XJ, Cao ZH. Effects of the interactions between selenium and phosphorus on the growth and selenium accumulation in rice (*Oryza sativa*). Environ Geochem Health. 2004;26(2):325–330.

11. Djanaguiraman M, Prasad PVV, Seppänen M. Selenium protects sorghum leaves from oxidative damage under high temperature stress by enhancing antioxidant defense system. Plant Physiol Bioch. 2010;48(12):999–1007.

12. Xu J, Zhu S, Yang F, Cheng L, Hu Y, Pan G, et al. The influence of selenium on the antioxidant activity of green tea. J Sci Food Agric. 2003;83(5):451–455.

13. Carey AM, Scheckel KG, Lombi E, Newville M, Choi Y, Norton G, et al. Grain accumulation of selenium species in rice (*Oryza sativa* L.). Environ Sci Technol. 2012;46 (10):5557–5564.

14. Pyrzynska K. Selenium speciation in enriched vegetables. Food Chem. 2009;114:1183–1191.

15. Finley JW. Reduction of cancer risk by consumption of selenium-enriched plants: enrichment of broccoli with selenium increases the anticarcinogenic properties of broccoli. J Med Food. 2003;6(1):19–26.

16. Lu XP, Gui YJ, Xiao BG, Li YP, Tong ZJ, Liu Y, et al. Development of DArT markers for a linkage map of flue-cured tobacco. Chin Sci Bull. 2013;58(6):641–648.

17. Jiang C, Zheng Q, Liu Z, Liu L, Zhao G, Long X, et al. Seawater-irrigation effects on growth, ion concentration, and photosynthesis of transgenic poplar overexpressing the Na$^+$/H$^+$ antiporter AtNHX^1. J Plant Nutr Soil Sci. 2011;174(2):301–310.

18. Jiang C, Zheng Q, Liu Z, Xu W, Liu L, Zhao G, et al. Overexpression of *Arabidopsis thaliana* Na$^+$/H$^+$ antiporter gene enhanced salt resistance in transgenic poplar (*Populus* × *euramericana* 'Neva'). Trees. 2012;26(3):685–694.

19. Li Y, Yang XX, Ren BB, Shen QR, Guo SW. Why nitrogen use efficiency decreases under high nitrogen supply in rice (Oryza sativa L.) seedlings. J Plant Growth Regul. 2012;31:47–52.

20. Gao J, Liu Y, Huang Y, Lin ZQ, Bañuelos GS, Lam MHW, et al. Daily selenium intake in a moderate selenium deficiency area of Suzhou, China. Food Chem. 2011;126(3):1088–1093.

21. Zhou XB, Shi WM, Zhang LH. Iron plaque outside roots affects selenite uptake by rice seedlings (*Oryza sativa* L.) grown in solution culture. Plant Soil. 2007;290(1–2):17–28.

22. Yao X, Chu J, Wang G. Effects of selenium on wheat seedlings under drought stress. Biol Trace Elem Res. 2009;130(3):283–290.

23. Wang YD, Wang X, Wong YS. Proteomics analysis reveals multiple regulatory mechanisms in response to selenium in rice. J Proteomics. 2012;75(6):1849–1866.

24. Chen TF, Zheng WJ, Luo Y, Yang F, Bai Y, Tu F. Effects of selenium stress on photosynthetic pigment contents and growth of *Chlorella vulgaris*. J Plant Physiol Mol Biol. 2005;31(4):369–373.

25. Azevedo Neto AD, Prico JT, Enéas-Filho J, Braga de Abreu CE, Gomes-Filho E. Effect of salt stress on antioxidative enzymes and lipid peroxidation in leaves and roots of salt-tolerant and salt-sensitive maize genotypes. Environ Exp Bot. 2006;56(1):87–94.

26. Cartes P, Gianfreda L, Mora ML. Uptake of selenium and its antioxidant activity in ryegrass when applied as selenate and selenite forms. Plant Soil. 2005;276 (1–2):359–367.

27. Cartes P, Gianfreda L, Paredes C, Mora ML. Selenium uptake and its antioxidant role in ryegrass cultivars as affected by selenite seed pelletization. J Soil Sci Plant Nut. 2011;11(4):1–14.

28. White PJ, Bowen HC, Marshall B, Broadley MR. Extraordinarily high leaf selenium to sulfur ratios define "Se-accumulator" plants. Ann Bot. 2007;100(1):111–118.

29. Galeas ML, Zhang LH, Freeman JL, Wegner M, Pilon-Smits EAH. Seasonal fluctuations of selenium and sulfur accumulation in selenium hyperaccumulators and related nonaccumulators. New Phytol. 2007;173:517–525.

30. Valdez Barillas JR, Quinn CF, Freeman JL, Lindblom SD, Fakra SC, Mar-cus MA, et al. Selenium distribution and speciation in the hyperaccumulator *Astragalus bisulcatus* and associated ecological partners. Plant Physiol. 2012;159(4):1834–1844.

31. Hasanuzzaman M, Hossain MA, Fujita M. Selenium in higher plants: physiological role, antioxidant metabolism and abiotic stress tolerance. J Plant Sci. 2010;5:354–375.

32. El Mehdawi AF, Pilon-Smits EAH. Ecological aspects of plant selenium hyperaccumulation. Plant Biol. 2012;14:1–10.

10

Acid α-galactosidase is involved in D-chiro-inositol accumulation during tartary buckwheat germination

Cai-Feng Jia, Wan-Hong Hu, Zhong-yi Chang, Hong-Liang Gao*

School of Life Sciences, East China Normal University, 500 Dongchuan Rd, Shanghai 200241, China

Abstract

Tartary buckwheat seed and especially its sprouts are rich in D-chiro-inositol (DCI). The research was to evaluate when DCI was most accumulated in tartary buckwheat sprouts. In addition, we explored the activity and expression pattern of α-galactosidase during tartary buckwheat seed germination. The results showed that DCI contents steadily increased at early stage of germination and reached the highest level of 33.42 µg/seed at 24 h during the 72 h trail. However, the total fagopyritol contents sharply decreased from 214.6 µg/seed to 46 µg/seed at the end of the germination. The activity of acid α-galactosidase increased gradually to the peak of 0.36 nkat/seed at 24 h after the primed seed imbibition. We cloned the gene fragment of α-galactosidase in tartary buckwheat for the first time. The deduced amino acid sequence is 93% identical to that of *P. vulgaris*. The quantitative PCR result of gene expression pattern was consistent with its enzyme activity during seed germination.

Keywords: tartary buckwheat; D-chiro-inositol; germination; α-galactosidase; gene expression

Introduction

Tartary buckwheat (*Fagopyrum tataricum*) is a functional food, widely cultivated all over the world including Asia and southwest of China. As a functional food *F. tataricum* has been receiving much attention for its healing effects over chronic diseases for the long time [1]. It has been illustrated that intragastric administration of buckwheat concentrate effectively lowered serum glucose concentrations in streptozotocin-diabetic rat [2]. In humans, buckwheat has a therapeutic potential against hyperglycemia and diabetes mellitus [3].

D-chiro-inositol (DCI), a naturally occurring isomer of myo-inositol, is the main active nutritional ingredient in buckwheat. It acts as a component of a putative insulin mediator, a galactosamine D-chiro-inositol with an insulin like bioactivity [4]. DCI increase insulin sensitivity and decrease plasma glucose in obese rhesus monkeys with spontaneous insulin resistance [5]. In humans, non-insulin-dependent diabetes mellitus (NIDDM) has also been associated with decreased urinary DCI excretion [6]. Therefore, DCI has great potential to work as an adjunctive drug in the treatment of insulin resistance ailments such as type 2 diabetes and polycystic ovary syndrome [7].

Buckwheat is an excellent dietary source of DCI in the form of its a-galactosides, fagopyritols, that accumulate in embryo tissues of seeds [8,9]. There are several form of fagopyritols accumulated in buckwheat seeds [10]. However, DCI exists as its galactosyl derivatives limits the nutritional value of buckwheat seed [8]. Previous research has demonstrated that germination may have the potential to improve the nutritional value of the grain and can effectively reduce antinutrients in cereals and legumes [11].

Seeds have a high demand for energy during early germination. Raffinose family oligosaccharides (RFOs), which are ubiquitous in plant seed and are rapidly mobilized by α-galactosidase during seed germination to provide energy [12]. RFOs are important for early germination of plant. The inhibition of raffinose oligosaccharide breakdown delayed pea seeds germination indicating that galactose is an important component during germination [13]. Uniquely, buckwheat seeds accumulate small amounts of RFOs but large amounts of fagopyritols, more than 40% percent of total soluble carbohydrates, which can be hydrolysed by the α-galactosidase in vivo, releasing galactose and the free DCI [14]. Seed priming is the technique which is commonly used to improve germination behavior and seedling emergence [15]. The application of seed priming will induce the synchronization of physiological and biochemical changes during seed germination [16].

Galactosidase can be divided into two types depending on their optimal pH for activity. The acidic α-galactosidases are most likely active in the acidic environment of vacuoles

---
* Corresponding author. Email: hlgao@bio.ecnu.edu.cn

Handling Editor: Grzegorz Jackowski

while the alkaline forms probably catalyze galactose release in the more neutral or alkaline cytoplasm [17]. The acidic α-galactosidases prefer raffinose as the substrate in comparison with the alkaline form which shows a higher affinity for stachyose [18,19]. In germinating legume seeds, α-galactosidase plays a role in the mobilization of RFOs [20]. The characterization and cloning of α-galactosidase during germination have been studied in other plant [21–23]. The activity of α-galactosidase and the expression pattern of the gene especially during tartary buckwheat germination remain unknown.

Although some studies have been carried out to improve DCI contents in buckwheat sprouts, the effects of seed priming on DCI accumulation in tartary buckwheat sprouts have never been reported. Additionally, we cloned the gene fragment of α-galactosidase of tartary buckwheat and further tested the gene expression during seed germination by the qPCR method.

Material and methods

Tartary buckwheat seed germination

Tartary buckwheat seeds were purchased from Sichuan province of Southwest China and stored at −20°C. The following germination method was adopted as we reported before [24]. The seeds were surface sterilized with 10% (v/v) of sodium hypochlorite for 3 h and then washed. Then the seeds were mixed with sand containing 4% (v/w) water, sealed in plastic box and primed in darkness at 15°C for 48 h. After treatment, seeds were washed under tap water and dried to the original moisture content determined by weighing with forced air under shade at 27 ±3°C for 2 days. The primed seeds were spread thinly on petri dishes containing layers of wet filter paper and initiated to germinate in the dark (25°C for 72 hours). Seed samples were collected at 12-hour intervals from 0 to 72 hours after imbibition, immediately frozen in liquid nitrogen and the samples were stored at −80°C for further use.

Analysis of DCI and total fagopyritol content

Phenyl R-D-glucoside, trimethylsilylimidazole (TMSI) and pyridine were purchased from Sigma-Aldrich (Shanghai, China), DCI standards were purchased from Wako Pure Chemicals Industries, Ltd. (Osaka, Japan). Fagopyritol standards were extracted from seed buckwheat (*Fagopyrum esculentum* L.). The DCI and total fagopyritol content were performed according to the procedure of Yang and Ren [25] with slight modification. Three replications of 10 seeds each were blended for 5 min using homogenizer with 20 ml of ethanol/water (1:1, v/v) containing 10 mg of phenyl R-D-glucoside as internal standard. The homogenate was centrifuged at 12 000 *g* for 10 min at room temperature, supernatant was removed, and the residue was re-extracted two times with 10 ml of ethanol/water (1:1, v/v) for 5 min and recentrifuged. An aliquot of the combined extracts was filtered through 0.22 µm film, transferred to silylation vials, and evaporated to dryness in a stream of nitrogen gas at 70°C water bath.

Extract residues were kept overnight in a desiccator over phosphorus pentaoxide to remove traces of water. Dry residues were derivatized with a silylation mixture (TMSI/pyridine, 1:1, v/v) in silylation vials at 70°C for 30 min, cooled, and analyzed by gas chromatograph system (GC-7900, Shanghai, China) equipped with a FID detector and a HP-5 capillary quartz column (50 m × 0.25 mm, 0.25 µm film thickness). The initial column temperature was 150°C, which increased gradually up to 200°C by the velocity of 3°C/min. Subsequently, it was gradually increased to 325°C by the velocity of 7°C/min, and then maintained for 20 min. The inlet temperature was 335°C and detector temperature was 350°C. The carrier gas nitrogen was at 1.0 ml/min (measured at 30°C). The injection volume was 1 µl. DCI and the fagopyritols were identified by GC retention times identical to the standard.

DCI and total fagopyritols contents were quantified based on standard curves: the ratios of the area of signals for each known compound to the area of the signal for phenyl R-D-glucoside, the internal standard, were plotted against known amounts of each compound [20]. Amounts below the level of detection are presented as zero.

Assay on α-galactosidase activity

Three replications of 10 seeds each were ground in a glass homogenizer in extraction buffer (50 mM HEPES-NaOH, pH 7.4). Homogenates were centrifuged at 10 000 *g* for 20 min at 4°C, the supernatants were stored at 4°C prior to assay of α-galactosidase activity. The enzyme activity of α-galactosidase was determined as previously described [26]. The quantitative analysis of its activity was measured by detection the p-nitrophenol released from p-nitrophenyl-α-D-galactopyranoside (pNPGal). The reaction mixtures consisted of 0.9 ml substrate (3 mM pNPGal in 100 mM NaAc buffer, pH 5.0) and 0.1 ml of suitably diluted enzyme preparation. The reaction was terminated by the addition of 3 ml of 3% Na_2CO_3 after incubation for 15 min at 37°C and the quantity of p-nitrophenol released was measured at 410 nm. Blanks were prepared by adding the enzyme extract after Na_2CO_3. One unit of enzymatic activity, nkatal (nkat) was defined as the amount of activity that released one nmole of p-nitrophenol per second at pH 5.0 and 37°C. The enzyme activity was expressed in units per seed (nkat/seed). The data presented for all α-galactosidase activity determinations are mean values of triplicate assays in which the standard deviations were always smaller than 10%.

Cloning and sequence analysis of FtaGAL in buckwheat

As the buckwheat belongs to dicotyledon , the alignment of the amino acid sequence of α-galactosidase among various dicotyledonous plant species was done with Clustal X software [27]. The conservative sequences were used for designing two degenerate PCR primers, GALF: 5'-TGGG(G/A)(A/G)T (G/A)GA(C/T)TA (C/T) T (G/A)AA(A/G)TATG-3' and GALR: 5'-TC(G/A)A(A/G)C AT (A/G) TC(G/ A) GG(A/G) TC (A/G) TTCC-3'.

Seed samples were powdered in liquid nitrogen with mortar and pestle and the total RNA was extracted using Trizol reagent according to the manufacturer's instructions (Invitrogen, Carlsbad, CA, USA). Less than 500 ng RNA was used for the RT-PCR as specified in the RNA PCR Kit (AMV) Ver.3.0 (TaKaRa, Japan). The oligo dT-adaptor primer was used for cDNA synthesis at 42°C, 60 min. The PCR amplification was carried out with above-described degenerate primers. The amplification profile was one cycle at 93°C for 3 min, followed by 35 cycles at 93°C for 15 s, 55°C for 30 s and 68°C for 5 min, and a final extension step at 72°C for 7 min. The amplified products were recovered from 2.0% agarose gels, cloned into the pGEM-T vector, and transformed into *Escherichia coli* DH5α. Sequencing was performed commercially and sequence analyses were performed using BLAST from the NCBI (http://www.ncbi.nlm.nih.gov/Tools/).

The expression pattern of α-galactosidase during tartary buckwheat germination

The primers for *FtαGAL* were designed based on the obtained cDNA sequence. The housekeeping gene β-actin was used as internal reference. The primer pairs for *FtαGAL* were FtαGAL-F: 5′-GATACCCTCCCATGCGTGATGC-3′ and FtαGA-R: 5′-GCATAGGCTGCC CACTTGTCAT-3′. The deduced amplification length was 203 bp. The primer pair for β-actin was actin-F: 5′-GCTGGATTTGCTG-GAGATGATGC-3′ and actin-R: 5′-CTTCTCCATGTCA TCCCAGTTGCT-3′ and the deduced amplification length was 196 bp.

Total RNA and first-stand complementary cDNA of buckwheat seeds at different germination stages were prepared as described above. The internal reference gene β-actin and target gene FtαGAL were analyzed in one plate, and each reaction was repeated three times to access the reproducibility. The cycling protocol consisted of denaturation at 95°C for 5 s, annealing at 58°C for 15 s, elongation at 72°C for 20 s, and the PCR reaction was run for 40 cycles. The fluorescence data was collected at 81°C for 20 s. The model $2^{-\Delta\Delta CT}$ for comparing relative expression results between samples in real-time PCR was applied. The expression of target gene, normalized to the reference control and relative to a calibrator (time-zero sample) is given by $R = 2^{-\Delta\Delta CT}$, where $\Delta\Delta CT = \Delta CT$ sample $- \Delta CT$ control. All samples were performed in triplicates. Positive and negative controls were performed on each plate.

Results

Seed germination of tartary buckwheat

Seed priming enhanced tartary buckwheat seed performance with respect to the speed and uniformity of germination (Fig. 1). The tartary buckwheat seed, after priming, began to germinate just 12 hours after incubation on wet paper layers. In the following time, seed germinated quickly and reach 73.2% at 24th h , then reached the maximum germinability as high as 94% at 36th h. It indicates that the fastest germination must take place between 12 h and 24 h and the complicated physiological and biochemical changes may occur in this period.

The contents of DCI and total fagopyritol during tartary buckwheat germination

The contents of DCI and total fagopyritol in tartary buckwheat seeds were determined at different germination stages (0–72 h; Fig. 2). The DCI contents increased dramatically during the early germination. At 24 h, it reached the maximum level of 33.4 μg/seed, which was about 2.3 times that of time-zero seeds. However, after 24 h, the DCI content in tartary buckwheat sprouts sharply decreased. After 60 h, the level was lower than that of time-zero seeds. Therefore, the optimum time for tartary buckwheat germination was 24 h, yielding the maximum DCI content. The total fagopyritol content revealed an opposite trend, decreased during the tested period from 214.6 (0 h) to 46 μg/seed (72 h). The breakdown of fagopyritol may release D-chiro-inositol and galactose, indicating that they may play a special role during early germination.

Enzyme activity assay of α-galactosidase during the germination

Acid α-galactosidase has great influence on seed development and germination, therefore, recent studies have paid much attention to it [28]. The activities of the alkaline α-galactosidase were very low and changed little during the buckwheat seed germination (data not shown). The activities of acidic α-galactosidase in tartary buckwheat seeds were measured at different germinating periods (Fig. 3). The activity of α-galactosidase exhibited high even at time-zero and increased slowly during the next 36 h of germination

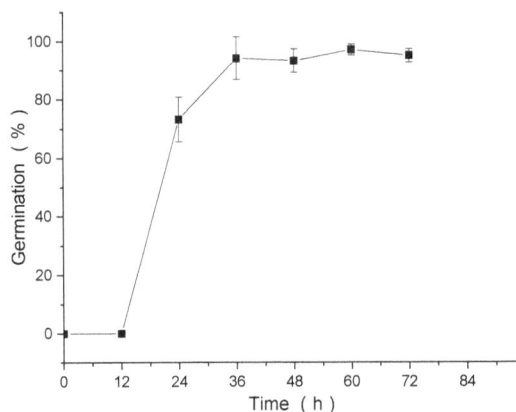

Fig. 1 Time course of germination of primed tartary buckwheat during the first 72 h. Data presented are means ±*SD* (*n* = 3).

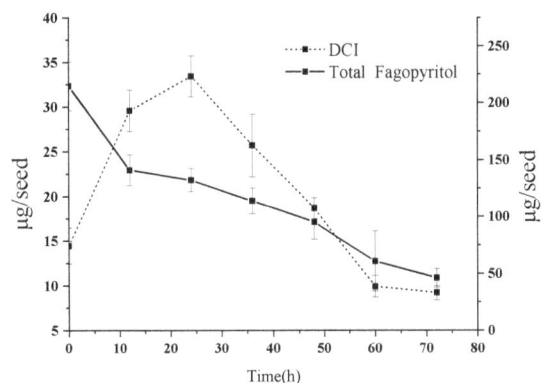

Fig. 2 Changes in DCI and total fagopyritol content of tartary buckwheat in germinating tartary buckwheat. Data presented are means ±*SD* (*n* = 3).

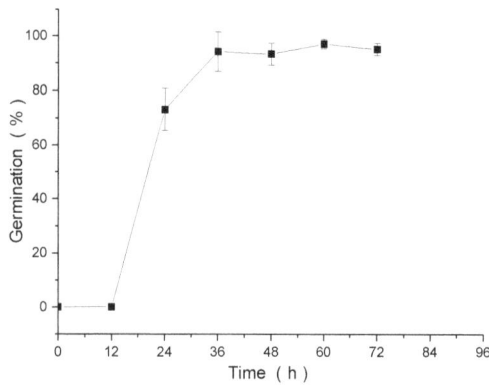

Fig. 3 α-galactosidase activity in primed tartary buckwheat seed during germination. Data presented are means ±SD (*n* = 3).

(from 0.30 to 0.36 nkat/seed). However, it decreased rapidly and at the end of the germination it was only 0.12 nkat/seed.

Cloning and identification of *FtaGAL* gene

The consecutive amino acid sequences of α-galactosidase in dicotyledonous were aligned and the protein exhibited a highly conservative sequence (Fig. 4). The degenerate primers were used and the cloned partial cDNA of *FtaGAL* is 293 nucleotides. Multiple alignments revealed a high degree of homology between deduced amino acid sequence of Ftα-GAL and α-Galactosidase of other plants. The highest identity was found to be 93% similarity with *P. vulgaris*. The deduced sequence covered 23% of the full length of the *P. vulgaris* α-galactosidase protein sequence. Ftα-GAL showed less homology with that of *Glycine max* (91%), *Cicer arietinum* (90%) and *Salvia mitiorrhiza* (90%).

Expression of α-galactosidase gene in tartary buckwheat seed during germination

Quantitative PCR was used to analyze the gene expression pattern during tartary buckwheat seed germination (Fig. 5). The mRNA level of α-galactosidase in buckwheat seeds increased steadily during early germination. At 24 h it reached the peak as 2.59 fold in comparison with that of time-zero seeds. Up-regulation of the genes responsible for the fagopyritol breakdown was expected, which was in accordance with the activity of α-galactosidase in tartary buckwheat seed. However the gene expression was found to be down-regulated at the later stages of germination, it declined to only 0.12 fold relative to control at the end of the germination. This correlated with the decrease of the activity of α-galactosidase in buckwheat germination.

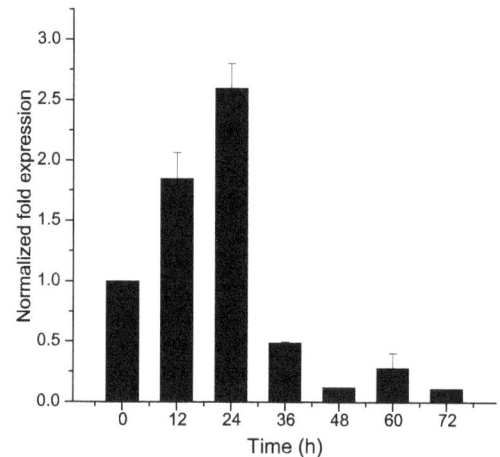

Fig. 5 Expression pattern of α-galactosidase gene during tartary buckwheat germination. The relative expression of α-galactosidase was quantified in comparison with the ACTIN using quantitative PCR with gene-specific primers. The values represent the average of three independent samples. Data presented are means ±SD (*n* = 3).

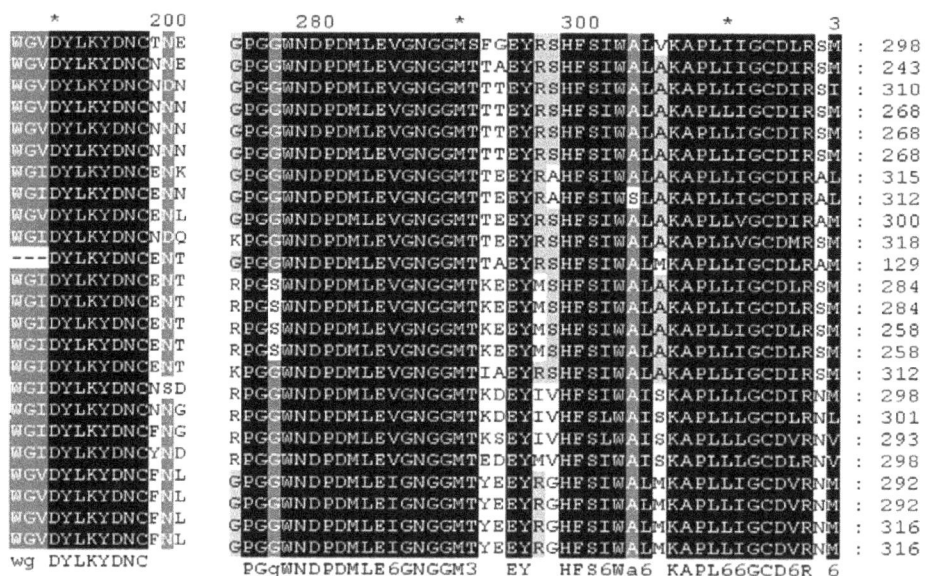

Fig. 4 The consecutive amino acid sequences of α-galactosidase gene in dicotyledon.

Discussion

Buckwheat has attracted increasing attention as a potential functional food for a long time. However, the best usage of it is rarely reported. Elicitor molecules such as salicylic acid (SA) and methyl jasmonate (MeJA) have been used to induce the high yield of free DCI in buckwheat [29]. DCI was significantly higher in the metallic additives treated samples than in the control during the tartary buckwheat germination [30]. The second metabolites have been demonstrated to be accumulated in some plants, by stimulating the activity of enzymes [31]. Germination has been reported to stimulate the production of second metabolites in some cereal seeds such as buckwheat [32]. Seed priming is a promising treatment to improve the rate and uniformity of germination and is associated with increase of enzyme activity in rice seed germination [33]. The seed priming method is used to induce the high yield of DCI during the germination process in this paper and the results show that it reaches the highest level of 33.42 μg/seed at 24 h after imbibition, which provides us the optimum time to make better use of the buckwheat seed for free DCI. These phenomena may attribute to the biochemical metabolism in seed germination [34]. Similarly Zalewski et al. observed that DCI content kept increasing in the first 48 h of seed germination in yellow lupin [35]. In addition, Pathak also found that soaking and early germination converted soybean seeds into an effective blood sugar regulator [36].

Galactosidase play an important role during plant germination. The initial enzymes in RFOs catabolism are α-galactosidases, which hydrolytically remove the terminal galactose moiety of RFOs. During soybean seed germination, the content of raffinose oligosaccharides decrease substantially, while the α-galactosidase activity increases [37]. In our paper we found that the total fagopyritol content decreased while the free DCI content increased during buckwheat germination, which indicated that DCI may be released from the fagopyritol. The analysis between α-galactosidase activities and DCI content revealed a significant positive correlation in germinated tartary buckwheat [30]. We employ the priming method in this paper and the results show that the activity of α-galactosidase exhibited high in time-zero seeds. This may be attribute to the priming treatments significantly mobilize the α-galactosidase activity even before imbibition.

The gene of α-galactosidase have been cloned and analysed in tomato seed, melon fruit and coffee bean [38–40]. Feurtado et al. [38] observed that α-galactosidase gene transcription gradually increased during tomato seed germination by using real-time fluorescence quantitative PCR. In this paper we cloned the partial sequence of α-galactosidase of tartary buckwheat and the gene expression pattern was also evaluated. The results were in close agreement with the enzyme activity of α-galactosidase and profiles of DCI accumulation.

Acknowledgments
The author acknowledges the financial support granted by University Innovation Foundation project (No. 78210259) of East China Normal University, People's Republic of China.

Authors' contributions
The following declarations about authors' contributions to the research have been made: contributed to the conducting experiments and writing the manuscript: CFJ; contributed to data acquisition: WHH; contributed to the analysis and interpretation of data: ZC; contributed to the conception and design of this paper: HLG.

Competing interests
No competing interests have been declared.

References
1. Zhang ZL, Zhou ML, Tang Y, Li FL,Tang YX, Shao JR, et al. Bioactive compounds in functional buckwheat food. Food Res Int. 2012;49(1):389–395.

2. Kawa JM, Taylor CG, Przybylski R. Buckwheat concentrate reduces serum glucose in streptozotocin-diabetic rats. J Agric Food Chem. 2003;51(25):7287–7291.

3. Zhang HW, Zhang YH, Lu MJ, Tong WJ, Cao GW. Comparison of hypertension, dyslipidaemia and hyperglycaemia between buckwheat seed-consuming and non-consuming Mongolian-Chinese populations in Inner Mongolia, China. Clin Exp Pharmacol Physiol. 2007;34(9):838–844.

4. Larner J, Price JD, Heimark D, Smith L, Rule G, Piccariello T, et al. Isolation, structure, synthesis, and bioactivity of a novel putative insulin mediator. A galactosamine chiro-inositol pseudo-disaccharide Mn²⁺ chelate with insulin-like activity. J Med Chem. 2003;46(15):3283–3291.

5. Ortmeyer HK, Bodkin NL, Lilley K, Larner J, Hansen BC. Chiroinositol deficiency and insulin resistance. I. Urinary excretion rate of chiroinositol is directly associated with insulin resistance in spontaneously diabetic rhesus monkeys. Endocrinology. 1993;132(2):640–645.

6. Suzuki S, Kawasaki H, Satoh Y, Ohtomo M, Hirai M, Hirai A, et al. Urinary chiro-inositol excretion is an index marker of insulin sensitivity in Japanese type II diabetes. Diabetes Care. 1994;17(12):1465–1468.

7. Yang N, Ren G. Application of near-infrared reflectance spectroscopy to the evaluation of rutin and D-chiro-inositol contents in tartary buckwheat. J Agric Food Chem. 2008;56(3):761–764.

8. Horbowicz M, Brenac P, Obendorf RL. Fagopyritol B1, O-α-D-galactopyranosyl-(1→2)-D-chiro-inositol, a galactosyl cyclitol in maturing buckwheat seeds associated with desiccation tolerance. Planta. 1998;205(1):1–11.

9. Szczecinski P, Gryff KA, Horbowicz M, Obendorf RL. NMR investigation of the structure of fagopyritol B1 from buckwheat seeds. Bull Pol Acad Sci Chem. 1998;46:9–13.

10. Ma JM, Horbowicz M, Obendorf RL. Cyclitol galactosides in embryos of buckwheat stem–leaf–seed explants fed d-chiro-inositol, myo-inositol or d-pinitol. Seed Sci Res. 2005;15(4):329–338.

11. Luo YW, Xie WH, Jin XX, Wang Q, Zai XM. Effects of germination on iron, zinc, calcium, manganese, and copper availability from cereals and legumes. Cyta J Food. 2013;11(4):318–323.

12. Blochl A, Peterbauer T, Richter A. Inhibition of raffinose oligosaccharide breakdown delays germination of pea seeds. J Plant Physiol. 2007;164(8):1093–1096.

13. Lahuta LB, Goszczyńska J. Inhibition of raffinose family oligosaccharides and galactosyl pinitols breakdown delays germination of winter vetch (Vicia villosa Roth.) seeds. Acta Soc Bot Pol. 2009;78(3):203–208.

14. Yang N. Ren G. Determination of D-chiro-inositol in tartary buckwheat

using high-performance liquid chromatography with an evaporative light-scattering detector. J Agric Food Chem. 2008;56(3):757–760.

15. Galhaut L, Lespinay A, Walker DJ, Bernal MP, Correal E, Lutts S. Seed priming of *Trifolium repens* L. improved germination and early seedling growth on heavy metal-contaminated soil. Water Air Soil Pollut. 2014;225(4):1–15.

16. Taie HA, Abdelhamid MT, Dawood MG, Nassar RM. Pre-sowing seed treatment with proline improves some physiological, biochemical and anatomical attributes of Faba bean plants under sea water stress. J Appl Sci Res. 2013;9(4):2853–2867.

17. Daldoul S, Toumi I, Reustle GM, Krczal G, Ghorbel A, Mliki A, et al. Molecular cloning and characterisation of a cDNA encoding a putative alkaline alpha-galactosidase from grapevine (*Vitis vinifera* L.) that is differentially expressed under osmotic stress. Acta Physiol Plant. 2012;34(3):891–903.

18. Gaudreault PR, Webb JA. Alkaline α-galactosidase activity and galactose metabolism in the family Cucurbitaceae. Plant Sci. 1986;45(2):71–75.

19. Schaffer A, Zhifang G. Alkaline alpha-galactosidase. US 6,607,901 B1 (Google Patents) 2003.

20. Guimaraes VM, Rezende ST, Moreira MA, Barros EG, Felix CR. Characterization of alpha-galactosidases from germinating soybean seed and their use for hydrolysis of oligosaccharides. Phytochemistry. 2001;58(1):67–73.

21. Atalide GM, Borges EE, Goncalves JF, Guimaraes VM, Bicalho EM, Flores AV. Activities of a-galactosidase and polygalacturonase during hydration of *Dalbergia nigra* [(Vell.) Fr All. ex Benth.] seeds at different temperatures. J Seed Sci. 2013;35(1):92–98.

22. Schillinger JA, Dierking EC, Bilyeu KD. Soybeans having high germination rates and ultra-low raffinose and stachyose content. US 8,471,107B2 (Google Patents) 2013.

23. Gajdhane S, Jadhav U, Dandge P. Biochemical study of α-galactosidase from cowpeas (*Vigna unguiculata*). Indian Streams Research Journal. 2014;4(1):1–5.

24. Jia CF, Zhao WC, Cai MM, Li J, Li J. Effect of sand priming treatment on seed germination of tartary buckwheat. Seed. 2011;30(1):96–98.

25. Steadman KJ, Burgoon MS, Schuster RL, Lewis BA, Edwardson SE, Obendorf RL. Fagopyritols, D-chiro-inositol, and other soluble carbohydrates in buckwheat seed milling fractions. J Agric Food Chem. 2000;48(7):2843–2847.

26. McCleary B, Matheson N. α-D-galactosidase activity and galactomannan and galactosylsucrose oligosaccharide depletion in germinating legume seeds. Phytochemistry. 1974;13(9):1747–1757.

27. Thompson JD, Gibson TJ, Plewniak F, Jeanmougin F, Higgins DG. The CLUSTAL X windows interface: flexible strategies for multiple sequence alignment aided by quality analysis tools. Nucleic Acids Res. 1997;25(24):4876–4882.

28. Keller F, Pharr DM. Metabolism of carbohydrates in sinks and sources: galactosyl-sucrose oligosaccharides. Photoassimilate distribution in plants and crops: source-sink relationships. New York, NY: Marcel Dekker; 1996.

29. Hu, YH, Yu YT, Piao CH, LiuJM, Yu HS. Methyl jasmonate-and salicylic acid-induced d-chiro-inositol production in suspension cultures of buckwheat (*Fagopyrum esculentum*). Plant Cell Tiss Org. 2011;106(3):419–424.

30. Wang L, Li XD, Niu M, Wang R, Chen ZX. Effect of additives on flavonoids, d-chiro-Inositol and trypsin inhibitor during the germination of tartary buckwheat seeds. J Cereal Sci. 2013;58(2):348–354.

31. Xu JG, Hu QP. Changes in γ-aminobutyric acid content and related enzyme activities in Jindou 25 soybean (*Glycine max* L.) seeds during germination. LWT – Food Sci Technol. 2014;55(1):341–346.

32. Makino Y, Soga N, Oshita S, Kawagoe Y, Tanaka A. Stimulation of γ-aminobutyric acid production in vine-ripe tomato (*Lycopersicon esculentum* Mill.) fruits under modified atmospheres. J Agric Food Chem. 2008;56(16):7189–7193.

33. Farooq M, Basra SM, Tabassm R, Afzal Z. Enhancing the performance of direct seeded fine rice by seed priming. Plant Prod Sci. 2006;9(4):446–456.

34. Randhir R, Lin YT, Shetty K. Phenolics, their antioxidant and antimicrobial activity in dark germinated fenugreek sprouts in response to peptide and phytochemical elicitors. Asia Pac J Clin Nutr. 2004;13(3):295–307.

35. Zalewski K, Nitkiewicz B, Lahuta LB, Głowacka K, Socha A, Amarowicz R. Effect of jasmonic acid–methyl ester on the composition of carbohydrates and germination of yellow lupine (*Lupinus luteus* L.) seeds. J Plant Physiol. 2010;167(12):967–973.

36. Pathak M. Soaked and germinated soybean seeds for blood sugar control: a preliminary study. Natural Product Radiance. 2005;4:405–409.

37. Viana SF , Guimarães VM, José IC, Oliveira MG, Costa NM,Barros EG, et al. Hydrolysis of oligosaccharides in soybean flour by soybean α-galactosidase. Food Chem. 2005;93(4):665–670.

38. Feurtado JA, Banik M, Bewley JD. The cloning and characterization of α-galactosidase present during and following germination of tomato (*Lycopersicon esculentum* Mill.) seed. J Exp Bot. 2001;52(359):1239–1249.

39. Carmi N, Zhang G, Petreikov M, Gao Z, Eyal Y, Granot D, et al. Cloning and functional expression of alkaline α-galactosidase from melon fruit: similarity to plant SIP proteins uncovers a novel family of plant glycosyl hydrolases. Plant J. 2003;33(1):97–106.

40. Zhu A, Goldstein J. Cloning and functional expression of a cDNA encoding coffee bean α-galactosidase. Gene. 1994;140(2):227–231.

11

Various scenarios of the cell pattern formation in *Arabidopsis* lateral root

Jolanta Białek[1], Izabela Potocka[2], Joanna Maria Szymanowska-Pułka[1]*

[1] Department of Biophysics and Plant Morphogenesis, University of Silesia in Katowice, Jagiellońska 28, 40-032 Katowice, Poland

[2] Laboratory of Cell Biology, University of Silesia in Katowice, Jagiellońska 28, 40-032 Katowice, Poland

Abstract

During lateral root (LR) development a coordinate sequence of cell divisions, accompanied by a change of the organ form takes place. Both the order of anatomical events and morphological features may vary for individual primordia. At early stages of LR primordia development oblique division walls are inserted in cells that are symmetrically located on both sides of the axis of the developing LR primordium, and thereby allow for the protrusion of the LR. We hypothesize that both oblique cell wall insertion and continuous changes in primordium form could be a consequence of a local change in stress distribution in the region of the LR initiation.

Keywords: lateral root formation; sequence of cell divisions; plant organ morphology

Introduction

Lateral roots (LRs) develop from the pericycle cells that locally acquire competence to form postembryonic meristems. Primordia are initiated in a left-right alternating pattern [1] and the process is auxin-regulated [2]. Eight stages of the LR formation in *Arabidopsis* have been identified [3], during which a sequence of cell divisions takes place. The initiation is morphologically detected when founder cells begin undergoing asymmetrical anticlinal divisions [4]. After the first periclinal divisions two cell layers are formed; an outer layer (OL) and an inner layer (IL). Further divisions lead to the formation of subsequent cell layers and to the increasing number of cells. A primordium expands by growing through the parent root tissues to emerge through the epidermis [3]. At this stage the geometry and cell patterning of the LR apex are fixed [5] and resemble those observed in the main root.

The proposed [3] scenario of events occurring during primordium development serves as a point of reference in studies on the LR formation. Based on these results we report and consider new morphological features that have hitherto not been described, with an emphasis on the occurrence of obliquely-oriented cell walls, irregularities in the cell division order, and primordia shapes. We show how a consideration of the reported features can provide new insights about LR formation and its possible relation to mechanical stress distribution.

Material and methods

Arabidopsis wild type Col-0 (WT) and transgenic lines DR5::GFP, DR5::GUS, pPIN1::PIN1:GFP, AUX1::YFP were used to analyze LR development. Seedlings were grown as described in [6]. For microscopic observations samples were used fresh or cleared in chloral hydrate. The roots were observed using phase contrast and Nomarski microscopy. Images of 1740 LR primordia of the mentioned accessions (Tab. 1) were analyzed in their axial planes. Identification of developmental stages of LR formation followed that of [3].

Results and discussion

A sequence of primordia at subsequent stages is shown in Fig. 1a–l. At the site of the LR initiation two founder cells are visible, separated by the wall perpendicular to the root axis, although in some primordia this wall is oblique (Fig. 1a). In such cases the wall orientation is preserved at further stages (Fig. 1c, Fig. 2a). In most primordia a stage of development could be identified and sequence of divisions followed after Malamy and Benfey [3], yet, in some cases a departure from this order was found. Very seldom (2 of the 1740 analyzed, one in DR5::GFP line and one in the AUX1::YFP line), a periclinal division of a single pericycle cell occurred (Fig. 1b), instead of a series of anticlinal divisions, typically preceding formation of the two-layer primordium. In this case a single cell might have become a founder cell. This observation is in accordance with the suggestion [7] that the number of pericycle cells capable of becoming founder cells is 1 to 3.

---

* Corresponding author. Email: jsp@us.edu.pl

Handling Editor: Beata Zagórska-Marek

Tab. 1 Quantitative comparison of morphological features observed in LR primordia in the analyzed lines.

	DR5::GFP	DR5::GUS	pPIN1::PIN1:GFP	AUX1::YFP	Col-0	All accessions
number of LRs	356	404	217	376	387	1740
relative number of LRs	20.5%	23.2%	12.5%	21.6%	22.2%	100.0%
changed division order	18 (5.1%)	29 (7.2%)	6 (2.8%)	12 (3.2%)	27 (7.0%)	number of cases (relative number of cases in the accession)
flattened apical part	38 (10.7%)	48 (11.9%)	14 (6.5%)	36 (9.6%)	40 (10.3%)	
asymmetrical shape	11 (3.1%)	21 (5.2%)	17 (7.8%)	36 (9.6%)	24 (6.2%)	

In later stages a change of division sequence was more frequent (5.29% of the 1740 analyzed cases) and it concerned mostly the insertion of anticlinal walls. In primordia at stages 2–4 anticlinal divisions preceding new layer formation occurred either in OL (Fig. 1d) or in IL (Fig. 1f,g,i). In some primordia a characteristic of stage 3 periclinal division took place in a cell of the OL neighboring an anticlinally divided cell of the IL (Fig. 1f). In other cases an anticlinally divided cell in the IL was found adjacent to a cell of the same layer that has undergone a regular periclinal division at stage 4 (Fig. 1i). Most of the changes were not exceptional and may be interpreted as a variation of the typical scenario. The early periclinal division of a single pericycle cell (Fig. 1b) is rather unique; more frequent are flip-flops at stages 2–4. The last may be associated with the overlapping of two subsequent stages and probably does not interfere with further development. Previous studies [8] showed that a program of cell divisions is not stereotypical and each individual primordium can pass through developmental stages with different numbers of cells.

From stage 2, walls perpendicular and parallel to the main root axis were inserted in the OL and IL (Fig. 1d–i). Orientation of these walls was adjusted to cell pattern of the main root. At stages 3 and 4 in cells located on the sides of the primordium axis, oblique walls were observed (Fig. 1g–i), which corresponded with a protrusion and emergence, and which led to formation of a dome-like shaped apex [6]. Derivatives of these cells are sharply pointed and situated close to the organ axis (Fig. 1l). In some cases a border between the cells derived from OL and IL is clearly visible (Fig. 1l) and a characteristic difference in the cell arrangement and cells' shapes occurs. The cells of OL are large and form layers aligned along the primordium outline, while the central cells of IL are elongated and form a column parallel to the organ axis. The cells of IL at the sides of the column are either arranged in accordance with the cells of OL or elongated and sharply tapered (Fig. 1l). Inclined cell walls have been reported in LR primordia of radish [9] in which oblique divisions at early stages led to formation of cuneiform cells, and in *Arabidopsis* [6], in which such walls were interpreted as a manifestation of a change of the principal directions of growth (PDG) [10] within the de novo formed meristem.

Under external mechanical stress plant cells divide in one of the principal directions of stress (PDS) [11]. As division walls are usually inserted along PDG [6], PDS and PDG may coincide [12]. Thus, the oblique divisions in a primordium may indicate a local reorientation of PDG resulting from redistribution of stress, as suggested by Lintilhac [13]. Such oblique walls are added no sooner than at stage 3 (Fig. 1g–i), which corresponds with the hypothesis [6] of a switch from a field of growth of the main root to the separate field of LR, and with the observation, that the isolated primordium is not capable of growth until it has developed 3 to 5 layers of cells [14].

On the sides of LR primordia strongly marked anticlinal cell walls are visible. First appearing in young primordia (Fig. 1e), they remain evident at later stages (Fig. 1h,i,k,l). Fates of cells in these side regions are similar either in observation [3] or in modeling [5,6]: they are formed at early stages and soon thereafter they stop dividing. The external walls of these cells seem to constitute LR boundaries separating the LR from the parent root. Such a separation may be related to the specific distribution of stress in the site of the LR formation. Most primordia show rather regular geometry; they are symmetrical in reference to their axes and they assume dome-like forms, yet, a number of deformed primordia also occur (see Tab. 1). In 10.11% of all analyzed primordia flattened apical portions with bent vascular bundles of the main root (Fig. 1j) were observed, whereas others manifested asymmetrical shapes (6.26% of the 1740 analyzed cases, Fig. 1k) or lack a regular shape at the tip (single cases, Fig. 1l). In LRs whose apical parts have emerged outside the parent root surface, such changes are hardly ever observed. This suggests that a changed geometry of primordia may be the result of mechanical stress from the surrounding tissues [15].

In few primordia at comparable stages, we observed significant differences in their shapes and dimensions. In Fig. 2a–c primordia consisting of two cell layers (stage 2) are shown. Two primordia (Fig. 2a,b) are rather flat, while in the third (Fig. 2c) a clear protrusion has appeared. Also, sizes and number of cells involved with LR formation changed from 4 cells (Fig. 2a) to about 14 cells (Fig. 2b,c). Moreover, the cells of the primordium in Fig. 2a are large in comparison to the two remaining cases. The variation in numbers of cells may be related to the varying number of the founder cells [7] and to the perturbed sequence of cell divisions.

In few cases (12 of the 1740 analyzed, single cases in particular accessions) atypical positioning of the LR primordia was observed. In Fig. 2d,e examples are shown in which two neighboring primordia are formed very close to each other. In some cases they appear on the same side

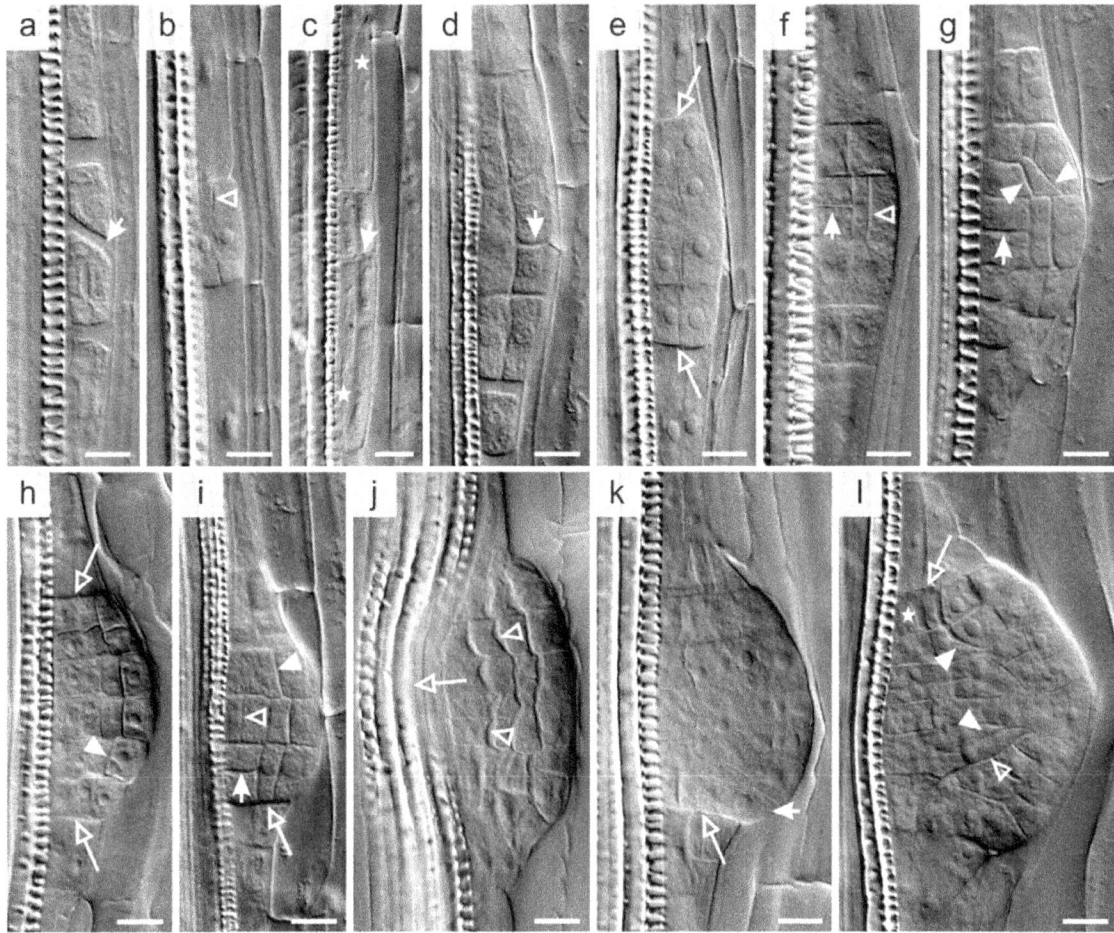

Fig. 1 LR primordia at consecutive stages: **a** – stage 1; **b** – stage 1/2; **c–e** – stage 2; **f,g** – stage 3; **h,i** – stage 4; **j–l** – presumably stages 5, 6, 7, respectively. **a** Oblique orientation of the central anticlinal wall (short arrow), preserved at stage 2 (**c**; short arrow). **c** Curved cell walls at the flanks (asterisks). A changed order of divisions: **b** Early periclinal division wall in pericycle (open arrowhead). Anticlinal division wall (short arrow) in OL (**d**) and in IL (**f,g,i**). **f** First periclinal division in OL (open arrowhead) characteristic of stage 3. **i** First periclinal division in IL (open arrowhead) characteristic of stage 4. **e,h,i,k,l** Strongly marked borders between the main root and primordium (arrows). **g–i** Oblique periclinal division walls (solid arrowheads) on the sides leading to sharply pointed cells (**l**; solid arrowheads) formation. **j** Local bending of vascular bands (arrow), curved periclinal walls (open arrowheads). A disturbed symmetry in the primordium shape: flattened apical part (**j**), stronger growth of the lower part (**k**; short arrow), conically shaped primordium with regular cell arrangement at the sides (**l**; asterisk), border between OL and IL (open arrowhead). Nomarski optics. Scale bars: 10 μm. **a,d,g,k** WT; **b,e,h,j** DR5::GFP; **c** AUX1::YFP; **f,i,l** DR5::GUS.

of the main root axis (Fig. 2d), while in other cases on the opposite sides (Fig. 2e). Typically, subsequent primordia are formed alternately on the left and right side [1] and they are distributed at more or less regular distances from each other, which is correlated with fluctuations in auxin distribution along the parent root [2]. However, LR density along the parent root was shown to be characteristic for individual accessions of Arabidopsis [1]. Root branching may be also controlled by mechanisms of lateral inhibition, caused by a competition between initiation and development for auxin [16]. As new LR initiation may be induced on the outside of the curve of the bent root [17], positioning may be also regulated by external mechanical signals.

In Tab. 1 a quantitative summary of the morphological features observed in LR primordia of all analyzed lines is presented. Slight differences in the relative number of cases between particular accessions may result from the fact that primordia were observed with the use of various microscopy techniques. For example, the roots of the pPIN1::PIN1:GFP line were mostly observed using phase contrast, which not allways allowed for clear identification of the cell pattern. That is why in this line there were only few primordia, in which a changed division order was observed in comparison with other lines. A sequence of events during the LR formation in Arabidopsis was described in [3]. Here, we have indicated and assessed additional features observed in the

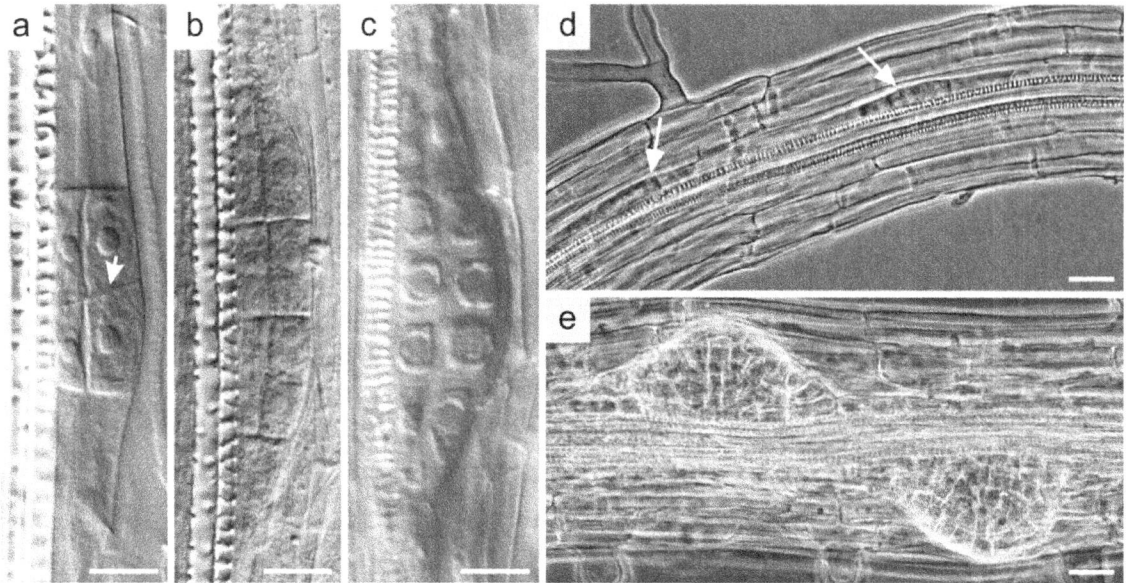

Fig. 2 Morphological features of the primordia. **a** Small primordium with few cells; oblique central wall in two layers (short arrow). **b** Large and flat primorium. **c** Small primordium forming a slight protrusion. Two closely formed LR primordia (arrows) on one (**d**) or opposite (**e**) sides of the root axis. **a–c** Nomarski optics; **d,e** phase contrast. Scale bars: **a–c** 10 μm; **d,e** 20 μm. **a,b,d** DR5::GFP; **c** DR5::GUS; **e** pPIN1::PIN1:GFP.

---

cell pattern of either WT or transgenic lines with comparable frequency. Some of these features are likely correlated with the stress distribution within the region of the LR initiation. In order to verify this hypothesis, further studies are required to investigate the effects of mechanical factors on morphology of *Arabidopsis* roots. The challenge for the future will also be to study the dynamic character of the mechanical stress distribution in the site of the LR formation.

Acknowledgments
We thank Prof. Lewis Feldman from UC Berkeley, USA for his helpful comments on the manuscript and revision of the English version. The research was supported in part by grant from the Polish Ministry of Science and Higher Education (grant No. N N303 333936).

Authors' contribution
The following declarations about authors' contributions to the research have been made: designed research: JSP, IP; performed experiments: JB, JSP; wrote the manuscript: JSP, IP.

References
1. Dubrovsky JG, Gambetta GA, Hernández-Barrera A, Shishkova S, González I. Lateral root initiation in *Arabidopsis*: developmental window, spatial patterning, density and predictability. Ann Bot. 2006;97(5):903–915.

2. de Smet I, Tetsumura T, de Rybel B, Frey NFD, Laplaze L, Casimiro I, et al. Auxin-dependent regulation of lateral root positioning in the basal meristem of *Arabidopsis*. Development. 2007;134(4):681–690.

3. Malamy JE, Benfey PN. Organization and cell differentiation in lateral roots of *Arabidopsis thaliana*. Development. 1997;124:33–44.

4. Casimiro I, Beeckman T, Graham N, Bhalerao R, Zhang H, Casero P, et al. Dissecting *Arabidopsis* lateral root development. Trends Plant Sci. 2003;8(4):165–171.

5. Szymanowska-Pułka J, Nakielski J. The tensor-based model for growth and cell divisions of the root apex. II. Lateral root formation. Planta. 2010;232(5):1207–1218.

6. Szymanowska-Pułka J, Potocka I, Karczewski J, Jiang K, Nakielski J, Feldman LJ. Principal growth directions in development of the lateral root in *Arabidopsis thaliana*. Ann Bot. 2012;110(2):491–501.

7. Dubrovsky JG, Rost TL, Colón-Carmona A, Doerner P. Early primordium morphogenesis during lateral root initiation in *Arabidopsis thaliana*. Planta. 2001;214(1):30–36.

8. Lucas M, Kenobi K, von Wangenheim D, Voss U, Swarup K, de Smet I, et al. Lateral root morphogenesis is dependent on the mechanical properties of the overlaying tissues. Proc Natl Acad Sci USA. 2013;110(13):5229–5234.

9. Blakely LM, Durham M, Evans TA, Blakely RM. Experimental studies on lateral root formation in radish seedling roots. I. General methods, developmental stages, and spontaneous formation of laterals. Bot Gaz. 1982;143(3):341–352.

10. Hejnowicz Z, Romberger JA. Growth tensor of plant organs. J Theor Biol. 1984;110(1):93–114.

11. Lynch TM, Lintilhac PM. Mechanical signals in plant development: a new method for single cell studies. Dev Biol. 1997;181(2):246–256.

12. Nakielski J, Hejnowicz Z. The description of growth of plant organs: a continuous approach based on the growth tensor. In: Nation J, Trofimova I, Rand JD, Sulis W, editors. Formal descriptions of developing systems. Dordrecht: Springer Netherlands; 2003. p. 119–136.

13. Lintilhac PM. The problem of morphogenesis: unscripted biophysical

control systems in plants. Protoplasma. 2014;251(1):25–36.

14. Laskowski MJ, Williams ME, Nusbaum HC, Sussex IM. Formation of lateral root meristems is a two-stage process. Development. 1995;121(10):3303–3310.

15. Szymanowska-Pułka J. Form matters: morphological aspects of lateral root development. Ann Bot. 2013;112(9):1643–1654.

16. Lucas M, Guédon Y, Jay-Allemand C, Godin C, Laplaze L. An auxin transport-based model of root branching in *Arabidopsis thaliana* PLoS ONE. 2008;3(11):e3673.

17. Ditengou FA, Teale WD, Kochersperger P, Flittner KA, Kneuper I, van der Graaff E, et al. Mechanical induction of lateral root initiation in *Arabidopsis thaliana*. Proc Natl Acad Sci USA. 2008;105(48):18818–18823.

Oxygenic photosynthesis: translation to solar fuel technologies

Julian David Janna Olmos[1], Joanna Kargul[2]*

[1] Faculty of Biology, University of Warsaw, Miecznikowa 1, 02-096 Warsaw, Poland

[2] Centre of New Technologies, University of Warsaw, Banacha 2c, 02-097 Warsaw, Poland (it is the present address of the first author)

Abstract

Mitigation of man-made climate change, rapid depletion of readily available fossil fuel reserves and facing the growing energy demand that faces mankind in the near future drive the rapid development of economically viable, renewable energy production technologies. It is very likely that greenhouse gas emissions will lead to the significant climate change over the next fifty years. World energy consumption has doubled over the last twenty-five years, and is expected to double again in the next quarter of the 21st century. Our biosphere is at the verge of a severe energy crisis that can no longer be overlooked. Solar radiation represents the most abundant source of clean, renewable energy that is readily available for conversion to solar fuels. Developing clean technologies that utilize practically inexhaustible solar energy that reaches our planet and convert it into the high energy density solar fuels provides an attractive solution to resolving the global energy crisis that mankind faces in the not too distant future. Nature's oxygenic photosynthesis is the most fundamental process that has sustained life on Earth for more than 3.5 billion years through conversion of solar energy into energy of chemical bonds captured in biomass, food and fossil fuels. It is this process that has led to evolution of various forms of life as we know them today. Recent advances in imitating the natural process of photosynthesis by developing biohybrid and synthetic "artificial leaves" capable of solar energy conversion into clean fuels and other high value products, as well as advances in the mechanistic and structural aspects of the natural solar energy converters, photosystem I and photosystem II, allow to address the main challenges: how to maximize solar-to-fuel conversion efficiency, and most importantly: how to store the energy efficiently and use it without significant losses. Last but not least, the question of how to make the process of solar energy conversion into fuel not only efficient but also cost effective, therefore attractive to the consumer, should be properly addressed.

Keywords: solar fuels; photosystem I; photosystem II; CO_2; photosynthesis; artificial leaf; artificial photosynthesis; solar-to-fuel nanodevices

Introduction

Finding renewable sources of energy is of paramount importance. It has been estimated that all the fossil fuel reserves are likely to be irreversibly exhausted by 2112 [1,2]. Coal reserves are estimated to be readily available up to 2112, and are likely to be the only fossil fuel remaining after 2042 [1]. Total global energy demand could rise by up to 80% by mid-21st century from its level in 2000 [3]. Sunlight, on the other hand, is by all means an infinite, inexhaustible and gargantuan source of energy. The power of 1 hour of solar energy reaching our planet is equivalent to the current annual global energy consumption of approximately 16.3 TW [4]. Crops, grasses, trees, cyanobacteria and algae capture solar energy via their photosynthetic machinery and store it in a wide range of feedstock in the form of chemical bonds (i.e., sugars, lipids and starch). The earliest form of photosynthesis is estimated to have evolved circa 3.5 billion years ago, occurring in an environment of low atmospheric oxygen and in the presence of high levels of reducing molecules including methane, hydrogen gas, hydrogen sulfide and other organic compounds which were all employed as the reducing agents necessary for anoxygenic photosynthesis [5]. As the reducing atmosphere was depleted over time, the need for using water as the source of reducing equivalents placed a mandatory selection pressure leading to the evolution of oxygenic photosynthesis where oxidation occurred by extracting hydrogen from water. The evolution of the water splitting reaction gave rise to oxygenic photosynthesis, whereby oxygen is produced as a by-product of water photo-oxidation. The first organisms to engage in such biologically revolutionary process were (proto)cyanobacteria, which are thought to have evolved from primitive anoxic photosynthetic bacteria [6].

* Corresponding author. Email: j.kargul@uw.edu.pl

Handling Editor: Beata Zagórska-Marek

Evolution of photosynthesis

It has been well established that photosynthetic organisms have transformed life on earth by their inherent ability to perform solar energy conversion. The earliest phototrophs were almost certainly non-oxygenic, and it is now known that photosynthesis evolved early in geological history of the Earth, in the early Archean eon circa 3.5 billion years ago. The enzymatic and energetic fundamentals of life were revolutionized approximately 2.4 billion years ago by the Great Oxidation Event, i.e., the rise of atmospheric oxygen due to oxygenic photosynthesis. The onset of oxygenic photosynthesis is often referred to as the "Big Bang of evolution" [7] as it drove evolutionary events in Earth's history on an unprecedented scale including appearance of aerobic respiration and colonization of terrestrial ecosystems by multicellular organisms due to the formation of the protective ozone layer. In parallel, development of the new photosynthetic pathways occurred through the modification of existing and alternative ways of carbon fixation. Phototrophs evolved further by means of lateral gene transfer leading to formation of the basic photosynthetic components, namely the photosynthetic reaction centers (RCs) as well as by some endosymbiotic events. Various types of evidence obtained from biochemical, physiological and biophysical studies of photosynthetic complexes together with comparative genomics data show a very rich complexity of photosynthetic features. At the same time, discovery of evolutionary similarity of structural features of the main photosynthetic components has allowed to formulate some hypotheses on the origin and evolution of photosynthesis, as overviewed below.

Molecular evolution occurs through DNA alterations (typically mutations) that create new proteins that offer alternative metabolic opportunities [8]. The building blocks for molecular innovation may be provided by an organism, gene or genome duplication [9,10]. It is thought that gene duplication, modification and divergence gave birth to the evolution of photosynthetic RCs (see [5] and references therein). It has been postulated that a single gene, which coded for a homodimeric protein, was duplicated subsequently to produce heterodimeric RCs. The mechanisms behind the evolution of the fused RC core and antennae in some species of higher plants and algae could be explained by gene fusion and splitting. On the other hand, the transfer of metabolic capacities between different organisms most likely occurred by lateral gene transfer [11] and could also explain the presently observed plethora of photosynthetic light-harvesting complexes (LHCs) associated with RCs that evolved in different families of phototrophs. As a matter of fact, certain genomes include gene clusters which themselves contain RC-coding genes together with certain photosynthetic pigments biosynthesis pathways [12], which most certainly provided the genetic platform for lateral transfer of photosynthetic traits between species.

Numerous structural, spectroscopic, thermodynamic, molecular sequence and genetic studies provided molecular evidence supporting the hypothesis that all RCs have evolved from a common ancestor and have been optimized so as to maximize the efficiency of solar energy conversion [5]. In all cases of various types of RCs, the fundamental mechanistic

principle is to achieve energy storage by ultrafast separation of the primary electron donors and acceptors during the primary charge transfer, thus avoid wasteful recombination reactions. This is achieved by optimal arrangement of the electron transport cofactors within the protein "smart matrix" of the RCs. However, it should be pointed out that the reversibility of charge separation (i.e. fast charge recombination processes) has also been demonstrated for various types of photosynthetic RCs (e.g., for photosystem I and photosystem II, PSI and PSII) as the regulatory mechanism for effective excitation quenching by closed (oxidized) RCs. The extent of this reversibility depends on the amount of free energy associated with the charge-separated state, which can be modulated by various modes of excitation dynamics that exist within the RCs and the associated antennae.

X-ray crystallographic analyses of photosynthetic RCs demonstrated that a three-dimensional protein and cofactor structure of RCs is remarkably conserved, despite only minimal amino acid sequence identity [5,13]. Two types of photosynthetic RCs can be distinguished depending on the type of the primary electron acceptor. These are type I and type II RCs which employ either (bacterio)chlorophyll (Chl) or (bacterio)pheophytin as the primary acceptor, respectively. Type I RCs are present in green sulfur bacteria and heliobacteria, while purple photosynthetic bacteria and green filamentous, non-sulfur bacteria have all type II RCs. In cyanobacteria, algae and higher plants one RC of each type is present, i.e., PSI (type I RC) and PSII (type II RC). These work in tandem to drive oxygenic photosynthesis using water as the primary source of reducing equivalents for subsequent reduction of CO_2 into sugars.

The common design scheme in all photosynthetic RCs is the charge-separation cofactor assembly that is bound within a protein dimer composed of 5-trans-membrane (TM) domain subunits. Each half of this common dimeric RC unit contains a nearly identical set of cofactors. The arrangements of these cofactors exhibit minimal variation in all the available crystal structures of RCs [5]. X-ray crystallographic studies revealed that all RCs are characterized by a pseudo-2 fold symmetry axis that relates both the cofactors and their protein scaffold. In cyanobacteria and eukaryotic phototrophs, the D1/D2 heterodimer of PSII RC, is flanked by two closely related 6-TM inner antenna complexes, CP43 and CP47 that both harbor additional core antenna pigments. These 6-TM antenna subunits are evolutionarily related to the N-terminal part of the type I RC, which forms the core antenna [14]. In contrast to the cyanobacterial type II RC of PSII in which 5 TMs and 6 TMs are separate proteins, all known type I RCs are 11-TM dimers, which consist of the C-terminal 5-TM RC complex and the N-terminal 6-TM core antenna, all of which belong to a single protein. In green sulfur bacteria and heliobacteria, type I RCs are homodimeric, whereas cyanobacterial, algal and higher plant type I RC represented by PSI is heterodimeric, not only structurally but also functionally, as the two electron transport branches that exist within the RC PsaA/PsaB heterodimer also exhibit different kinetic characteristics [5].

Primary photosynthetic events: unparalleled efficiency

As discussed above, oxygenic photosynthesis represents the major process of solar energy conversion on Earth today. Most significantly, this is without a doubt the most important process for the input of organic energy into the biosphere: thermal, radioactive and chemical inputs being by comparison negligible. Unfortunately, photosynthesis as an overall process is rather inefficient. Massive losses in light absorption, transmission, reflection, and scattering, as well as losses of absorbed photon energy as heat limit the energy storage in biomass to a mere 5–7% (as an upper limit) for plants and algae, with an average of 0.2% [15], as shown in Fig. 1.

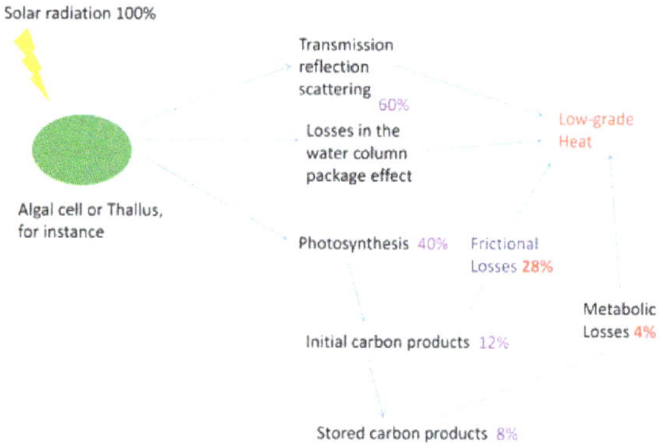

Fig. 1 Energy losses and dissipation in photosynthesis. Much of the incident energy is lost as low-grade heat either directly or indirectly. Only a small amount of energy is stored as chemical free energy in new organic material. Adapted from Larkum [95].

Taking most factors into consideration, including the light absorption processes, trans-membrane electron/proton transport and any losses due to radiative and non-radiative dissipation the overall energy conversion efficiency decreases from the primary photosynthetic light capture and charge separation events (up to 95%) to mere 30% for the process of the Calvin–Benson cycle [16]. In terms of efficiency of solar energy conversion, the rates of the even best crop plants are less than 1% [17]. It is then unsurprising that scientists have invested considerable time and resources to study the structure and function of the primary solar energy converters, PSII and PSI which both exhibit remarkably high power conversion efficiencies. The first molecular steps of photosynthesis, the so-called light reactions, work at an impressive quantum yield close to unity, and no man-made energy-converting device can reach such a remarkable efficiency. At low-to-medium light levels PSII and PSI operate at a quantum efficiency of above 95% [18]. Quantum yield of PSI is particularly impressive at unity, making it an almost perfect Einstein photoelectric device. It is then obvious that in order to maximize the solar energy conversion efficiency for production of solar fuels we must short-circuit the photosynthetic energy conversion processes and use directly the reducing equivalents generated through the primary

photosynthetic events, without proceeding to generation of biomass where the most significant energy losses occur (see Fig. 2).

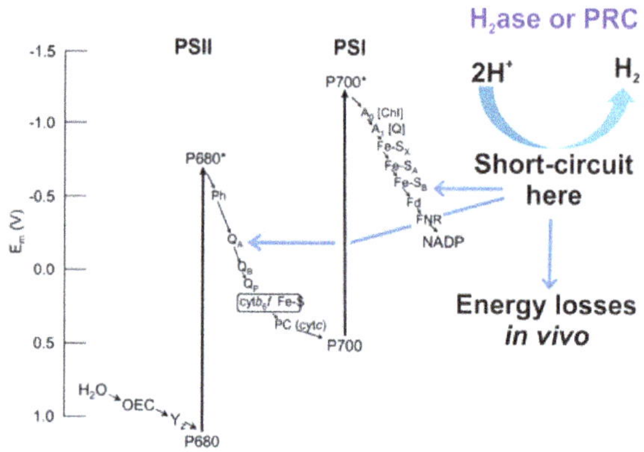

Fig. 2 Short-circuiting the light reactions to maximize power conversion efficiency and minimize energy losses. The dark reactions of photosynthesis are energetically expensive, leading to the loss of energy conversion efficiency. Short-circuiting of primary photosynthetic event for proton reduction and photocatalysis may considerably enhance solar-to-fuel conversion efficiency. PSI and PSII can be short-circuited at the terminal F_B cluster ([18] and references therein) or the Q_A site ([96] and references therein), respectively.

PSI and PSII capture quanta of solar light which are transferred to specialized pairs of chlorophyll (Chl) a molecules forming the photochemical RCs, P700 and P680, respectively. Upon capture of solar light PSII and PSI work in tandem to generate vectorial electron flow (with water molecules as the primary source of electrons and protons), as well as forming the trans-membrane proton gradient across the thylakoid membrane through the action of the third macromolecular protein complex also embedded in the thylakoid membranes between PSII and PSI, cytochrome b_6f. The electromotive force of proton gradient powers the thylakoid-bound ATP synthase to produce ATP and enables reduction of NADP, being the final recipient of photogenerated electrons (see Fig. 3).

In a different route of electron flow, directed mainly around PSI, electrons may be reused from either NADPH to plastoquinone or reduced ferredoxin, and subsequently to the cytochrome b_6f complex [19–21]. A pH gradient is generated by such cyclic electron flow, leading to formation of exclusively ATP without accumulation of the reduced NADP (NADPH). The precise physiological role of cyclic electron flow around PSI is still somewhat controversial, but it is known that in higher plants this type of photosynthetic electron flow consists of two partially redundant pathways, i.e., NADPH dehydrogenase (NDH)- and PGRL1/PGR5-dependent pathways, both of which are fuelled with electrons from reduced ferredoxin [22–24]. Recently, it has been shown that cyclic electron flow is important for regulation of the photosynthetic yield, under various stress and metabolic conditions (e.g., [25,26]).

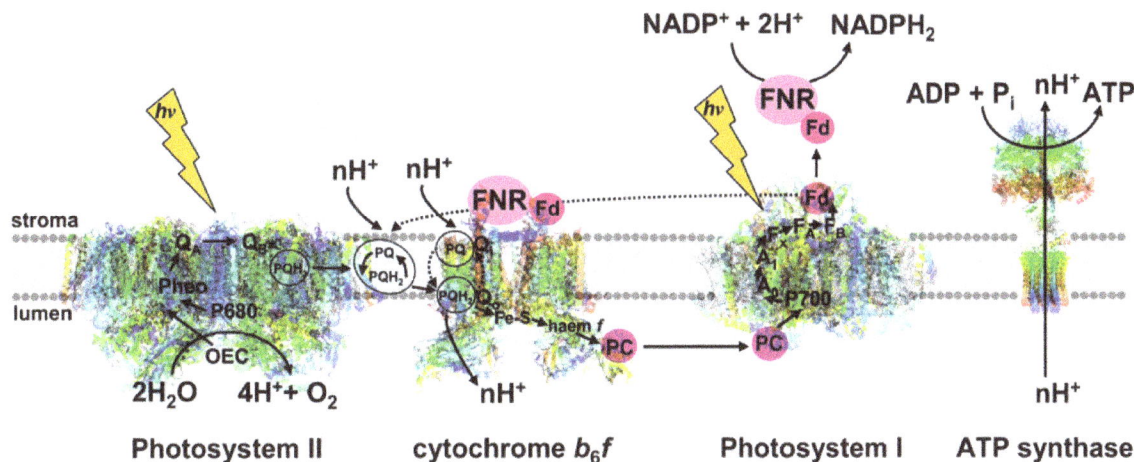

Fig. 3 Structural representation of the linear photosynthetic electron transfer path in higher plants. X-ray crystal structures of major protein complexes are highlighted, photosystem II (3ARC), cytochrome $b_6 f$ (1VF5), and photosystem I (1JB0), and their arrangement in the thylakoid membrane in order to conduct a linear photosynthetic electron transport and generate reducing equivalents and NADPH. The proton gradient generated powers ATP production by proton transport via the ATP synthase. Both ATP and NADPH are employed for fixation and reduction of CO_2 into carbohydrates. The OEC (oxygen evolving complex) is indicated by an arrow, and is comprised of a $Mn_4 CaO_5$ cluster (see body of the text) which oxidizes substrate water molecules. Pheo – pheophytin: a primary electron acceptor; Q_A, Q_B – primary and secondary quinone electron acceptors; PQ/PQH_2 – plastoquinone/plastoquinol: oxidized and reduced plastoquinone; PC – plastocyanin; A_0 – Chla; A_1 – phylloquinione; F_x, F_A, F_B – [4Fe-4S] clusters; Fd – ferredoxin; FNR – ferredoxin: NADP reductase. Under certain metabolic ATP-depleted conditions, cyclic electron flow is induced (dotted arrow) which generates ATP without production of NADPH. Adapted from Kargul and Barber [5].

Water splitting at the oxygen-evolving center of photosystem II

PSII is the molecular nanomachine responsible for generation of oxygenic life on Earth through its catalytic activity of water photooxidation. Whilst most other photosynthetic type II RCs use small organic molecules as electron donors, the presence of the oxygen-evolving center (OEC) in close proximity to the P680 photochemical RC ingeniously permits for the use of H_2O as the electron donor [27,28]. The $Mn_4 CaO_5$ cluster of the OEC contains a $Mn_3 CaO_4$ distorted cubane structure, in which with three Mn atoms and one Ca cation occupy four corners, whilst the four O atoms occupy the other corners (see Fig. 4, inset) [29]. The fourth Mn ion (Mn4 or the "dangler" Mn) is covalently linked to the cubane by a di-μ-oxo bridge involving the O5 and O4 atoms of the cubane [29]. The metals within the cluster are surrounded by a number of amino acid residues, of which seven (mostly originated from the D1 subunit) provide the direct ligands to the metals through their carboxylate groups in a monodentate or bidentate mode [29]. Importantly, the latest near-atomic (1.9 Å) X-ray crystal structure of PSII purified from a cyanobacterium *Thermosynechococcus vulcanus* provides the first direct evidence on the location of the primary water substrate molecules which are believed to be coordinated to both the Ca ion and the dangler Mn [29]. The detailed mechanism of water oxidation, especially the formation of molecular oxygen, remains hotly debated, but high-valent Mn species are almost certainly involved in the activation of oxo-ligands to form O-O bond during the water oxidation cycle [28,30].

The process of water splitting requires overall two water molecules and yields four electrons, four protons, and one molecule of O_2. The process, which requires capture of four subsequent photons in the P680 RC, involves the charge separated states and the consequent accumulation of 4 positive holes on the OEC which are then reduced with 4 electrons derived from 2 substrate water molecules. PSII has been estimated to conduct the water splitting reaction at a turnover frequency (TOF) of 300 s^{-1} in vivo [31]. The overall structure of the OEC catalytic site and the mechanistic knowledge that has been superbly advanced by the latest achievements of PSII crystallography greatly inspires a quest for the development of synthetic catalysts for mimicking the energetically demanding water splitting reaction in artificial photoelectrochemical systems to generate solar fuels (as discussed below) at a minimal overpotential, as does PSII.

Reducing power of photosystem I

Upon absorption of red photons by P700 RC of PSI (E_m +0.43 eV) composed of 2 Chla molecules, the highly reducing excited state P700* is formed (E_m −1.3 eV), which then rapidly undergoes oxidization to P700+ followed by electron transfer by the conserved redox cofactors system involving 4 additional Chla molecules, 2 phylloquinones and 3 iron-sulfur [4Fe-4S] clusters. The final electron acceptor, F_B is formed by the iron-sulfur cluster coordinated by the Cys residues of PsaC subunit on the acceptor side of PSI. Reduction of a positive hole P700+ on the donor side of PSI occurs through the specific interaction with the mobile electron carriers, plastocyanin or cytochrome c_6. The potential difference between the primary electron donor P700* and the terminal electron acceptor F_B^- is about 1 V, corresponding to a thermodynamic efficiency of nearly 60% [32].

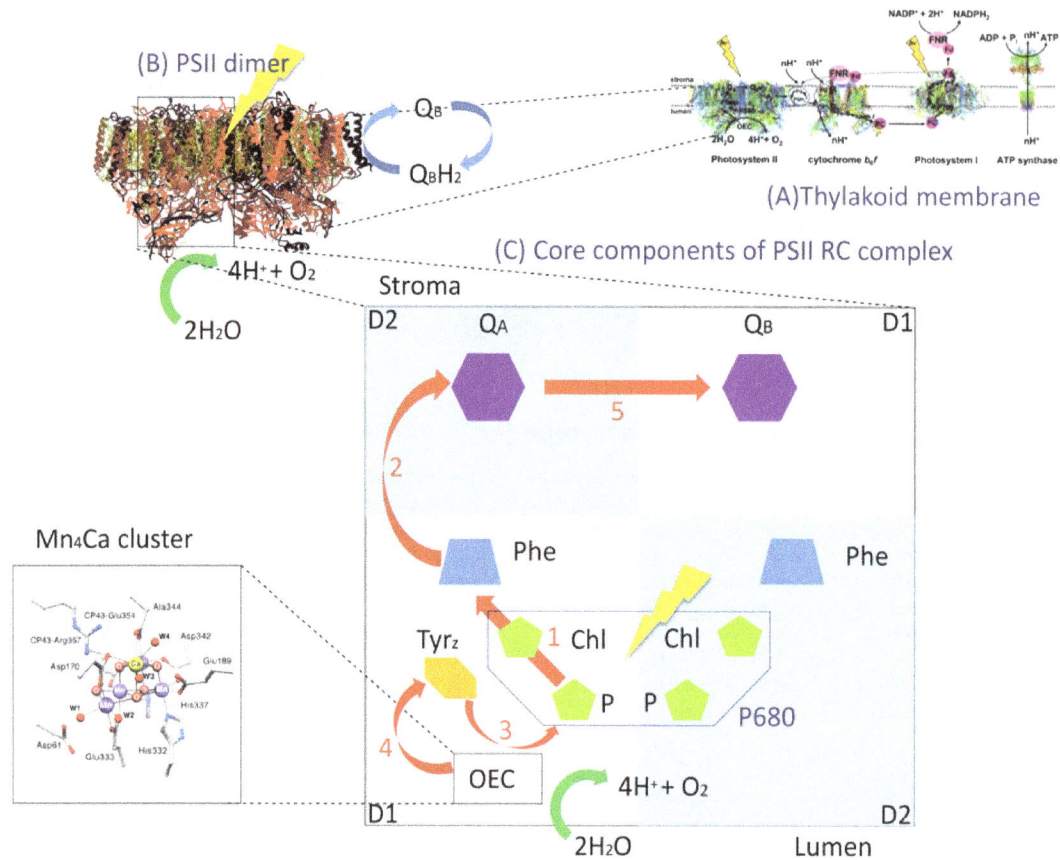

Fig. 4 From structure to function of photosystem II. **a** Structural representation of the linear photosynthetic electron transfer path in the thylakoid membrane. **b** 1.9 Å crystal structure of PSII dimer from *Thermosynechococcus vulcanus* (PDB: 3ARC) [29]. Water splitting at the OEC results in the release of H^+, e^- and O_2, concomitant with the double reduction of the terminal quinone, Q_B, in the Q_B pocket to liberate doubly reduced PQ (PQH_2, plastoquinol) into the thylakoid space. **c** Diagrammatic representation of the electron transfer cofactor arrangement within the RC D1/D2 dimer. Red arrows indicate the direction of the forward electron transfer. Step 1: photoexcitation leads to formation of the charge separated state $P680^+Phe_{D1}^-$. Step 2: transfer of electrons from Phe_{D1}^- to Q_A. Step 3: the positive hole of $P680^+$ is filled by an electron from Tyr_z. Step 4: hole migration from Tyr_z^+ to the Mn_4CaO_5 cluster. Step 5: electron transfer from Q_A^- to the terminal acceptor Q_B. Insert: the latest structure of Mn_4CaO_5 cluster together with coordinating ligands at 1.9 Å [29].

Despite structural complexity (12–17 protein subunits and 127–178 cofactors for the cyanobacterial and higher plant PSI [33–35], PSI operates with a quantum yield of unity and so far, no artificial system gained such a remarkable efficiency. As a consequence, this robust and highly abundant photoactive protein complex constitutes the favorite biological component for construction of the biohybrid photovoltaic (PV) and photoelectrochemical (PEC) devices ([18,36] and references therein). Due to an exceptionally long-lived charge-separated state $P700^+F_B^-$ (~60 ms) and exceptionally low redox potential of the F_B cluster (E_m −0.58 eV), PSI in principle generates sufficient driving force to reduce protons to H_2 at neutral pH [18,37]. As PSI is capable of being both an acceptor ($P700^+$) and a donor (F_B^-) of electrons, this complex has been used in semi-artificial assemblies for generation of both anodic and cathodic photocurrents, as well as solar fuels [18,36,37].

Photosynthesis and reduction of carbon footprint

Microalgae-based technologies to reduce carbon footprint

The term "carbon footprint" has evolved as an important measure of greenhouse gas intensity for diverse activities and products. The carbon footprint is perhaps best expressed as the amount of greenhouse gasses (primarily CO_2) that is produced per capita [38]. The most advanced industrial societies can have enormous values, such as 19–22 tons per capita for the USA, and even more elevated for a few small Arab and European nations, and can be as low to 0.1 or even less for non-industrial nations (e.g., Ethiopia and Somalia). In the light of proven anthropogenic climate change it is important to consider lowering the contribution of various sources towards carbon footprint per capita, particularly for the modern economies which are predominantly powered by burning fossil fuels. This fact places our biosphere in a

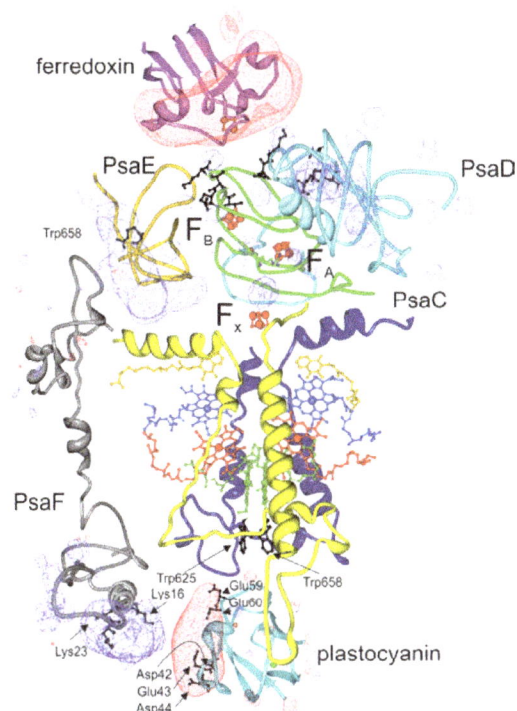

Fig. 5 Electron transport chain in PSI. Depicted are 2 branches of electron transfer cofactors including P700 primary electron donor Chls (green), accessory Chls (red), primary electron acceptor Chls (blue), secondary electron acceptor phylloquinones (yellow) and 3 [4Fe-4S] clusters, F_X, F_A and F_B (red balls). Interaction domains between PSI and plastocyanin and ferredoxin including critical amino acid residues are also shown. Positively and negatively charged regions are colored in as blue and red mesh, respectively. Coordinates are: 3LW5 [35] for PSI, 2PLT for plastocyanin [97] and 3AV8 for ferredoxin [98]. Reproduced with permission from Kargul et al. [18].

precarious situation: not only are our fuel reserves being irreversibly depleted but also man-made climate changes have been altering our biosphere on the scale that is only beginning to be fully comprehended. For these reasons, CO_2 capture and utilization, a concept of turning a greenhouse gas into a useful feedstock, has gained much attention in recent years [39].

Microalgae-based CO_2 photosynthetic fixation is regarded as a potential way to not only reduce CO_2 emission but also achieve energy utilization within microalgal biomass (Fig. 6). However, in this approach culture processing of microalgae plays an important role as it is directly related to the mechanism of microalgal CO_2 fixation and characteristics of microalgal biomass production [40]. According to the capture mechanism, post-combustion CO_2 capture methods can be roughly categorized as chemical absorption, physicochemical adsorption, membrane filtration, cryogenics, chemical looping combustion and biotechnology (e.g., with the use of hydroponic algae or terrestrial vegetations). From a technical point of view all of these methods are feasible in spite of the differences in CO_2 capture efficiency and capture capability of various microalgae.

An alternative feedstock to eukaryotic microalgae is cyanobacteria. The simpler structure of cyanobacteria allows them to be more efficient at carbon fixation from the atmosphere than eukaryotic algae [41]. However, they yield a much smaller photosynthetic conversion efficiency compared to algae [42]. Regardless, cyanobacteria are still being investigated for their inorganic carbon fixation capabilities as they possess a much simpler genetic make-up which can be easily engineered for an enhanced biomass yield, RuBisCO's increased affinity for CO_2 and production of high value products (HVPs) [43–45]. Moreover, several model cyanobacterial species have their genomes fully sequenced. The latter, combined with the fact that genetic manipulation of cyanobacteria is well established allowed for application of high throughput systems biology approaches to engineer cyanobacterial strains producing HVPs [41–46].

Soybean, sesame seeds, wheat, sunflower, switchgrass, rapeseed and peanuts comprise the most common feedstocks used in the first and second generations of biofuels. Moreover, different liquid fuels such as vegetable oil and alcohols (ethanol, propanol and butanol) are generated from these sources. Unfortunately, the competition with arable land for food production is the major limitation of these energy crops. To overcome this limitation, the so-called "third generation" feedstock is postulated to use non-food biomass sources for energy generation [46]. Cyanobacteria are the most promising candidates for the production of such HVPs including biofuels for the following reasons: (*i*) they contain a significant amount of lipids, present mostly in the thylakoid membranes, (*ii*) their photosynthetic and growth rates are higher than those of algae and higher plants, (*iii*) they grow easily and with addition of minimum amount of nutrients and water, employing light as the only energy source for many strains, (*iv*) cultivation is simple and inexpensive.

High value products as a route for carbon footprint reduction

BIOHYDROGEN. Hydrogen gas combustion produces only water as a side product. Hydrogen combustion has an extremely low enthalpy of combustion, at −286 kJ/mol. Moreover, combustion of hydrogen gas releases 572 kJ of energy. This is quite impressive, thermodynamically speaking. Hydrogen, by far, has the highest energetic value of all HVPs [47]. The reversible activity of hydrogenases allows many strains of cyanobacteria to produce hydrogen, often termed the "green fuel". H_2 may be formed as a by-product of N_2 fixation by nitrogenases when cyanobacteria are exposed to N_2 starvation [47]. Additionally, non-heterocystous cyanobacteria seem to be less efficient in H_2 production than heterocystous organisms. Several excellent reviews have reported cyanobacterial species capable of evolving H_2 gas due to the activity of chloroplast reversible hydrogenase that is directly coupled to the linear photosynthetic electron flow [47–49]. One of the problems of employing cyanobacteria (and some hydrogen producing green algae, such as *Chlamydomonas reinhardtii*) is that the most active hydrogenases responsible for H_2 production (Fe-Fe hydrogenases) are irreversibly poisoned by O_2 inherently evolved during photosynthetic light reactions. Moreover, the availability of reducing agents such as NADPH and ferredoxin is an additional limitation as these compounds are also involved

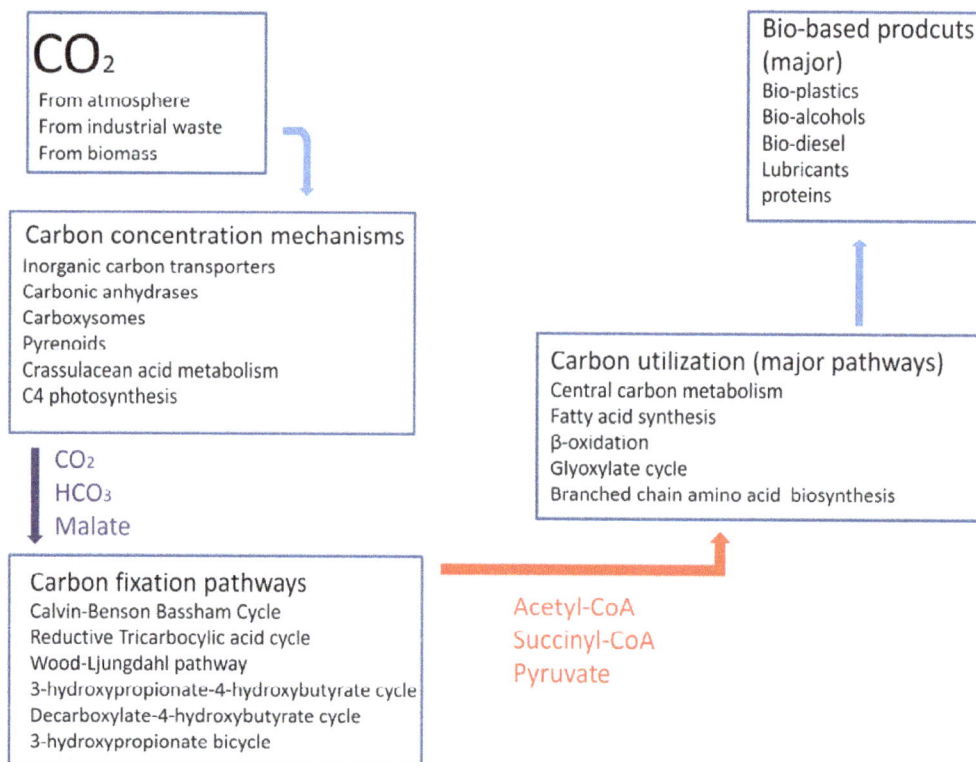

Fig. 6 Carbon concentrating mechanisms, carbon fixation pathways and carbon utilization strategies adopted by various autotrophic organisms, leading to bio-based products. Photosynthetic organisms (cyanobacteria, algae and higher plants) are the main but not exclusive autotrophs that are capable of CO_2 assimilation through the photosynthetic dark reactions. Research on autotrophic bacteria over the last decade showed that these CO_2-utilizing microorganisms could also in principle be applied for reducing carbon footprint on an industrial scale. Scheme adapted from Jajesniak et al. [39].

in other competing metabolic pathways, such as respiration. To overcome these problems Anastasios Melis and colleagues have developed a smart strategy for H_2 production in a green alga *C. reinhardtii* by inducing anaerobiosis in the light following sulfur starvation of the cells, which causes reversible PSII inactivation and thus, inhibition of photosynthetic O_2 evolution. Lowering the ratio of photosynthetic rate below oxidative respiration rate induces anaerobiosis in the cells, which is sufficient for the induction of reversible hydrogenase and cellular H_2 production in the light using photogenerated electrons originated from photoreduced ferredoxin [50]. The economic viability of this approach is somewhat limited by a relatively poor yield. Nevertheless, over the last few years a good progress has been made in this field mainly by isolating the specific mutants of *C. reinhardtii* altered in the rate of cyclic electron transport and degree of non-photochemical quenching (fluorescence and heat) that showed an improved yield of solar-to-H_2 conversion upon sulfur deprivation and increased irradiance compared to the pioneering work of Melis and colleagues [50].

Thus, Kruse and colleagues [51] have developed a mutagenesis-based approach to increase H^+ and e^- supply to the hydrogenases in anaerobic conditions in order to improve H_2 evolution in *C. reindhardtii*. Initially, they isolated the mutants impaired in state 1-to-state 2 transition (e.g., *stm6*

[51]) that were locked in linear photosynthetic electron flow. State transition is a rapid regulatory mechanism of balancing photosynthetic electron flow under fluctuating light which in *Chlamydomonas* is also associated with induction of cyclic electron flow around PSI upon overreduction of the plastoquinone pool [52]. As an example of such mutant, *stm6* was not only blocked in cyclic electron flow, but also exhibited excessive starch accumulation, and increased O_2 consumption (due to photofermentation of starch and upregulation of alternative oxidase activity) [51]. All these factors activated endogenous hydrogenase to catalyze molecular hydrogen production, whilst minimizing O_2-induced irreversible inactivation of this enzyme. The reported hydrogen production rates of *stm6* were over 10-fold higher compared to wild type, with ~540 ml of H_2 produced per liter culture over 10–14 days at a maximal rate of 4 ml per hour [51]. More recently, the same group reported another mutant [53] with the truncated light harvesting antenna of PSII, the LHCII system. This mutant exhibited even higher H_2 production rates compared to *stm6* due to better penetration of light into the dense culture inside the bioreactor and minimization of solar energy conversion losses due to the diminished non-photochemical quenching and photoinhibition in this LHCII-depleted mutant under increased light intensities promoting biomass production [53]. Therefore, truncation

of the LHCII antenna results in balancing the rate of photosynthetic O_2 production with the rate of respiration-based O_2 consumption. This in turn facilitates rapid induction of endogenous hydrogenase allowing for the continuous photobiological H_2 production using reducing equivalents produced by PSII from water [53].

PHOTANOL AND PHOTOALCOHOLS. Several model cyanobacterial species have been used as "biosolar cell factories" [54] for metabolic engineering by combining photosynthetic and fermentation pathways using C_3 sugars (glyceraldehyde-3-phosphate and pyruvate) as the central linking intermediates. These include *Synechocystis* sp. PCC 6803, *Synechococcus elongatus* sp. PCC 7492, *Synechococcus* sp. PCC 7002 and *Anabaena* sp. PCC 7120 [55]. To this end, the given fermentation pathways have been metabolically engineered into the cyanobacterial phototrophic metabolism by inserting the appropriate gene cassettes encoding heterologous enzymes required for conversion of photosynthetically produced C_3 sugars into the particular fermentation end product [55,56]. Thus, photofermentative pathways engineered in cyanobacteria resulted in the production of various short-chain alcohols as the end products including isobutanol, 2,3-butanediol, lactate, and ethanol (photanol) [55,57,58]. Of all these HVPs the highest yield for a direct solar-to-biofuel conversion was thus far obtained for synthesis of ethanol (at a rate of 212 mg/l/day) by genetic engineering of heterologous pyruvate decarboxylase from *Zymomonas mobilis* and overexpressing homologous alcohol dehydrogenase in *Synechocystis* sp. PCC6803 [58]. The main challenge of the metabolic engineering approach is to re-direct the great majority of the fixed carbon from biomass into the desired energy product. A promising strategy towards this goal is to remove carbon sinks, e.g., those employed for the storage compounds polyhydroxybutyrate and glycogen [59].

BIODIESEL. Biodiesel is an environmentally friendly, non-toxic biodegradable fuel with low CO_2 emissions obtained from various feedstocks, including oleaginous plants, microalgae and animal fat [60–63]. It consists of methyl esters of long-chain (C_{12}-C_{18}) fatty acids derived from triglycerides that are present in biological feedstocks. As an example, soybean oil is principally employed for biodiesel production in the USA, illustrating at the same time the global "food vs. fuel" problem, i.e., competition for arable land and water supplies as a result of crop-based biodiesel production [2,62]. In this context, the use of microalgae for biodiesel production provides an attractive option as cultivating algae in minimal nutrient requirements does not exert pressure on arable land or water supplies, it can be scaled up and it is not subjected to seasonal fluctuations of solar-to-biomas conversion efficiency. There are numerous examples of microalgae used for biodiesel production including the genera *Botryococcus*, *Chlorella*, *Scenedesmus*, *Chlamydomonas*, *Dunuliella*, and *Nannochloropsis* ([60] and references therein).

The other obvious advantages of using microalgae for conversion of their biomass to biodiesel are their high growth rate and often high lipid content. Importantly for the high combustion quality of algae-derived biodiesel, the lipids synthesized by microalgae upon exposure to environmental stress factors contain mainly neutral fatty acids of a low degree of unsaturation, making algal biodiesel suitable for partial replacement for fossil diesel [61–63].

METHANE. An alternative to reduce climate-associated emissions could be methane and photocatalytic methanol production using CO_2 captured from coal-fired power stations [64]. Under anoxic conditions, biomass can be converted to CH_4 through a process referred to as anaerobic digestion (AD). This process allows for the remaining material after lipid extraction from cyanobacterial biomass to be converted into CH_4, considerable enhancing the total energy recovery. Apart from decreasing the overall costs of the process for bioenergy production, this might lead to a more favorable or positive energetic balance of the overall cyanobacterial biofuel production. Moreover, the oil recovery results in a 21% of the energetic value when the algae accumulate less than 40% of lipids, despite the fact that the energetic costs of lipid recovery involved are higher than 30% [65]. AD thus could represent a good choice for total energetic recovery of biomass. Unfortunately, the research on the matter is quite limited, particularly for CH_4 production by cyanobacterial biomass. Because of this lack of information in the literature it is quite challenging to compare the yield of different cyanobacterial strains. Economically speaking, cyanobacterial methane production is not viable compared to methane derived from fossil fuels. If the current technology is made more cost-effective in the upcoming years the situation might change. Needless to say, the process of methane production by itself or in conjunction to other bioenergy producing processes needs more investment in research [66].

Artificial photosynthesis: a multidisciplinary approach to produce solar fuels

The translation of fundamental mechanistic knowledge, which has been tremendously advanced in recent years from the X-ray crystallographic and state-of-the-art biophysical studies of PSII and PSI, into man-made solar-to-fuel devices is an ongoing key aspect of the artificial photosynthesis research. The main scientific challenge in these endeavors is to construct an "artificial leaf" capable of efficient capturing and conversion of solar energy into chemical bonds of a high-energy density liquid fuel (such as gasoline, kerosene, formate, methanol etc.), while simultaneously generating oxygen from splitting water (Fig. 7).

To date, the main focus has been to design and synthesize water splitting and proton reducing catalysts that can be potentially connected to artificial light-driven charge separation RCs [67]. Various organic dyes have been employed for the latter (e.g. porphyrins, Zn-chlorins, perylenes) [68,69], whereas inorganic semiconductors (e.g., WO_3, TiO_2, Fe_2O_3) are currently the material of choice to split water or power reductive chemistry within the photoelectrochemical cells (Fig. 8) [4,70].

One of the greatest challenges in this field is nanoengineering of suitable semiconductors to efficiently transfer the captured energy of sunlight into multi-electron reduction reactions that are required to convert CO_2 into fuels. The other major challenge is to efficiently interface stable molecular catalysts with these semiconductors that would

Artificial photosynthesis

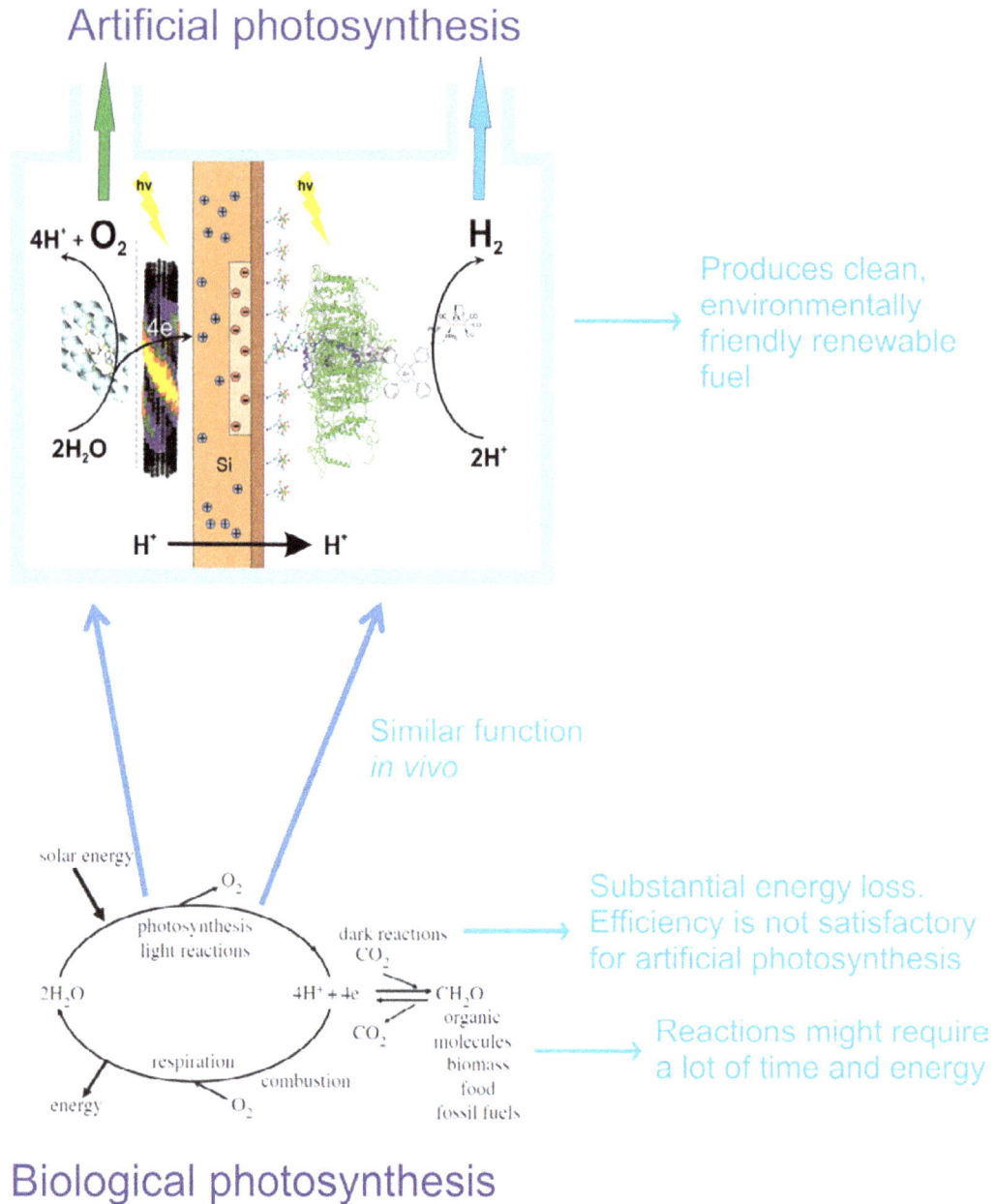

Biological photosynthesis

Fig. 7 Artificial versus natural photosynthesis. A comparison between photosynthesis in vivo and artificial photosynthesis explains why it is more energetically viable to employ artificial photosynthesis for clean fuel production. Natural photosynthesis is very inefficient in terms of solar-to-biomass conversion. Artificial photosynthesis short-circuits the natural process by utilizing the most energetically efficient primary events of light capture, charge separation and charge transfer. Lower panel of $H_2/O_2/H_2O$ cycle driven by solar light is reproduced with permission from Barber and Tran [4].

efficiently split water and reduce water-derived protons upon absorption of visible light, as do the natural photosystems. There has been an explosion of reports describing novel synthetic molecular catalysts, some of which are characterized by high turn-over numbers (TON) and turn-over frequencies (TOF), in some cases matching or exceeding those of the natural photosystems. As an example of some of the most remarkable advances in this field is a rationally designed Ru-based catalyst with a TOF of 300 s^{-1}, which is comparable with that of PSII [71]. Another spectacular catalysts for water splitting are based on a relatively abundant element, cobalt. Kanan and Nocera have reported a cheap self-assembling catalyst composed of Co-oxo-borate or phosphate complexes [72,73], which can efficiently split water with a low overpotential at neutral pH, similar to the OEC of PSII. Remarkably, the Co-oxo-phosphate catalyst

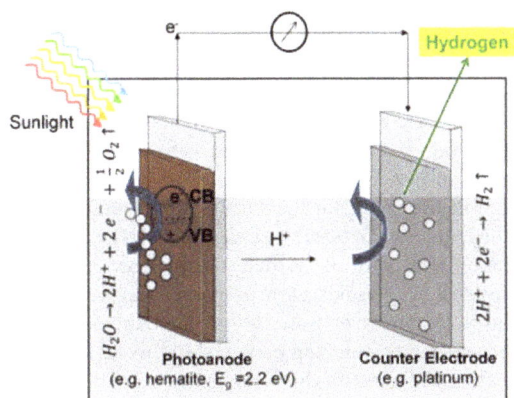

Fig. 8 Principles of the water-splitting and hydrogen-evolving photoelectrochemical cell. Light induces a charge separation within the semiconductor photoanode (CB – conduction band; VB – valence band), and this leads to photocatalytic water splitting into H+, molecular oxygen and electrons. The protons migrate to the counter electrode (cathode) where they are reduced to H_2 with the photogenerated electrons. In the majority of PEC cells, a certain external overpotential (bias) of 0.5–1 mV needs to be applied in order reach significant H_2 production rates. Figure reproduced with permission from Ihssen et al. [70].

adopts a cubane-like structure, which is very similar to the OEC of PSII [74]. Many more water splitting and hydrogen evolving synthetic catalysts have been reported that have promising TON and TOF numbers. These are overviewed in several excellent recent reviews [4,75–78] and are beyond the scope of the present review.

The advantage of solar-to-H_2 conversion is that molecular hydrogen can be utilized directly as a fuel but can also be employed to reduce CO_2 to simple carbon-based fuels (e.g., methane, methanol, formic acid or carbon monoxide) as molecular precursors for higher molecular weight carbon compounds [76–78]. Efforts to generate high-energy carbon containing fuels, such as methanol, have been made, although progress in the field is rather slow due to the technical challenges associated with the multi-electron character of the CO_2 reduction, as opposed to a single-step reduction of protons into molecular hydrogen. However, the biggest challenge is to functionally integrate robust and highly active individual components (photocatalysts, light harvesting antenna and photoelectrodes) into a complete device operating at high solar power conversion efficiencies in which water splitting and hydrogen production are spatially and temporarily separated. To this end, several exciting advances have been made over the last 3 years that have started turning this highly ambitious goal into realistic technology. These are overviewed below.

Recent advances in biomimetic solar-to-fuel devices

Among various semiconducting substrates used for engineering of artificial leaves hematite (α-Fe_2O_3) has recently become the material of choice for photoanodes due to its inherent material characteristics, such as low cost, environmentally benign nature, corrosion resistance over a

wide pH range (and particularly robustness towards photocorrosion), its narrow band gap (2.1 eV), and importantly, its photocatalytic activity in the visible part of the spectrum (approximately 40% solar light is absorbed by this n-type semiconductor) [79]. Many consider hematite essential and even critical for improvement of power conversion efficiency in an artificial leaf, hence, its bulk and surface electronic properties have been investigated in depth for many decades [79]. Recently, a novel photoanode heterostructure of n-type hematite and p-type NiO/α-Ni(OH)$_2$ has been reported by Bora et al. [80] which is particularly efficient in terms of solar energy conversion. This pn-type water-splitting heterojunction assembly was engineered in order to lower bias potential and to maximize current density to overcome the kinetic barrier of the water oxidation reaction. Indeed, such a heterojunction system exhibited an impressive current density which was several times higher than that of pristine hematite film (at 16 mA/cm²), with the oxidized Ni(OH)$_2$ layer producing the sudden increase in current density [80]. Furthermore, once exposed to AM 1.5 (standard solar illumination) the system displayed charge-storing capacity, along with electrochromic behavior. This is not present in the case of pn-junction-like devices made by mere deposition of NiO on hematite by simple thermal annealing. Moreover, no such behavior was observed for hematite alone. In this way, this novel type of electrode offers a simple low-cost option for both water splitting and charge storage [80]. A related approach features an electron hole doped film in the α-Fe_2O_3 photoanode upon electrochemical oxidation [81].

In contrast to such a fully synthetic water-splitting photoanode, Kargul and colleagues very recently described the successful nanoengineering of nanocrystalline hematite/FTO with the robust red algal PSI-LHCI supercomplex, thus forming a truly green biohybrid photoanode (Fig. 9) [37]. The individual synthetic and biological photoactive components were interfaced to construct a highly organized biophotoanode, which was subsequently used for assembly of the biohybrid dye-sensitized solar cell (DSSC) using Pt counter electrode for reduction of water-derived protons [37]. Electron microscopy and X-ray diffraction analyses showed that red algal PSI-LHCI was immobilized as a structured multilayer over highly ordered nanocrystalline arrays of hematite. Compared to a related tandem system based on TiO$_2$/PSI-LHCI material (unpublished), the α-Fe_2O_3/PSI-LHCI biophotoanode generated the largest open circuit photocurrent and operated at the highest power conversion efficiency due to a better electronic tuning between the PSI layer and the conductive band of hematite [37]. The nanostructuring of the PS-LHCI multilayer in which the subsequent layers of PSI were organized in the head-to-tail orientation was accomplished by surface charge manipulation at various pH, which enabled immobilization of the PSI-LHCI complex with its reducing side towards the hematite surface [37]. Importantly, upon illumination with visible light above 590 nm, the biohybrid PSI-LHCI-DSSC was capable of sustained photoelectrochemical H_2 production at a rate of 744 μmoles H_2/mg Chl/h, representing one of the best performing biohybrid "green" solar-to-fuel nanodevices capable of sustained H_2 production under standard solar illumination [37].

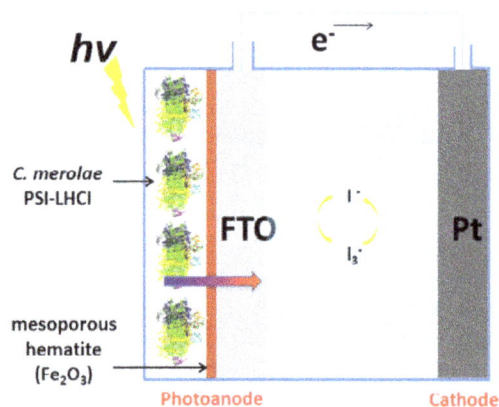

Fig. 9 Diagrammatic representation of biohybrid PSI-based DSSC incorporating photosystem I from a red microalga *Cyanidioschyzon merolae*. Ultrastable red algal PSI-LHCI was nanostructured in head-to-tail oriented multilayer over hematite/FTO electrode [37]. In this way, PSI acts as an efficient photosensitizer/charge separator for hematite, allowing this substrate to split water upon absorption of red/far red light and generate a photocurrent density of up to 57 µA/cm². The protons derived from water are reduced on the Pt counter electrode to molecular hydrogen that is sustainably produced by this assembly under standard red-shifted solar illumination.

Braun and colleagues have recently demonstrated a promising strategy for enhancing photocurrent density on semiconductor substrates using other rather simpler in structure light harvesting proteins. Low-cost biomaterials such as cyanobacterial phycocyanin [82] and enzymatically produced melanin [70] were both employed to functionalize electrode surfaces constructed with hematite, thus increasing the overall performance of such cheap metal oxide photoanodes in a biohybrid PEC system. The overall stability and performance of such biohybrid photoanode assemblies was further improved by application of environmentally benign and pH-neutral electrolyte systems [70,82].

Imitating the Z scheme in an artificial leaf

An important benchmark in the quest for a fully operational viable artificial leaf has been very recently provided by the biohybrid approach using natural photosystems for construction of anodic and cathodic half-cells within PEC devices. As overviewed above, PSII and PSI are very efficient as electron generators for the reduction of protons, due to their high quantum efficiency of solar conversion to charge separation, and especially as their substrates are only water and visible light. Photocatalytic properties of photosystems such as the light-driven water splitting conducted by PSII and generation of reducing power by PSI make them the invaluable candidates for the development of semi-artificial devices which convert light energy into stable HVPs such as molecular hydrogen and simple carbon-based fuels [16,18,36].

Recently, an elegant proof-of-concept approach to serially couple PSII and PSI with an Au electrode has yielded one of the first operational semi-artificial PECs producing photocurrents so as to imitate the Z scheme of natural photosynthesis (Fig. 10) [83]. This full Z-scheme mimic resulted in the PEC device in which the two electrochemical half-cells were separated by the Au electrode with PSII or PSI on either side smarty interfaced by the two types of redox compatible osmium complex hydrogel to facilitate direct electron transfer. In this way, the complete biohybrid PEC made a significant advance to previous studies by the same group on construction of individual PSII- and PSI-based half cells [84,85]. Importantly, in contrast to these previous studies the complete photovoltaic cell operated as a closed circuit without any sacrificial electron donors or mediators, using water and solar light as the only substrates [83].

Overall, such serial coupling of two independent processes of light capturing by PSII and PSI in two separate interconnected compartments yielded a fully autonomous biophotovoltaic cell, albeit operating at very low power conversion efficiency. This type of cell is different from previously published biophotovoltaic devices utilising PSI as a photosensitizing light harvesting/charge separating

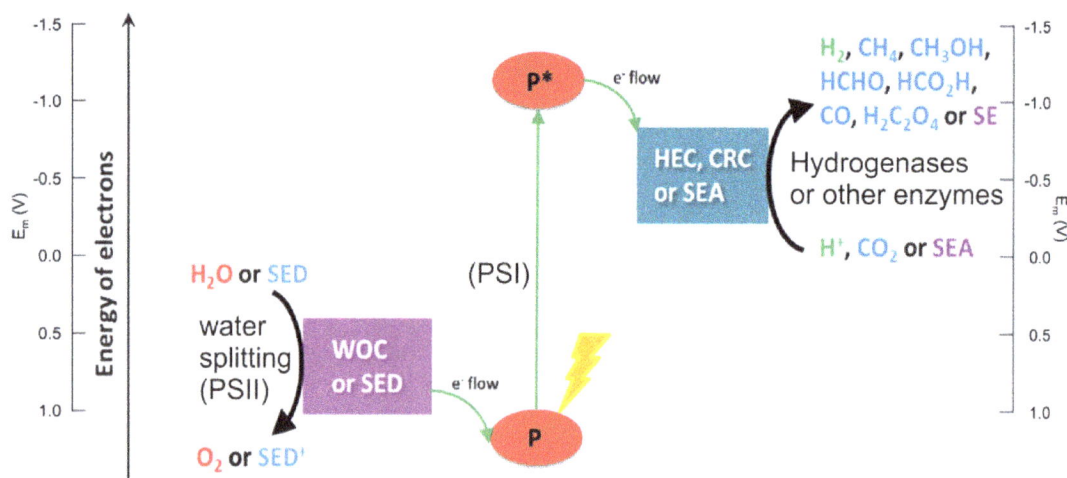

Fig. 10 Reconstruction of the photosynthetic Z scheme in biohybrid solar-to-HVP devices. Diagrammatic representation of artificial photosynthesis. Light converting modules (P), water oxidation catalyst (WOC), sacrificial electron donor (SED), hydrogen evolving catalyst (HEC), CO₂ reduction catalyst (CRC), sacrificial electron acceptor (SEA). Adapted from Berardi et al. [16].

photoanodic component [37,86] as it provides the blueprint for the potential application of such a "biobattery" in conjunction with various synthetic and biological catalysts, since the photoactivated electrons originating from water splitting may be employed for chemical energy conversion rather than reducing oxygen by methyl viologen (as in the case of the Kothe et al. assembly [83]). Of course, PSI has been used in numerous studies as the cathodic component of the H_2-producing half-cell devices, yet all of such devices required an artificial electron donor and metal/hydrogenase catalysts to operate [18,36]. The separation of the oxygen evolving PSII photoanode from the PSI photocathode opens up the possibility to couple PSI with oxygen-sensitive enzymes of HVP pathways such as nitrogenases, CO_2 dehydrogenases or hydrogenases [16,87], as shown schematically in Fig. 10. However, in such prospective assemblies, PSII will likely have to be replaced by a more robust, ideally self-renewing molecular water splitting catalyst to bypass the problem of PSII degradation.

In a similar approach which also utilizes redox-active polymers, Yehezkeli et al. [88] employed polyN,N′-dibenzyl-4,4′-bipyridinium (poly-benzyl viologen, PBV^{2+}) which was employed as the smart matrix for nanostructured assemblies of PSI and/or PSII layers, deposited on indium tin oxide (ITO) electrodes (Fig. 11). As expected, the PSII-functionalized electrode evolved oxygen producing an anodic photocurrent upon illumination with visible light, without the necessity of an artificial electron donor/mediator. In contrast, the layered assembly of PSI generated a cathodic photocurrent only in the presence of a sacrificial electron donor/mediator system (ascorbate/dichlorophenol indophenol). The photocurrent was generated by electron transfer from PSII to the redox polymer and subsequently to the electrode via the PBV^{2+}. The concomitant evolution of molecular oxygen was accompanied by injection of electrons into the electrode by the charge-separated species. Compared to the anodic photocurrent generated by a PSII-modified electrode, the system composed of a PSII layer assembled on a layer of PSI within the matrix of PBV^{2+} showed a 2-fold higher photocurrent compared to the system engineered with PSII alone, most likely due to better charge separation within the PSII/PSI tandem system [88]. Vectorial electron transfer to the electrode was elegantly demonstrated by further nanoengineering of the two photosystems on the electrode using two different redox polymers better tuned in with the redox properties of the photoactivated photosystems, resulting in further (~6-fold) enhancement in the photocurrent density [88], demonstrating the importance of a rational smart matrix-based design for better overall performance of such biohybrid PEC devices.

Although photooxidation of water has been optimized by Nature by means of water splitting cycle catalyzed by the OEC of PSII, mimicking this energetically demanding reaction in an artificial leaf remains a major challenge, not only in terms of operation at a minimal overpotential, exhibiting high TON and TOF numbers of the water splitting catalyst, but also managing the proton coupled electron transfer during the catalysis. One of the more successful assemblies fulfilling these criteria has been recently reported for the water splitting/H_2 generating hybrid system, in which an inorganic H_2-producing photocatalysts CdS and Rh-doped $Ru/SrTiO_3$:Rh (Rh-doped perovskite titanate) and plant O_2-evolving PSII were functionally coupled and integrated with an inorganic redox system [$Fe(CN)_6^{3-}/Fe(CN)_6^{4-}$] in an aqueous solution [89], yielding both O_2 and H_2 upon illumination with visible light. The overall water splitting achieved under standard solar irradiation and the H_2-production rate of this hybrid system reached the impressive values of 1334 mol O_2/mol PSII/h and 2489 mol H_2/mol PSII/h, respectively. In this hybrid system, PSII particles were self-assembling onto the inorganic photocatalysts' surface and the electron transport from PSII to the synthetic photocatalyst was determined to be the rate-limiting step. Importantly, the biohybrid PSII-perovskite system was characterized by exceptionally high TON for H_2 production (over 3751), without the loss of activity for over 3 hours in direct sunlight. In this way, an autonomous biohybrid operational solar-to-fuel nanodevice of PSII and the perovskite catalyst was generated that used solely water and solar energy as substrates for hydrogen production, as does PSII [89].

Recently, Reisner and colleagues reported a hybrid photoanode consisting of *T. elongatus* PSII particles deposited on a high surface area mesoporous ITO electrode [90]. The particular three-dimensional architecture of this material permits favorable high protein coverage and direct electron transfer from PSII to the mesoporous ITO surface. The authors reported the TOF of water oxidation of 0.18 ±0.04 mol O_2/mol PSII/s and a current density of 1.6 ±0.3 μA/cm² [90]. Moreover, the study demonstrated great enhancement in photocurrent density upon addition of two external electron mediators, potassium 1,4-naphthoquinone-2-sulfonate (NQS) and 2,6-dichloro-1,4-benzoquinone (DCBQ), indicating somewhat non-favorable interfacing of PSII with ITO which has to be overcome by the addition of external electron transfer mediators. Furthermore, mechanistic studies show electron transfer from the terminal quinone Q_A in addition to the terminal quinone Q_B to the ITO surface [90], something that does not occur in vivo as only Q_B is involved in electron relay in the thylakoid membrane.

In a follow-up work, the same group reported an improvement of the above biohybrid system by both electrostatic and covalent immobilization of the PSII particles on a self-assembled monolayer (SAM)-modified ITO electrode [91]. Significantly, the directed immobilization of PSII, which placed the terminal quinones (Q_A and Q_B) in close proximity of the ITO surface not only enhanced the current density generated by the assembly but also required much less sample deposition (1.1 pmol PSII) [91]. Moreover, the system resulted in enhanced electronic interfacial communication between the PSII and the ITO surface and an increased stability [91] compared to the previous study [90]. Addition of external electron mediators resulted in no significant increase of photocurrent density upon illumination, suggesting that the proximity of the terminal quinones with the ITO surface was good enough for an optimal electron transfer. The work is an excellent proof-of-concept example illustrating that the oriented immobilization of photosynthetic modules with the appropriate electrode is a prerequisite for enhanced direct electron transfer and improved robustness of the biohybrid PEC system.

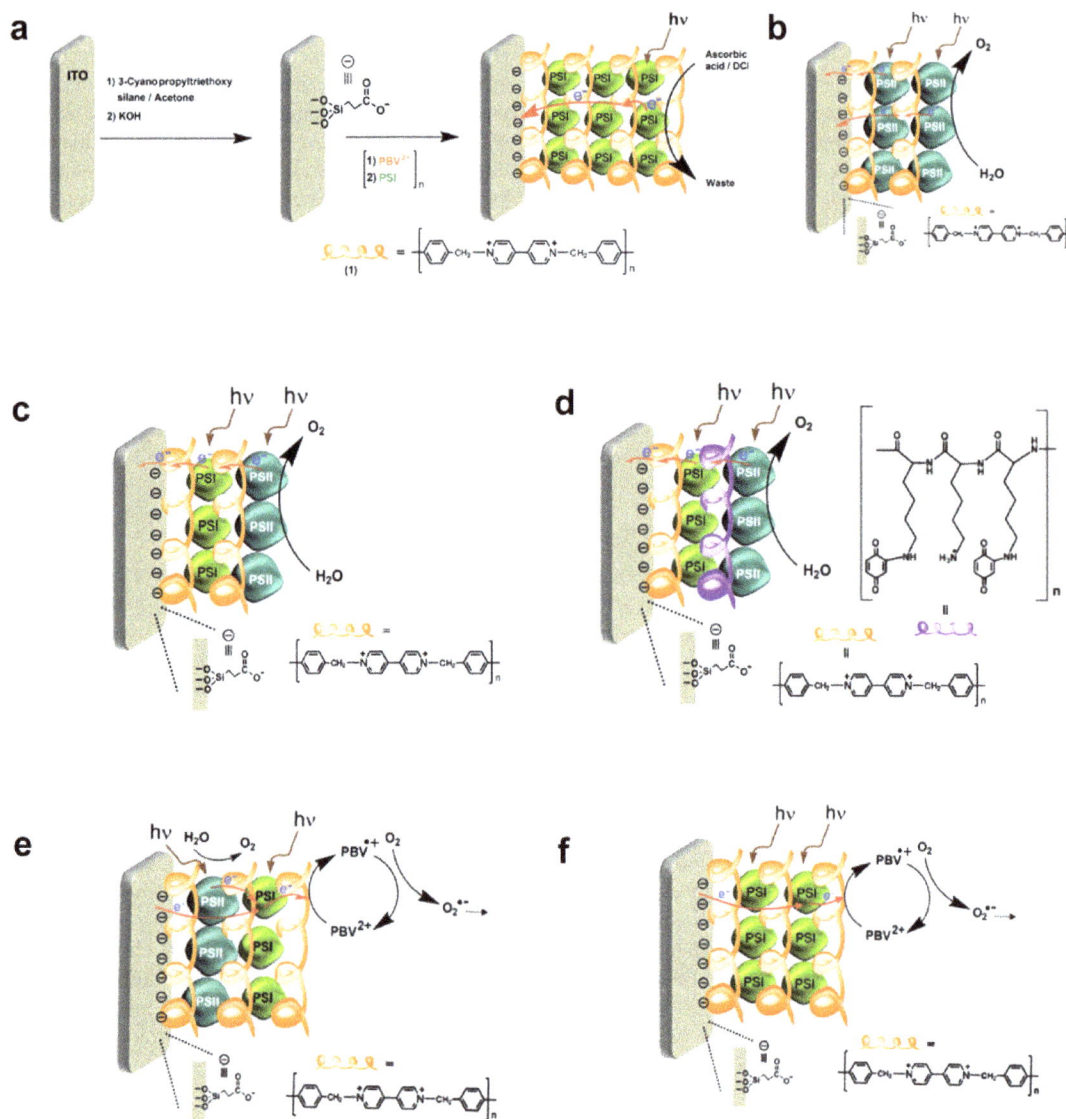

Fig. 11 Nanoengineering the PSII/PSI photoelectrodes using redox active polymers as the smart matrix. Various organizations of PSI and PSII layers redox coupled to each other and/or the electrode surface according to Yehezkeli et al. [88]. **a** Diagrammatic synthesis of the layered PBV^{2+}/PSI photoactive conjunction on the ITO electrode. **b** Diagrammatic assembly of the layered PBV^{2+}/PSII photoactive conjunction on an ITO electrode. **c** Diagrammatic assembly of the layered PBV^{2+}/PSI/PBV^{2+}/PSII photoactive conjunction deposited on an ITO electrode. **d** Diagrammatic assembly of the layered PBV^{2+}/PSI/PBQ/PSII photoactive conjunction on an ITO electrode. **e** Diagrammatic assembly of the layered PSII/PBV^{2+}/PSI photoactive conjunction deposited on an ITO electrode. **f** Diagrammatic representation of the assembly composed of the layered PSI/PBV^{2+}/PSI photoactive conjunction deposited on an ITO electrode. Structures of the interfacing matrices are also shown. Reproduced with permission from Yehezkeli et al. [88].

An interesting approach in the construction of the bio-hybrid H_2-producing nanodevices is based on application of nanostructured hollow nanospheres sized in the optical range and composed of Pt-doped carbon nitride organic semiconductor exhibiting both water splitting or proton reduction activity [92]. These small, hollow semiconducting nanospheres (Fig. 12) function not only as nanostructured

scaffold for the Pt co-catalyst but also as the light-harvesting antennae that facilitate photoreduction catalysis. Significantly, such sophisticated highly nanostructured assemblies of optimized thickness have been shown to operate at an impressive solar-to-hydrogen quantum efficiency of 7.5% [92]. Hollow polymerized nanospheres with controlled surface functionalities and shell thickness provide an important

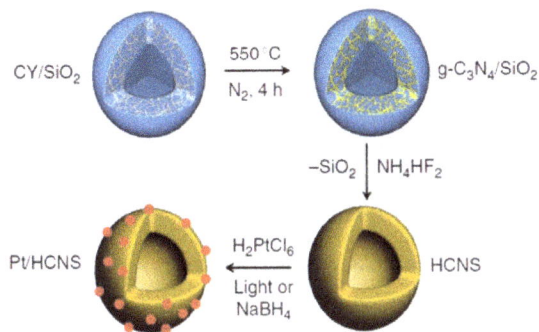

Fig. 12 Synthetic scheme of nanosphere synthesis. A schematic diagrammatic illustration of the hollow carbon nitride nanosphere (HCNS) and Pt co-catalyst/HCNS nanosphere, highlighting the chemical synthesis pathway. This light-harvesting nanosphere of the precisely controlled thickness serves as a photocatalytically active scaffold for immobilization of the metal co-catalyst resulting in both water splitting and molecular hydrogen production under visible light illumination and reaching a solar-to-hydrogen quantum efficiency of 7.5%. Reproduced with permission from Sun et al. [92].

development in artificial photosynthesis field as they can act as a platform for assembling various functionalities into complex nanostructures, while at the same time maintaining "interior" and "exterior" compartments, with a "membrane" separating both of them. In natural photosynthesis, the equivalent of such nanospheres is represented by the thylakoid membrane, where both the water-oxidation catalyst (PSII) and reduction center (PSI) are spatially separated and ordered in tandem for highly efficient energy transfer and photoconversion within the nanosized structures of the reaction centers [5,93].

Recently, Zhou et al. [94] reported a unique design strategy for solar-to-fuel conversion using a real leaf's three-dimensional hierarchical matrix for immobilization of nanostructured crystalline perovskite titanate clusters ($ATiO_3$, with A being Sr, Ca, and Pb) acting as catalysts for both water splitting and photoreduction of CO_2 into simple hydrocarbons (CH_4 and CO). This novel approach allowed for highly efficient, improved gas diffusion and light harvesting within such a photocatalytical semi-artificial system, in an analogous fashion to a well-defined leaf anatomy and physiology. Remarkably, the system operated with no external bias potential (Fig. 13) for water splitting, whilst the solar-to-hydrocarbon quantum yield was further improved by doping with the Au co-catalyst acting as the electron reservoir and

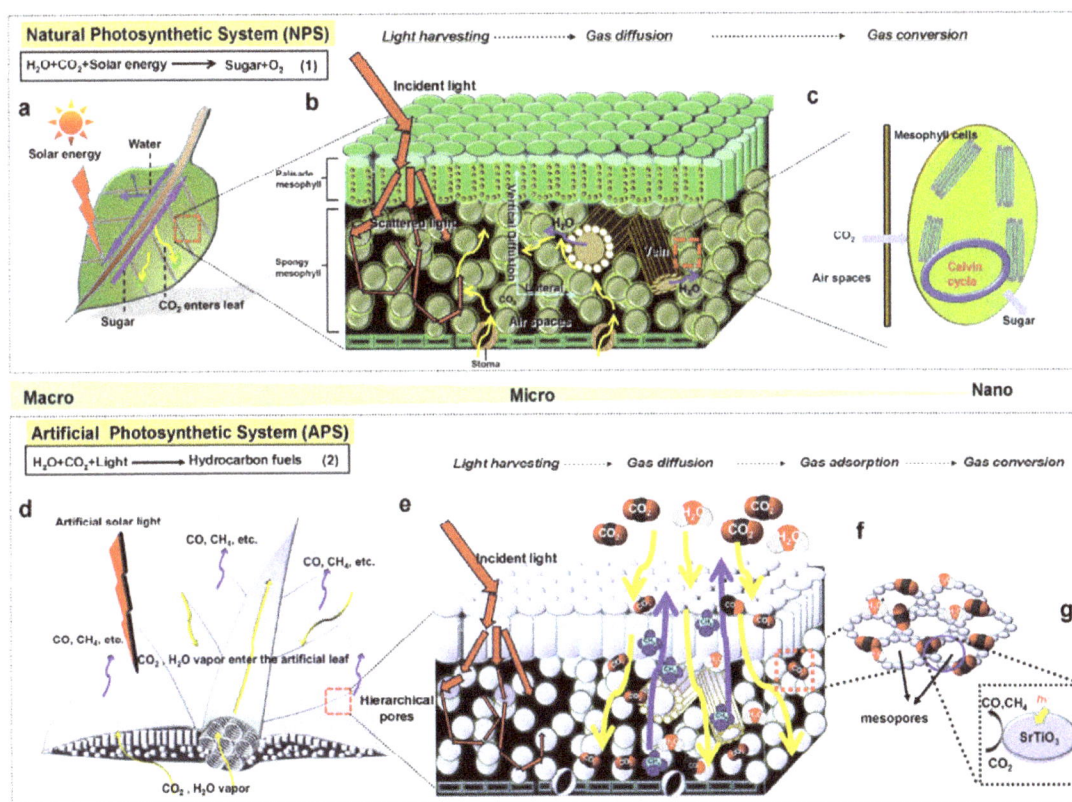

Fig. 13 Schematic diagrammatic illustration and comparison of the essential processes in natural photosynthetic system (NPS) and artificial photosynthetic system (APS). **a** Essential process of photosynthesis in NPS in the leaf. **b** Gas diffusion and light harvesting processes in NPS at the mesophyllic level. **c** Process of gas conversion in mesophyll cells at the chloroplastic level. **d** Artificial photosynthesis basic process in APS at the "artificial leaf" scale. **e** Gas diffusion and light harvesting processes in APS at the mesophyllic level. **f** Process of gas adsorption in APS at the supramolecular level. **g** Processs of gas conversion in APS at the molecular level. Reproduced with permission from Zhou et al. [94].

facilitating proton-coupled electron transfer [94]. The authors postulate that this approach may be extended to construction of other 3D scaffolds with other interesting multi-metallic oxides, such as niobates (e.g. $NaNbO_3$) or tantalates (e.g. $NaTaO_3$, $KTaO_3$, $Sr_2Ta_2O_7$) [94]. Moreover, this approach may also be applied to polymeric metal-free photocatalysts (e.g., graphitic carbon nitride) based on earth-abundant elements. This is the first example that employs leaf architecture as a scaffold for the nanoengineering of artificial leaves for CO_2 photoreduction, and thus in our view, represents a major breakthrough in the field of artificial photosynthesis.

Conclusions and future outlook

The inherent depletion of fossil fuels and relying entirely on the use of fossil fuels to power our economies must be skillfully addressed in the coming years to sustain economic growth. An attractive alternative to burning fossil fuels is provided through introduction of carbon-neutral, transportable energy carriers made from renewable resources. To date, biomass-based fuels are still quite restricted in terms of scale and their production remains controversial in many cases mainly due to the "food or fuel" problem. The limitation is global, and unfortunately the scale of fossil and biomass-based fuels cannot be compared, with the latter being present on our planet in much lesser quantities. Moreover, when the energy demand will grow in the not-too-distant future, from the current annual use of 16 TW to projected 50 TW by 2100 [1], the situation is likely to worsen. As mentioned previously, solar energy seems to be the most abundant source of renewable energy available on our planet. The terms "solar fuel" and "artificial leaf" have become well established within the scientific community seeking to provide clean, green alternatives to fossil fuels for energy production.

In order to be able to adapt solar energy on a global scale, scientific breakthroughs must demonstrate that this is a viable technology, i.e., the energy must be harnessed, stored and used at the maximum possible efficiency, whilst minimizing energy losses or unwanted back reactions. In contrast to other renewable energy sources there is more than enough of solar energy falling on Earth each hour to meet global energy demand of mankind for an entire year. Therefore, transition from fossil fuel-based economies to this renewable practically limitless energy source, even at modest power conversion efficiencies seems the only viable option for generation of solar fuels or photocurrents in long term. It is also important to emphasize that in order to maximize the efficiency of solar energy conversion, the energy ideally should be stored in the form of high energy density liquid fuels. Realistically, the efficiency target for "artificial leaves" discussed in this review must reach 10 per cent or better in order to compete with traditional fossil fuel-based technologies of energy production and to fulfill the need of realistic alternative technologies addressing the increasing global energy demand. Smart matrix engineering seems to be the key aspect towards this goal, whereby intelligent nanoengineering of robust catalytic and light harvesting modules together with charge separation modules and management of proton relay network is skillfully applied to produce fully integrated working devices that produce fuel from water upon capture of solar energy. Just as the leaf does...

Acknowledgments
JDJO and JK are grateful for the support of the Polish Ministry of Science and Higher Education and the European Science Foundation (grant No. 844/N-ESF EuroSolarFuels/10/2011/0 to JK) and all the fruitful collaborations within the EUROCORES/EuroSolarFuels/Solarfueltandem network. JDJO is supported by the PRELUDIUM grant awarded by the Polish National Science Center (grant No. UMO-2013/11/N/NZ1/02390 to JDJO).

Authors' contributions
The following declarations about authors' contributions to the research have been made: co-wrote the article and prepared the majority of the figures: JDJO; co-wrote, revised the article and prepared some figures: JK.

Competing interests
No competing interests have been declared.

References
1. Shafiee S, Topal E. When will fossil fuel reserves be diminished? Energy Policy. 2009;37(1):181–189.

2. Stephens E, Ross IL, Mussgnug JH, Wagner LD, Borowitzka MA, Posten C, et al. Future prospects of microalgal biofuel production systems. Trends Plant Sci. 2010;15(10):554–564.

3. Industries [Internet]. Shell global. 2014 [cited 2014 Sep 9].

4. Barber J, Tran PD. From natural to artificial photosynthesis. Interface Focus. 2013;10(81):20120984.

5. Kargul J, Barber J. Structure and function of photosynthetic reaction centres. In: Wydrzynski TJ, Hillier W, editors. Molecular solar fuels. Cambridge: Royal Society of Chemistry; 2011. p. 107–142.

6. Larkum AWD. Evolution of the reaction centers and photosystems. In: Renger G, editor. Primary processes of photosynthesis: principles and apparatus. Cambridge: Royal Society of Chemistry; 2008. p. 489–521.

7. Barber J. Engine of life and big bang of evolution: a personal perspective. Photosynth Res. 2004;80(1–3):137–155.

8. Hohmann-Marriott MF, Blankenship RE. Evolution of photosynthesis. Annu Rev Plant Biol. 2011;62(1):515–548.

9. Hurles M. Gene duplication: the genomic trade in spare parts. PLoS Biol. 2004;2(7):e206.

10. Pennisi E. Genome duplications: the stuff of evolution? Science. 2001;294(5551):2458–2460.

11. Raymond J, Blankenship RE. Horizontal gene transfer in eukaryotic algal evolution. Proc Natl Acad Sci USA. 2003;100(13):7419–7420.

12. Igarashi N, Harada J, Nagashima S, Matsuura K, Shimada K, Nagashima KV. Horizontal transfer of the photosynthesis gene cluster and operon rearrangement in purple bacteria. J Mol Evol. 2001;52(4):333–341.

13. Sadekar S. Conservation of distantly related membrane proteins: photosynthetic reaction centers share a common structural core. Mol Biol Evol. 2006;23(11):2001–2007.

14. Murray JW, Duncan J, Barber J. CP43-like chlorophyll binding proteins: structural and evolutionary implications. Trends Plant Sci. 2006;11(3):152–158.

15. Blankenship RE, Tiede DM, Barber J, Brudvig GW, Fleming G, Ghirardi M, et al. Comparing photosynthetic and photovoltaic efficiencies and recognizing the potential for improvement. Science. 2011;332(6031):805–809.

16. Berardi S, Drouet S, Francàs L, Gimbert-Suriñach C, Guttentag M, Richmond C, et al. Molecular artificial photosynthesis. Chem Soc Rev. 2014;43(22):7501–7519.

17. Field CB. Primary production of the biosphere: integrating terrestrial and oceanic components. Science. 1998;281(5374):237–240.

18. Kargul J, Janna Olmos JD, Krupnik T. Structure and function of photosystem I and its application in biomimetic solar-to-fuel systems. J Plant Physiol. 2012;169(16):1639–1653.

19. Munekage Y, Hashimoto M, Miyake C, Tomizawa KI, Endo T, Tasaka M, et al. Cyclic electron flow around photosystem I is essential for photosynthesis. Nature. 2004;429(6991):579–582.

20. Johnson GN. Physiology of PSI cyclic electron transport in higher plants. Biochim Biophys Acta. 2011;1807(3):384–389.

21. Joliot P, Johnson GN. Regulation of cyclic and linear electron flow in higher plants. Proc Natl Acad Sci USA. 2011;108(32):13317–13322.

22. Hertle AP, Blunder T, Wunder T, Pesaresi P, Pribil M, Armbruster U, et al. PGRL1 is the elusive ferredoxin-plastoquinone reductase in photosynthetic cyclic electron flow. Mol Cell. 2013;49(3):511–523.

23. DalCorso G, Pesaresi P, Masiero S, Aseeva E, Schünemann D, Finazzi G, et al. A complex containing PGRL1 and PGR5 is involved in the switch between linear and cyclic electron flow in *Arabidopsis*. Cell. 2008;132(2):273–285.

24. Peng L, Fukao Y, Fujiwara M, Takami T, Shikanai T. Efficient operation of NAD(P)H dehydrogenase requires supercomplex formation with photosystem I via minor LHCI in *Arabidopsis*. Plant Cell. 2009;21(11):3623–3640.

25. Yamori W, Sakata N, Suzuki Y, Shikanai T, Makino A. Cyclic electron flow around photosystem I via chloroplast NAD(P)H dehydrogenase (NDH) complex performs a significant physiological role during photosynthesis and plant growth at low temperature in rice. Plant J. 2011;68(6):966–976.

26. Kukuczka B, Magneschi L, Petroutsos D, Steinbeck J, Bald T, Powikrowska M, et al. Proton gradient regulation5-like1-mediated cyclic electron flow is crucial for acclimation to anoxia and complementary to nonphotochemical quenching in stress adaptation. Plant Physiol. 2014;165(4):1604–1617.

27. Nelson N, Yocum CF. Structure and function of photosystems I and II. Annu Rev Plant Biol. 2006;57(1):521–565.

28. Cardona T, Sedoud A, Cox N, Rutherford AW. Charge separation in photosystem II: a comparative and evolutionary overview. Biochim Biophys Acta. 2012;1817(1):26–43.

29. Umena Y, Kawakami K, Shen JR, Kamiya N. Crystal structure of oxygen-evolving photosystem II at a resolution of 1.9 Å. Nature. 2011;473(7345):55–60.

30. Kanady JS, Tsui EY, Day MW, Agapie T. A synthetic model of the Mn_3Ca subsite of the oxygen-evolving complex in photosystem II. Science. 2011;333(6043):733–736.

31. Ananyev G, Dismukes GC. How fast can photosystem II split water? Kinetic performance at high and low frequencies. Photosynth Res. 2005;84(1–3):355–365.

32. Badura A, Kothe T, Schuhmann W, Rögner M. Wiring photosynthetic enzymes to electrodes. Energy Environ Sci. 2011;4(9):3263.

33. Jordan P, Fromme P, Witt HT, Klukas O, Saenger W, Krauss N. Three-dimensional structure of cyanobacterial photosystem I at 2.5 A resolution. Nature. 2001;411(6840):909–917.

34. Ben-Shem A, Frolow F, Nelson N. Crystal structure of plant photosystem I. Nature. 2003;426(6967):630–635.

35. Amunts A, Drory O, Nelson N. The structure of a plant photosystem I supercomplex at 3.4 A resolution. Nature. 2007;447(7140):58–63.

36. Nguyen K, Bruce BD. Growing green electricity: progress and strategies for use of photosystem I for sustainable photovoltaic energy conversion. Biochim Biophys Acta. 2014;1837(9):1553–1566.

37. Ocakoglu K, Krupnik T, van den Bosch B, Harputlu E, Gullo MP, Olmos JDJ, et al. Photosystem I-based biophotovoltaics on nanostructured hematite. Adv Funct Mater. 2014 (in press).

38. Pandey D, Agrawal M. Carbon footprint estimation in the agriculture sector. In: Muthu SS, editor. Assessment of carbon footprint in different industrial sectors. Singapore: Springer; 2014. p. 25–47. (vol 1).

39. Jajesniak P, Ali H, Wong TS. Carbon dioxide capture and utilization using biological systems: opportunities and challenges. J Bioprocess Biotech. 2014;4(155).

40. Zhao B, Su Y. Process effect of microalgal-carbon dioxide fixation and biomass production: a review. Renew Sustain Energy Rev. 2014;31:121–132.

41. Oliver JWK, Machado IMP, Yoneda H, Atsumi S. Combinatorial optimization of cyanobacterial 2,3-butanediol production. Metab Eng. 2014;22:76–82.

42. Machado IMP, Atsumi S. Cyanobacterial biofuel production. J Biotech. 2012;162(1):50–56.

43. Rabinovitch-Deere CA, Oliver JWK, Rodriguez GM, Atsumi S. Synthetic biology and metabolic engineering approaches to produce biofuels. Chem Rev. 2013;113(7):4611–4632.

44. Smith KS, Ferry JG. Prokaryotic carbonic anhydrases. FEMS Microbiol Rev. 2000;24(4):335–366.

45. Rosgaard L, de Porcellinis AJ, Jacobsen JH, Frigaard NU, Sakuragi Y. Bioengineering of carbon fixation, biofuels, and biochemicals in cyanobacteria and plants. J Biotech. 2012;162(1):134–147.

46. Quintana N, van der Kooy F, Van de Rhee MD, Voshol GP, Verpoorte R. Renewable energy from cyanobacteria: energy production optimization by metabolic pathway engineering. Appl Microbiol Biotechnol. 2011;91(3):471–490.

47. Das D. Hydrogen production by biological processes: a survey of literature. Int J Hydrog. Energy. 2001;26(1):13–28.

48. Abed RMM, Dobretsov S, Sudesh K. Applications of cyanobacteria in biotechnology. J Appl Microbiol. 2009;106(1):1–12.

49. Dutta D, De D, Chaudhuri S, Bhattacharya SK. Hydrogen production by cyanobacteria. Microb Cell Fact. 2005;4(1):36.

50. Melis A, Zhang L, Forestier M, Ghirardi M, Seibert M. Sustained photobiological hydrogen gas production upon reversible inactivation of oxygen evolution in the green alga *Chlamydomonas reinhardtii*. Plant Physiol. 2000;122(1):127–136.

51. Kruse O, Rupprecht J, Bader K-P, Thomas-Hall S, Schenk PM, Finazzi G, et al. Improved photobiological H_2 production in engineered green algal cells. J Biol Chem. 2005;280(40):34170–34177.

52. Kargul J, Barber J. Photosynthetic acclimation: structural

reorganisation of light harvesting antenna - role of redox-dependent phosphorylation of major and minor chlorophyll *a/b* binding proteins. FEBS J. 2008;275(6):1056–1068.

53. Oey M, Ross IL, Stephens E, Steinbeck J, Wolf J, Radzun KA, et al. RNAi knock-down of LHCBM1, 2 and 3 increases photosynthetic H$_2$ production efficiency of the green alga *Chlamydomonas reinhardtii*. PLoS ONE. 2013;8(4):e61375.

54. Angermayr SA, Hellingwerf KJ, Lindblad P, Teixeira de Mattos MJ. Energy biotechnology with cyanobacteria. Curr Opin Biotechnol. 2009;20(3):257–263.

55. Savakis P, Hellingwerf KJ. Engineering cyanobacteria for direct biofuel production from CO$_2$. Curr Opin Biotechnol. 2015;33:8–14.

56. van der Woude AD, Angermayr SA, Puthan Veetil V, Osnato A, Hellingwerf KJ. Carbon sink removal: increased photosynthetic production of lactic acid by *Synechocystis* sp. PCC6803 in a glycogen storage mutant. J Biotech. 2014;184:100–102.

57. Wijffels RH, Kruse O, Hellingwerf KJ. Potential of industrial biotechnology with cyanobacteria and eukaryotic microalgae. Curr Opin Biotechnol. 2013;24(3):405–413.

58. Gao Z, Zhao H, Li Z, Tan X, Lu X. Photosynthetic production of ethanol from carbon dioxide in genetically engineered cyanobacteria. Energy Environ Sci. 2012;5(12):9857.

59. Qi F, Yao L, Tan X, Lu X. Construction, characterization and application of molecular tools for metabolic engineering of *Synechocystis* sp. Biotechnol Lett. 2013;35(10):1655–1661.

60. Ho SH, Ye X, Hasunuma T, Chang JS, Kondo A. Perspectives on engineering strategies for improving biofuel production from microalgae – a critical review. Biotechnol Adv. 2014;32(8):1448–1459.

61. PC Lai E. Biodiesel: environmental friendly alternative to petrodiesel. J Pet Env. Biotechnol. 2014;5(1).

62. Ragauskas AME, Ragauskas AJ. Re-defining the future of FOG and biodiesel. J Pet Environ Biotechnol. 2013;4(1).

63. Talebi AF, Mohtashami SK, Tabatabaei M, Tohidfar M, Bagheri A, Zeinalabedini M, et al. Fatty acids profiling: a selective criterion for screening microalgae strains for biodiesel production. Algal Res. 2013;2(3):258–267.

64. Trudewind CA, Schreiber A, Haumann D. Photocatalytic methanol and methane production using captured CO$_2$ from coal power plants. Part II – well-to-wheel analysis on fuels for passenger transportation services. J Clean Prod. 2014;70:38–49.

65. Sialve B, Bernet N, Bernard O. Anaerobic digestion of microalgae as a necessary step to make microalgal biodiesel sustainable. Biotechnol Adv. 2009;27(4):409–416.

66. Rittmann BE. Opportunities for renewable bioenergy using microorganisms. Biotechnol Bioeng. 2008;100(2):203–212.

67. Thapper A, Styring S, Saracco G, Rutherford AW, Robert B, Magnuson A, et al. Artificial photosynthesis for solar fuels – an evolving research field within AMPEA, a joint programme of the european energy research alliance. Green. 2013;3(1):43–57.

68. Ocakoglu K, Joya KS, Harputlu E, Tarnowska A, Gryko DT. A nanoscale bio-inspired light-harvesting system developed from self-assembled alkyl-functionalized metallochlorin nano-aggregates. Nanoscale. 2014;6(16):9625.

69. Llansola-Portoles MJ, Bergkamp JJ, Tomlin J, Moore TA, Kodis G, Moore AL, et al. Photoinduced electron transfer in perylene-TiO$_2$ nanoassemblies. Photochem Photobiol. 2013;89(6):1375–1382.

70. Ihssen J, Braun A, Faccio G, Gajda-Schrantz K, Thöny-Meyer L. Light harvesting proteins for solar fuel generation in bioengineered photoelectrochemical cells. Curr Protein Pept Sci. 2014;15(4):374–384.

71. Duan L, Bozoglian F, Mandal S, Stewart B, Privalov T, Llobet A, et al. A molecular ruthenium catalyst with water-oxidation activity comparable to that of photosystem II. Nat Chem. 2012;4(5):418–423.

72. Kanan MW, Nocera DG. In situ formation of an oxygen-evolving catalyst in neutral water containing phosphate and CO^{2+}. Science. 2008;321(5892):1072–1075.

73. Reece SY, Hamel JA, Sung K, Jarvi TD, Esswein AJ, Pijpers JJH, et al. Wireless solar water splitting using silicon-based semiconductors and earth-abundant catalysts. Science. 2011;334(6056):645–648.

74. Kanan MW, Yano J, Surendranath Y, Dincă M, Yachandra VK, Nocera DG. Structure and valency of a cobalt–phosphate water oxidation catalyst determined by in situ X-ray spectroscopy. J Am Chem Soc. 2010;132(39):13692–13701.

75. Tran PD, Wong LH, Barber J, Loo JSC. Recent advances in hybrid photocatalysts for solar fuel production. Energy Environ Sci. 2012;5(3):5902.

76. Bensaid S, Centi G, Garrone E, Perathoner S, Saracco G. Towards artificial leaves for solar hydrogen and fuels from carbon dioxide. ChemSusChem. 2012;5(3):500–521.

77. Kim JH, Nam DH, Park CB. Nanobiocatalytic assemblies for artificial photosynthesis. Curr Opin Biotechnol. 2014;28:1–9.

78. Benson EE, Kubiak CP, Sathrum AJ, Smieja JM. Electrocatalytic and homogeneous approaches to conversion of CO$_2$ to liquid fuels. Chem Soc Rev. 2009;38(1):89.

79. Bora DK, Braun A, Constable EC. "In rust we trust". Hematite – the prospective inorganic backbone for artificial photosynthesis. Energy Environ Sci. 2013;6(2):407–425.

80. Bora DK, Braun A, Erni R, Müller U, Döbeli M, Constable EC. Hematite–NiO/α-Ni(OH)$_2$ heterostructure photoanodes with high electrocatalytic current density and charge storage capacity. Phys Chem Chem Phys. 2013;15(30):12648.

81. Gajda-Schrantz K, Tymen S, Boudoire F, Toth R, Bora DK, Calvet W, et al. Formation of an electron hole doped film in the α-Fe$_2$O$_3$ photoanode upon electrochemical oxidation. Phys Chem Chem Phys. 2013;15(5):1443.

82. Bora DK, Rozhkova EA, Schrantz K, Wyss PP, Braun A, Graule T, et al. Functionalization of nanostructured hematite thin-film electrodes with the light-harvesting membrane protein C-phycocyanin yields an enhanced photocurrent. Adv Funct Mater. 2012;22(3):490–502.

83. Kothe T, Plumeré N, Badura A, Nowaczyk MM, Guschin DA, Rögner M, et al. Combination of a photosystem 1-based photocathode and a photosystem 2-based photoanode to a Z-scheme mimic for biophotovoltaic applications. Angew Chem Int Ed Engl. 2013;52(52):14233–14236.

84. Badura A, Guschin D, Esper B, Kothe T, Neugebauer S, Schuhmann W, et al. Photo-induced electron transfer between photosystem 2 via cross-linked redox hydrogels. Electroanalysis. 2008;20(10):1043–1047.

85. Badura A, Guschin D, Kothe T, Kopczak MJ, Schuhmann W, Rögner M. Photocurrent generation by photosystem 1 integrated in crosslinked redox hydrogels. Energy Environ Sci. 2011;4(7):2435.

86. Mershin A, Matsumoto K, Kaiser L, Yu D, Vaughn M, Nazeeruddin MK, et al. Self-assembled photosystem-I biophotovoltaics on nanostructured TiO$_2$ and ZnO. Sci Rep. 2012;2:1–7.

87. Wenk SO, Qian DJ, Wakayama T, Nakamura C, Zorin N, Rögner M,

et al. Biomolecular device for photoinduced hydrogen production. Int J Hydrog. Energy. 2002;27(11–12):1489–1493.

88. Yehezkeli O, Tel-Vered R, Michaeli D, Nechushtai R, Willner I. Photosystem I (PSI)/photosystem II (PSII)-based photo-bioelectrochemical cells revealing directional generation of photocurrents. Small. 2013;9(17):2970–2978.

89. Wang W, Chen J, Li C, Tian W. Achieving solar overall water splitting with hybrid photosystems of photosystem II and artificial photocatalysts. Nat Commun. 2014;5:4647.

90. Kato M, Cardona T, Rutherford AW, Reisner E. Photoelectrochemical water oxidation with photosystem II integrated in a mesoporous indium–tin oxide electrode. J Am Chem Soc. 2012;134(20):8332–8335.

91. Kato M, Cardona T, Rutherford AW, Reisner E. Covalent immobilization of oriented photosystem II on a nanostructured electrode for solar water oxidation. J Am Chem Soc. 2013;135(29):10610–10613.

92. Sun J, Zhang J, Zhang M, Antonietti M, Fu X, Wang X. Bioinspired hollow semiconductor nanospheres as photosynthetic nanoparticles. Nat Commun. 2012;3:1139.

93. Engel GS, Calhoun TR, Read EL, Ahn TK, Mančal T, Cheng YC, et al. Evidence for wavelike energy transfer through quantum coherence in photosynthetic systems. Nature. 2007;446(7137):782–786.

94. Zhou H, Guo J, Li P, Fan T, Zhang D, Ye J. Leaf-architectured 3D hierarchical artificial photosynthetic system of perovskite titanates towards CO_2 photoreduction into hydrocarbon fuels. Sci Rep. 2013;3:1667.

95. Larkum AWD. Harvesting solar energy through natural or artificial photosynthesis: scientific, social, political and economic implications. In: Wydrzynski TJ, Hillier W, editors. Molecular solar fuels. Cambridge: Royal Society of Chemistry; 2011. p. 1–19.

96. Kato M, Zhang JZ, Paul N, Reisner E. Protein film photoelectrochemistry of the water oxidation enzyme photosystem II. Chem Soc Rev. 2014;43(18):6485.

97. Redinbo MR, Cascio D, Choukair MK, Rice D, Merchant S, Yeates TO. The 1.5-.ANG. crystal structure of plastocyanin from the green alga *Chlamydomonas reinhardtii*. Biochemistry. 1993;32(40):10560–10567.

98. Kameda H, Hirabayashi K, Wada K, Fukuyama K. Mapping of protein-protein interaction sites in the plant-type [2Fe-2S] ferredoxin. PloS One. 2011;6(7):e21947.

13

Identification and expression analysis of a novel phytocystatin in developing and germinating seeds of triticale (×*Triticosecale* Wittm.)

Joanna Simińska*, Wiesław Bielawski

Department of Biochemistry, Faculty of Agriculture and Biology, Warsaw University of Life Sciences – SGGW, Nowoursynowska 159, 02-776 Warsaw, Poland

Abstract

In this paper the complete cDNA sequence of a newly identified triticale phytocystatin, TrcC-7, was analyzed. Because *TrcC-7* transcripts were present in seeds, we hypothesized that it may regulate storage protein accumulation and degradation. Therefore, changes in mRNA and protein levels during the entire period of seed development and germination were examined. Expression of *TrcC-7* increased during development and decreased at the end of maturation and subsequently increased during seed germination. Based on these results, TrcC-7 likely regulates cysteine proteinase activity during the accumulation and mobilization of storage proteins.

Keywords: phytocystatin; cysteine proteinase inhibitor; seed development; germination

Introduction

In cereal seeds, germinating embryos use accumulated storage materials, which are primarily starch, proteins and lipids. Most protein accumulation occurs during the middle and late maturation stages. The largest group of proteinases responsible for degradation and mobilization of storage proteins during germination and seedling growth are cysteine proteinases [1]. One mechanism of controlling the activity of these enzymes involves specific inhibitors, phytocystatins (PhyCys).

To date, 5 PhyCys have been identified in triticale (×*Triticosecale* Wittm.), and one (TrcC-4) has been shown to have inhibitory activity against endogenous cysteine proteinase EP8, what may be related to pre-harvest sprouting tolerance [2–4]. Therefore, we examined another triticale phytocystatin. Because the transcripts of *TrcC-7* were present in developing and germinating seeds, we postulated that it may be involved in seed development and germination. To verify this hypothesis, gene expression analysis was performed.

Material and methods

Plant material

Two cultivars of triticale that differ in their resistance to pre-harvest sprouting, Hortenso (more resistant) and Leontino (less resistant), were analyzed. The seeds were provided by Danko Plant Breeders Ltd. (Laski, Poland).

RNA extraction

The total RNA from seeds was extracted according to the Chomczynski and Sacchi method [5], which was preceded by an extraction in 50 mM Tris-HCl pH 9.0, 200 mM NaCl, 1% sarcosyl, 20 mM EDTA, 5 mM DTT and a further extraction in phenol:chloroform:isoamyl alcohol (24.5:24.5:1). Total RNA was treated with RNase-free DNase (Applied Biosystems, USA) according to the manufacturer's protocol.

Sequence identification

First-strand cDNA was synthesized with a Reverse Transcription System (Promega, USA) according to the manufacturer's instructions. PCR primers were designed with the gene sequence of barley HvCPI-8 phytocystatin (Gen Bank: CAG38129), which do not have known homologues in wheat or rye. This sequence was aligned with known triticale phytocystatin sequences: TrcC-1 (GU395200); TrcC-4 (GU395201); TrcC-5 (GU395202); TrcC-8 (JX003861); TrcC-9 (HO068312) and regions characteristic exclusively for HvCPI-8 were selected. PCR with primers complementary to those regions resulted in one-band product. After obtaining full *TrcC-7* gene sequence single nucleotide change (A to G) in the region recognized by forward primer was revealed. However, this variation did not prevent primer hybridization. The 5′ and 3′ ends of *TrcC-7* sequence were amplified using a GeneRacer Kit (Invitrogen, USA) with specific and GeneRacer primers (Tab. 1).

* Corresponding author. Email: joanna_siminska@sggw.pl

Handling Editor: Elżbieta Bednarska-Kozakiewicz

Tab. 1 PCR primers used for identification and expression analysis of *TrcC-7*.

Gene	Product size (bp)	Type of reaction	Forward primer (5′–3′)	Reverse primer (5′–3′)
TrcC-7	215	PCR; rqRT-PCR	ATCCCGGACGTGAAGGAC	GTCCAGGACTGCTCGTAG
EF1α	109	PCR; rqRT-PCR	GATCAGCAACGGCTATGCC	CTCAATCTCCTTGCCAGACC
TrcC-7	470	5′RACE	GeneRacer 5′ Primer	GGTCCAGGACTGCTCGTAGACCTC
TrcC-7	534	3′RACE	CGAGCAGCAGGTCGTCTCCG	GeneRacer Nested 3′ Primer

Relative Quantitative RT-PCR

The mRNA level of *TrcC-7* was quantified with a Titanium One-Step RT-PCR kit (Clontech Laboratories Inc., USA). In all reactions, 20 ng of RNA was used as a template. As an internal control for RNA quantity, the *EF1α* (*elongation factor 1α*) gene was amplified. (Tab. 1).

Bioinformatic analysis

Primer sequences were designed with Primer3 v.0.4.0 software [6]. The nucleotide and amino acid sequences of TrcC-7 were analyzed with the following tools: MAFFT v7.130b [7], EMBOSS Transeq [8] and ProtParam [9]. To identify signal peptide, SignalP 4.1 was used [10].

Western Blotting

12 µg of protein extracts from seeds were used for SDS-PAGE. Proteins were transferred onto PVDF membrane (0.2 µm; Merck, Germany). For detection of TrcC-7, rabbit anti-TrcC-7 polyclonal antibodies against 16 aa peptide H-VALGGRGARVGGWGPI-NH₂ (Eurogentec, Belgium) which distinguishes TrcC-7 from other known triticale PhyCys were used. Anti-rabbit IgG alkaline phosphatase conjugated secondary antibodies (Sigma-Aldrich, USA) were used for visualization with BCIP/NBT.

Results

Identification and sequence analysis of new phytocystatin

In embryos of the Hortenso and Leontino varieties of winter triticale, unique phytocystatin transcripts are present. The cDNA fragment (215 bp) of *TrcC-7* was obtained by reverse transcription of mRNA extracted from embryos at 8 hours after imbibition. This sequence was extended in the 5′ and 3′ directions using RACE. The complete gene sequence of phytocystatin, along with non-coding regions was 826 bp long. The open reading frame (ORF) was 369 bp. The resulting phytocystatin gene sequence was named *TrcC-7* (GenBank: KJ209713). Gene sequence alignment with other Poaceae phytocystatins is shown in supplementary material (Fig. S1). This sequence was identical in the Hortenso and Leontino varieties. Neither gene had introns, as the PCR products from cDNA and genomic DNA templates were the same lengths. The predicted amino acid sequence of the new triticale phytocystatin was used for further bioinformatic analysis. *TrcC-7* cDNA encodes a protein of 123 amino acids with a molecular mass of 13.0 kDa and a theoretical pI 9.50, which shows the highest identity to HvCPI-8 (89.34%) from barley. Sequence analysis (Fig. 1) demonstrated that

this inhibitor have 3 characteristic cystatin motifs, which are responsible for enzyme-inhibitor interactions and a motif with unknown function, LARFAVxEHN-like, which is characteristic of plant cystatins. Also TrcC-7 most likely has signal peptide. Phylogenetic analysis showed that TrcC-7 belongs to phylogenetic group C, identified by Martinez et al. [11], along with another triticale PhyCys TrcC-5. However, TrcC-7 was assigned to subgroup C1 and TrcC-5 to subgroup C2 (Fig. S2).

Analysis of expression

TrcC-7 mRNA levels were examined during seed development, which lasts approximately 50 days. The gene was expressed during the entire period, but transcript levels changed depending on the stage of seed development (Fig. 2a). *TrcC-7* expression pattern showed a rapid increase in expression between 4 and 10 DAP, maximum at 14 DAP after which the transcripts progressively decreased until the seeds reached maturity. There were no significant differences in gene expression in the plant varieties examined. These results were confirmed by the presence of TrcC-7 protein in developing seeds during the same stages (Fig. 3a). Changes in gene expression were also observed during seed germination in both embryos (Fig. 2b, Fig. 3b) and endosperm (Fig. 2c, Fig. 3c). In the Hortenso cultivar, the initially low *TrcC-7* mRNA level increased starting at 12 HAI, decreased at approximately 24 HAI and reached its maximum at the last examined time point. A similar expression pattern was observed in the Leontino cultivar, but the increase in transcripts started earlier, at approximately 8 HAI. Protein levels also increased during germination, but starting at 36 HAI for Hortenso and 24 for Leontino. *TrcC-7* expression in the endosperm was low and constant for both mRNA and protein.

Discussion

Numerous phytocystatins are present in several Poaceae family species. There are 13 known phytocystatins in barley [11], 12 in rice [12,13], 10 in maize [14] and 6 in wheat [15,16], 5 in triticale [2]. The new PhyCys described in this paper, TrcC-7, is characterized by conserved regions that are present in most cystatin superfamily inhibitors and most likely has signal peptide (Fig. 1). Bioinformatic analysis also suggested that the signal peptide directs the PhyCys to ER and, eventually, to the extracellular space. This is consistent with the results presented for barley [17], wheat [18] and rice [19] PhyCys.

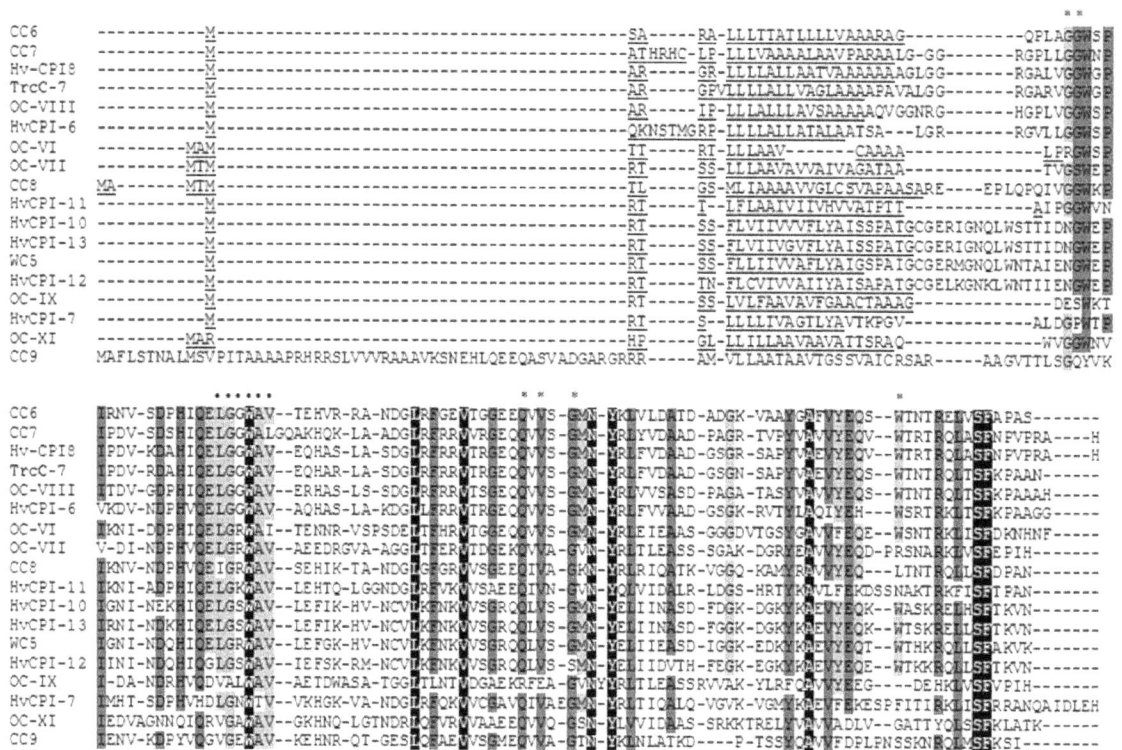

Fig. 1 Alignment of deduced amino acid sequence of TrcC-7 with phytocystatins from phylogenetic group C (subgroup C1 and C2) created by MAFFT. The conserved sequences of the cystatin superfamily are marked by asterisks. Region characteristic for phytocystatin family is marked by dots. Highly conserved amino acids (100% identity) are white letters on black, less conserved (more than 80% identity) are black letters shaded in dark gray. Amino acids less conserved are shaded in light gray. Putative signal peptides are underlined.

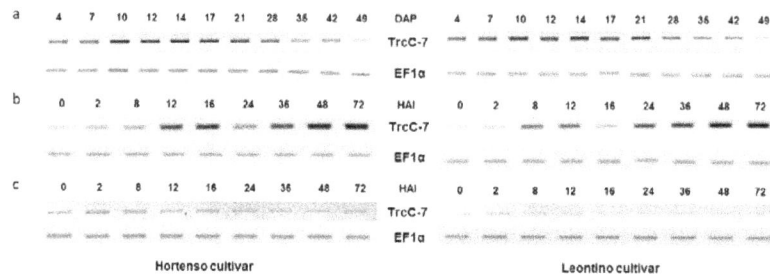

Fig. 2 *TrcC-7* expression patterns. The constitutively expressed *EF1α* was used as a control. **a** Whole seeds during development. **b** Embryos during germination. **c** Endosperms during germination. DAP – day after pollination; HAI – hour after imbibition.

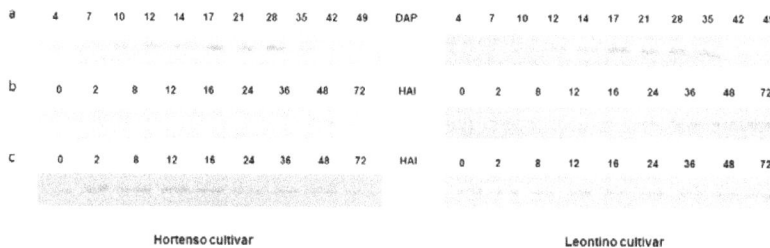

Fig. 3 TrcC-7 protein expression patterns. **a** Whole seeds during development. **b** Embryos during germination. **c** Endosperms during germination. DAP – day after pollination; HAI – hour after imbibition.

TrcC-7 is expressed throughout seed development (Fig. 2a, Fig. 3a). Its transcripts increased during seed development between 10 and 28 DAP. Thus, TrcC-7 may possibly control proteolysis during embryo development and the accumulation of storage proteins. In the final stages of seed maturation, *TrcC-7* mRNA and protein decreased. They were low also in mature seeds, in both embryo and endosperm, and through the first 8 HAI (mRNA) and first 24–36 HAI (protein). This result is similar to *CC6* and *CC7* (subgroup C1) and *CC8* and *CC9* (subgroup C2) [14]. These genes are expressed during seed development, but their expression decreases during filling and maturation stages. This indicates they are not crucial for pre-harvest sprouting tolerance, but for protein accumulation. Other expression patterns are observed for barley PhyCys: *HvCPI-6* and *HvCPI-8* from subgroup C1

[11]. In conclusion, PhyCys in group C (subgroups C1 and C2) exhibit considerable variation in expression during seed development and most likely play various functions during this process. During germination, *TrcC-7* mRNA and protein levels increased only in embryos and remained unchanged in endosperms. Such expression patterns are similar to barley *HvCPI-6* and *HvCPI-8* from subgroup C1 [11]. Although their expression in embryos peaked at 8 HAI and then significantly decreased, it gradually increased from 16 HAI.

These results suggest that the newly identified triticale phytocystatin, TrcC-7, may be involved in the control of cysteine proteinase activity during embryo development and the accumulation and processing of storage proteins in developing seeds. It is also most likely essential during germination, when storage proteins degradation occurs.

Acknowledgments

This work was supported by the National Science Centre (Poland) grant No. DEC-2011/03/N/NZ9/04115.

Authors' contributions

The following declarations about authors' contributions to the research have been made: identification of a new phytocystatin, bioinformatic analysis, mRNA expression analysis, western blotting: JS; writing the manuscript: JS; revising and final approval of the manuscript: WB.

Competing interests

No competing interests have been declared.

Supplementary material

The following supplementary material for this article is available online at http://pbsociety.org.pl/journals/index.php/asbp/rt/suppFiles/asbp.2015.011/0:
1. Fig. S1: comparison of gene sequences encoding phytocystatins from phylogenetic subgroup C1 with *TrcC-7*.
2. Fig. S2: unrooted phylogenetic tree of Poaceae phytocystatins, including 6 from triticale.

References

1. Grudkowska M, Zagdańska B. Multifunctional role of plant cysteine proteinases. Acta Biochim Polon. 2004;51:609–624.

2. Szewińska J, Zdunek-Zastocka E, Pojmaj M, Bielawski W. Molecular cloning and expression analysis of triticale phytocystatins during development and germination of seeds. Plant Mol Biol Rep. 2012;30:867–877.

3. Prabucka B, Drzymała A, Grabowska A. Molecular cloning and expression analysis of the main gliadin-degrading cysteine endopeptidase EP8 from triticale. J Cereal Sci. 2013;58:284–289.

4. Szewińska J, Prabucka B, Krawczyk M, Mielecki M, Bielawski W. The participation of phytocystatin TrcC-4 in the activity regulation of EP8, the main prolamin degrading cysteine endopeptidase in triticale seeds. Plant Growth Regul. 2013;69:131–137.

5. Chomczynski P, Sacchi N. Single-step method of total RNA isolation by acid guanidinium thiocyanate-phenol-chloroform extraction. Anal Biochem. 1987;162:156–159.

6. Untergrasser A, Cutcutache I, Koressaar T, Ye J, Faircloth BC, Remm M, Rozen SG. Primer3 – new capabilities and interfaces. Nucleic Acids Res. 2012;40(15):e115.

7. Katoh K, Standley DM. MAFFT multiple sequence alignment software version 7: improvements in performance and usability. Mol Biol Evol. 2013;30:772–780.

8. Goujon M, McWilliam H, Li W, Valentin F, Squizzato S, Paern J, Lopez R. A new bioinformatics analysis tools framework at EMBL-EBI. Nucleic Acids Res. 2010;38(2 suppl):W695–W699.

9. Gasteiger E, Hoogland C, Gattiker A, Duvaud S, Wilkins MR, Appel RD, Bairoch A. Protein identification and analysis tools on the ExPASy server. In: Walker JM, editor. The proteomics protocols handbook. New York, NY: Humana Press; 2005. p. 571–607.

10. Petersen TN, Brunak S, von Heijne G, Nielsen H. SignalP 4.0: discriminating signal peptides from transmembrane regions. Nat Methods. 2011;8:785–786.

11. Martinez M, Cambra I, Carrillo L, Diaz-Mendoza M, Diaz I. Characterization of the entire cystatin gene family in barley and their target cathepsin L-like cysteine-proteases, partners in the hordein mobilization during seed germination. Plant Physiol. 2009;151:1531–1545.

12. Abe K, Emori Y, Kondo H, Suzuki K, Arai S. Molecular cloning of a cysteine proteinase inhibitor of rice (oryzacystatin). Homology with animal cystatins and transient expression in the ripening process of rice seeds. J Biol Chem. 1987;262:16793–16797.

13. Kondo H, Abe K, Nishimura I, Watanabe H, Emori Y, Arai S. Two distinct cystatin species in rice seeds with different specificities against cysteine proteinases. Molecular cloning, expression, and biochemical studies on oryzacystatin-II. J Biol Chem. 1990;265:15832–15837.

14. Massonneau A, Condamine P, Wisniewski JP, Zivy M, Rogowsky PM. Maize cystatins respond to developmental cues, cold stress and drought. Biochim Biophys Acta. 2005;1729:186–199.

15. Corre-Menguy F, Cejudo FJ, Mazubert C, Vidal J, Lelandais-Brière C, Torres G, Rode A, Hartmann C. Characterization of the expression of a wheat cystatin gene during caryopsis development. Plant Mol Biol. 2002;50:687–698.

16. Kuroda M, Kiyosaki T, Matsumoto I, Misaka T, Arai S, Abe K. Molecular cloning, characterization, and expression of wheat cystatins. Biosci Biotechnol Biochem. 2001;65:22–28.

17. Abraham Z, Martinez M, Carbonero P, Diaz I. Structural and functional diversity within the cystatin gene family of *Hordeum vulgare*. J Exp Bot. 2006;57:4245–4255.

18. Dutt S, Singh VK, Marla SS, Kumar A. In silico analysis of sequential, structural and functional diversity of wheat cystatins and its implication in plant defense. Genomics Proteomics Bioinformatics. 2010;8:42–56.

19. Womack JS, Randall J, Kemp JD. Identifcation of a signal peptide for oryzacystatin-I. Planta. 2000;210:844–847.

14

Extraordinary accuracy in floret position of *Helianthus annuus*

Takuya Okabe*

Graduate School of Integrated Science and Technology, Shizuoka University, 3-5-1 Johoku, Naka-ku, 432-8561 Hamamatsu, Japan

Abstract

Divergence angles were measured for inflorescences of *Helianthus annuus* with several hundreds to more than a thousand disk flowers. Quantitative analysis showed that the angles are robustly fixed in the vicinity of the ideal golden angle 137.508° as accurately as ~0.001°. The mean deviation from the ideal value varies for each sample. The results have important implications for phyllotaxis models, which are discussed by referring to a necessary modification proposed by Roberts.

Keywords: phyllotaxy; Asteraceae; contact pressure model; expanding apex model; Hofmeister's rule

Introduction

Phyllotaxis patterns of leaves or modified leaves in general on a plant stem are classified into several types and each type as well as its transformation are rich in diversity. Distichy and simple spiral patterns have one leaf at each node, while decussate and whorled patterns have multiple leaves at a node. Each spiral direction occurs with equal probability [1]. Some plants may have interestingly variable patterns, e.g., a female catkin of *Betula alba*, an ear of *Plantago major*, bracts of *Pl. media* [1], a vegetative shoot of *Abies balsamea* [2], gynoecium of *Magnolia acuminata* [3,4], capitulum of *Carlina acaulis* [5], primary vascular system of *Anagallis arvensis* [6], areoles of the family Cactaceae [7]. Thus there is no doubt that not all phyllotaxis properties are genetically determined. Independently of the diversity problems, simple spiral phyllotaxis in itself is especially noted not only for its predominant occurrence but because divergence angles between successive leaves are precisely fixed at a very special value. The spiral arrangement is so precise that Hirmer particularly called it "spiral phyllotaxis with precise divergences" to distinguish from variable counterparts related to distichy and decussate systems [8]. It is interesting to note that, in the family Cactaceae, the tribe Cacteae in which the simple spiral pattern dominates has more precise divergence angles than other tribes in which the decussate pattern occurs with high frequency [7].

Normally, the divergence angle of a simple spiral pattern is intriguingly close to the golden angle $\alpha_{gold} = 360/(1 + \tau) \cong$ 137.508°, which is the smaller one of the two angles created by sectioning the circumference of a full circle according to the golden proportion: $1:\tau = \tau - 1:1 \cong 1:1.618$.

The invariability of divergence angle was already remarked by Bravais and Bravais [9] (p. 69, 106): "By direct measurements, we find constantly an angle differing very little from 137° or 138°, and this precision may be very surprising if we note that on a stem of average size, for example a centimeter in diameter, the arc of 1° does not exceed the 8 hundredth of a millimeter, a quantity almost inappreciable to the naked eye.". "The genetic spiral extends to the underground stem by keeping the same invariable divergence, and even to the organs of the flower.". The accurate occurrence of the limit divergence α_{gold} was underscored by van Iterson [10]. Comparing with his theory, he wrote (p. 247): "However, especially in cases of the contact 1 and 2 at the apex, the trouble becomes apparent. While the theory allows all divergences between 180° and 128.5° as possible for this contact, one finds only divergences between 137° and 142°, namely a great approach to the limit divergence of the main sequence.". The uniqueness of the golden angle was strongly supported by Hirmer and his colleagues (see references cited in [11]). To cite an example, for composite flowers of *Rudbeckia laciniata*, *Lepachys pinnatifida*, *Chrysanthemum millefoliatum*, *Bidens leucantha*, *Galinsoga parvifiora*, *Cineraria lobata* and others, it was argued that the number of ray florets tends to be Fibonacci numbers like 5, 8, 13, 21, 34 because the ray florets are regularly arranged with the golden angle [8]. The universality of the golden angle was directly demonstrated by Fujita [11], who measured divergence angles in bud sections of thirty species of flowering plants with one degree accuracy. Showing distribution curves conspicuously peaked at 137° or 138°, he remarked: "From these data we can see that the peak point of the variation curve of divergence comes strongly close to the limit value, though with a large variation width. In other words, in the regular spiral arrangement, the divergence follows the limit value, as Hirmer claims, independently of the numbers of conjugate parastichies.". Snow suggested that the golden

---

* Email: okabe.takuya@shizuoka.ac.jp

Handling Editor: Beata Zagórska-Marek

angle might have some biological advantage by referring to the difficulty of van Iterson underlined by this result of Fujita [12]. As for sample variance, it was reported for young shoots of *Erigeron sumatrensis* that each shoot has an individual deviation from the golden angle and the mean average was 137.499° ±0.212° [13]. The first quantitative analysis of floret patterns of *Helianthus* was made by Ryan et al. [14]. They reported systematic and persistent fluctuations of ±5° about the ideal value and concluded that capitula do not possess a uniquely defined growth center. As reported for *Helianthus annuus* [15], divergence angles of mature leaves are not measured with a better accuracy because of secondary twist of the stem. As a matter of fact, secondary deformation of the phyllotaxis pattern generally causes apparent fluctuations in divergence angle [16]. Therefore, measured angles may depend on the procedure to determine the center position of the phyllotaxis pattern [17]. Accordingly, a true deviation from the ideal value α_{gold} is almost always bound to be hidden beneath various practical errors by chance. Nevertheless, by whatever mechanism it occurs, there is no doubt that any particular real life system has a more or less definite deviation from the mathematical ideality $\alpha_{gold} = 137.50776\ldots$. The present study makes a first attempt at detecting and assessing a statistically significant deviation from the ideality for a representative system of spiral phyllotaxis. To suppress statistical errors in a systematic manner, a large number of divergence angles have to be collected from an individual sample. Inflorescences of the Asteraceae are the best suited to this end. In fact, as it turns out, they are perhaps the one and only subject with which to achieve the purpose. This paper reports that accuracy in divergence angle of *Helianthus annuus* amounts to as high as ~0.001°. Although this may seem implausible at first, it is immediately obvious by observing how regularly Fibonacci numbers (1, 2, 3, 5, 8, 13, 21, 34, 55, 89, 144, 233, 377, 610, 987, etc.) are arranged in Fig. 1 and Fig. 2. The measurement results of this study reinforce evidence for the accurate constancy of divergence angle to an unquestionable degree and may shed new light on the fundamental problem of phyllotaxis.

Fig. 1 A sunflower head with divergence angle $\alpha \simeq 137.513$ ±0.003°. Disk florets are numbered algorithmically [16]. They are formed counterclockwise inward. Adapted from Yellow sunflower 001/Wikimedia Commons.

Material and methods

The divergence angle α is evaluated by measuring the angle $\angle P_n O P_{n+q}$ subtended by two arbitrary florets P_n and P_{n+q}, where O denotes a fixed center [9]. Solving

$$\angle P_n O P_{n+q} = q\alpha - 360p \qquad (1)$$

for α,

$$\alpha_n(q) = (\angle P_n O P_{n+q} + 360p)/q. \qquad (2)$$

This is the main formula used in this study. The angle $\angle P_n O P_{n+q}$ is measured in the direction of the genetic spiral, that is, the imaginary spiral connecting all the florets in increasing order of the index n, so that the divergence angle $\alpha_n(q)$ is a positive number. On the right-hand side of Equation (1), an integer multiple of full turns, $360p$ in degrees of arc, is subtracted to reduce the total angle $q\alpha$ (>360°) to

a net angle within $\pm 180°$. The angle $\angle P_n O P_{n+q}$ is calculated from the xy-coordinates of two vectors OP_n and OP_{n+q}. The coordinates of the floret position P_n are digitally read from photographic images published in media file repositories, which are now of ideal quality incomparable with any figures in published papers. For indexing and data collection, the algorithm described previously was applied [16]. As the angle $\angle P_n O P_{n+q}$ comprises q divergence angles, the effect of secondary disturbances on each divergence angle is averaged out when the integer q is large enough. As a rough guide, the standard error in measuring $\angle P_n O P_{n+q}$ is typically less than 1 degree. Accordingly, accuracy of $\alpha_n(q)$ is less than about $1/q$ degree. As $\alpha_n(q)$ turns out to be very close to the golden angle, it is not necessary but convenient to employ Fibonacci numbers for q and p in the above formula, namely $(q, p) = (1,0), (2,1), (3,1), (5,2), (8,3), (13,5)$ and so on.

Fig. 2   A large sunflower head with $\alpha \simeq 137.507 \pm 0.001°$. The phyllotaxis fraction is as high as 377/987. Florets are formed clockwise inward. Adapted from Sunflower Closeup Hungary/Wikimedia Commons.

Results

The divergence angle $\alpha_n(q)$ was evaluated for disk florets of Fig. 1. The n-dependence of $\alpha_n(q)$ is qualitatively similar for any value of q, whereas the variation range depends on q. In Fig. 3, $\alpha_n(233)$ is plotted against the floret index n representatively. The mean and standard deviation of $\alpha_n(233)$ are 137.5139 ±0.0041, 137.5145 ±0.0038, 137.5080 ±0.0060, 137.5000 ±0.0155 for $n \le 100$, $100 < n \le 200$, $200 < n \le 300$, $300 < n$, respectively. Thus, $\alpha_n(q)$ is robustly constant over a wide range of n. This is surprising and not at all self-evident because the hundreds of florets were formed under varying conditions. The mean variation is about the same order as statistical errors, namely 0.01° at most. To suppress statistical noise including measurement error in $\angle P_n O P_{n+q}$, florets near the center were excluded from the statistical analysis. Tab. 1 presents the mean $\langle\alpha\rangle$ and standard deviation σ of $\alpha_n(q)$ evaluated by using the first 400 florets of Fig. 1. The constant mean indicates that the florets are formed in accordance with a fixed divergence angle $\simeq 137.51°$. On the other side, the standard deviation σ decreases as q increases, as expected. If the deviation σ is interpreted as a result of q independent random variations of individual divergence angles $\delta\alpha$ [cf. Equation (1)], then

$$\sigma \approx \delta\alpha/\sqrt{q}. \qquad (3)$$

This is a standard result of statistical theory. The results for large q in Tab. 1 are consistent with this relation for $\delta\alpha \simeq 0.06°$, which is also remarkable. The deviation σ for a small q represents the persistent fluctuations mentioned in Introduction, which therefore does not follow this relation. Owing to σ being suppressed systematically, the mean value begins to show a clear sign of deviation from the ideal value, that is, $\langle\alpha\rangle - \alpha_{gold} > 1.5\sigma$. See Fig. 3 and Tab. 1. This is supported by consistent results in Tab. 2 and Tab. 3 for other independent samples. Tab. 4 presents the results for

a larger sample of Fig. 2. Tab. 5 is for a packing pattern of mature seeds. The seed pattern has an order of magnitude larger values of σ and $\delta\alpha$, while the deviation $\langle\alpha\rangle - \alpha_{gold}$ is not affected comparatively. Thus it appears that the floret patterns are not altered substantially by the growth that follows their establishment.

These results indicate that the deviation $\langle\alpha\rangle - \alpha_{gold}$ is robust and typically of the order of a 10^{-3} degree. To suppress σ to this order, the sample size, or q for Equation (3), has to be larger than ~100 even when $\delta\alpha$ is as small as 0.05°. Thus, in hindsight, the necessary condition for detecting the deviation is barely satisfied for outer florets of the sunflower capitula. It is not asserted from the present study alone whether the higher accuracy of $\langle\alpha\rangle - \alpha_{gold} \simeq 10^{-4}$ in Tab. 4 is due to a large capitulum size or just a coincidence. It is interesting to see whether the minute individual difference identified in this study is used as a fingerprint of each plant. In fact, the flower head of Fig. 1 is distinguished from that of Fig. 2 by their different divergence angles, though the difference of 0.007° is too small to be noticed without special attention.

Discussion

Schwendener, Church and van Iterson attempted to explain the tendency to the limit divergence α_{gold} by postulating that contact parastichies tend to cross at right angles (p. 248 [10]; p. 340 [18]; see also [19]). They were criticized because the predicted angles are too close to α_{gold} to test empirically [8] and when the predicted deviation is large enough it is disproved by observation [20]. As a matter of fact, the crossing angles of contact parastichies are not even constant. The angle $\angle P_{n+34} P_n P_{n+55}$ in Fig. 1 varies from 116° for $n = 0$ to 49° for $n = 444$. In fact, divergence angles remain constant independently of the crossing angles of parastichies.

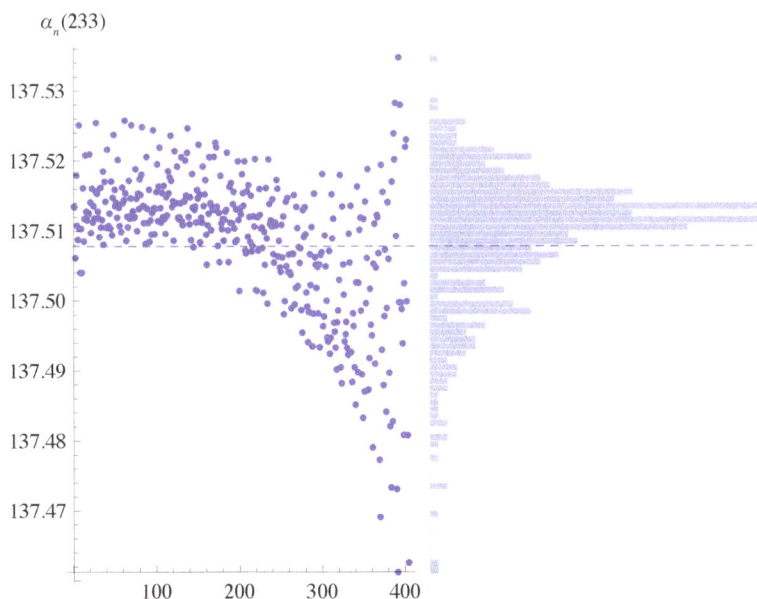

$\alpha_n(233)$

Fig. 3 The divergence angle $\alpha_n(233)$ for the pattern of Fig. 1 is plotted against the floret index n. The ideal golden angle of 137.508° is indicated by a dashed line.

Tab. 1 Divergence angles of disk flowers of a sunflower.

q	$\langle\alpha\rangle$	σ	$\sqrt{q}\sigma$	$\langle\alpha\rangle - \alpha_{gold}$	$(\langle\alpha\rangle - \alpha_{gold})/\sigma$
1	137.5054	2.2337	2.234	−0.0024	
2	137.5115	1.1829	1.673	0.0038	0.003
3	137.5130	0.6575	1.139	0.0052	0.08
5	137.5132	0.2970	0.664	0.0055	0.02
8	137.5126	0.1337	0.378	0.0048	0.04
13	137.5121	0.0605	0.218	0.0043	0.07
21	137.5130	0.0294	0.135	0.0053	0.18
34	137.5135	0.0147	0.086	0.0057	0.39
55	137.5136	0.0089	0.066	0.0059	0.66
89	137.5139	0.0064	0.061	0.0062	0.96
144	137.5142	0.0047	0.056	0.0064	1.4
233	137.5142	0.0040	0.061	0.0064	1.6

The mean $\langle\alpha\rangle$ and standard deviation σ of the divergence angle $\alpha_n(q)$ are evaluated for first 400 disk flowers of Fig. 1. The ideal golden angle is given by $\alpha_{gold} = 137.50776°$. The first row for $q = 1$ corresponds to the standard method of evaluating the angle between successive florets.

Tab. 2 Divergence angles of disk flowers of a sunflower.

q	$\langle\alpha\rangle$	σ	$\sqrt{q}\sigma$	$\langle\alpha\rangle - \alpha_{gold}$	$(\langle\alpha\rangle - \alpha_{gold})/\sigma$
1	137.5143	2.9275	2.928	0.0065	
34	137.5113	0.0182	0.106	0.0035	0.19
55	137.5117	0.0113	0.084	0.0039	0.34
89	137.5121	0.0078	0.073	0.0043	0.56
144	137.5125	0.0053	0.064	0.0048	0.90
233	137.5126	0.0042	0.064	0.0048	1.1

The mean $\langle\alpha\rangle$ and standard deviation σ of the divergence angle $\alpha_n(q)$ for 400 disk flowers of a sunflower head with about 580 florets and a parastichy pair of (34,55).

Tab. 3 Divergence angles of disk flowers of a sunflower.

q	$\langle\alpha\rangle$	σ	$\sqrt{q}\sigma$	$\langle\alpha\rangle - \alpha_{gold}$	$(\langle\alpha\rangle - \alpha_{gold})/\sigma$
1	137.4909	2.7422	2.742	−0.0169	
34	137.4988	0.0251	0.147	−0.0089	−0.36
55	137.4984	0.0171	0.127	−0.0094	−0.55
89	137.4986	0.0134	0.127	−0.0092	−0.68
144	137.4988	0.0096	0.115	−0.0090	−0.93
233	137.4987	0.0047	0.072	−0.0091	−1.9

The mean $\langle\alpha\rangle$ and standard deviation σ of the divergence angle $\alpha_n(q)$ for 330 disk flowers of a sunflower head with about 380 florets and a parastichy pair of (34,55).

Tab. 4 Divergence angles of disk flowers of a sunflower.

q	$\langle\alpha\rangle$	σ	$\sqrt{q}\sigma$	$\langle\alpha\rangle - \alpha_{gold}$	$(\langle\alpha\rangle - \alpha_{gold})/\sigma$
1	137.5067	2.2778	2.278	−0.0010	
55	137.5072	0.0067	0.050	−0.0006	−0.09
89	137.5072	0.0043	0.041	−0.0006	−0.14
144	137.5072	0.0028	0.034	−0.0006	−0.21
233	137.5071	0.0022	0.034	−0.0006	−0.29
377	137.5070	0.0016	0.032	−0.0008	−0.46
610	137.5069	0.0013	0.031	−0.0009	−0.67
987	137.5071	0.0010	0.031	−0.0006	−0.64

The mean $\langle\alpha\rangle$ and standard deviation σ of the divergence angle $\alpha_n(q)$ for 1161 florets among the total number of about 1700 in Fig. 2.

Tab. 5 Divergence angles of sunflower seeds.

q	$\langle\alpha\rangle$	σ	$\sqrt{q}\sigma$	$\langle\alpha\rangle - \alpha_{gold}$	$(\langle\alpha\rangle - \alpha_{gold})/\sigma$
1	137.5117	5.077	5.077	0.0039	
34	137.5112	0.0256	0.149	0.0034	0.13
55	137.5112	0.0204	0.151	0.0034	0.17
89	137.5119	0.0175	0.165	0.0041	0.24
144	137.5125	0.0163	0.196	0.0047	0.29
233	137.5126	0.0153	0.234	0.0049	0.32

For a seed packing. The mean $\langle\alpha\rangle$ and standard deviation σ of the divergence angle $\alpha_n(q)$ for 400 seeds of a sunflower head with about 550 florets and a parastichy pair of (34,55).

Roberts [21] proposed to make up deficiencies of phyllotaxis models [22,23]. One of them concerns the accuracy problem. His remark applies to any model based on the premise that the position of a new primordium is determined by the existing primordia. The difficulty has been left unaddressed in recent models [24,25]. The pattern in which P_n has contact with P_{n-q} and $P_{n-q'}$ is called a (q,q') system. In any pattern, including a $(1,2)$ system, a correct model should put P_3 at the golden section point in the largest gap between P_1 and P_2 (Appendix S1). To account for this, Roberts proposed to modify the premise such that the position of a new primordium (e.g., P_3) is determined not by its near neighbors (e.g., P_1 and P_2) but by older primordia remotely located from it (e.g., P_0 and P_{-2}). In a word, it is equivalent to postulating that internal or "chemical" contact of the organs is of higher order than it appears from the outside. To meet the accuracy observed in this study, the chemical contact has to be supposed as high as (89,144), even if the pattern appears to be (34,55). Take Fig. 1 for example. The mean and standard deviation of $\angle P_{n-89}OP_n$ and $\angle P_{n-144}OP_n$ are $-1.3 \pm 0.6 < 0$ and $2.0 \pm 0.7 > 0$, respectively. The fact that the former (latter) is negative (positive) means that P_n lies on the negative (positive) side of OP_{n-89} (OP_{n-144}). Accordingly,

P_n lies between the narrow angle $\angle P_{n-144}OP_{n-89} \cong 3°$ (in the gap $\angle P_{n-55}OP_{n-34} \cong 8°$). See Fig. 1. P_{377} lies between P_{233} and P_{288}. For the sake of argument, let it be assumed that P_n is determined by P_{n-89} and P_{n-144} (i.e., P_{377} by P_{288} and P_{233}). Then their interaction, of whatever sort, has to be such as to meet the following requirements. (*i*) **Constancy problem.** For each sample, the angle $\angle P_{n-q}OP_n$ is fixed independently of the floret *n*. (*ii*) **Individuality problem.** The fixed angle $\angle P_{n-q}OP_n$ varies for each sample. The former refers to the robustness of $\alpha_n(q)$ – cf. Fig. 3. This is evidenced by $\angle P_{n-34}OP_n = -4.5 \pm 0.5$, $\angle P_{n-55}OP_n = 3.2 \pm 0.5$ and $\angle P_{n-89}OP_n = -1.3 \pm 0.6$ for Fig. 1. The latter refers to the sample dependence of $\alpha_n(q)$. Remark that $\alpha_n(q)$ by Equation (2) is given by the angle $\angle P_n OP_{n+q}$, which is determined by the assumed interaction. To illustrate (*ii*), $\angle P_{n-89}OP_n = -1.3 \pm 0.6$ for Tab. 1 is contrasted with $\angle P_{n-89}OP_n = -2.8 \pm 1.2$ for Tab. 3. For reference, the pattern of Fig. 2 has $\angle P_{n-89}OP_n = -1.86 \pm 0.38$, whereas the exact golden angle α_{gold} gives $\angle P_{n-89}OP_n = -1.8090$. Furthermore, (*iii*) the angles satisfy the following relation:

$$q\angle P_{n-q'}OP_n - q'\angle P_{n-q}OP_n = \pm 360° \qquad (4)$$

where, for the normal phyllotaxis, q and q' are successive Fibonacci numbers. The above (*i*) signifies that $\alpha_n(q)$ is independent of *n*. This relation signifies that $\alpha_n(q)$ is independent of q, which is obvious from Tab. 1–Tab. 5. Indeed, for the left-hand side of the last equation, the pattern of Fig. 1 gives 360.0 ± 6.2 for $(q,q') = (2,3)$, 359.8 ± 28.8 for $(34,55)$, 355.6 ± 62.8 for $(89,144)$ and so on. To sum up, the observed results indicate that all angles $\angle P_{n-q}OP_n$ are determined by a single constant $\alpha_n(q) = \alpha$, which is minutely specific to an individual sample. Thus, Roberts' proposal creates more difficult problems.

The riddle of phyllotaxis is why the constant α is $\alpha_{gold} = 137.508°$. The theoretical difficulty stems from the fundamental premise that divergence angles are determined by the existing primordia, or that the phyllotaxis system is "a dynamical system" (in the technical sense of mathematical physics) in that the angle $\angle P_n OP_{n-1}$ is determined by the preceding P_m's (i.e., $m < n$). This has been an unchallenged, working hypothesis since early times. There is a contrasting approach of explaining special traits of living things, according to which the robustness of divergence angles may be explained without understanding the molecular mechanism of developmental processes of primordia formation. It can be simply that plants have evolved so as to generate lateral organs at fixed intervals of angle and so the divergence angle is pre-determined independently of the existing primordia. Hence, it would be instructive to revisit the original motivation of the phyllotaxis models, which is stated in the opening paragraphs of van Iterson [10].

"The times are over when we content ourselves with describing and classifying manifestations of the plant world with great accuracy. Since the theory of evolution has taught us to consider a number of peculiarities of inner and outer constructions as functional for the plants, we have made efforts to explain morphological facts as manifestations of adaptation more and more. Although we are taken certainly too far in this endeavor, it cannot be denied that this way of treatment has produced the finest results and stimulated many new studies. In addition to this line of research, now a second has developed, in which, setting aside the question of functionality, one seeks to bring various manifestations into causal relationship, and whereby one seeks to interpret a property as the mechanical necessary consequence of certain others. Thereby the first way of treatment loses none of its importance, as natural selection will retain only a preservable one from different forms which are possible for mechanical reasons. Therefore those special cases in that functionality is least pronounced and different possibilities make their appearance in most complete forms will be most suitably carried through with the mechanical way of understanding.

Now, there are few morphological facts showing such regularity and peculiarity as the phenomena of phyllotaxis, as they became well-known to us particularly by the works of Schimper and Braun, A. and L. Bravais, and just right here the functionality is not at all obvious[1]. So it is understandable that the most thorough attempts of carrying out a mechanical explanation have been made in this field. Here is pointed out only phyllotaxis theories of Hofmeister, Airy, Delpino, Schwendener and Church, among which Schwendener's theory is the most well-known and surely the most important.".

The footnote mentions a conceivable function of phyllotaxis: "[1] Although Wiesner believes to see in the preference of the main sequence an adaptation to light environment, but it seems to me the proof is not given in his work.".

Thus, the apparent lack of functionality was an initial motivation for pursuing the second, causal approach, a mechanical explanation of phyllotaxis. However, the first, evolutionary approach was recently revisited with a suggestion that the arrangement with the golden angle is the most functional when it comes to be rearranged in ranked patterns complying with Schimper–Braun's rule [26] (for an animated demonstration see [27]). At any rate, the presented results are rather compatible with a conventional view that the divergence angle is an inherited characteristic with which natural selection has to do. Evolutionary perspectives should not be readily dismissed on superficial grounds.

Last but not least, although the present study was devoted exclusively to the key problem of the most typical case, qualitative diversity in phyllotaxis should not be disregarded [1–7]. From a broad perspective, there is also good reason to consider that the golden angle is a necessary consequence of developmental dynamics of the shoot apical meristem. Although phyllotaxis patterns are species-specific, the pattern often changes for no apparent reason. Qualitative diversity of phyllotaxis patterns due to ontogenetic changes are well explained by geometrical considerations in terms of variable-sized primordia [28] and dislocations in regular patterns [3]. The causal approach since Hofmeister, in which evolution plays, if any, only a minor role, can provide a parsimonious explanation for all phyllotaxis phenomena, including the most striking phenomenon of the golden angle in spiral phyllotaxis.

Acknowledgments
The author received no financial support for this study.

Competing interests
No competing interests have been declared.

Supplementary material
The following supplementary material for this article is available online at http://pbsociety.org.pl/journals/index.php/asbp/rt/suppFiles/asbp.2015.007/0:
1. Appendix S1: the riddle in terms of the golden section.

References
1. Braun A. Vergleichende Untersuchung über die Ordnung der Schuppen an den Tannenzapfen als Einleitung zur Untersuchung der Blattstellung. Nov Acta Ac CLC. 1831;15:195–402.
2. Zagórska-Marek B. Phyllotactic patterns and transitions in *Abies balsamea*. Can J Bot. 1985;63:1844–1854.
3. Zagórska-Marek B. Phyllotaxic diversity in *Magnolia* flowers. Acta Soc Bot Pol. 1994;63:117–137.
4. Wiss D, Zagórska-Marek B. Geometric parameters of the apical meristem and the quality of phyllotactic patterns in *Magnolia* flowers. Acta Soc Bot Pol. 2012;81:203–216.
5. Szymanowska-Pułka J. Phyllotactic patterns in capitula of *Carlina acaulis* L. Acta Soc Bot Pol. 1994;63:229–245.
6. Kwiatkowska D. Ontogenetic changes in the shoot primary vasculature of *Anagallis arvensis* L. Acta Soc Bot Pol. 1995;64:213–222.
7. Gola EM. Phyllotactic pattern formation in early stages of cactus ontogeny. Acta Soc Bot Pol. 2006;75:271–279.
8. Hirmer M. Zur Kenntnis der Schraubenstellungen im Pflanzenreich. Planta. 1931;14:132–206.
9. Bravais L, Bravais A. Essai sur la disposition des feuilles curvisériées, Annales des Sciences Naturelles Botanique 1837;7:42–110.
10. van Iterson G. Mathematische und Mikroskopisch-Anatomische Studien über Blattstellungen. Gustav Fischer, Jena; 1907.
11. Fujita T. Statistische Untersuchungern über den Divergenzwinkel bei den schraubigen Organstellungen. Bot Mag Tokyo. 1939;53:194–199.
12. Snow R. Problems of phyllotaxis and leaf determination. Endeavour. 1955;14:190–199.
13. Kumazawa M, Kumazawa M. Periodic variations of the divergence angle, internode length and leaf shape, revealed by correlogram analysis. Phytomorphology. 1971;21:376–389.
14. Ryan G, Rouse J, Bursill L. Quantitative analysis of sunflower seed packing. J Theor Biol. 1991;147:303–328.
15. Couder Y. Initial transitions, order and disorder in phyllotactic patterns: the ontogeny of *Helianthus annuus*. A case study. Acta Soc Bot Pol. 1998;67:129–150.
16. Okabe T. Systematic variations in divergence angle. J Theor Biol 2012;313:20–41.
17. Matkowski A, Karwowski R, Zagórska-Marek B. Two algorithms of determining the middle point of the shoot apex by surrounding organ primordia positions and their usage for computer measurements of divergence angles. Acta Soc Bot Pol. 1998;67:151–159.
18. Church AH. On the relation of phyllotaxis to mechanical laws. London: Williams & Norgate; 1904.
19. van Iterson G. New studies on phyllotaxis. Proceedings of Koninklijke Nederlandse Akademie van Wetenschappen. Series C. 1960;63:137–150.
20. Fujita T. Zur Kenntnis der Organstellungen im Pflanzenreich. Jpn J Bot. 1942;12:1–55.
21. Roberts DW. The origin of Fibonacci phyllotaxis – an analysis of Adler's contact pressure model and Mitchison's expanding apex model. J Theor Biol 1978;74:217–233.
22. Adler I. A model of contact pressure in phyllotaxis. J Theor Biol. 1974;45:1–79.
23. Mitchison GH. Phyllotaxis and the Fibonacci series. Science. 1977;196:270–275.
24. Smith RS, Guyomarc'h S, Mandel T, Reinhardt D, Kuhlemeier C, Prusinkiewicz P. A plausible model of phyllotaxis. Proc Natl Acad Sci USA. 2006;103:1301–1306.
25. Jönsson H, Heisler MG, Shapiro BE, Meyerowitz EM, Mjolsness E. An auxin-driven polarized transport model for phyllotaxis. Proc Natl Acad Sci USA. 2006;103:1633–1638.
26. Okabe T. Physical phenomenology of phyllotaxis. J Theor Biol. 2011;280:63–75.
27. Okabe T. Phyllotaxis of plant shoots [Internet]. 2014 [cited 2014 Mar 12].
28. Zagórska-Marek B, Szpak M. Virtual phyllotaxis and real plant model cases. Funct Plant Biol. 2008;35:1025–1033.

15

Regulation of abscisic acid metabolism in relation to the dormancy and germination of cereal grains

Justyna Fidler, Edyta Zdunek-Zastocka*, Wiesław Bielawski

Department of Biochemistry, Warsaw University of Life Sciences – SGGW, Nowoursynowska 159, 02-776 Warsaw, Poland

Abstract

Seed dormancy is of particular importance in the cultivation of cereals, as it directly affects the quality of crop yield. If the dormancy period is too short, this may lead to pre-harvest sprouting, whereas a dormancy period that is too long may cause uneven germination; both of these scenarios are associated with economic losses. Most enzymes engaged in the metabolism of abscisic acid (ABA) have been identified, and significant progress has been made in understanding the role of this phytohormone in the induction and maintenance of dormancy, mainly as a result of research conducted in *Arabidopsis*. Much less is known about the metabolism and function of ABA in cereal grains, especially in relation to dormancy and germination. This review focuses on the regulation of ABA metabolism in dormant and non-dormant cereal grains, in both the dry state and upon imbibition. Moreover, this review describes the influence of factors such as after-ripening, light, temperature, nitric oxide, and reactive oxygen species (ROS) on the dormancy and germination of cereal grains. These factors, with the exception of ROS, appear to affect the level of dormancy and germination of grains through regulation of ABA metabolism.

Keywords: 9-*cis*-epoxycarotenoid dioxygenase; ABA 8′-hydroxylase; abscisic acid; abscisic acid metabolism; dormancy; germination

Introduction

Dormancy is one of the most intensively studied aspects of seed biology. Primary dormancy of seeds is initiated during the seed maturation period, and it is characterized by the inability of intact viable seeds to germinate under favorable conditions [1–4]. The level of seed dormancy in many species of cultivated plants, cereal grains in particular, directly affects the quality of crop yield. Short and shallow dormancy, which is characteristic of numerous varieties of cereal species that are important for agriculture, may result in the harmful phenomenon of pre-harvest sprouting, in which the seeds gain the ability to germinate while they are still on the mother plant [5,6]. Wheat, rye, and triticale are particularly prone to this unfavorable process. In contrast, the dormancy of barley at harvest can be too strong, which disrupts the fast and uniform germination required in the malting process; as a result, the technological costs are increased due to the long after-ripening period (dry storage of mature seeds after harvest). Seed dormancy and germination are collectively controlled by numerous genes and environmental factors, particularly the prevailing conditions during seed development and storage after harvest [7]. The environmental factors that are of particular importance include light quality, temperature, soil moisture, and the length of the after-ripening period. Although the mechanisms associated with release and breaking of dormancy still remain largely unexplained, it is now generally accepted that abscisic acid (ABA) is the primary mediator of seed dormancy; however, other participating hormones, such as gibberellins, ethylene, and brassinosteroids, are also very important [6,8]. ABA plays a central role not only in the acquisition of primary dormancy during seed maturation but also in maintaining dormancy in imbibed seeds. Dormancy release is accompanied by a decrease in embryo ABA content and/or a decrease in the sensitivity of embryos to ABA in parallel with a simultaneous increase in gibberellin levels. However, gibberellins are increasingly considered to promote germination after dormancy release, rather than to participate in breaking seed dormancy [6,9–13].

Recently, there have been reports suggesting that the capacity of seeds to metabolize ABA, as a result of the regulation of synthesis and catabolism of this phytohormone, undergoes changes during the after-ripening period, and this may be one of the key factors that determines the breaking of cereal grain dormancy [14–16]. Additionally, factors such as the quality of light, temperature, and donors of nitric oxide (NO) present during imbibition of seeds appear to

* Corresponding author. Email: edyta_zdunek_zastocka@sggw.pl

Handling Editor: Grzegorz Jackowski

regulate the state of dormancy and germination via changes in ABA content and modifications in the expression of genes participating in its metabolism. Most information regarding the regulation of ABA metabolism and its effect on dormancy has been obtained from studies on *Arabidopsis*. However, due to the immense economic significance of cereals, the intensification of research that would enhance understanding of the mechanisms responsible for breaking dormancy in cereal grains becomes necessary. This paper presents a review of the available literature regarding identification of genes participating in the biosynthesis and catabolism of ABA in cereals as well as the literature exploring the effect of factors such as after-ripening, light, and temperature on ABA content and metabolism in dry and imbibed grains of cereals.

ABA metabolism

The ABA biosynthesis and catabolism pathways have been clarified to a large extent mainly through the genetic modifications that caused ABA deficiency or ABA accumulation in *Arabidopsis* plants (Fig. 1), and are summarized in many earlier and recent reviews [17–23]. However, only some of the potential genes participating in ABA biosynthesis and

catabolism have been identified so far in cereals (Tab. 1). Although the expression products of these genes have not yet been characterized biochemically, the phenotypic analyses of mutants and transgenic plants, have proved that the potential genes encoding two ABA biosynthetic enzymes, zeaxanthin epoxidase (ZEP) and 9-*cis*-epoxycarotenoid dioxygenase (NCED), and an enzyme catalyzing the predominant ABA catabolic pathway, ABA 8'-hydroxylase (ABA8'OH), are involved in regulation of the seed ABA content and dormancy also in cereals (Tab. 2). In imbibed seed, the expression of these genes is also influenced by factors such as after-ripening, light quality, temperature, nitric oxide, and is described in subsequent sections.

Effect of after-ripening on ABA metabolism

In cereals, after-ripening does not alter ABA content or ABA metabolism in dry grains in most experiments; however, the situation changes diametrically when dormant and after-ripened grains undergo imbibition. Freshly harvested mature dry grains of wheat or barley that were still dormant (D grains) had ABA content similar to those of grains that underwent 3–4 months of after-ripening to release dormancy (AR grains, after-ripened grains). Although no differences

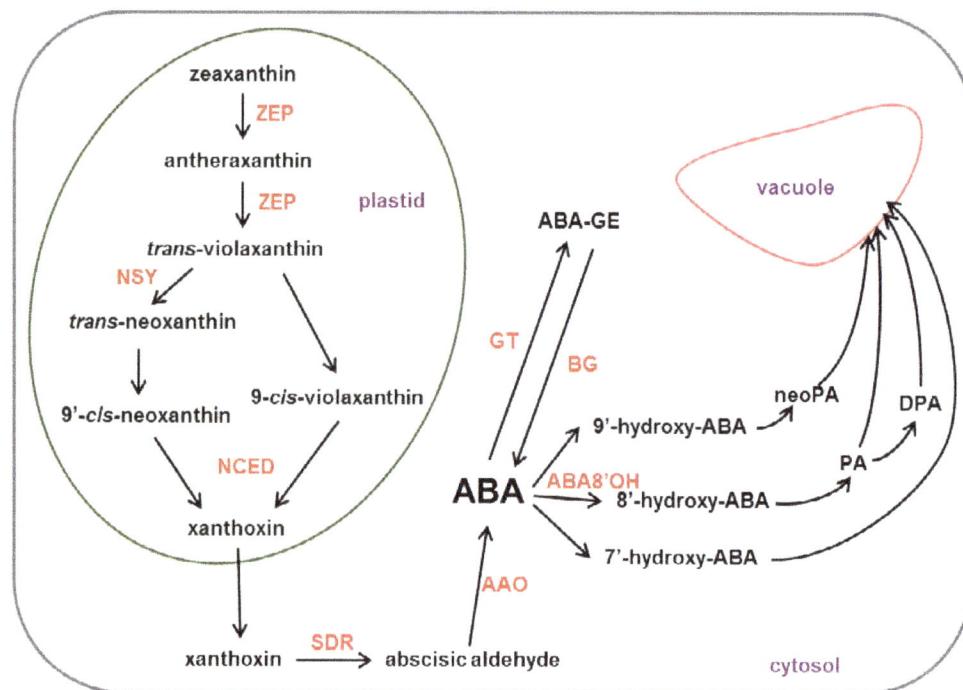

Fig. 1 Abscisic acid biosynthesis and inactivation pathways in higher plants. In plastids, the immediate precursor of ABA, zeaxanthin, is converted to trans-violaxanthin in reactions catalyzed by zeaxanthin epoxidase (ZEP). Trans-violaxanthin is then converted, to cis-violaxanthin by an unknown isomerase and to cis-neoxanthin in reaction catalyzed by neoxanthin synthase (NSY). Oxidative cleavage of cis-violaxanthin and cis-neoxanthin to xanthoxin is catalyzed by 9-*cis*-epoxycarotenoid dioxygenase (NCED). Xanthoxin is transported to the cytosol and is then converted into abscisic acid in reactions catalyzed by a short-chain dehydrogenase/reductase (SDR) and an abscisic aldehyde oxidase (AAO). ABA catabolism occurs by either hydroxylation or conjugation. The 8'-hydroxylation of ABA is catalyzed by ABA 8'-hydroxylase (ABA8'OH), a cytochrome P-450 monooxygenase. ABA hydroxylation may also take place at the C-7' and C-9' positions. ABA and its oxidative catabolites, phaseic acid (PA), neophaseic acid (neoPA), and dihydrophaseic acid (DPA) are the potential targets for conjugation. The conjugation of ABA with glucose is catalyzed by glucosyltransferases (GTs). ABA conjugation is reversible, and the hydrolysis of ABA glucosyl ester (ABA-GE) is catalyzed by β-glucosidases (BG).

Tab. 1 Potential genes participating in ABA synthesis and catabolism in cereals. The complete coding sequences are shown in bold.

Enzyme	Species	Gene	References
Zeaxanthin epoxidase	*Oryza sativa*	**OsABA1**	[63]
	Hordeum vulgare	**HvZEP1**	[66]
		HvZEP2	
		HvZEP3	
		HvZEP4	
		HvZEP5	
	Triticum aestivum	TaZEP1	[67]
	Zea mays	ZEP1	[68,69]
		ZEP2	
Nine-*cis*-epoxycarotenoid dioxygenase	*Zea mays*	**vp14/ZmNCED1**	[69,70]
		ZmNCED2	
		ZmNCED3	
		ZmNCED4	
		ZmNCED5	
		ZmNCED6	
	Hordeum vulgare	**HvNCED1**	[14,25]
		HvNCED2	
	Oryza sativa	**OsNCED1**	[32]
		OsNCED2	
		OsNCED3	
		OsNCED4	
		OsNCED5	
	Brachypodium distachyon	**BdNCED1**	[28]
		BdNCED2	
	Triticum aestivum	**TaNCED1**	[16,65]
		TaNCED2	
Short-chain dehydrogenase/ reductase	*Hordeum vulgare*	**HvSDR1**	[66]
		HvSDR2	
		HvSDR3	
		HvSDR4	
		HvSDR5	
		HvSDR6	
		HvSDR7	
	Oryza sativa	SDR1	[71]
		SDR2	
		SDR3	
Abscisic aldehyde oxidase	*Hordeum vulgare*	HvAO1	[66]
		HvAO2	
		HvAO3	
		HvAO4	
		HvAO5a	
		HvAO5b	
		HvAO6	
		HvAO7	
	Oryza sativa	AAO1	[71]
		AAO2	
		AAO3	
β-glucosidase	*Hordeum vulgare*	**HvBg1**	[66]
		HvBg2	
		HvBg3	
		HvBg4	
		HvBg5	
		HvBg6	
		HvBg7	
		HvBg8	
		HvBg9	
		HvBg10	
ABA 8'-hydroxylase	*Hordeum vulgare*	**HvCYP707A1/HvABA8'OH1**	[14,25,66]
		HvABA8'OH2	
		HvABA8'OH3	
	Triticum aestivum	**TaCYP707A1/TaABA8'OH1**	[16,31,72]
		TaABA8'OH2	

Tab. 1 (continued)

Enzyme	Species	Gene	References
	Triticum monococcum	**TmABA8'OH2**	[73]
	Oryza sativa	**CYP707A5**	[74]
		CYP707A6/	
		OsCYP707A5	[32]
		OsCYP707A6	
		OsCYP707A7/	
		OsABA8ox1	[75]
		OsABA8ox2	
		OsABA8ox3	
	Zea mays	**ABAOx1a**	[68]
		ABAOx1b	
		ABAOx2	
		ABAOx3a	
		ABAOx3b	
	Brachypodium distachyon	**BdABA8'OH1**	[28]
		BdABA8'OH2	

Tab. 2 Mutants and transgenic lines in which the potential genes involved in ABA metabolism in cereal grains are silenced or overexpressed.

Source of the gene	Gene	Mutant/transgenic line	Phenotypic effects	References
Oryza sativa	OsABA1(ZEP)	*Oryza sativa Osaba1* mutant	A strong viviparous mutant with wilty phenotype, precocious germination.	[63]
Zea mays	vp14 (NCED)	*Zea mays vp 14* mutant	Lower ABA content in embryos of developing seeds, reduced seed dormancy.	[64]
Oryza sativa	OsNCED3	Overexpression of *OsNCED3* in *Arabidopsis thaliana*	Higher seed ABA content, delayed seed germination.	[33]
Triticum aestivum	TaNCED1	Overexpression of *TaNCED1* in *Nicotiana tabacum*	No obvious differences in germination rates between the transgenic and WT plants, however higher germination rate in transgenic lines under drought treatment.	[65]
Hordeum vulgare	HvABA8'OH1	*Hordeum vulgare HvABA8'OH1* RNAi transgenic lines	Higher ABA level in embryos of dry seeds, increased seed dormancy.	[15]
Triticum aestivum	TaABA8'OH1	A double *Triticum aestivum* mutant in *TaABA8'OH1* on the A and D genome	Higher ABA content in embryos of developing and mature seeds, lower germination rate.	[31]

in embryo ABA content were found between dry D and AR grains, the AR grains germinated much earlier [10,14–16,24]. In turn, significant differences in ABA levels were observed between D and AR embryos after the imbibition. During the early hours of imbibition, the ABA content rapidly decreased in both D and in AR grains; however, in the subsequent hours, the decrease was observed only in AR grains, whereas in D grains the ABA content stabilized or even increased in certain varieties [10,14–16,25]. Thus, the direct indicator of the depth of dormancy is not always the level of ABA in the grains before imbibition, but rather the varied capacity of grains to catabolize and synthesize ABA upon imbibition [26]. It is assumed that the decrease in ABA content

observed during grain imbibition results from intensified catabolism of this phytohormone, as it was accompanied by an increase in phaseic acid (PA) content, an oxidative catabolite of ABA [10,13]. On the other hand, studies using norflurazon, which disturbs the synthesis of ABA, due to inhibition of carotenoid synthesis, revealed that de novo ABA synthesis is responsible, in some extent, for maintaining the dormancy of imbibed grains of deep dormant rice cultivar. It has been observed that dehusked grains of deep dormant, medium dormant and non-dormant rice cultivars, in the presence of norflurazon, showed similar germination rates [27]. However, inhibitor of ABA biosynthesis did not change germination rate of the intact grains of deep dormant rice cultivar, which suggests that not only ABA biosynthesis during imbibition but also some compounds in the husk might affect seed dormancy [27].

Analysis of the expression of genes potentially encoding regulatory enzymes involved in ABA metabolism in correlation with ABA level and dormancy release during imbibition of monocots seeds was conducted for barley, *Brachypodium distachyon*, wheat, and rice [14–16,25,27,28]. In barley, an increase in *HvNCED1* and *HvNCED2* transcripts, either transitory or stabilizing after a few hours following the start of imbibition, was observed in embryos of both D and AR grains [14,15]. However, the expression profiles of these genes observed upon imbibition as a result of after-ripening differ and in the case of *HvNCED1* it also depends on the variety. In Proctor barley variety, the increase in the *HvNCED1* transcript level was more obvious in imbibed D grains than in AR grains, while there were no significant differences between D and AR grains during imbibition of Betzes variety [15]. Surprisingly, the *HvNCED2* transcript level was higher in the embryos of AR than D grains in all varieties, analyzed so far [14,15,25]. Thus, the expression of genes participating in the biosynthesis of ABA seems not to affect (or has a limited effect on) the differences in the ABA content between D and AR grains observed during imbibition of barley grains. Although after-ripening has ambiguous effect on the expression of genes participating in ABA biosynthesis, it always considerably increases the expression of *HvABA8′OH1*, the gene engaged in ABA catabolism. The expression of *HvABA8′OH1* increased in both AR and D grains within the first few hours after the start of imbibition and decreased thereafter; however, it was always considerably higher in the embryos of AR versus D grains [14,15,25]. The *HvABA8′OH2* expression level was very low during the imbibition of both D and AR grains. It is therefore suggested that the higher decrease in ABA content, observed in the embryos of AR versus D barley grains during first hours of imbibition is due to increased ABA catabolism, through increased expression of *HvABA8′OH1*. RNAi silencing of *HvABA8′OH1* expression in barley plants supported an involvement of *HvABA8′OH1* in dormancy release. Reduced expression of *HvABA8′OH1* in RNAi plants resulted in higher ABA content of embryos isolated from dry as well imbibed grains, and increased dormancy compared to grains of wild-type and null segregant plants [15]. However, silencing of *HvABA8′OH1* only slightly affected the after-ripening time compared to wild-type grains, which indicates that the loss of seed dormancy due to after-ripening is not solely caused by increased ABA catabolism; rather, it may also result from decreased sensitivity of embryos to ABA in AR grains [15,29]. In situ localization of *HvABA8′OH1* in imbibed D and AR grains of barley revealed that in the embryos of AR grains, expression was observed only in the coleorhiza (tissue surrounding the primary root of the cereal embryo), whereas in the embryos of D grains, the expression of this gene was undetectable [14]. The authors suggest that the coleorhiza plays a crucial role in the regulation of grain dormancy, and the control mechanism may be based on the fact that a high ABA level prevents the coleorhiza from weakening and growing, leading to a blockage of radicle elongation, whereas intensive ABA catabolism removes this limitation [14,30].

Changes in the expression level of the genes engaged in ABA metabolism during imbibition were analyzed with reference to changes in the content of this phytohormone also in the embryos of wheat grains. In imbibed grains of wheat, after-ripening not only increased the expression of genes engaged in ABA catabolism but also resulted in significant changes in the expression of genes participating in ABA biosynthesis [16]. During the first few hours of imbibition, a considerable increase in the *TaNCED1* transcript level was observed in the embryos of both D and AR grains. Subsequently, the expression of this gene in AR embryos rapidly decreased, while in the embryos of D grains, it was stably maintained at a relatively high level. *TaNCED2* expression did not demonstrate any clear correlation with the ABA content in the embryos of D grains, whereas in the embryos of AR grains, despite the initial relatively high level, it rapidly decreased after six hours following the start of imbibition [16]. The *TaABA8′OH1* transcript level increased in the embryos of both types of grains, although it increased to a significantly higher level in the embryos of AR grains. Similar to what was observed in imbibed barley grains, *TaABA8′OH2* expression remained at a very low and stable level in both in D and AR grains [16]. The above expression analysis conducted upon imbibition of AR and D grains of wild wheat as well as the examination of the double wheat mutant in the *TaABA8′OH1* gene demonstrates that research involving manipulation of the *TaABA8′OH1* gene may contribute to solving the problem of pre-harvest sprouting of cereal grains [15,31]. However, recent report indicate that AR mediated wheat dormancy release might be associated with changes in ABA signaling and sensitivity without affecting the metabolism of this phytohormone, since there were no significant differences in expression of ABA metabolic genes between D and AR wheat grains, even after imbibition [24]. Differences in the results obtained by different research groups may arise from analysis of the different varieties of wheat, various tissues (grains/separated embryos), as well as various conditions of after-ripening. However, it should be assumed that the resolution of dormancy by after-ripening is rather the result of both, changes in the metabolism of ABA and in the grain responsiveness to ABA. Perhaps, depending on the tissue and the conditions of after-ripening, one of these processes might be more or less substantial.

Similarly to the results obtained for other cereal species, which showed differences in the content of ABA in imbibed D and AR grains, in imbibed grains of deep dormant rice cultivar ABA accumulation had been observed, while in medium dormant and non-dormant rice cultivars increasing in ABA content had not been observed [27]. It is also suggested that genes engaged in the synthesis and catabolism of ABA in rice, *OsNCED3* and *OsCYP707A5*, respectively, participate in the regulation of seed dormancy and germination. *OsNCED3* expression in embryos of dry grains was obviously higher, and *OsCYP707A5* expression was much lower in the *PA64s* cultivar, which exhibited stronger dormancy than the *9311* cultivar with weaker dormancy [32]. Differences in the expression of potential regulatory genes of ABA metabolism reflected higher ABA content in the embryos of *PA64s* grains compared with those of *9311* grains. In addition, *OsNCED3* overexpression in *Arabidopsis thaliana* resulted in increased accumulation of ABA and a delay in germination of transgenic seeds compared to wild-type seeds [33].

Effect of light on ABA metabolism

Light quality is a key environmental factor that regulates dormancy and germination of numerous species of plants, including cereals and other Poaceae [34,35]. White light, similarly to blue light, represses the germination of barley, *Lolium rigidum*, and *Brachypodium distachyon*, whereas red light stimulates the germination of *Brachypodium distachyon* but does not affect the dormancy of barley [15,28,36–38]. In barley, imbibition of primary dormant grains in the dark resulted in considerable dormancy release, similarly to the results of after-ripening, and in both cases it was associated with a significant, similar decrease in ABA content as well as an increase in PA in the embryos over the initial 12-h period of imbibition [10].

The expression of genes encoding enzymes engaged in ABA metabolism was examined to determine whether the decrease of the embryo ABA content, and the consequent dormancy release during imbibition in the dark, is due to the differential expression of these genes. It appeared that imbibition for 24 hours under white light or blue light strongly induced *HvNCED1* expression in the embryos of both D and AR barley grains, whereas after-ripening had a limited effect on expression of this gene. Expression of a gene encoding ABA 8′-hydroxylase, *HvABA8′OH1*, was strongly induced by AR; however, it was not affected by white or blue light [14,15]. It have been also shown, that longer incubation (more than 24 hours) of primary dormant barley grains at low temperature, under blue light inhibits germination, which was associated with an increase in expression of both *HvNCED1* and *HvNCED2*, while expression of *HvABA8′OH1* was not significantly changed [38]. These results indicate that the effect of white or blue light on maintaining dormancy of D barley grains are correlated with increase in embryo ABA content and is the result of intensified biosynthesis of this phytohormone rather than reduced ABA catabolism. However, this effect might also result from changed sensitivity of embryos to ABA in response to white or blue light [15,38]. In barley, blue light significantly increased embryo sensitivity to abscisic acid [38]. Such a situation may occur during germination of *Brachypodium distachyon* grains in white light [28]. Exposure to white light during imbibition had very little effect on the grain ABA concentration as well as on the expression of ABA metabolism genes, despite the fact that a reduction in the *BdNCED1* transcript level and an increase in *BdABA8′OH* were observed as a result of after-ripening [28]. The results indicate that after-ripening considerably affects ABA metabolism in *Brachypodium distachyon* grains, although the effect of white light on grain germination in this species cannot be explained by changes in ABA metabolism [28].

It have been suggested that inhibition of germination under blue light is cryptochrome (blue light photoreceptor) dependent, and this was recently confirmed by Barrero et al.

[36]. RNAi silencing of cryptochrome 1 (CRY1) expression in barley resulted in increased mRNA level of *ABA8′OH1* in first few hours after imbibition under blue light and significant reduction of *NCED1* expression after 18 hours imbibition, which was consistent with lower embryo ABA content. These results indicate that Hv-CRY1 is engaged in blue light dependent maintenance of high ABA content in barley [36].

Effect of temperature on ABA metabolism

In cereals and other Poaceae growing in moderate climates, a germination temperature above 15–20°C gradually deepens the primary dormancy observed in mature grains [7]. For example, primary dormant grains of barley germinated easily at 20°C, while imbibition at 30°C resulted in almost complete suppression of grain germination [39,40]. Incubation of primary dormant grains at 30°C can even induce thermodormancy; i.e., after incubation of primary dormant grains at high temperatures the grains lose the ability to germinate at lower temperatures (e.g., 20°C) [40,41]. This phenomenon may be considered as an induction of secondary dormancy, and it was correlated with maintaining a relatively high embryo ABA content, increased sensitivity of the embryos to ABA, and changes in the expression of genes engaged in the metabolism of this phytohormone [42–45].

In barley, relatively high embryo ABA content during imbibition of primary dormant grains at 30°C and during imbibition of secondary dormant grains at 20°C, was correlated with the maintenance of a high *HvNCED1* transcript level. Instead, during imbibition of primary dormant grains at 20°C, the *HvNCED1* expression level was considerably lower in comparison to that in dry grains, which in turn was correlated with a sudden decrease in embryo ABA content [39,40]. *HvNCED2* expression in embryos was constantly decreasing during imbibition of primary dormant grains at 30°C; however, interestingly, it increased when grains with induced by 30°C secondary dormancy were imbibed at 20°C [40]. The *HvABA8′OH1* expression level did not change significantly in the embryos of primary dormant grains imbibed at 30°C; however, its expression increased in the embryos of grains after induction of secondary dormancy. Therefore, participation of this gene in the induction and regulation of secondary dormancy induced by temperature is rather dubious [40–44]. It seems that in barley, regulation of ABA synthesis during secondary dormancy takes place with the participation of *HvNCED1* and *HvNCED2*, while *HvABA8′OH1* can be considered as a key gene regulating primary dormancy.

H₂O₂ and ABA have antagonistic effects on germination

Reactive oxygen species (ROS) are constantly produced in plants, as by-products of many metabolic pathways. Beside their well-documented toxicity, they are considered as important signaling molecules in many processes related to plant growth, development, and stress responses. In seed biology, they are believed to participate in regulation of dormancy, after-ripening, and germination [46–48]. In dry seeds, ROS are probably generated in non-enzymatic reactions such as lipid peroxidation or Amadori and Maillard reactions. In hydrated seeds, ROS may originate from mitochondria, peroxisomes, glyoxysomes and chloroplasts, or can be produced through the activity of NADPH oxidase, amine oxidase, peroxidase or cytochrome P450 [46,48]. The production of hydrogen peroxide, which is one of the reactive forms of oxygen, primarily with the participation of NADPH oxidase was observed during the early imbibition stages in the grains of wheat and barley [49,50].

Exogenously applied H_2O_2 promotes seed germination of many plant species, including cereals [51,52]. It has been demonstrated that H_2O_2 may stimulate germination by participating in programmed cell death of the aleurone layer in cereal grains, and its production is accelerated by gibberellins and inhibited by ABA [53,54]. Despite the fact that H_2O_2 stimulates germination, a slight increase in ABA content was observed in the embryos of barley treated with exogenous H_2O_2. This increase was associated with the induction of expression of genes participating in the biosynthesis of this phytohormone, *HvNCED1* and *HvNCED2*, whereas no changes in the expression level of the gene encoding ABA 8′-hydroxylase were observed [55]. It is assumed that the increase of ABA in embryos under the influence of exogenous H_2O_2 may be related to the role of this phytohormone in activation of the antioxidative system, thus preventing oxidative stress; this phenomenon was observed in vegetative tissues of *Cynodon dactylon* grass [55,56]. Moreover, germination in the presence of H_2O_2 with simultaneous high ABA content may be explained by changes in the balance between ABA and gibberellins [55].

Is NO involved in the regulation of ABA metabolism in cereal grains?

Treatment of seeds with a NO donor, such as sodium nitroprusside (SNP), results in reduction or release of the dormancy of *Arabidopsis thaliana*, barley and wheat seeds, whereas the use of c-PTIO [2-(4-carboxyphenyl)-4,4,5,5-tetramethylimidazoline-1-oxyl-3 oxide], an NO scavenger, effectively promotes the maintenance of seed dormancy [16,57,58]. In imbibed *Arabidopsis thaliana* seeds, NO induces a rapid ABA decrease, which is correlated with an increase in the transcript level of one of the genes encoding ABA 8′-hydroxylase, *CYP707A2*. In addition, the germination rate of the *cyp707a2* mutant seeds was not elevated by SNP treatment, contrary to what was observed in the wild-type seeds [59,60]. These results indicate a significant role for the *CYP707A2* gene in the NO-mediated control of ABA levels during seed germination of *Arabidopsis* seeds; however, it has yet to be determined whether NO also affects seed dormancy in cereal grains through the regulation of genes participating in ABA metabolism.

Conclusions and perspectives

The length and depth of seed dormancy plays a key role in crop cultivation, thus affecting the economic aspect of agricultural production. Weak seed dormancy in plants such as cereals is important from an economic perspective, as it causes pre-harvest sprouting, which is a serious problem for growers in many regions of the world. Because of the scale and scope of pre-harvest sprouting, improving cereal resistance to this phenomenon has become one of the most difficult tasks that breeders are currently facing. Recently, there have been numerous reports exploring seed dormancy and germination at the molecular level. Many of these studies provided very promising results, which may be important for cultivation of plants in the future. However, most of the studies conducted thus far have been performed using a model plant, *Arabidopsis thaliana*, and the published data regarding cereals are still limited. Incomplete knowledge about the molecular mechanisms regulating dormancy and germination of cereals results primarily from difficulties in methodology and problems caused by the complexity of genome and polyploid nature of these plants [4]. Use of the wild grass *Brachypodium distachyon* in molecular studies seems to be promising, as it may become a diploid model of cereals. *Brachypodium distachyon*, apart from its mentioned diploidy, is also characterized by a small genome size and a short life cycle, which makes it a perfect model for studies at the molecular level [61,62]. Expanding our knowledge of mechanisms regulating dormancy and germination of cereal grains is therefore becoming one of the most important directions of research in the upcoming years.

Acknowledgments

This work was financially supported by the Department of Biochemistry, Warsaw University of Life Sciences – SGGW.

Authors' contributions

The following declarations about authors' contributions to the research have been made: created the concept, searched the literature and wrote the manuscript (contributed equally to the preparation of this article): JF, EZZ; critically reviewed the paper and proposed some useful suggestions: WB.

Competing interests

No competing interests have been declared.

References

1. Bewley JD. Seed germination and dormancy. Plant Cell. 1997;9:1055–1066.

2. Finch-Savage WE, Leubner-Metzger G. Seed dormancy and the control of germination. New Phytol. 2006;171:501–523.

3. Graeber K, Nakabayashi K, Miatton E, Leubner-Metzger G, Soppe WJ. Molecular mechanisms of seed dormancy. Plant Cell Environ. 2012;35:1769–1786.

4. Kumar S, Hirani AH, Asif M, Goyal A. Molecular mechanisms controlling dormancy and germination in barley. In: Asif M, Goyal A, editors. Crop production. ??: InTech; 2013. p. 69–98.

5. Gerjets T, Scholefield D, Foulkes MJ, Lenton JR, Holdsworth MJ. An analysis of dormancy, ABA responsiveness, after-ripening and pre-harvest sprouting in hexaploid wheat (*Triticum aestivum* L.) caryopses. J Exp Bot. 2010;61:597–607.

6. Gubler F, Millar AA, Jacobsen JV. Dormancy release, ABA and pre-harvest sprouting. Curr Opin Plant Biol. 2005;8:183–187.

7. Corbineau F, Come D. Barley seed dormancy. Bios. 1996;261:113–119

8. Koornneef M, Bentsink L, Hilhorst H. Seed dormancy and germination. Curr Opinion Plant Biol. 2002;5:33–36.

9. Finkelstein R, Gampala SSL, Rock CD. Abscisic acid signaling in seeds and seedlings. Plant Cell. 2002;14:S15-S45.

10. Jacobsen JV, Pearce DW, Poole AT, Pharis RP, Mander LN. Abscisic acid, phaseic acid and gibberellin contents associated with dormancy and germination in barley. Physiol Plant. 2002;115:428–441.

11. Nambara E, Okamoto M, Tatematsu K, Yano R, Seo M, Kamiya Y. Abscisic acid and the control of seed dormancy and germination. Seed Sci Res. 2010;20:55–67.

12. Rodriguez MV, Mendiondo GM, Cantoro R, Auge GA, Luna V, Masciarelli O, et al. Expression of seed dormancy in grain sorghum lines with contrasting pre-harvest sprouting behavior involves differential regulation of gibberellin metabolism genes. Plant Cell Physiol. 2012;53:64–80.

13. Rodriguez-Gacio MC, Matilla-Vazquez MA, Matilla AJ. Seed dormancy and ABA signaling: the breakthrough goes on. Plant Signal Behav. 2009;4:1035–1048.

14. Millar A, Jacobsen J, Ross J, Helliwell C, Poole A, Scofield G, et al. Seed dormancy and ABA metabolism in *Arabidopsis* and barley: the role of ABA 8′-hydroxylase. Plant J. 2006;45:942–954.

15. Gubler F, Hughes T, Waterhouse P, Jacobsen J. Regulation of dormancy in barley by blue light and after-ripening: effects on abscisic acid and gibberellin metabolism. Plant Physiol. 2008;147:886–896.

16. Jacobsen JV, Barrero JM, Hughes T, Julkowska M, Taylor JM, Xu Q, et al. Roles for blue light, jasmonate and nitric oxide in the regulation of dormancy and germination in wheat grain (*Triticum aestivum* L.). Planta. 2013;238:121–138.

17. Finkelstein R. Abscisic acid synthesis and response. Arabidopsis Book. 2013;11:1–36.

18. Nambara E, Marion-Poll A. Abscisic acid biosynthesis and catabolism. Annu Rev Plant Biol. 2005;56:165–185.

19. Schwartz SH, Qin X, Zeevaart JAD. Elucidation of the indirect pathway of abscisic acid biosynthesis by mutants, genes, and enzymes. Plant Physiol. 2003;131:1591–1601.

20. Schwartz SH, Zeevaart JAD. Abscisic acid biosynthesis and metabolism. In: Davies PJ, editor. Plant hormones. Dordrecht: Springer Netherlands; 2010. p. 137–155.

21. Seo M, Koshiba T. Complex regulation of ABA biosynthesis in plants. Trends Plant Sci. 2002;7:41–48.

22. Taylor IB, Sonneveld T, Bugg TD, Thompson AJ. Regulation and manipulation of the biosynthesis of abscisic acid, including the supply of xanthophyll precursors. J Plant Growth Regul. 2005;24:253–273.

23. Xu ZY, Kim DH, Hwang I. ABA homeostasis and signaling involving multiple subcellular compartments and multiple receptors. Plant Cell Rep. 2013;32:807–813.

24. Liu A, Gao F, Kanno Y, Jordan MC, Kamiya Y, Seo M, et al. Regulation of wheat seed dormancy by after-ripening is mediated by specific transcriptional switches that induce changes in seed hormone metabolism and signaling. PLoS ONE. 2013;8:1–18.

25. Chono M, Hondo I, Shinoda S, Kushiro T, Kamiya Y, Nambara E, et al. Field studies in the regulation of abscisic acid content and germinability during grain development of barley: molecular and chemical analysis of pre-harvest sprouting. J Exp Bot. 2006;57:2421–2434.

26. Kermode AR. Role of abscisic acid in seed dormancy. J Plant Growth Regul. 2005;24:319–344.

27. Liu Y, Fang J, Xu F, Chu J, Yan C, Schlappi MR, et al. Expression patterns of ABA and GA metabolism genes and hormone levels during rice seed development and imbibition: a comparison of dormant and non-dormant rice cultivars. J Genet Genomics. 2014;41:327–338.

28. Barrero JM, Jacobsen JV, Talbot M, White R, Swain M, Garvin D, et al. Grain dormancy and light quality effects on germination in the model grass *Brachypodium distachyon*. New Phytol. 2012;193:376–386.

29. Walker-Simmons M. ABA levels and sensitivity in developing wheat embryos of sprouting resistant and susceptible cultivars. Plant Physiol. 1987;84:61–66.

30. Barrero JM, Talbot MJ, White RG, Jacobsen JV, Gubler F. Anatomical and transcriptomic studies of the coleorhiza reveal the importance of this tissue in regulating dormancy in barley. Plant Physiol. 2009;150:1006–1021.

31. Chono M, Matsunaka H, Seki M, Fujita M, Kiribuchi-Otobe C, Oda S, et al. Isolation of a wheat (*Triticum aestivum* L.) mutant in ABA 8′-hydroxylase gene: effect of reduced ABA catabolism on germination inhibition under field condition. Breed Sci. 2013;63:104–115.

32. Liu F, Zhang H, Wu G, Sun J, Hao L, Ge X, et al. Sequence variation and expression analysis of seed dormancy- and germination-associated ABA- and GA-related genes in rice cultivars. Front Plant Sci. 2011;2:1–13.

33. Hwang SG, Chen HC, Huang WY, Chu YC, Shii CT, Cheng WH. Ectopic expression of rice *OsNCED3* in *Arabidopsis* increases ABA level and alters leaf morphology. Plant Sci. 2010;178:12–22.

34. Sawada Y, Aoki M, Nakaminami K, Mitsuhashi W, Tatematsu K, Kushiro T, et al. Phytochrome and gibberellin-mediated regulation of abscisic acid metabolism during germination of photoblastic lettuce seeds. Plant Physiol. 2008;146:1386–1396.

35. Seo M, Hanada A, Kuwahara A, Endo A, Okamoto M, Yamauchi Y, et al. Regulation of hormone metabolism in *Arabidopsis* seeds: phytochrome regulation of abscisic acid metabolism and abscisic acid regulation of gibberellin metabolism. Plant J. 2006;48:354–366.

36. Barrero JM, Downie AB, Xu Q, Gubler F. A Role for Barley CRYPTOCHROME1 in light regulation of grain dormancy and germination. Plant Cell. 2014;26:1094–1104.

37. Goggin D, Steadman K, Powles S. Green and blue light photoreceptors are involved in maintenance of dormancy in imbibed annual ryegrass (*Lolium rigidum*) seeds. New Phytol. 2008;148:81–89.

38. Hoang HH, Sechet J, Bailly C, Leymarie J, Corbineau F. Inhibition of germination of dormant barley (*Hordeum vulgare* L.) grains by blue light as related to oxygen and hormonal regulation. Plant Cell Environ. 2014;37:1393–1403.

39. Benech-Arnold RL, Gualano N, Leymarie J, Come D, Corbineau F. Hypoxia interferes with ABA metabolism and increases ABA sensitivity in embryos of dormant barley grains. J Exp Bot. 2006;57:1423–1430.

40. Leymarie J, Robayo-Romero ME, Gendreau E, Benech-Arnold RL, Corbineau F. Involvement of ABA in induction of secondary dormancy in barley (*Hordeum vulgare* L.) seeds. Plant Cell Physiol. 2008;49:1830–1838.

41. Corbineau F, Black M, Come D. Induction of thermodormancy in *Avena sativa* seeds. Seed Sci Res. 1993;3:111–117.

42. Argyris J, Dahal P, Hayashi E, Still DW, Bradford KJ. Genetic variation for lettuce seed thermoinhibition is associated with temperature sensitive expression of abscisic acid, gibberellin, and ethylene biosynthesis, metabolism, and response genes. Plant Physiol. 2008;148:926–947.

43. Corbineau F, Come D. Dormancy of cereal seeds as related to embryo sensitivity to ABA and water potential. In: Viemont JD, Crabbe J, editors. Dormancy in plants: from whole plants behaviour to cellular control. Oxon: CAB International; 2000.

44. Leymarie J, Benech-Arnold RL, Farrant JM, Corbineau F. Thermo-dormancy and ABA metabolism in barley grains. Plant Signal Behav. 2009;4:205–207.

45. Toh S, Imamura A, Watanabe A, Nakabayashi K, Okamoto M, Jikumaru Y, et al. High temperature induced abscisic acid biosynthesis and its role in the inhibition of gibberellin action in *Arabidopsis* seeds. Plant Physiol. 2008;146:1368–1385.

46. Bailly C, Kranner I. Methods for analyses of reactive oxygen species and antioxidants in relation to seed longevity and germination. Methods Mol Biol. 2011;773:343–367.

47. Ye N, Zhu G, Liu Y, Zhang A, Li Y, Liu R, et al. Ascorbic acid and reactive oxygen species are involved in the inhibition of seed germination by abscisic acid in rice seeds. J Exp Bot. 2011;63:1809–1822.

48. El-Maarouf-Bouteau H, Bailly C. Oxidative signaling in seed germination and dormancy. Plant Signal Behav. 2008;3:175–182.

49. Caliskan M, Cuming AC. Spatial specificity of H_2O_2-generating oxalate oxidase gene expression during wheat embryo germination. Plant J. 1998;15:165–171.

50. Ishibashi Y, Tawaratsumida T, Zheng SH, Yuasa T, Iwaya-Inoue M. NADPH oxidases act as key enzyme on germination and seedling growth in barley (*Hordeum vulgare* L.). Plant Prod Sci. 2010;13:45–52.

51. Ishibashi Y, Yamamoto K, Tawaratsumida T, Yuasa T, Iwaya-Inoue M. Hydrogen peroxide scavenging regulates germination ability during wheat (*Triticum aestivum* L.) seed maturation. Plant Signal Behav. 2008;3:183–188.

52. Barba-Espin G, Diaz-Vivancos P, Clemente-Moreno MJ, Albacete A, Faize L, Faize M, et al. Interaction between hydrogen peroxide and plant hormones during germination and the early growth of pea seedlings. Plant Cell Environ. 2010;33:981–994.

53. Fath A, Bethke P, Beligni V, Jones R. Active oxygen and cell death in cereal aleurone cells. J Exp Bot. 2002;53:1273–1282.

54. Ishibashi Y, Tawaratsumida T, Kondo K, Kasa S, Sakamoto M, Aoki N, et al. Reactive oxygen species are involved in gibberellin/abscisic acid signaling in barley aleurone cells. Plant Physiol. 2012;158:1705–1714.

55. Bahin E, Bailly C, Sotta B, Kranner I, Corbineau F, Leymarie J. Crosstalk between reactive oxygen species and hormonal signaling pathways regulates grain dormancy in barley. Plant Cell Environ. 2011;34:980–993.

56. Lu S, Su W, Li H, Guo Z. Abscisic acid improves drought tolerance of triploid bermuda grass and involves H_2O_2- and NO-induced antioxidant enzyme activities. Plant Physiol Biochem. 2009;47:132–138.

57. Bethke PC, Gubler F, Jacobsen JV, Jones RL. Dormancy of *Arabidopsis* seeds and barley grains can be broken by nitric oxide. Planta. 2004;219:847–855.

58. Bethke PC, Libourel IGL, Reinöhl V, Jones RL. Sodium nitroprusside, cyanide, nitrite, and nitrate break *Arabidopsis* seed dormancy in a nitric oxide–dependent manner. Planta. 2006;223:805–812.

59. Liu Y, Shi L, Ye N, Liu R, Jia W, Zhang J. Nitric oxide-induced rapid decrease of abscisic acid concentration is required in breaking seed dormancy in *Arabidopsis*. New Phytol. 2009;183:1030–1042.

60. Matakiadis T, Albores A, Jikumaru Y, Tatematsu K, Pichon O, Renou JP, et al. The *Arabidopsis* abscisic acid catabolic gene *CYP707A2* plays a key role in nitrate control of seed dormancy. Plant Physiol. 2009;149:949–960.

61. Barrero JM, Jacobsen J, Gubler F. Seed dormancy: approaches for finding new genes in cereals. In: Pua EC, Davey MR, editors. Plant developmental biology – biotechnological perspectives. Berlin: Springer; 2010. p. 361–381.

62. Vain P. *Brachypodium* as a model system for grass research. J Cereal Sci. 2011;54:1–7.

63. Agrawal GK, Yamazaki M, Kobayashi M, Hirochika R, Miyao A, Hirochika H. Screening of the rice viviparous mutants generated by endogenous retrotransposon *Tos17* insertion: tagging of a zeaxanthin epoxidase gene and a novel *OsTATC* gene. Plant Physiol. 2001;125:1248–1257.

64. Tan BC, Schwartz SH, Zeevaart JAD, McCarty DR. Genetic control of abscisic acid biosynthesis in maize. Proc Natl Acad Sci USA. 1997;94:12235–12240.

65. Zhang SJ, Song GQ, Li YL, Gao J, Liu JJ, Fan QQ, et al. Cloning of 9-*cis*-epoxycarotenoid dioxygenase gene (*TaNCED1*) from wheat and its heterologous expression in tobacco. Biol Plant. 2014;58:89–98.

66. Seiler C, Harshavardhan VT, Rajesh K, Reddy PS, Strickert M, Rolletschek H, et al. ABA biosynthesis and degradation contributing to ABA homeostasis during barley seed development under control and terminal drought-stress conditions. J Exp Bot. 2011;62:2615–2632.

67. Ji X, Dong B, Shiran B, Talbot MJ, Edlington JE, Hughes T, et al. Control of abscisic acid catabolism and abscisic acid homeostasis is important for reproductive stage stress tolerance in cereals. Plant Physiol. 2011;156:647–662.

68. Vallabhaneni R, Wurtzel ET. From epoxycarotenoids to ABA: the role of ABA 8′-hydroxylases in drought-stressed maize roots. Arch Biochem Biophys. 2010;504:112–117.

69. Capelle V, Remoue C, Moreau L, Reyss A, Mahe A, Massonneau A, et al. QTLs and candidate genes for desiccation and abscisic acid content in maize kernels. BMC Plant Biol. 2010;10:2.

70. Schwartz SH, Tan BC, Gage DA, Zeevaart JA, McCarty DR. Specific oxidative cleavage of carotenoids by VP14 of maize. Science. 1997;276:1872–1874.

71. Chen QF, Ya HY, Feng YR, Jiao Z. Expression of the key genes involved in ABA biosynthesis in rice implanted by ion beam. Appl Biochem Biotechnol. 2014;137:239–247.

72. Zhang CL, He XY, He ZH, Wang LH, Xia XC. Cloning of *TaCYP707A1* gene that encodes ABA 8′-hydroxylase in common wheat (*Triticum aestivum* L.). Agric Sci China. 2009;8:902–909.

73. Nakamura S, Chono M, Abe F, Miura H. Mapping a diploid wheat abscisic acid 8′-hydroxylase homologue in the seed dormancy QTL region on chromosome 5Am. Euphytica. 2010;171:111–120.

74. Yang SH, Choi D. Characterization of genes encoding ABA 8′-hydroxylase in ethylene-induced stem growth of deepwater rice (*Oryza sativa* L.). Biochem Biophys Res Commun. 2006;350:685–690.

75. Saika H, Okamoto M, Miyoshi K, Kushiro T, Shinoda S, Jikumaru Y, et al. Ethylene promotes submergence-induced expression of *OsABA8ox1*, a gene that encodes ABA 8′-hydroxylase in rice. Plant Cell Physiol. 2007;48:287–298.

16

Facing global markets – usage changes in Western Amazonian plants: the example of *Euterpe precatoria* Mart. and *E. oleracea* Mart.

Rainer W. Bussmann[1]*, Narel Y. Paniagua Zambrana[2]

[1] Missouri Botanical Garden, Box 299, St. Louis, Missouri 63166-0299, USA

[2] Instituto de Ecología, Universidad Mayor de San Andrés, Box 10077, Correo Central, La Paz, Bolivia

Abstract

Palms (Arecaceae) are one of the most important families of useful plants, and indigenous societies have developed very distinct ways of utilizing this resource. The clonal *Euterpe oleracea* Mart. has long been used for the preparation of frothy beverages in the eastern Amazon, in particular by colonists and caboclos, but to a much lesser extent by the indigenous population. *Euterpe precatoria* Mart., which grows in the western Amazon, is traditionally reported as resource for construction and thatch, but not as important species in alimentation. Our recent work indicates that the use of both species has dramatically shifted in the recent past. Prices for *Euterpe* products have increased dramatically due to the global commodization first of palm hearts and "Açaí berry juice" as nutritional supplement. This is especially evident in western Amazonia: In Bolivia and Peru, where older indigenous informants mostly reported thatch and houseposts as regular use for *E. precatoria* and did not know *E. oleracea*. Younger informants most commonly reported to a large extent on *E. precatoria* being used for the production of palm hearts, but less for other, while the youngest informants in many cases only knew *E. precatoria* fruits as source of beverages, as commercial fruit, and as source for handicrafts, and indicated that *E. oleracea* was being introduced because the species yielded higher harvests. In addition, many mid-age and younger informants reported *Euterpe* sp. as medicinal species, a less frequently mentioned by older informants. The local mestizo population in contrast had a broader distributed knowledge with regard to "food" uses of *Euterpe* sp., and mentioned the species as source of construction material less frequently.

Keywords: Açaí, *Euterpe*, ethnobotany, Amazon, plant use, cultural change, globalization

Introduction

Palms (Arecaceae) are one of the most important families of useful plants, and indigenous as well as mestizo societies have developed very distinct ways of utilizing this resource. For this reason the family can serve as an excellent tool to depict the change of use knowledge over time, in particular because a considerable number of palm products has entered the global market in recent decades.

Various species of the genus *Euterpe* are widely known under their main vernacular name Açaí. *Euterpe oleracea* Mart. (Fig. 1) is a clonal species that occurs naturally in periodically inundated areas in Northern South America, in particular the Brazilian Amazon, the Orinoco basin, and in costal swamps of Columbia and Ecuador (Kahn). In contrast *Euterpe precatoria* Mart. (Fig. 2) grows solitary, on tierra firme, and covers much the same geographic region as the previous species, but occurs also in the western Amazon basin of Peru and Bolivia [1].

The use of *Euterpe oleracea* has long been documented, in particular in Eastern Amazonia. There, the fruits are traditionally immersed in water, mashed, and the resulting frothy mixture is strained and drunk, or added to rice or cassava (*Manihot esculenta*) flour. This can provide up to 42% of local diets [2,3]. In addition, the juice is traditionally used to treat pain, flu, and fever [4], while oil extracted from the fruit is known to treat diarrhea [5]. During the last decades, *E. oleracea* has become a global commodity however [6]. Originally only harvested from flooded areas, the palm is now being grown in large plantations [7–17]. *E. oleracea* was initially commercially important as source of palm hearts [3,18], but the species has more recently attracted fame for its supposed health-promoting benefits [4,19–33]. The much wider distribution of Açaí beverages has led to a tremendous increase in price, as well as the use of quicker, non-traditional preparation techniques, depending on the fancy of the end-users [34–36]. This in turn is suspected to have led to outbreaks of orally transmitted Chagas disease [37–40]. *Euterpe prectoria* in contrast has been reported much more frequently for its traditional use in house construction, e.g. as posts and for thatch. During the last decades reports on the use of *E. precatoria* adventive roots for the preparation of medicines to treat malaria, hepatitis and other ailments have become more frequent [41].

* Corresponding author. Email: rainer.bussmann@mobot.org

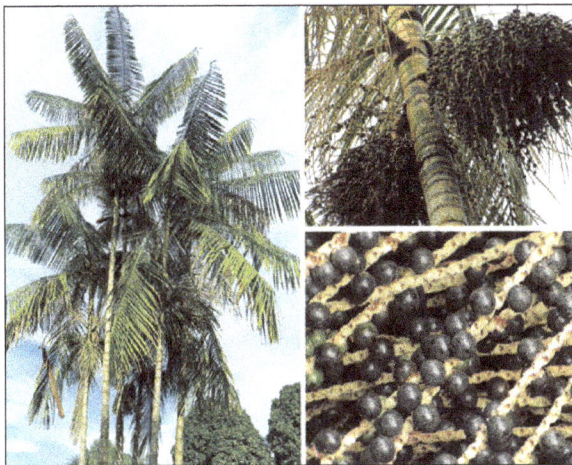

Fig. 1 *Euterpe oleracea* Mart.

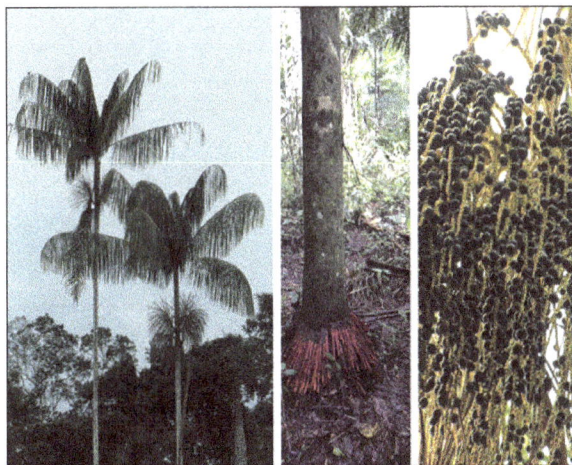

Fig. 2 *Euterpe precatoria* Mart.

Our presented research focused on the uses of both species in Western Amazonia (Bolivia and Peru), in order to evaluate change of usage by local communities as possible effect of the global boom in *Euterpe* products.

The changing use of *Euterpe* sp. in the Western Amazon

As late as 2010 Macia et al. [42] outlined the broader use records of palm species in Western Amazonia based on a large-scale literature review. According to this overview *Euterpe oleracea* had been used in the Amazonian areas of Peru and Bolivia, only by mestizo communities, and mostly for the preparation of beverages and palm hearts [43,44], and only the roots were infrequently applied as medicine, mostly for endocrineurinary system disorders.

Recent research [45,46] indicates however that the use of *Euterpe oleracea* has increased in recent years, and is slowly spreading from mestizo to indigenous communities, e.g. the Cocama and Awajun in Peru. In addition to the uses described above, *E. oleracea* is also infrequently used for construction purposes (thatch and house-posts), and the roots serve to treat

anemia, as well as a galactogogue. While older participants showed a trend to know more food uses, younger participants had a tendency to mention more medicinal uses (Tab. 1).

In contrast, *Euterpe precatoria* has traditionally been much more widely used in Western Amazonia, both by indigenous and mestizo communities. However, usage centered mostly on construction [44,47–62], household utensils and other artifacts [44,48–50,55,58–62], as well as medicinal uses [43,44,47–51,53–65]. It is however important to note that fruits hardly play any role in the latter category – if at all, fruits were infrequently mentioned as "vitamins", but in no case whatsoever as cure for any ailment. This could be corroborated with through our own research. Almost all communities mentioned the medicinal use of *E. precatoria* roots for a wide range of mostly parasitic and viral ailments (malaria, "parasites", hepatitis, yellow fever, anemia related to malaria, hepatitis), as well as to treat pain, urinary infections, and as galactogogue. This medicinal knowledge was spread equally across all ages. However, not a single informant mentioned to use *Euterpe* fruits as medicine. In addition, older participants were more inclined to know the species as source of material for utensils, and significantly, for construction. Younger participants were more familiar with food uses. This was particular significant in case of young mestizos. Ethnicity had virtually no influence on use-knowledge in any of the use categories observed [45,46,66–70] (Tab. 1).

The use of *Euterpe precatoria* fruits as food also showed striking differences between indigenous and mestizo communities. Literature indicated a much more frequent use of the species for the production of palm hearts amongst mestizos [7,43,49,52,53,56–61,64], versus indigenous communities [48,55,59,60,63,71]. A similar trend holds true for the use of the species to produce beverages: again, mestizo communities are much more inclined to use *E. precatoria* fruits for this purpose [7,44,49,53,56–58,61], than indigenous communities [48,50,51,54,72–83].

This trend however seems to have changed dramatically in the more recent past. Our own data indicate a widely spread use of *Euterpe precatoria* amongst all communities interviewed, without regard to ethnic background. However, we did encounter a significant difference related to participant age – in case of food uses, older participants did most frequently not mention *E. precatoria* as food, while younger participants often only knew the species as potential source of palm hearts and beverages. However, many of these participants indicated that they were not themselves consuming the material, but rather collected it for sale [45,46,66–70] (Tab. 1).

Conclusions

The plant use knowledge of local communities in the Western Amazon of Peru and Bolivia clearly depends little on their ethnic background. Differences in knowledge can rather be linked to the age of the users. Young mestizo and indigenous inhabitants are most familiar with the use of *E. precatoria* fruits and palm hearts or food, and know about medicinal uses, while older informants focus on the use of the species for construction and utensils. A reason for this discrepancy might be that the harvest of palm hearts inevitably destroys the resource, and thus if a species is needed as construction material, it makes little sense to destroy the resource for short-term gain, in particular if market access is very limited. The global boom

Tab. 1 Use differences of *Euterpe olearacea* and *Euterpe precatoria* in the present study.

Ethnic group	Food – fruits	Food – palm heart	Construction	Medicine – roots	Medicine – fruits	Utensils	Art
Euterpe oleracea							
Ese Eja							
Arazaeri							
Awajun	eaten		leaves	galactagogue			
Mestizo Riberalta	infrequent	sale	stem, leaves	anemia			
Chacobo							
Mestizo Iquitos	eaten	eaten	stem, leaves	anemia			
Cocama	eaten	eaten	stem, leaves	anemia			
lecos							
Yuracare							
Euterpe precatoria							
Lamas	eaten		stem, leaves	hepatitis, yellow fever, malaria, kidney			used
Ese Eja	eaten	eaten	stem, leaves	Infections, pain, hepatitis			used
Arazaeri	eaten	eaten	stem, leaves	anemia, infections, prostata			used
Awajun	eaten		stem, leaves	galactagogue, hepatitis, anemia			
Mestizo Riberalta	eaten	sold only	stem, leaves	anemia			used
Chacobo		sold only	stem, leaves			used	
Mestizo Iquitos	eaten	eaten	stem, leaves	malaria, urinary, anemia, infections			used
Cocama	eaten	eaten	stem, leaves	malaria, urinary, anemia, infections			used
Lecos	eaten	eaten	stem				
Yuracare	eaten	eaten	stem, leaves	parasites			

of palm hearts and later açaí fruit products, and the associated influx of quick cash, led to a replacement of the originally used products and species with others that can be sold commercially. In case of *Euterpe oleracea*, still only few communities in Western Amazonia know the palm at all, indicating that this species has relatively recently been introduced as reaction to the global boom of açaí products.

The fact that the older local population does hardly ever eat açaí fruits (or use them for the preparation of beverages), while essentially any other palm with soft fruits is consumed in this way, might show some relation to the Chagas disease risk mentioned above (Health, Oksen, Signori Pereira, Velente). No other palm species have been reported as source of food-borne Chagas. This could mean that the indigenous population originally avoided *Euterpe* sp. as food source, precisely to avoid potential illness. This hypothesis would merit further investigation.

Overall it is clear that the global commodity boom has led to a shift in the usage of *Euterpe* sp. from construction and utility use to various forms of food usage. This has inevitably increased the pressure on natural populations, even in the relatively remote Western Amazon, and has led to the introduction and cultivation of *Euterpe oleracea*.

Acknowledgments

Special thanks to all the inhabitants of the communities in Bolivia and Peru who freely shared their knowledge and hospitality with the authors.

We also thank the Russell E. "Train education for nature program" of WWF, the Chatham-Fellowship of the Garden Club of America and the William L. Brown Center at Missouri Botanical Garden for financial and logistical support.

References

1. Kahn F. Palms as key swamp forest resources in Amazonia. Forest Ecol Manag. 1991;38(3–4):133–142.

2. Lee R, Balick MJ. Palms, people, and health. Explore. 2008;4(1):59–62.

3. Goulding M, Smith NJH. Palms: sentinels for Amazon conservation. Saint Louis MO: Missouri Botanical Garden Press; 2007.

4. Matheus ME, de Oliveira Fernandes SB, Silveira CS, Rodrigues VP, de Sousa Menezes F, Fernandes PD. Inhibitory effects of *Euterpe oleracea* Mart. on nitric oxide production and iNOS expression. J Ethnopharmacol. 2006;107(2):291–296.

5. Schauss AG, Wu X, Prior RL, Ou B, Patel D, Huang D, et al. Phytochemical and nutrient composition of freeze-dried Amazonian palm berry *Euterpe oleracea* Mart. (açaí). J Agric Food Chem. 2006;54(22):8598–8603.

6. Rogez H. Açaí: prepare, composicao, e melhoramiento da conservacao. Belem: Edufpa; 2000.

7. Anderson AB. Use management of native forest dominated by açaí palm (*Euterpe oleracea* Mart.) in the Amazon estuary. Adv Econ Bot. 1988;6:144–154.

8. Bobbio FO, Druzian JI, AbrãO PA, Bobbio PA, Fadelli S. Identificação e quantificação das antocianinas do fruto do açaizeiro (*Euterpe oleracea* Mart.). Ciênc Tecnol Aliment. 2000;20(3):388–390.

9. Brondízio E. Agriculture intensification, economic identity, and shared invisibility in Amazonian peasantry: Caboclos and Colonists in comparative perspective [Internet]. In: Adams C, Murrieta R, Neves W, Harris M, editors. Amazon peasant societies in a changing environment. Dordrecht: Springer Netherlands; 2009. p. 181–214.

10. Brondízio ES, Safar CAM, Siqueira AD. The urban market of açaí fruit (*Euterpe oleracea* Mart.) and rural land use change: ethnographic insights

into the role of price and land tenure constraining agricultural choices in the Amazon estuary. Urban Ecosyst. 2002;6(1):67–97.

11. Brondízio ES. From staple to fashion food: shifting cycles and shifting opportunities in the development of the açaí palm fruit economy in the Amazon estuary. In: Zarin DJ, editor. Working forests in the neotropics. New York: Columbia University Press; 2004. p. 339–365.

12. Hiraoka M. Land use changes in the Amazon estuary. Global Environ Chang. 1995;5(4):323–336.

13. Lewis JA. The power of knowledge: information transfer and açaí intensification in the peri-urban interface of Belém, Brazil. Agroforest Syst. 2008;74(3):293–302.

14. Moegenburg SM, Levey DJ. Prospects for conserving biodiversity in Amazonian extractive reserves. Ecol Let. 2002;5(3):320–324.

15. Muñiz-Miret N, Vamos R, Hiraoka M, Montagnini F, Mendelsohn RO. The economic value of managing açaí (Euterpe precatoria Mart.) in the floodplains of the Amazon estuary, Pará, Brazil. For Ecol Man. 1996;87:163–173.

16. Pokorny B, Cayres G, Nunes W. Participatory extension as basis for the work of rural extension services in the Amazon. Agric Human Values. 2005;22(4):435–450.

17. Weinstein S, Moegenburg S. Acai palm management in the Amazon estuary: course for conservation or passage to plantations? Conservat Soc. 2004;2(2):315–346.

18. Galetti M, Fernandez JC. Palm heart harvesting in the Brazilian Atlantic forest: changes in industry structure and the illegal trade. J Appl Ecol. 1998;35(2):294–301.

19. Heinrich M, Dhanji T, Casselman I. Açai (Euterpe oleracea Mart.) – a phytochemical and pharmacological assessment of the species' health claims. Phytochem Let. 2011;4(1):10–21.

20. Chin YW, Chai HB, Keller WJ, Kinghorn AD. Lignans and other constituents of the fruits of Euterpe oleracea (Açaí) with antioxidant and cytoprotective activities. J Agric Food Chem. 2008;56(17):7759–7764.

21. Oliveira de Souza M, Silva M, Silva ME, de Paula Oliveira R, Pedrosa ML. Diet supplementation with acai (Euterpe oleracea Mart.) pulp improves biomarkers of oxidative stress and the serum lipid profile in rats. Nutrition. 2010;26(7–8):804–810.

22. Del Pozo-Insfran D, Brenes CH, Talcott ST. Phytochemical composition and pigment stability of açaí (Euterpe oleracea Mart.). J Agric Food Chem. 2004;52(6):1539–1545.

23. Desmarchelier C. Neotropics and natural ingredients for pharmaceuticals: why isn't South American biodiversity on the crest of the wave? Phytother Res. 2010;24(6):791–799.

24. dos Santos G, Maia G, de Sousa P, da Costa J, de Figueroa R, do Prado G. Correlation between antioxidant activity and bioprotective compounds in açaí (Euterpe oleracea Mart.) chemical pulps. Arch Latinoam Nutr. 2008;58:187–192.

25. Engels G. Açai – Euterpe oleracea (herb profile). Herbalgram. 2010;86:1–2.

26. Gallori S, Bilia AR, Bergonzi MC, Barbosa WLR, Vindicieri FF. Polyphenolic constituents of anthocyanins from açaí fruit (Euterpe oleracea Mart.). Cienc Tecnol Aliment. 2004;20:388–390.

27. Gruenwald J. Novel botanical ingredients for beverages. Clin Dermatol. 2009;27(2):210–216.

28. Iaderoza M. Anthocyanins from fruits of acai (Euterpe oleracea, Mart) and jucara (Euterpe edulis, Mart). Trop Sci. 1992;32:41–46.

29. Kinghorn AD, Chai HB, Sung CK, Keller WJ. The classical drug discovery approach to defining bioactive constituents of botanicals. Fitoterapia. 2011;82(1):71–79.

30. Lichtenthäler R, Rodrigues RB, Maia JGS, Papagiannopoulos M, Fabricius H, Marx F. Total oxidant scavenging capacities of Euterpe oleracea Mart. (açaí) fruits. Int J Food Sci Nutr. 2005;56(1):53–64.

31. Pacheco-Palencia LA, Duncan CE, Talcott ST. Phytochemical composition and thermal stability of two commercial açaí species, Euterpe oleracea and Euterpe precatoria. Food Chem. 2009;115(4):1199–1205.

32. Pompeu DR, Silva EM, Rogez H. Optimisation of the solvent extraction of phenolic antioxidants from fruits of Euterpe oleracea using Response Surface Methodology. Bioresour Technol. 2009;100(23):6076–6082.

33. Rodrigues RB, Lichtenthäler R, Zimmermann BF, Papagiannopoulos M, Fabricius H, Marx F, et al. Total oxidant scavenging capacity of Euterpe oleracea Mart. (açaí) seeds and identification of their polyphenolic compounds. J Agric Food Chem. 2006;54(12):4162–4167.

34. Menezes E, Deliza R, Chan HL, Guinard JX. Preferences and attitudes towards açaí-based products among North American consumers. Food Res Int. 2011;44(7):1997–2008.

35. Sabbe S, Verbeke W, Deliza R, Matta V, Van Damme P. Effect of a health claim and personal characteristics on consumer acceptance of fruit juices with different concentrations of açaí (Euterpe oleracea Mart.). Appetite. 2009;53(1):84–92.

36. Sabbe S, Verbeke W, Deliza R, Matta VM, Van Damme P. Consumer liking of fruit juices with different açaí (Euterpe oleracea Mart.) concentrations. J Food Sci. 2009;74(5):S171–S176.

37. Health articles [Internet]. Açaí (berry) palm (Euterpe oleracea) the good, the bad and the warning. 2012 [cited 2012 Jun 21].

38. Oksen DV. An epidemiological overview on oral outbreaks of Chagas disease in South America [Master thesis]. Umeå: Umeå University; 2011.

39. Pereira KS, Schmidt FL, Guaraldo AMA, Franco RMB, Dias VL, Passos LAC. Chagas' disease as a foodborne illness. J Food Prot. 2009;72(2):441–446.

40. Valente SADS, da Costa Valente V, das Neves Pinto AY, de Jesus Barbosa César M, dos Santos MP, Miranda CO, et al. Analysis of an acute Chagas disease outbreak in the Brazilian Amazon: human cases, triatomines, reservoir mammals and parasites. Trans R Soc Trop Med Hyg. 2009;103(3):291–297.

41. Hajdu Z, Hohmann J. An ethnopharmacological survey of the traditional medicine utilized in the community of Porvenir, Bajo Paraguá Indian Reservation, Bolivia. J Ethnopharmacol. 2012;139(3):838–857.

42. Macía MJ, Armesilla PJ, Cámara-Leret R, Paniagua-Zambrana N, Villalba S, Balslev H, et al. Palm uses in northwestern South America: a quantitative review. Bot Rev. 2011;77(4):462–570.

43. Gutiérrez-Vásquez CA, Peralta R. Palmas comunes de Pando, Santa Cruz de la Sierra, Bolivia. La Paz: PANFOR/OIMT/BOLFOR; 2001.

44. Vásquez R. Sistemática de las plantas medicinales de uso frecuente en el area de Iquitos. Fol Amazon. 1992;4:65–80.

45. Paniagua Zambrana NY, Bussmann RW, Vega C, Téllez C, Macía MJ. Kampanak se usa para el techo pero ya no hay – uso y conservación de palmeras entre los Awajun, Amazonas, Perú. Trujillo: Graficart; 2012.

46. Paniagua Zambrana NY, Bussmann RW, Blacutt E, Macía MJ. Conservando nuestros bosques: conocimiento y uso de las palmeras en las comunidades campesinas del norte de Bolivia. Trujillo: Graficart; 2012.

47. Alexiades MN. Ethnobotany of the Ese Ej: plants, health, and change in an Amazonian society [PhD thesis]. New York: City University of New York; 1999.

48. Armesilla PJ. Usos de las palmeras (Arecaceae) en la Reserva de la Biosfera-Tierra Comunitaria de Origen Pilon Lajas, (Bolivia) [Master thesis]. Madrid: Universidad Autonoma de Madrid; 2006.

49. Balslev H, Grandez C, Paniagua NY, Møller AL, Hansen SL. Useful palms (Arecaceae) near Iquitos, Peruvian Amazon. Rev Peru Biol. 2008;15(1 suppl):121–132.

50. Boom BM. The Chacobo Indians and their palms. Principes. 1986;30(2):63–70.

51. Bourdy G, Valadeau C, Albán J. Yato' ramuesh: plantas medicinales yaneshas. Lima: Institute of Research for Development; 2008.

52. Henkemans AB. Tranquilidad and hardship in the forest: livelihoods and perceptions of "Camba" forest dwellers in the northern Bolivian Amazon. Riberalta: Utrecht University; 2001. (Programa Manejo de Bosques de la Amazonía Boliviana; vol 5).

53. Kvist LP, Andersen MK, Stagegaard J, Hesselsøe M, Llapapasca C. Extraction from woody forest plants in flood plain communities in Amazonian Peru: use, choice, evaluation and conservation status of resources. For Ecol Man. 2001;150(1–2):147–174.

54. Langevin M. Mapajo – nuestra selva y cultura. La Paz: PRAIA (FIDA/CAF); 2002.

55. Macia MJ. Multiplicity in palm uses by the Huaorani of Amazonian Ecuador. Bot J Linn Soc. 2004;144(2):149–159.

56. Mejía K. Las palmeras en los mercados de Iquitos. Bull Inst Fr Etudes Andines. 1992;21(2):755–769.

57. Moraes M, Sarmiento J, Oviedo E. Richness and uses in a diverse palm site in Bolivia. Biodivers Conserv. 1995;4(7):719–727.

58. Moraes M. Flora de palmeras de Bolivia. La Paz: Herbario Nacional de Bolivia, Instituto de Ecología, Carrera de Biología, Universidad Mayor de San Andrés; 2004.

59. Paniagua Zambrana NY. Diversidad, densidad, distribución y uso de las palmas en la región del Madidi, noreste del departamento de La Paz (Bolivia). Ecol Bolivia. 2005;40:265–280.

60. Paniagua Zambrana NY. Guía de plantas útiles de la comunidad de San José de Uchupiamonas. La Paz: FUND-ECO/LIDEMA/Herbario Nacional de Bolivia; 2001.

61. Paniagua Zambrana NY. Expedición de la Universidad de Oxford a Bolivia. Investigación etnobotánica de las Palmae en el noroeste del departamento de Pando. 25 junio-7 de septiembre de 1992. Oxford: Oxford University Press; 1992.

62. Thomas E, Vandebroek I. Guía plantas medicinales de los Yuracarés y Trinitarios del Territorio Indígena Parque Nacional Isidoro-Sécure. Santa Cruz: Industrias Gráficas Sirena; 2006.

63. Aguirre G. Plantas medicinales utilizadas por los indigenas Moseten-Tsimane' de la comunidad Asuncion del Quiquibey en la RB-TCO Pilon Lajas, Beni, Bolivia [Master thesis]. La Paz: Higher University of San Andrés; 2006.

64. Moreno Suárez LR, Moreno Suárez OI. Colecciones de las palmeras de Bolivia: Palmae – Arecaceae. Santa Cruz de la Sierra: Editorial FAN; 2006.

65. Pérez D. Etnobotánica medicinal y biocidas para malaria en la región Ucayali. Fol Amazon. 2002;13:87–108.

66. Paniagua Zambrana NY, Bussmann RW, Blacutt Rivero E, Macia MJ. Los

Chácobo y las palmeras. Trujillo: Graficart; 2011.

67. Paniagua Zambrana NY, Bussmann RW, Macia MJ. Conocimientos de los ancestros – los Ese Eja y sus usos de palmeras Madre de Dios, Peru. Trujillo: Graficart; 2012.

68. Paniagua Zambrana NY, Bussmann RW, Macia MJ. El bosque SI tiene valor – el uso de palmeras en las comunidades campesinas e indígenas de la región de Inambari, Madre de Dios, Perú. Trujillo: Graficart; 2012.

69. Paniagua Zambrana NY, Bussmann RW, Vega C, Téllez C, Macia MJ. Nuestro conocimiento y uso de las palmeras – una herencia para nuestros hijos. Comunidades Llaquash, San Martín, Peru. Trujillo: Graficart; 2012.

70. Bussmann RW, Paniagua Zambrana NY. Traditional knowledge in a changing world – new insights from the Chacobo in Bolivia. Healing the planet: medicinal plants and the legacy of Richard E. Schultes. In: Ponman BE, Bussmann RW, editors. Medicinal plants and the legacy of Richard E. Schultes: proceedings of the Botany 2011 Symposium Honoring Dr. Richard E. Schultes. Trujillo: Graficart; 2012. p. 23–24.

71. Sánchez Sáenz M. Use of tropical rain forest biodiversity by indigenous communities in northwestern Amazonia. Amsterdam: University of Amsterdam; 2005.

72. Albán J. La mujer y las plantas útiles silvestres en la comunidad Cocama-Cocamilla de los ríos Samiria y Marañón. Informe de Proyecto. Lima: Wild World Life Foundation – Biodiversity Support Program 7560; 1994.

73. Cárdenas D, Marín CA, Suárez LS. Plantas útiles en dos comunidades del departamento de Putumayo. Bogotá: Instituto Amazónico de Investigaciones Científicas, SINCHI; 2002.

74. Castaño-Arboleda N, Cárdenas D, Otavo E. Ecología, aprovechamiento y manejo sostenible de nueve especies de plantas del departamento del Amazonas, generadoras de productos maderables y no maderables. Bogotá: Instituto Amazónico de Investigaciones Científicas, SINCHI; 2007.

75. Cerón CE, Montalvo CG. Etnobotánica de los huaorani de quehueiri-ono, Napo-Ecuador. Quito: Abya-Yala; 1998.

76. Forero MC. Aspectos etnobotánicos de uso y manejo de la familia Arecaceae (palmas) en la comunidad indígena Ticuna de Santa Clara de Tarapoto, del resguardo Ticoya del municipio de Puerto Nariño, Amazonas, Colombia [Master thesis]. Bogotá: Pontificia Universidad Javeriana; 2005.

77. Glenboski LL. Ethnobotany of the Tukuna Indians, Amazonas, Colombia. Bogotá: National University of Colombia; 1983.

78. Mondragón ML, Smith R. Bete quiwiguimamo. Salvando el bosque para vivir sano: algunas plantas y árboles utilizados por la nacionalidad huaorani de la amazonía ecuatoriana. Quito: Abya-Yala; 1997.

79. Huertas B. Nuestro territorio Kampu Piyawi (Shawi). Lima: Terra Nuova; 2007.

80. Ortiz R. Uso, conocimiento y manejo de algunos recursos naturales en el mundo yucuna, Mirití-Paraná, Amazonas, Colombia. Quito: Abya-Yala; 1994.

81. Ponce M. Etnobotánica de palmas de Jatun Sacha. Memorias del Tercer Simposio Colombiano de Etnobotánica. Bogotá: INCIVA; 1992.

82. Vargas L. Vida y medicina tradicional de los Mosetenes de Muchanes. Bogotá: Instituto Amazónico de Investigaciones Científicas, SINCHI; 1997.

83. Kronik J. Fééjahisuu: fééne jatimejé hiyaachi suunu: palmas de los nietos de la tierra y montaña verde del centro. København: Centre for Development Research; 1999.

17

Paulinella chromatophora – rethinking the transition from endosymbiont to organelle

Eva C. M. Nowack*

Department of Biology, Heinrich-Heine-Universität Düsseldorf, Universitätsstrasse 1, 40225 Düsseldorf, Germany

Abstract

Eukaryotes co-opted photosynthetic carbon fixation from prokaryotes by engulfing a cyanobacterium and stably integrating it as a photosynthetic organelle (plastid) in a process known as primary endosymbiosis. The sheer complexity of interactions between a plastid and the surrounding cell that started to evolve over 1 billion years ago, make it challenging to reconstruct intermediate steps in organelle evolution by studying extant plastids. Recently, the photosynthetic amoeba *Paulinella chromatophora* was identified as a much sought-after intermediate stage in the evolution of a photosynthetic organelle. This article reviews the current knowledge on this unique organism. In particular it describes how the interplay of reductive genome evolution, gene transfers, and trafficking of host-encoded proteins into the cyanobacterial endosymbiont contributed to transform the symbiont into a nascent photosynthetic organelle. Together with recent results from various other endosymbiotic associations a picture emerges that lets the targeting of host-encoded proteins into bacterial endosymbionts appear as an early step in the establishment of an endosymbiotic relationship that enables the host to gain control over the endosymbiont.

Keywords: organellogenesis; plastid evolution; endosymbiosis; cyanobacterium; photosynthesis; chromatophore; protein targeting; Rhizaria

The evolution of plastids

Primary plastids, the photosynthetic organelles of Plantae (i.e. green algae/land plants, red algae, and glaucophytes) evolved more than 1 billion years ago (gya) through the endosymbiotic uptake of a cyanobacterium by a heterotrophic protist host [1–3]. Photosynthetic carbon fixation from the newly acquired plastid relieved the host from its dependency on the continuous uptake of organic carbon molecules from the environment. In addition to photosynthetic carbon fixation, plastids provide other beneficial metabolic functions to the plant/algal cell such as the assimilation of ammonia and sulfate into amino acids [4], assembly of iron-sulfur clusters [5], de novo biosynthesis of fatty acids [6], isopentenyl diphosphate (IPP) [7], and aromatic amino acids [8]; and they contribute to pathways that are distributed over several compartments, such as photorespiration and biosynthesis of pyrimidines and folates [9–11].

Evolution of plastids was accompanied by a strong reduction of the size of the cyanobacterial genome and the transfer of thousands of cyanobacterial genes into the host nuclear genome, a process termed endosymbiotic gene transfer (EGT) [12,13]. This resulted in proteins localized to the cyanobacterium, now a full-fledged plastid, being synthesized in the cytoplasm and then imported into the plastid through a sophisticated import machinery (the TIC/TOC complex; short for: translocon of the inner and outer chloroplast membranes) targeted by N-terminal transit peptides [14]. A multitude of solute transporters underlie the intricate metabolic crosstalk between plastid and the surrounding cell by controlling fluxes of metabolites and ions through the double membrane system that delimits the plastid [15–17]. Signaling pathways from and to the plastid communicate environmental cues and the metabolic and developmental status of different compartments within the plant/algal cell [18,19]. An interplay of eukaryote- and cyanobacterium-derived proteins coordinate plastid division with the host cell cycle [20].

Following the establishment of primary plastids in Plantae, photosynthetic ability spread to other eukaryotic lineages through secondary endosymbioses [21,22]. In secondary endosymbioses a red or green alga served as an endosymbiont, leading to the establishment of secondary plastids in heterokontophytes (stramenopiles), cryptophytes, haptophytes, dinoflagellates, apicomplexa, chlorarachniophytes, and euglenophytes. In most cases the only remnant of the eukaryotic symbiont is one or two extra membranes that surround the secondary plastid. The nuclear genome of the

* Email: e.nowack@uni-duesseldorf.de

Handling Editor: Andrzej Bodył

eukaryotic endosymbiont was reduced and genes important to operate the plastid were transferred into the host nucleus. However, in the cryptophytes and chlorarachniophytes, a residual eukaryotic nucleus is still present in the periplastidial compartment [23,24].

Since plastids became dependent on the import of nuclear-encoded proteins for metabolism, growth, and proliferation, they cannot be regarded as independent organisms, but rather as semiautonomous parts of the host cell. Thus, the endosymbiont and surrounding eukaryote merged into one chimeric organism. This makes the endosymbiotic acquisition of organelles one of the most transformative forces of eukaryotic evolution. Understanding the molecular mechanisms that underlie this transformation process is fascinating in its own right. Additionally, it might gain an important practical significance for engineering custom-made organelles with novel functions in the context of synthetic biology. However, it is challenging to deduce the intermediate steps in the evolution of the complex and highly optimized interactions between the plastid and the surrounding plant/algae cell from studies of extant primary plastids that are the product of over 1 billion years of evolutionary tinkering. Therefore, it is of pivotal interest to identify cyanobacteria-derived photosynthetic organelles at earlier stages of evolution and dissect their biology. Recently, the photosynthetic amoeba *Paulinella chromatophora* (Fig. 1a,b), was identified as such a much sought-after intermediate stage [25–27].

Paulinella chromatophora

Paulinella chromatophora was first described in 1895 by Robert Lauterborn from an old riverbed of the upper river Rhine in Germany [28,29] and has been reported since from freshwater (and sometimes brackish [30]) habitats around the world, including sites in Austria [31], Switzerland [32], the Netherlands [33], Great Britain [34], the United States [35], Japan [36], and Canada [37]. Its preferred habitat seems to be shady sediments of freshwater bodies that are microaerophilic, rich in organic compounds, with increased salinity, and relatively low pH [38]. Like its heterotrophic

marine relatives (*Paulinella carsoni, P. gigantica, P. indentata, P. intermedia, P. lauterborni, P. multipora, P. ovalis,* and *P. suzukii*) [37], *P. chromatophora* is a thecate amoeba that lives inside an oval-shaped, lucid shell (or "theca") consisting of five rows of silicate scales with a terminal aperture (or "mouth opening") for the naked filopodia that enable the cells to perform sluggish movements (Fig. 1b).

In contrast to its heterotrophic relatives that feed on bacteria, including cyanobacteria [39,40], digestive vacuoles have never been observed in *P. chromatophora* [28,32,33,41], suggesting that the amoeba has dispensed with phagotrophic nutrition. Instead *P. chromatophora* carries two prominent blue-green sausage-shaped photosynthetic units in its cytoplasm that seem to support a phototrophic lifestyle and are referred to as "chromatophores" or "cyanelles". Already Lauterborn indicated in his original description that chromatophores are morphologically very distinct from typical plastids, but resemble cyanobacteria – in particular members of the genus *Synechococcus*. In contrast to free-living *Synechococcus* spp. which are rod-shaped cells approximately 0.5–1.5 μm long, chromatophores are approximately 15–20 μm long with a 3.5–4 μm diameter. Electron microscopic studies of *P. chromatophora* by Ludwig Kies in 1974 substantiated the similarity to cyanobacteria on an ultrastructural level [41]: the chromatophores, which are located in the host cytoplasm are surrounded by two membranes and a peptidoglycan wall between inner and outer membrane. The latter is interpreted as a host-derived vesicle by Kies. In their cytoplasm, chromatophores contain concentrically arranged thylakoids covered with phycobilisomes (the light-harvesting complexes of cyanobacteria) and polyhedral carboxysomes (proteinaceous micro-compartments involved in the carbon concentration mechanisms in cyanobacteria). Despite their morphological similarity to cyanobacteria, chromatophores cannot survive independently of their host.

Within the host, chromatophores are strictly vertically inherited [33]. Before cell division, the mother cell assembles a new theca from silica scales that are produced in cytoplasmic vacuoles. Filopodia are used to arrange the secreted scales into a new theca [42]. The daughter cell, including one of the chromatophores, is then squeezed through the mouth

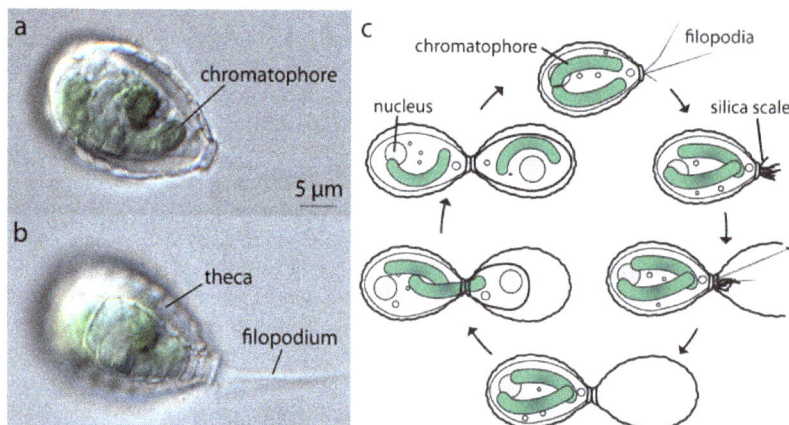

Fig. 1 *Paulinella chromatophora.* **a,b** Different focal planes of micrographs of *P. chromatophora* (differential interference contrast). **c** Schematic representation of cell division in *P. chromatophora.*

opening into the new theca [33,42]. The chromatophores then grow in length taking on a horseshoe shape and finally divide by binary fission, restoring the original state of two chromatophores per cell (Fig. 1c).

$H^{14}CO_3^-$ labeling experiments by Kies and Kremer confirmed in 1979 that *P. chromatophora* photosynthetically assimilates radiocarbon. After 3 h of labeling in the light, radiocarbon had been incorporated into a wide variety of assimilates including glucose, organic phosphates, amino acids, TCA cycle intermediates, and lipids. However, over 40% of the radiolabel ended up in the ethanol-insoluble fraction, probably as proteins and polyglucans [43].

Chromatophores originated independently of plastids

Initially, a rigorous dissection of *P. chromatophora* biology was hampered by the inability to establish a culture. All experiments had to be performed on individuals collected from the environment. It was not until 1990 that the establishment of a clonal culture of *P. chromatophora* in the lab of Michael Melkonian from an isolate from a small pond in the Spessart Forest, Germany (strain M 0880 and later strain CCAC 0185, a further purified subisolate thereof) [38] and advances in DNA sequencing technology allowed for exploring this unusual protist in more depth.

Sequence analysis of the 18S rDNA placed the *P. chromatophora* host cell within the Euglyphidae [44], an order of filose testate amoebae within the Cercozoa, Rhizaria [45,46]. Aside from *P. chromatophora*, there is one other photosynthetic clade among the Cercozoa: the chlorarachniophytes. These net-forming amoebae, which are only distantly related to the Euglyphida, obtained a plastid by secondary endosymbiosis of a green alga [47]. This raised the question whether the chromatophore of *P. chromatophora* was also product of a secondary endosymbiosis, possibly involving a glaucophyte (whose plastids contain a peptidoglycan wall and concentrically arranged thylakoid membranes with phycobilisomes) [48], or whether it arose independently of plastids.

Phylogenetic analyses of the ribosomal and the carboxysomal operon of the chromatophore and numerous cyanobacteria and plastids placed the chromatophore at the base of the *Synechococcus/Prochlorococcus/Cyanobium* clade of cyanobacteria, distant from plastids [25,49]. This group of cyanobacteria is also referred to as α-cyanobacteria because they contain a type 1A ribulose-1,5-bisphosphate carboxylase/oxygenase (RubisCO), in contrast to the remaining β-cyanobacteria and plastids (with the exception of red algal plastids [50,51]) that contain a type 1B RubisCO [49]. The α-cyanobacterial origin of the chromatophore was later confirmed by phylogenetic analyses of multiple protein coding genes [36,52] and genomic data (see below). Taken together, these data yielded overwhelming evidence that the chromatophore originated independently from plastids. Since *P. chromatophora* was the first – and so far only – permanently photosynthetic eukaryote that evolved its photosynthetic ability independently of the event that gave rise to the plastids, this finding resulted in considerable scientific interest in *P. chromatophora*.

Reductive genome evolution in the chromatophore

To understand in which way the chromatophore was integrated into the metabolic networks of the host cell and to determine whether its genome had been reduced with respect to free-living cyanobacteria, the chromatophore genome sequence of *P. chromatophora* CCAC 0185 was completed in 2008 [27]. With a size of only 1.02 Mbp comprising 867 protein-coding genes, the chromatophore genome had been reduced to about 1/3 of the size of genomes of basally branching free-living α-cyanobacteria (e.g. *Synechococcus* WH5701 with a genome size of approximately 3 Mbp and 3346 protein-coding genes). Nevertheless, the chromatophore genome is still about 5–10 times larger than a typical plastid genome, which rarely exceed 200 kbp and generally encode less than 200 genes. Genome reduction in the chromatophore went along with an increase in A+T content to 62%, as frequently observed for bacterial endosymbionts (for a discussion of this phenomenon see e.g. [53]). Gene losses were most pronounced for genes without assigned functions, likely constituting a reservoir for functional responses to environmental changes and, thus, possibly no longer needed in the homogeneous intracellular environment. Also capabilities for DNA repair and regulation of transcription were limited. Furthermore, biosynthetic capabilities have been clearly reduced: pathways for the biosynthesis of several amino acids (Glu, Arg, His, Try, and Met) and cofactors (NAD+, riboflavine, thiamine, biotin, cobalamine, pantothenate, and coenzyme A) were missing. For other amino acids and purine nucleotides, precursors cannot be formed, although later biosynthetic steps are retained. Moreover, all genes encoding enzymes of the TCA cycle, which provides primary building blocks for the synthesis of important cellular compounds, are missing.

Conversely, the chromatophore seems to be much more autonomous than plastids. While in plants, plastid division proteins of cyanobacterial origin are encoded on the nuclear genome, most of the cell-division genes commonly found in α-cyanobacteria (*ftsZ, ftn2, minC, minD, minE, cdv1,* and *cdv2*) are encoded on the chromatophore genome. All genes essential for light harvesting, the function of photosystem (PS) I and II, the cytochrome b_6/f complex, and the F-type ATPase, are present on the chromatophore genome. Only two genes that encode low molecular weight subunits of PSI, *psaE* and *psaK*, were lost. Besides its photosynthetic machinery, the chromatophore retained other biosynthetic capabilities that are potentially beneficial for the host, such as assimilatory sulfate reduction, biosynthesis of fatty acids, IPP, cofactors such as lipoic acid and folate, and some amino acids (Ala, Val, Ile, and Leu).

The gene complement encoded on the chromatophore genome suggests an intricate metabolic crosstalk between host and chromatophore. This would require efficient metabolite transport systems in both directions. However, transport capabilities encoded on the chromatophore genome are also strongly reduced (e.g. uptake systems for any form of nitrogen are missing). This suggests that nuclear-encoded transport systems are involved in metabolite exchange between host and chromatophore. In agreement with this idea, it has been reported that the majority of plastid solute transporters in

Plantae were derived from the host. Authors hypothesized that the insertion of host-derived transporters into the plastid envelope membranes provided a mechanism to rapidly establish metabolic control over the evolving plastid [54].

Age of the chromatophore and speciation

Determining the age of an endosymbiotic association without direct fossil record is inherently difficult. However, using a Bayesian relaxed molecular clock calibrated with multiple calibration points from the microfossil record (dinoflagellates, diatoms, and coccolithophorids), divergence between *P. chromatophora* and the non-photosynthetic relative *Euglypha rotunda* was estimated to be 200 mya, delimiting the maximum age of the chromatophore [55]. The minimum age of the chromatophore was estimated to be 60 mya, based on timing estimates for the complete disintegration of inactivated genes in endosymbiotic bacteria [27]. An interesting question concerns the diversity that evolved within the photosynthetic *Paulinella* population within the 60–200 million years since chromatophore acquisition.

In 2009, a second *P. chromatophora* strain (FK01 or NIES-2635) was brought into culture by Takeshi Nakayama from an isolate from Daigo, Ibaraki Prefecture, Japan [36]. FK01 differs from CCAC 0185 with regard to cell size, and number and fine-structure of the silica scales. Phylogenetic analyses of the 16S rDNA of the chromatophores of CCAC 0185 and FK01 plus four further *P. chromatophora* individuals collected from the environment indicate monophyly of all chromatophores that form two distinct sister clades, one containing CCAC 0185, the other FK01 [36]. Additionally, 18S rDNA and actin phylogenies support monophyly of the CCAC 0185 and FK01 host cells [36]. This phylogenetic pattern is most easily explained by a single origin of chromatophores followed by diversification. Based on morphological and molecular differences observed between CCAC 0185 and FK01 it is likely that these two strains represent distinct species [36]. The establishment of strain FK01 opened the exciting possibility to study divergent evolution of this endosymbiotic association after uptake of the cyanobacterium.

The completion of the chromatophore genome sequence from strain FK01 revealed a genome with a size of 0.977 Mbp, which is slightly smaller than the chromatophore genome of CCAC 0185 (1.02 Mbp). Gene order appears to be largely conserved with respect to CCAC 0185. Nevertheless, 27 genes that are encoded in FK01 are absent from CCAC 0185, whereas 39 genes encoded in CCAC 0185 are absent from FK01 [56]. The majority of these differentially lost genes seem to be outright lost rather than transferred into the nucleus.

Cyanobacterial genes in the nuclear genome of *P. chromatophora*

From the chromatophore genome sequences alone it was not clear in which way the host cell compensates for lost chromatophore functions. An important question was whether in *P. chromatophora* some of the genes that have been lost from the chromatophore were transferred to the host nuclear genome, similar to the evolution of plastids. To answer this question, cDNA libraries from strain FK01 and CCAC 0185 were generated and analyzed. The first two likely chromatophore-derived genes identified in the nuclear transcripts of strain FK01 were genes coding for low molecular weight subunits of PSI, PsaE [57] and PsaI [56]. The corresponding chromatophore genes were missing in FK01. A broader survey of EGT in *P. chromatophora* CCAC 0185 revealed more than 30 nuclear genes of possible chromatophore origin [58]. Most of these EGT genes encoded very short proteins with a function in photosynthesis or light-acclimation of the cell, including the low molecular weight subunits of PSI – PsaE and two divergent copies of PsaK. The predicted functions strongly suggested that at least a subset of the products encoded by EGT genes function within the chromatophore. A G+C content similar to host genes but distinct from the A+T-rich chromatophore genes, the presence of spliceosomal introns, and polyA tails in the mRNAs confirmed that EGT genes were located in, and expressed from the nuclear genome [56–58].

Intriguingly, also 10 genes of β-cyanobacterial or unclear cyanobacterial origin were identified in the transcriptome of CCAC 0185 [58]. These genes had more diverse metabolic functions and did not share the bias towards short genes. Whether or not their products function within the chromatophore is unclear. In phagotrophic protists, horizontal gene transfer (HGT) from food organisms is well documented [59–61]. Thus, these genes presumably represent HGTs from cyanobacteria that served as food sources to *P. chromatophora*, before it gave up phagotrophy.

Larkum et al. [62] argued that organellogenesis does not necessarily involve only one host engulfing a single target organism, but that a host might form a series of transient endosymbioses before a stable relationship is finally achieved. In their "shopping bag model", the resulting organelle is regarded as a chimera of products collected through EGT from different predecessors [62]. In the same line, recently, genes derived from probably co-housed *Chlamydia*, intracellular bacterial pathogens, were proposed to play an important part in the establishment of primary plastids in the Plantae [63–66]. Both *P. chromatophora* cultures available for the transcriptome analyses contained co-cultured bacteria (however, no free cyanobacteria). Some of these co-cultured bacteria live attached to *P. chromatophora* cells and seem to promote the growth of the amoebae (own observations); hence, they are difficult to remove from the culture. As a consequence, the transcriptome data derived from these non-axenic cultures could not easily be analyzed for HGT from bacteria other than cyanobacteria. Nevertheless, HGT from other bacteria might have contributed to forge the nuclear genome of *P. chromatophora*.

Our upcoming genomic/transcriptomic studies from a recently established axenic *P. chromatophora* culture indicate that several bacterial genes might play critical roles in integrating the chromatophore. Thus, it would be reasonable to extend the shopping bag model [62] from genes collected from transient symbionts to genes obtained from pathogens and prey organisms that all might contribute to establishing an endosymbiont and integrating it into a eukaryotic host as an organelle.

A mechanistic view on gene transfers

EGT from plastids and mitochondria is thought to involve the direct movement of DNA following organelle degradation [67–70] and non-homologous end joining at double-strand breaks in the nucleus [71]. In *P. chromatophora*, a stable number of two chromatophores per cell established, and no developmental stages are known that involve the degradation of chromatophores. Damage and lysis of one of the chromatophores would probably represent a rather traumatic event that occurs very rarely and is even more rarely survived. Thus, if indeed lysis of organelles/endosymbionts is necessary for EGT to occur, it is possible that most EGTs in *P. chromatophora* either predate the acquisition of chromatophores or originate from a time when the number of chromatophores per cell was not fixed yet. This view is supported by the fact that also in a heterotrophic *P. ovalis*-like cell a nuclear gene of α-cyanobacterial origin was identified encoding a diaminopimelate epimerase [72].

It is plausible to assume that the symbiotic association between host and α-cyanobacterium established gradually from a predator/prey relationship (Fig. 2). Delaying digestion of the cyanobacterial prey initially would have increased the food supply for the amoeba by allowing its prey to grow and proliferate, using CO_2 and light energy, i.e. nutrient and energy sources that could not be directly used by the host. This scenario, would have given plenty of opportunity to transfer genes from lysed prey/early symbiotic cells and a "practice ground" for establishing protein targeting. However, there are indications that some of the EGTs found in *P. chromatophora* might be more recent events. Two examples are the nuclear genes *csoS4A* and *psaI*: (*i*) in the case of *csoS4A*, encoding a short carboxysomal shell polypeptide, there is – in addition to an expressed nuclear copy – a second copy of the gene in the chromatophore genome. While the amino acid sequence of the chromatophore-encoded copy is highly conserved with respect to cyanobacterial orthologues, the nuclear copy is relatively divergent [58]; (*ii*) in the case of *psaI* that has been identified in the nuclear transcripts of FK01 (see above), the gene is still identifiable in the FK01 chromatophore genome but has been silenced by two nonsense mutations [56]; in CCAC 0185, the chromatophore-encoded copy of *psaI* is still intact. These two cases illustrate the different possible fates of genes that managed to get expressed from the nuclear genome following EGT: as long as a protein encoded by the nuclear copy did not achieve targeting to its site of function, the gene is initially under strongly reduced functional constraints and hence prone to accumulate mutations. This can lead to the nuclear gene either evolving a new function or becoming non-functional (i.e. a pseudogene) and finally erode [73]. If the gene product, however, – probably much more rarely – becomes targeted to its site of function, loss of either copy could be fixed by selection (or stochastically).

Doolittle described very convincingly that even if achieving expression and correct targeting appears to be highly unlikely for a transferred gene, over evolutionary times, the repeated possibility for EGTs after organelle/endosymbiont lysis leads to a "gene transfer ratchet" by which most organellar genes functionally integrate in the nucleus [60]. For *psaI* and *csoS4A* this gene transfer ratchet seems to be caught in the act, with *psaI* getting fixed in the nucleus and *csoS4A* likely being in the process of obtaining a new function or getting lost again from the nucleus either stochastically or because it failed to obtain the correct targeting information.

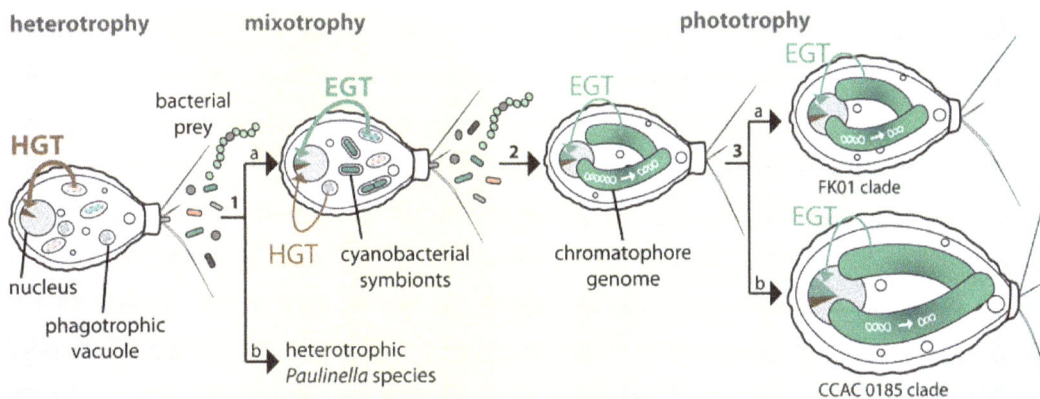

Fig. 2 Proposed model for the evolution of phototrophy in *P. chromatophora*. Heterotrophic and phototrophic *Paulinella* species share a common bacterivorus predecedor (left). **1a** A mixotrophic cell evolved by initially delaying digestion of a particular type of α-cyanobacteria and exploiting their photosynthetic ability. During this stage EGTs and HGTs were possible. The host started to insert transporters into the symbiont-surrounding membranes and target protein into the bacterial symbionts. **1b** Heterotrophic *Paulinella* species did not acquire the ability to house cyanobacterial endosymbionts. **2** As the host cell continues to target proteins into the endosymbiont and regulate metabolic fluxes between the cytoplasm and the endosymbiont, it gains control over the endosymbiont's growth and division. This underpins a stable vertical inheritance. Efficient metabolite exchange makes phagotrophy dispensable (although organic solutes might still be taken up from the environment) and functional constraints relax for hundreds of chromatophore genes leading to massive chromatophore genome reduction. Lysis of chromatophores becomes less frequent and the rate of EGT slows down. **3a,b** Divergent evolution results in different phototrophic *Paulinella* species (represented by strain CCAC 0185 and FK01). Speciation is accompanied by continued chromatophore genome reduction that leads to differential gene losses (and probably differential EGTs) in CCAC 0185 vs. FK01. Colored sections in the nucleus represent contribution of HGT (brown) and EGT (green) to the nuclear genome, the thickness of the arrows represents prevalence of the particular type of gene transfer during different evolutionary stages. At which point the number of chromatophore-surrounding membranes was reduced from 3 to 2 is uncertain.

It is tempting to speculate that fixation of the nuclear copy of a transferred gene is not purely stochastic. Rather, by giving the host transcriptional and translational control over some cellular compounds that are critical to endosymbiont function, coordination of host and endosymbiont metabolism, growth, and proliferation is facilitated and the association between the two partners becomes stabilized.

Another interesting observation was that several EGT genes were present in multiple, in sequence divergent copies, suggesting that the genes duplicated – likely after transfer into the nuclear genome. The largest gene family of EGT genes with at least 15 copies encode a class of proteins designated high-light inducible proteins, or Hlips. These small proteins are proposed to be the progenitors of the light-harvesting chlorophyll a, b binding proteins in green algae and land plants and have been shown to be critical for light acclimation of cyanobacteria [74,75].

Import of nuclear-encoded proteins into the chromatophore

The observation that EGT genes were conserved in amino acid sequence while adjusting their codon usage to the host and acquiring spliceosomal introns strongly suggests that the cellular function of EGT gene products was retained. Experimental evidence that EGT genes in *P. chromatophora* indeed yield products which function in the chromatophore came from immunogold analyses using antibodies raised against *P. chromatophora* PsaE and compositional analysis of PSI isolated from *P. chromatophora* CCAC 0185 [26]. Furthermore, autoradiography of resolved PSI subunits, isolated from *P. chromatophora* cells that were radiolabeled in the presence of chloramphenicol or cycloheximide (that selectively inhibit protein biosynthesis in chromatophore and cytoplasm, respectively), demonstrated that PsaE, PsaK1, and PsaK2 are synthesized in the cytoplasm of the amoeba.

Various possible routes to import protein into the chromatophore have been proposed [76]. Some genes that evolved in Plantae into components of the TIC/TOC complex are conserved on the chromatophore genome (e.g. Toc12, Tic21, and Tic32) and it has been speculated that the corresponding gene products might be involved in protein import into the chromatophore; e.g. a good candidate for a protein transport channel in the inner chromatophore membrane is the homologue of Tic21 [76]. However, homologues of other essential TIC/TOC subunits are missing. Most importantly Omp85, the cyanobacterial homologue of Toc75, which forms the major protein translocating pore in the outer chloroplast membrane, is neither encoded on the chromatophore genome nor was it found in nuclear transcripts of CCAC 0185. Detection of a significant amount of PsaE in the Golgi by immunogold analysis suggested that protein import into the chromatophore involves vesicular transport through the Golgi [26]. Surprisingly, however, all three imported proteins seem to be devoid of cleavable N-terminal signal peptides (SPs) that usually govern co-translational protein import into the endoplasmic reticulum [58]. Also bioinformatic analyses showed that there is no common pre-sequence feature associated with the EGT gene sequences [58,77].

Understanding the mechanism that targets nuclear-encoded proteins into the chromatophore in more detail is difficult because no genetic transformation system is available for *P. chromatophora*, which would enable experimental exploration of targeting signals. The expression of transgenes in *P. chromatophora* might be difficult to establish because the cells grow extremely slowly [78], the cells do not grow on solid media (own observation), and the prominent silica theca might impede delivery of a transgene into the *P. chromatophora* nucleus.

Paulinella chromatophora as a model for early plastid evolution?

Evolution is not necessarily repetitive. Thus, an important question is in how far chromatophores recapitulate the same or similar steps as early primary plastids. Comparing chromatophores and primary plastids, it is fascinating that several 100s of millions of years after the establishment of plastids in the Plantae, a symbiosis involving a cyanobacterium and a protist host that are both very different from the partner organisms that gave rise to the Plantae, converged into a strikingly similar result (a photosynthetic organelle surrounded by two envelope membranes; genome reduction; EGT; protein import; control of metabolite fluxes by host-encoded transport systems). Despite the astonishing similarities between chromatophores and primary plastids, details of the molecular mechanisms employed to integrate the cyanobacterial symbiont might differ.

In primary plastids the outer membrane is believed to be mainly homologous to the outer cyanobacterial membrane but has acquired some eukaryotic features [79]. In contrast, the outer chromatophore membrane might be rather host-derived based on its morphology and the lack of genes encoding porins on the chromatophore genome; however also a chimeric nature is possible (for a discussion of this issue see [80]). Although the exact route of protein import into chromatophores is still elusive, the localization of chromatophore-targeted proteins in the Golgi implies that – other than in primary plastids, where most proteins pass the outer envelope membrane through a translocator pore – the outer chromatophore membrane is likely passed through vesicle fusion. Intriguingly, also in primary plastids a few specific proteins were found to be imported through a TIC/TOC-independent, Golgi-mediated pathway [81,82]. This finding led to the question whether a protein secretion mechanism pre-existing in the host could have been recruited initially for protein targeting to the evolving plastid and was only later replaced by the TIC/TOC system [83]. However, other authors argued that gradual evolution of the TIC/TOC multi-subunit translocon from a simpler system composed of cyanobacterial components is the more parsimonious and thus the more likely scenario, because it would not require the evolution of SPs and subsequently their modification into plastid transit peptides for plastid-targeted proteins; in this case, Golgi-dependent protein import would be regarded as a derived feature [79]. Also, phylogenetic analyses of proteins targeted into primary plastids via the endomembrane system supported the view of Golgi-dependent protein import as an evolutionary derived feature [84].

However, irrespective of the mechanism by which nuclear-encoded proteins traffic into the chromatophore and whether or not a similar targeting mechanism has been used by evolving primary plastids, *P. chromatophora* has the potential to provide important insights germane to organelle evolution. These insights concern in particular: (*i*) early stages in the merging of genetic systems in response to endosymbiotic interactions; (*ii*) contribution of genes obtained by other organisms through HGT to the integration of an evolving organelle; (*iii*) molecular mechanisms that can provide a eukaryotic host with control over growth and division of an endosymbiont early on; (*iv*) origin and characteristics of transport systems for exchange of metabolites with the surrounding cell; (*v*) acquisition of regulatory and detoxifying capacity that allow a previously heterotrophic host to control gene expression/translation in response to light signals and/or redox state of the evolving organelle and to counteract increased levels of oxidative stress that are associated with a phototrophic lifestyle.

There are other examples of highly reduced cyanobacterial endosymbionts, such as *Candidatus* Atelocyanobacterium thalassa living symbiotically with a haptophyte alga [85] or the "spheroid bodies" in rhopalodiacean diatoms [86,87]. However, these symbionts lost their ability to perform oxygenic photosynthesis and their main function is thought to be nitrogen fixation. Hence, the chromatophore of *P. chromatophora* is the closest model to an early primary plastid found so far.

Transitions from endosymbionts to organelles

Until recently, the difference between organelle and endosymbiont appeared to be clear cut: while endosymbionts might lose many genes and depend on their hosts for providing metabolites, an endosymbiont codes for and synthesizes all its own proteins; the import of functional protein was regarded as an exclusive feature of canonical organelles, i.e. mitochondria and plastids [88,89]. However, with growing knowledge on bacterial endosymbionts, it becomes increasingly difficult to pinpoint when exactly the transition into an organelle occurs (for a discussion see e.g. [26,83,90–96]). There is accumulating evidence that, in fact, in many endosymbiotic systems host-encoded proteins get targeted into the bacterial partners that have been regarded as endosymbionts and not as organelles for decades.

Facultative endosymbiotic nitrogen-fixing bacteria (Rhizobia), living in the nodules of the legume *Medicago truncatula*, were reported to import plant-encoded peptides that evolved from effectors of the plant's innate immune system [97]. In the weevil *Sitophilus* spp., a host-derived antimicrobial peptide was found to get selectively targeted into relatively recently acquired (20 mya) γ-proteobacterial endosymbionts that reside in specialized symbiont-harboring cells, termed "bacteriocytes" [98]. In both cases, the immune system-derived peptides trigger dramatic morphological and physiological changes upon import into the endosymbiont [97,98]. And finally, the pea aphid *Acyrthosiphon pisum*

contains the γ-proteobacterial endosymbiont *Buchnera aphidicola* in bacteriocytes. Although the symbionts are transovarially transmitted (i.e. *Buchnera* cells are exocytosed from maternal bacteriocytes and endocytosed by syncytial blastulae) and thus have access to the host germ line [99], functional EGTs have not been found [100]. Nevertheless, *A. pisum* acquired several genes from other bacteria via HGT [100,101]. The mechanism of HGT in aphids is not known, but would likely involve other bacteria reaching the blastula and getting lysed there. Expression of some of these bacteria-derived genes was found to be highly upregulated in bacteriocytes [100]. Using antibodies raised against the predicted protein sequence of one of these bacterium-derived nuclear *A. pisum* genes, *rlpA4*, Nakabachi et al. show that this protein is specifically targeted into *B. aphidicola* [94]. The function of this protein is not yet known. In all three examples, endosymbionts are enclosed in host-derived membrane vesicles and targeting of the proteins into the symbionts is initiated through the secretory pathway mediated by N-terminal SPs. Interestingly, also protein import into complex plastids is initiated through the secretory pathway, mediated by N-terminal SPs [22].

Although a large diversity of bacterial endosymbionts has been described and there is a plethora of genomic data available (for reviews see e.g. [53,102–105]), most endosymbiotic associations have not been studied at the biochemical level. Therefore, whether or not host-encoded protein gets targeted into the symbionts is unknown. However, genomes of several bacterial endosymbionts in various insects are reduced to organellar sizes (e.g. *Hodgkinia cicadicola*, 0.144 Mbp; *Carsonella ruddii*, 0.160 Mbp; *Sulcia muelleri*, 0.246 Mbp) [106–108] and in some cases genome reduction involves the loss of genes involved in DNA replication, transcription, and translation (i.e. functions that occur in the endosymbiont but that are not readily compensated for at the metabolite level) [109]. This finding strongly suggests that an existence as autoreduplicating units is impossible without the import of functional protein from the host.

Since eukaryotic host cells were found to target functional proteins into – even facultative – endosymbionts, the criterion of import of functional protein does not seem to be valid to differentiate organelles from endosymbionts. Furthermore, the transfer of symbiont genes into the nucleus of the host is often regarded as a first required step in the origin of an import apparatus for nuclear-encoded proteins and thus as a required step in the origin of organelles [88,89]. This view is problematic since the examples discussed above demonstrate that protein import is not subordinate to preceding EGT but can apply to proteins deriving from HGTs or the host itself. In fact, with the findings described in this text section, it seems possible that the import of host-encoded proteins into endosymbiotic bacteria – rather than being the final result of a long co-evolution – represents an early step in the establishment of an endosymbiotic relationship. Targeting of protein into the endosymbiont might enable the host to control and manipulate the endosymbiont according to its needs. So, theoretically, it would be feasible that essential functions that are lost, due to genome reduction in an endosymbiont, are replaced by cytoplasmically-synthesized proteins that are imported into the endosymbiont, but that are encoded by genes that do not originate from the endosymbiont. Irrespective of the origin of the imported protein, conceptually this system would be the same as an organelle.

Thus, a good line to draw between an endosymbiont and an organelle seems to be the moment when the endosymbiont – as a consequence of gene loss – becomes dependent for survival and proliferation on the import of nuclear-encoded proteins. These proteins can be derived from EGT, HGT, or the host itself. In this situation, the symbiont cannot be regarded as an independent organism anymore, but is genetically integrated into the host organism and thus has to be regarded as a part of the host. Further criteria that an organelle has to fulfill are vertical inheritance and providing a benefit to the host. If we regard these three criteria as defining an organelle, chromatophores clearly have to be addressed as organelles, even though the coding capacity of their genome is still much larger than the one of plastids and mitochondria. It is possible that some other bacterial entities that are so far regarded as endosymbionts (e.g. some nutritional endosymbionts in insects) will have to be re-classified as early stages of symbiont-derived organelles; but, before this classification can be made, further data on protein import and the cellular functions of imported proteins are required.

Summary and future directions

Although *P. chromatophora* was described over 100 years ago, initiating a rigorous dissection of its biology was only possible after establishment of clonal cultures. Phylogenetic analyses of these cultures clearly established an independent and more recent origin of the chromatophore with respect to primary plastids. After integration of the chromatophore, divergent evolution resulted in at least two different photosynthetic *P. chromatophora* lineages – likely representing distinct species. Genomic and biochemical studies characterized the chromatophore as a bona fide, but early stage photosynthetic

organelle. This makes *P. chromatophora* the closest model for early primary plastids identified so far and hence a unique model to explore the transformation of a cyanobacterium into a genetically integrated photosynthetic organelle.

The nuclear genome sequence of *P. chromatophora* would represent a valuable informational foundation to explore mechanism and dynamics of EGT and the contribution of genes acquired by HGT to chromatophore integration. Experimental determination of the full extent of protein import into the chromatophore would foster our understanding of the molecular mechanisms through which integration occurs. It would be important to determine (*i*) whether import is restricted to short polypeptides like the three PSI subunits that have been described so far; (*ii*) whether also HGT and host-derived proteins get imported; (*iii*) what functions are fulfilled by imported proteins; (*iv*) and whether imported proteins share common structural characteristics that might provide clues about a common targeting mechanism.

Targeting of functional host-encoded proteins into bacterial endosymbionts is likely more frequent than currently assumed. In all bacterial endosymbionts in which protein import has been described (plus secondary plastids), the endomembrane system seems to be involved in targeting. Thus, it is reasonable to assume that endomembrane system-mediated protein targeting into bacterial endosymbionts represents a general mechanism by which a eukaryotic host cell gains control over an endosymbiont at relatively early stages in the establishment of an endosymbiotic relationship.

In the light of recent data, the import of functional host-encoded protein appears not to be a suitable criterion for defining an organelle anymore. Thus, the following criteria were suggested as defining an organelle: (*i*) dependence on import of host-encoded protein for growth and proliferation; (*ii*) vertical inheritance; and (*iii*) benefit for the host. In order to find out if more bacterial endosymbionts have reached the status of early organelles, biochemical studies of further endosymbiotic associations would be necessary.

Acknowledgments
I am grateful to Dr. Michael Melkonian, Dr. Ru Zhang, Dr. Andrzej Bodył as the editor, and to three anonymous reviewers for proof-reading the manuscript and providing many helpful comments. Work in this area is supported in our lab by Deutsche Forschungsgemeinschaft Grant NO 1090/1-1.

Competing interests
No competing interests have been declared.

References
1. Dorrell RG, Howe CJ. What makes a chloroplast? Reconstructing the establishment of photosynthetic symbioses. J Cell Sci. 2012;125(8):1865–1875.
2. Dyall SD, Brown MT, Johnson PJ. Ancient invasions: from endosymbionts to organelles. Science. 2004;304(5668):253–257.
3. Parfrey LW, Lahr DJG, Knoll AH, Katz LA. Estimating the timing of early eukaryotic diversification with multigene molecular clocks. Proc Natl Acad Sci USA. 2011;108(33):13624–13629.
4. Weber A, Flügge UI. Interaction of cytosolic and plastidic nitrogen metabolism in plants. J Exp Bot. 2002;53(370):865–874.
5. Balk J, Lobréaux S. Biogenesis of iron–sulfur proteins in plants. Trends Plant Sci. 2005;10(7):324–331.
6. Wang Z, Benning C. Chloroplast lipid synthesis and lipid trafficking through ER–plastid membrane contact sites. Biochem Soc Trans. 2012;40(2):457–463.
7. Lichtenthaler HK, Schwender J, Disch A, Rohmer M. Biosynthesis of isoprenoids in higher plant chloroplasts proceeds via a mevalonate-independent pathway. FEBS Lett. 1997;400(3):271–274.
8. Herrmann KM, Weaver LM. The shikimate pathway. Annu Rev Plant Physiol Plant Mol Biol. 1999;50(1):473–503.
9. Hanson AD, Gregory III JF. Synthesis and turnover of folates in plants. Curr Opin Plant Biol. 2002;5(3):244–249.
10. Witz S, Jung B, Fürst S, Möhlman T. De novo pyrimidine nucleotide synthesis mainly occurs outside of plastids, but a previously undiscovered nucleobase importer provides substrates for the essential

salvage pathway in *Arabidopsis*. Plant Cell. 2012;24(4):1549–1559.

11. Maurino VG, Peterhansel C. Photorespiration: current status and approaches for metabolic engineering. Curr Opin Plant Biol. 2010;13(3):248–255.

12. Martin W, Rujan T, Richly E, Hansen A, Cornelsen S, Lins T, et al. Evolutionary analysis of *Arabidopsis*, cyanobacterial, and chloroplast genomes reveals plastid phylogeny and thousands of cyanobacterial genes in the nucleus. Proc Natl Acad Sci USA. 2002;99(19):12246–12251.

13. Timmis JN, Ayliffe MA, Huang CY, Martin W. Endosymbiotic gene transfer: organelle genomes forge eukaryotic chromosomes. Nat Rev Genet. 2004;5(2):123–135.

14. Schleiff E, Becker T. Common ground for protein translocation: access control for mitochondria and chloroplasts. Nat Rev Mol Cell Biol. 2011;12(1):48–59.

15. Flügge UI, Häusler RE, Ludewig F, Gierth M. The role of transporters in supplying energy to plant plastids. J Exp Bot. 2011;62(7):2381–2392.

16. Weber AP. Solute transporters as connecting elements between cytosol and plastid stroma. Curr Opin Plant Biol. 2004;7(3):247–253.

17. Pick TR, Weber APM. Unknown components of the plastidial permeome. Front Plant Sci. 2014;5:410.

18. Chi W, Sun X, Zhang L. Intracellular signaling from plastid to nucleus. Annu Rev Plant Biol. 2013;64(1):559–582.

19. Barkan A. Expression of plastid genes: organelle-specific elaborations on a prokaryotic scaffold. Plant Physiol. 2011;155(4):1520–1532.

20. Miyagishima S. Mechanism of plastid division: from a bacterium to an organelle. Plant Physiol. 2011;155(4):1533–1544.

21. Archibald JM. The puzzle of plastid evolution. Curr Biol. 2009;19(2):R81–R88.

22. Gould SB, Waller RF, McFadden GI. Plastid evolution. Annu Rev Plant Biol. 2008;59(1):491–517.

23. Douglas S, Zauner S, Fraunholz M, Beaton M, Penny S, Deng LT, et al. The highly reduced genome of an enslaved algal nucleus. Nature. 2001;410(6832):1091–1096.

24. Gilson PR, Su V, Slamovits CH, Reith ME, Keeling PJ, McFadden GI. Complete nucleotide sequence of the chlorarachniophyte nucleomorph: nature's smallest nucleus. Proc Natl Acad Sci USA. 2006;103(25):9566–9571.

25. Marin B, Nowack ECM, Melkonian M. A plastid in the making: evidence for a second primary endosymbiosis. Protist. 2005;156(4):425–432.

26. Nowack ECM, Grossman AR. Trafficking of protein into the recently established photosynthetic organelles of *Paulinella chromatophora*. Proc Natl Acad Sci USA. 2012;109(14):5340–5345.

27. Nowack ECM, Melkonian M, Glöckner G. Chromatophore genome sequence of *Paulinella* sheds light on acquisition of photosynthesis by eukaryotes. Curr Biol. 2008;18(6):410–418.

28. Lauterborn R. Protozoenstudien. Z Wiss Zool. 1895;59:537–544.

29. Melkonian M. Robert Lauterborn (1869–1952) and his *Paulinella chromatophora*. Protist. 2005;156(2):253–262.

30. Pankow H. *Paulinella chromatophora* Lauterb., eine bisher nur im Süßwasser nachgewiesene Thekamöbe, in den Boddengewässern des Darß und des Zingst (südliche Ostsee). Arch Protistenkd. 1982;126(3):261–263.

31. Geitler L. Bemerkungen zu *Paulinella chromatophora*. Zool Anz. 1927;72:333–334.

32. Penard E. Notes sur quelques Sarcodinés – 12. *Paulinella chromatophora* Lauterborn. Rev Suisse Zool. 1905;13:585–616.

33. Hoogenraad HR. Zur Kenntnis der Fortpflanzung von *Paulinella chromatophora* Lauterb. Zool Anz. 1927;72:140–150.

34. Brown JM. Freshwater rhizopods from the English lake district. Zool J Linn Soc. 1910;30(201):360–368.

35. Lackey JB. Some fresh water protozoa with blue chromatophores. Biol Bull. 1936;71(3):492–497.

36. Yoon HS, Nakayama T, Reyes-Prieto A, Andersen RA, Boo SM, Ishida KI, et al. A single origin of the photosynthetic organelle in different *Paulinella* lineages. BMC Evol Biol. 2009;9(1):98.

37. Nicholls KH. Six new marine species of the genus *Paulinella* (Rhizopoda: Filosea, or Rhizaria: Cercozoa). J Mar Biol Assoc UK. 2009;89(07):1415–1425.

38. Melkonian M, Surek B. Famous algal isolates from the Spessart forest (Germany): the legacy of Dieter Mollenhauer. Algol Stud. 2009;129(1):1–23.

39. Hannah F, Rogerson A, Anderson OR. A description of *Paulinella indentata* n. sp. (Filosea: Euglyphina) from subtidal coastal benthic sediments. J Eukaryot Microbiol. 1996;43(1):1–4.

40. Johnson PW, Hargraves PE, Sieburth JM. Ultrastructure and ecology of *Calycomonas ovalis* Wulff, 1919, (Chrysophyceae) and its redescription as a testate rhizopod, *Paulinella ovalis* n. comb. (Filosea: Euglyphina). J Protozool. 1988;35(4):618–626.

41. Kies L. Electron microscopical investigations on *Paulinella chromatophora* Lauterborn, a thecamoeba containing blue-green endosymbionts (Cyanelles). Protoplasma. 1974;80(1):69–89.

42. Nomura M, Nakayama T, Ishida K. Detailed process of shell construction in the photosynthetic testate amoeba *Paulinella chromatophora* (Euglyphid, Rhizaria). J Eukaryot Microbiol. 2014;61(3):317–321.

43. Kies L, Kremer BP. Function of cyanelles in the thecamoeba *Paulinella chromatophora*. Naturwissenschaften. 1979;66(11):578–579.

44. Bhattacharya D, Helmchen T, Melkonian M. Molecular evolutionary analyses of nuclear-encoded small subunit ribosomal RNA identify an independent rhizopod lineage containing the Euglyphina and the Chlorarachniophyta. J Eukaryot Microbiol. 1995;42(1):65–69.

45. Cavalier-Smith T, Chao EEY. Phylogeny and classification of phylum Cercozoa (Protozoa). Protist. 2003;154(3–4):341–358.

46. Moreira D, von der Heyden S, Bass D, López-García P, Chao E, Cavalier-Smith T. Global eukaryote phylogeny: combined small- and large-subunit ribosomal DNA trees support monophyly of Rhizaria, Retaria and Excavata. Mol Phylogenet Evol. 2007;44(1):255–266.

47. McFadden GI, Gilson PR, Hofmann CJ, Adcock GJ, Maier UG. Evidence that an amoeba acquired a chloroplast by retaining part of an engulfed eukaryotic alga. Proc Natl Acad Sci USA. 1994;91(9):3690–3694.

48. Raven JA. Carboxysomes and peptidoglycan walls of cyanelles: possible physiological functions. Eur J Phycol. 2003;38(1):47–53.

49. Marin B, Nowack EC, Glöckner G, Melkonian M. The ancestor of the *Paulinella* chromatophore obtained a carboxysomal operon by horizontal gene transfer from a *Nitrococcus*-like γ-proteobacterium. BMC Evol Biol. 2007;7(1):85.

50. Delwiche CF, Palmer JD. Rampant horizontal transfer and duplication of rubisco genes in eubacteria and plastids. Mol Biol Evol. 1996;13(6):873–882.

51. Valentin K, Zetsche K. Structure of the Rubisco operon from the unicellular red alga *Cyanidium caldarium*: evidence for a polyphyletic origin of the plastids. Mol Gen Genet. 1990;222(2–3):425–430.

52. Yoon HS, Reyes-Prieto A, Melkonian M, Bhattacharya D. Minimal

plastid genome evolution in the *Paulinella* endosymbiont. Curr Biol. 2006;16(17):R670–R672.

53. McCutcheon JP, Moran NA. Extreme genome reduction in symbiotic bacteria. Nat Rev Microbiol. 2012;10:13–26.

54. Tyra HM, Linka M, Weber AP, Bhattacharya D. Host origin of plastid solute transporters in the first photosynthetic eukaryotes. Genome Biol. 2007;8(10):R212.

55. Berney C, Pawlowski J. A molecular time-scale for eukaryote evolution recalibrated with the continuous microfossil record. Proc R Soc B. 2006;273(1596):1867–1872.

56. Reyes-Prieto A, Yoon HS, Moustafa A, Yang EC, Andersen RA, Boo SM, et al. Differential gene retention in plastids of common recent origin. Mol Biol Evol. 2010;27(7):1530–1537.

57. Nakayama T, Ishida KI. Another acquisition of a primary photosynthetic organelle is underway in *Paulinella chromatophora*. Curr Biol. 2009;19(7):R284–R285.

58. Nowack ECM, Vogel H, Groth M, Grossman AR, Melkonian M, Glöckner G. Endosymbiotic gene transfer and transcriptional regulation of transferred genes in *Paulinella chromatophora*. Mol Biol Evol. 2011;28(1):407–422.

59. Andersson JO. Lateral gene transfer in eukaryotes. Cell Mol Life Sci. 2005;62(11):1182–1197.

60. Doolittle WF. You are what you eat: a gene transfer ratchet could account for bacterial genes in eukaryotic nuclear genomes. Trends Genet. 1998;14(8):307–311.

61. Keeling PJ, Palmer JD. Horizontal gene transfer in eukaryotic evolution. Nat Rev Genet. 2008;9(8):605–618.

62. Larkum AWD, Lockhart PJ, Howe CJ. Shopping for plastids. Trends Plant Sci. 2007;12(5):189–195.

63. Ball S, Colleoni C, Cenci U, Raj JN, Tirtiaux C. The evolution of glycogen and starch metabolism in eukaryotes gives molecular clues to understand the establishment of plastid endosymbiosis. J Exp Bot. 2011;62(6):1775–1801.

64. Becker B, Hoef-Emden K, Melkonian M. Chlamydial genes shed light on the evolution of photoautotrophic eukaryotes. BMC Evol Biol. 2008;8(1):203.

65. Facchinelli F, Colleoni C, Ball SG, Weber APM. Chlamydia, cyanobiont, or host: who was on top in the ménage à trois? Trends Plant Sci. 2013;18(12):673–679.

66. Huang J, Gogarten J. Did an ancient chlamydial endosymbiosis facilitate the establishment of primary plastids? Genome Biol. 2007;8(6):R99.

67. Fuentes I, Karcher D, Bock R. Experimental reconstruction of the functional transfer of intron-containing plastid genes to the nucleus. Curr Biol. 2012;22(9):763–771.

68. Hanekamp T, Thorsness PE. Inactivation of YME2/RNA12, which encodes an integral inner mitochondrial membrane protein, causes increased escape of DNA from mitochondria to the nucleus in *Saccharomyces cerevisiae*. Mol Cell Biol. 1996;16(6):2764–2771.

69. Lister DL, Bateman JM, Purton S, Howe CJ. DNA transfer from chloroplast to nucleus is much rarer in *Chlamydomonas* than in tobacco. Gene. 2003;316:33–38.

70. Sheppard AE, Ayliffe MA, Blatch L, Day A, Delaney SK, Khairul-Fahmy N, et al. Transfer of plastid DNA to the nucleus is elevated during male gametogenesis in tobacco. Plant Physiol. 2008;148(1):328–336.

71. Hazkani-Covo E, Zeller RM, Martin W. Molecular poltergeists: mitochondrial DNA copies (numts) in sequenced nuclear genomes. PLoS Genet. 2010;6(2):e1000834.

72. Bhattacharya D, Price DC, Yoon HS, Yang EC, Poulton NJ, Andersen RA, et al. Single cell genome analysis supports a link between phagotrophy and primary plastid endosymbiosis. Sci Rep. 2012;2:356.

73. Selosse MA, Albert B, Godelle B. Reducing the genome size of organelles favours gene transfer to the nucleus. Trends Ecol Evol. 2001;16(3):135–141.

74. Havaux M, Guedeney G, He Q, Grossman AR. Elimination of high-light-inducible polypeptides related to eukaryotic chlorophyll *a/b*-binding proteins results in aberrant photoacclimation in *Synechocystis* PCC6803. Biochim Biophys Acta. 2003;1557:21–33.

75. He Q, Dolganov N, Björkman O, Grossman AR. The high light-inducible polypeptides in *Synechocystis* PCC6 expression and function in high light. J Biol Chem. 2001;276(1):306–314.

76. Bodyl A, Mackiewicz P, Stiller JW. Comparative genomic studies suggest that the cyanobacterial endosymbionts of the amoeba *Paulinella chromatophora* possess an import apparatus for nuclear-encoded proteins. Plant Biol. 2010;12:639–649.

77. Mackiewicz P, Bodyl A, Gagat P. Possible import routes of proteins into the cyanobacterial endosymbionts/plastids of *Paulinella chromatophora*. Theory Biosci. 2012;131(1):1–18.

78. Nowack ECM. *Paulinella chromatophora* – a model for the acquisition of photosynthesis by eukaryotes [PhD thesis]. Cologne: University of Cologne; 2009.

79. Bodyl A, Mackiewicz P, Stiller JW. Early steps in plastid evolution: current ideas and controversies. Bioessays. 2009;31(11):1219–1232.

80. Mackiewicz P, Bodyl A. A hypothesis for import of the nuclear-encoded PsaE protein of *Paulinella chromatophora* (cercozoa, rhizaria) into its cyanobacterial endosymbionts/plastids via the endomembrane system. J Phycol. 2010;46(5):847–859.

81. Nanjo Y, Oka H, Ikarashi N, Kaneko K, Kitajima A, Mitsui T, et al. Rice plastidial N-glycosylated nucleotide pyrophosphatase/phosphodiesterase is transported from the ER-Golgi to the chloroplast through the secretory pathway. Plant Cell. 2006;18(10):2582–2592.

82. Villarejo A, Burén S, Larsson S, Déjardin A, Monné M, Rudhe C, et al. Evidence for a protein transported through the secretory pathway en route to the higher plant chloroplast. Nat Cell Biol. 2005;7(12):1224–1231.

83. Bhattacharya D, Archibald JM, Weber APM, Reyes-Prieto A. How do endosymbionts become organelles? Understanding early events in plastid evolution. Bioessays. 2007;29(12):1239–1246.

84. Gagat P, Bodyl A, Mackiewicz P. How protein targeting to primary plastids via the endomembrane system could have evolved? A new hypothesis based on phylogenetic studies. Biol Direct. 2013;8(1):18.

85. Thompson AW, Foster RA, Krupke A, Carter BJ, Musat N, Vaulot D, et al. Unicellular cyanobacterium symbiotic with a single-celled eukaryotic alga. Science. 2012;337(6101):1546–1550.

86. Nakayama T, Kamikawa R, Tanifuji G, Kashiyama Y, Ohkouchi N, Archibald JM, et al. Complete genome of a nonphotosynthetic cyanobacterium in a diatom reveals recent adaptations to an intracellular lifestyle. Proc Natl Acad Sci USA. 2014;111(31):11407–11412.

87. Prechtl J, Kneip C, Lockhart P, Wenderoth K, Maier UG. Intracellular spheroid bodies of *Rhopalodia gibba* have nitrogen-fixing apparatus of cyanobacterial origin. Mol Biol Evol. 2004;21(8):1477–1481.

88. Cavalier-Smith T, Lee JJ. Protozoa as hosts for endosymbioses and the conversion of symbionts into organelles. J Protozool. 1985;32(3):376–379.

89. Theissen U, Martin W. The difference between organelles and

endosymbionts. Curr Biol. 2006;16(24):R1016–R1017.

90. Bhattacharya D, Archibald JM. The difference between organelles and endosymbionts – response to Theissen and Martin. Curr Biol. 2006;16(24):R1017–R1018.

91. Bodył A, Mackiewicz P, Stiller JW. The intracellular cyanobacteria of *Paulinella chromatophora*: endosymbionts or organelles? Trends Microbiol. 2007;15(7):295–296.

92. Bodył A, Mackiewicz P, Gagat P. Organelle evolution: *Paulinella* breaks a paradigm. Curr Biol. 2012;22(9):R304–R306.

93. Wernegreen JJ. Strategies of genomic integration within insect-bacterial mutualisms. Biol Bull. 2012;223(1):112–122.

94. Nakabachi A, Ishida K, Hongoh Y, Ohkuma M, Miyagishima SY. Aphid gene of bacterial origin encodes a protein transported to an obligate endosymbiont. Curr Biol. 2014;24(14):R640–R641.

95. Archibald JM. Back to the future. In: One plus one equals one: symbiosis and the evolution of complex life. Oxford: Oxford University Press; 2014. p. 157–172.

96. Reyes-Prieto M, Latorre A, Moya A. Scanty microbes, the "symbionelle" concept. Environ Microbiol. 2014;16(2):335–338.

97. van de Velde W, Zehirov G, Szatmari A, Debreczeny M, Ishihara H, Kevei Z, et al. Plant peptides govern terminal differentiation of bacteria in symbiosis. Science. 2010;327(5969):1122–1126.

98. Login FH, Balmand S, Vallier A, Vincent-Monegat C, Vigneron A, Weiss-Gayet M, et al. Antimicrobial peptides keep insect endosymbionts under control. Science. 2011;334(6054):362–365.

99. Koga R, Meng XY, Tsuchida T, Fukatsu T. Cellular mechanism for selective vertical transmission of an obligate insect symbiont at the bacteriocyte-embryo interface. Proc Natl Acad Sci USA. 2012;109(20):E1230–E1237.

100. Nikoh N, McCutcheon JP, Kudo T, Miyagishima SY, Moran NA, Nakabachi A. Bacterial genes in the aphid genome: absence of functional gene transfer from *Buchnera* to its host. PLoS Genet. 2010;6(2):e1000827.

101. Nikoh N, Nakabachi A. Aphids acquired symbiotic genes via lateral gene transfer. BMC Biol. 2009;7(1):12.

102. Moran NA, McCutcheon JP, Nakabachi A. Genomics and evolution of heritable bacterial symbionts. Annu Rev Genet. 2008;42(1):165–190.

103. Moya A, Peretó J, Gil R, Latorre A. Learning how to live together: genomic insights into prokaryote–animal symbioses. Nat Rev Genet. 2008;9(3):218–229.

104. Nowack ECM, Grossman AR. Evolutionary pressures and the establishment of endosymbiotic associations. In: Bakermans C, editor. Microbial evolution under extreme conditions. Berlin: De Gruyter; 2015 (in press). p. 223–246.

105. Nowack ECM, Melkonian M. Endosymbiotic associations within protists. Phil Trans R Soc B. 2010;365(1541):699–712.

106. McCutcheon JP, McDonald BR, Moran NA. Origin of an alternative genetic code in the extremely small and GC-rich genome of a bacterial symbiont. PLoS Genet. 2009;5(7):e1000565.

107. McCutcheon JP, Moran NA. Parallel genomic evolution and metabolic interdependence in an ancient symbiosis. Proc Natl Acad Sci USA. 2007;104(49):19392–19397.

108. Nakabachi A, Yamashita A, Toh H, Ishikawa H, Dunbar HE, Moran NA, et al. The 160-kilobase genome of the bacterial endosymbiont *Carsonella*. Science. 2006;314(5797):267–267.

109. Tamames J, Gil R, Latorre A, Peretó J, Silva FJ, Moya A. The frontier between cell and organelle: genome analysis of *Candidatus* Carsonella ruddii. BMC Evol Biol. 2007;7(1):181.

18

Primary endosymbiosis: have cyanobacteria and Chlamydiae ever been roommates?

Philippe Deschamps*

Unité d'Ecologie, Systématique et Evolution, CNRS UMR 8079, Université Paris-Sud, 91405 Orsay, France

Abstract

Eukaryotes acquired the ability to process photosynthesis by engulfing a cyanobacterium and transforming it into a genuine organelle called the plastid. This event, named primary endosymbiosis, occurred once more than a billion years ago, and allowed the emergence of the Archaeplastida, a monophyletic supergroup comprising the green algae and plants, the red algae and the glaucophytes. Of the other known cases of symbiosis between cyanobacteria and eukaryotes, none has achieved a comparable level of cell integration nor reached the same evolutionary and ecological success than primary endosymbiosis did. Reasons for this unique accomplishment are still unknown and difficult to comprehend. The exploration of plant genomes has revealed a considerable amount of genes closely related to homologs of Chlamydiae bacteria, and probably acquired by horizontal gene transfer. Several studies have proposed that these transferred genes, which are mostly involved in the functioning of the plastid, may have helped the settlement of primary endosymbiosis. Some of these studies propose that Chlamydiae and cyanobacterial symbionts coexisted in the eukaryotic host of the primary endosymbiosis, and that Chlamydiae provided solutions for the metabolic symbiosis between the cyanobacterium and the host, ensuring the success of primary endosymbiosis. In this review, I present a reevaluation of the contribution of Chlamydiae genes to the genome of Archaeplastida and discuss the strengths and weaknesses of this tripartite model for primary endosymbiosis.

Keywords: primary endosymbiosis; Archaeplastida; Chlamydiae; horizontal gene transfers

Introduction

Oxygenic photosynthesis in eukaryotes arose more than one billion years ago from the association of a cyanobacterial endosymbiont (or cyanobiont) with a heterotrophic host [1–3]. This event, called primary endosymbiosis, is considered to have happened with success only once during the evolution of eukaryotes. It allowed the emergence and diversification of a monophyletic supergroup, the Archaeplastida (or Plantae), composed of three lineages: Chloroplastida (green algae and land plants), Rhodophyta (red algae) and Glaucophyta (Fig. 1) [4–6]. From the original endosymbiotic relationship to the diversification of Archaeplastida, major metabolic and genomic rearrangements occurred. First, the genome of the cyanobiont was highly reduced and a significant part of it was transferred to the host nucleus [7,8]. Secondly, a specific targeting system evolved to export nucleus encoded proteins to the plastid, based on the combination of a N-terminus transit peptide addition to the protein to be exported, and on a specific multi-protein transporter located in the plastid membrane [9,10]. Finally, some regulatory

and feedback pathways were implemented to regulate the organelle activity as well as protein import [11,12]. All these features are shared and processed by homologous genes in all three Archaeplastida lineages, indicating that all of them were tuned in their common ancestor before diversification. Many other eukaryotic phyla acquired the ability to perform photosynthesis by "recycling" the primary plastid by secondary endosymbioses, which involve a heterotrophic host and an Archaeplastida symbiont (either a green or a red alga). Red alga-like secondary plastids are found in cryptophytes, haptophytes, stramenopiles and alveolates while euglenids and chlorarachniophytes bear green alga-like plastids. There are also cases of tertiary endosymbiosis, where the symbiont is a photosynthetic species derived from secondary endosymbiosis; for instance, some dinoflagellates obtained their plastid from cryptophytes or haptophytes. The number of secondary and tertiary endosymbioses that occurred during the evolution of eukaryotes is still strongly debated (for a complete review, refer for example to [13]).

Thought it seems very likely that events similar to primary endosymbiosis happened during the evolution of eukaryotes, none have left any observable living lineage. The sole comparable case of endosymbiosis involving a heterotrophic eukaryote and a cyanobacterium is observed in the cercozoan genus *Paulinella* [14]. *Paulinella chromatophora* is a testate

---

* Corresponding author. Email: philippe.deschamps@u-psud.fr

Handling Editor: Andrzej Bodył

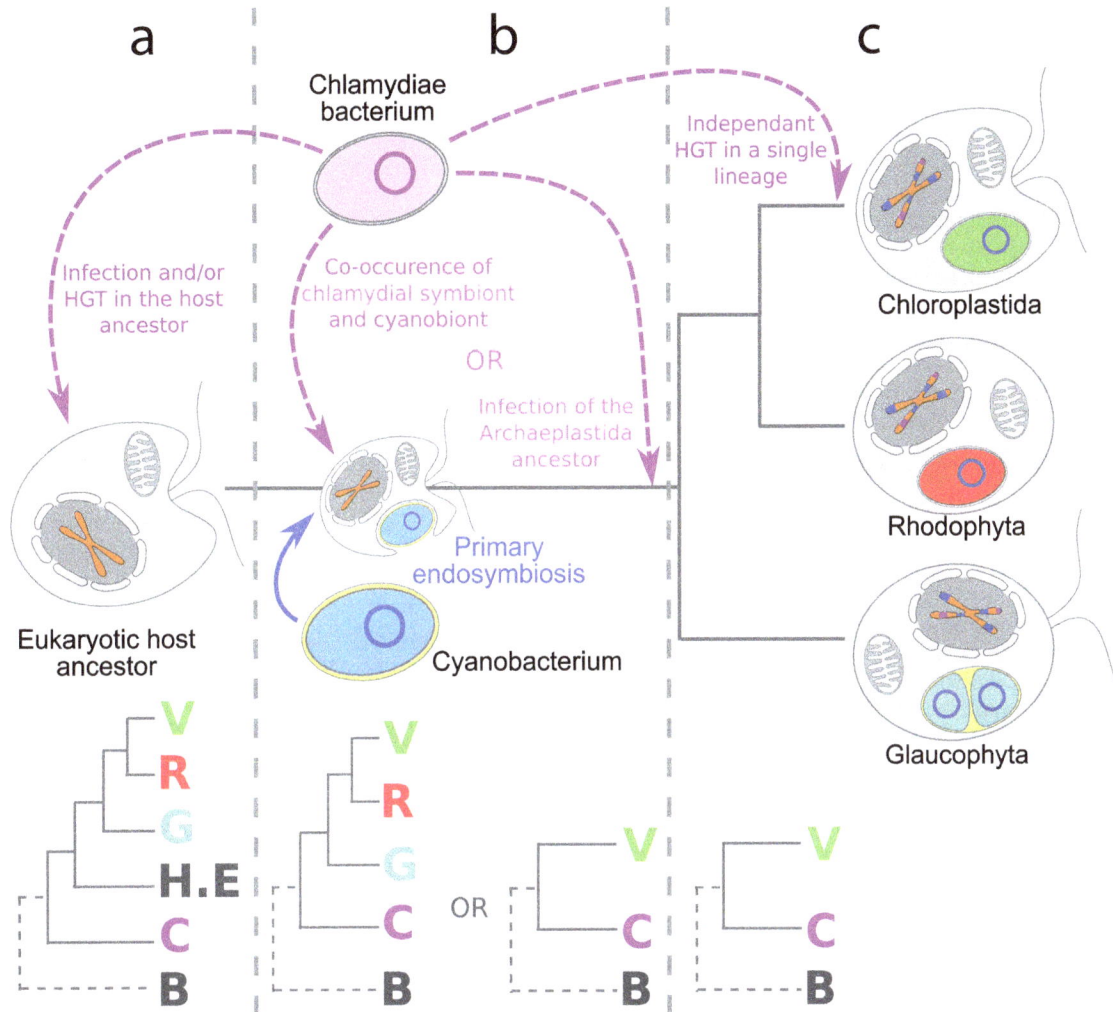

Fig. 1 Time-line of the evolution of Archaeplastida, from their common heterotrophic ancestor to extant Archaeplastida lineages. This line is split in three parts corresponding to three "moments" when HGT from Chlamydiae could have occurred: in the host ancestor before primary endosymbiosis (**a**); in the common ancestor of Archaeplastida, after the engulfment of the cyanobiont but before their diversification (**b**); in a single lineage after the diversification of Archaeplastida (**c**). For each of these moments, the corresponding theoretical phylogenetic tree supporting a HGT from Chlamydiae is displayed. V – Viridiplantae; R – Rhodophyta; G – Glaucophyta; H.E – Heterotrophic eukaryotes; C – Chlamydiae; B – Other bacteria.

amoeba-like organism carrying symbionts called "chromatophores" that are related to *Synechoccocus/Prochlorococcus* cyanobacteria and were probably acquired ~60 million years ago. *Paulinella chromatophora* cells display a lesser level of cytological integration compared to the plastids of Archaeplastida [15]. Indeed, very few endosymbiotic gene transfers could be detected in the nuclear genome of *Paulinella* (0.3% ~ 0.8% of the genome) [16] and no targeting system comparable to the one found in Archaeplastida seems to exist, even if sequences similar to targeting peptides have been detected [17]. Nowack and Grossman have speculated that proteins may use the Golgi secretory pathway for trafficking toward the chromatophores [18]. Finally, there is so far no description of regulation/feedback processes between the host and the chromatophores [16]. Thereby, primary endosymbiosis of the Archaeplastida plastid remains a unique event with

respect to the conversion of the symbiont into a genuine organelle, and to the diversification and spreading of the descent phyla over the planet. Thus, one can wonder what were the specific circumstances that made this event more successful than any other.

Genome mosaicism in Archaeplastida

Like for any other eukaryote, Archaeplastida genomes are mosaics of genes of various prokaryotic origins [19,20]. Some are related to archaeal genes and support functions related to the maintenance, replication and expression of the genome [21,22]; some others are related to alphaproteobacterial genes and are mainly involved in operating the mitochondria [23]. These components are a testimony of the involvement of the corresponding prokaryotes in shaping the genome, if not the cell of the ancestor of all eukaryotes [2,24–26]. Additionally, many enzymes involved in central metabolism are encoded by genes acquired from a wide range of bacterial sources [23,27,28] probably via horizontal gene transfers (HGT). Finally, a collection of genes is restricted to eukaryotes and represent specific innovations of this domain. In addition to that common set, Archaeplastida (and derived photosynthetic eukaryotes) carry genes of other prokaryotic origin. The majority of them are related to cyanobacterial genes and were acquired via endosymbiotic gene transfers (EGT) [7,8] during primary endosymbiosis. Previous studies have assayed the amount of cyanobacterial genes in Archaeplastida and ended up with results ranging from 10 to 20% of the nuclear genome [29–32]. Many of these genes encode plastid-located enzymes, but a significant part of them were alternatively relocated to other compartments of the cell, providing an incredible source of metabolic reshuffling and innovation. Whole genome surveys of genes having prokaryotic homologs as closest relatives in phylogenetic trees (a proxy of their evolutionary origin) [32] provide a mixed estimate of the Archaeplastida genome mosaic composition, which is composed of HGT potentially shared by all eukaryotes as well as H/EGT restricted to Archaeplastida species. To have a more precise view on the H/EGT that specifically contributed to Archaeplastida genomes, an interesting yet limited strategy is to apply the same survey procedures to plastid proteomes (excluding plastid-encoded proteins). This analysis, conducted several times independently, led to comparable results where cyanobacteria were the main contributors to plastid proteomes, followed by proteobacteria and Chlamydiae, as well as many other bacterial phyla [33–35]. While we start to understand how EGT from the cyanobiont have physically happened [36], it is still difficult to determine the chronology of their (probably continuous) occurrence in the interval between the engulfment of the cyanobiont and the diversification of Archaeplastida. Similarly, we know almost nothing about the mechanism and timing underlying HGT from other bacterial sources [27,37].

The intriguing phylogenetic link between plants and Chlamydiae

The Chlamydiae is a bacterial phylum composed of obligate intracellular specialized parasites and/or symbionts of eukaryotic cells. The whole group is considered to have diverged between one and two billion years ago [38,39]. The main order of the phylum, the Chlamydiales, is composed of the Chlamydiaceae (animal pathogens) and of several groups of symbionts of free-living Amoebozoa (for a complete review on Chlamydiae, refer to [40]). All Chlamydiae share a biphasic life cycle where one cell type is specialized in the infection process (the elementary body), and another one (the reticulate body) is dedicated to hijacking the host metabolism and producing daughter cells in an inclusion vesicle for future infection. Chlamydiae were believed to be "energy parasites" because of their ability to uptake ATP from their host. For this purpose, they use an ATP/ADP exchanger

which is homologous to the one found in Rickettsiales (another group of intracellular parasitic bacteria), Microsporidia and photosynthetic eukaryotes [41]. However, this view has been challenged by the observation of a potentially active ATP synthesis complex encoded in the genome of some Chlamydiae species [40]. All Chlamydiae also possess a large repertoire of transporters able to import nucleotides [42], glucose-6-phosphate [43], NAD or amino acids [44]. Finally, all Chlamydiae have the ability to secret proteins into the cytoplasm of their host [40]. For this purpose, they use a Sec-independent protein secretion pathway, called the type III secretion system (TTSS), composed of about 20 components. The exported proteins, called effectors, are responsible for hijacking the central metabolism of the host cell in order to redirect certain metabolites toward the symbiont/parasite.

Starting from the 90's, several publications have reported the existence of close phylogenetic relationships between plant genes and chlamydial genes [45–48]. These observations were first interpreted as HGTs from plants towards infectious bacteria, because Chlamydiae were sister group to plants or were nested within plants in phylogenetic trees [49]. The latter topology was probably induced by low species sampling in gene trees (restricted to Chlamydiaceae parasites) and to the very long branches of their genes due to an acceleration of their evolutionary rate (like in many parasites). Brinkman et al. [48] in 2002 observed that many of these genes were also present in cyanobacterial genomes and/or encode plastidial functions. They proposed that these genes were a testimony of an ancestral relationship between Cyanobacteria, Chlamydiae and the ancestor of the chloroplast. New genome data, in particular from "environmental" Chlamydiae [39,50] contributed to detect additional genes related to plants and to enhance the resolution of the phylogenetic trees. Indeed, Chlamydiae symbionts of Amoebozoa have larger genomes than those of the species parasitizing animals, as well as genes displaying a smaller evolutionary rate [50]. Recent publications have reported between 20 and 55 genes relating plants and Chlamydiae [45–47,51]. Gene trees supporting this relationship recover plants and Chlamydiae as sister groups or, when the gene is present in other photosynthetic lineages, Archaeplastida is found as the closest group to Chlamydiae.

Many of these genes encode important enzymes located in the plastid (involved in lipid and amino-acid synthesis or tRNA synthesis) as well as many transporters, including the ADP/ATP transporter mentioned above, which is found in the plastid membrane of all photosynthetic eukaryotes. It has been demonstrated that, at least in plants, this transporter is involved in fueling enzymatic reactions in the dark, when the plastidial production of ATP is lower or disabled. This includes mechanisms of resorption of oxidative molecules from the chloroplast stroma [52] and mutants lacking this transporter suffer and die from oxidative stress. Given that several essential genes for the functioning of the plastid were acquired from Chlamydiae, many authors hypothesized that, maybe, primary endosymbiosis that led to Archaeplastida was successful thanks to a co-infection of the host by a Chlamydiae bacterium.

A little help from Chlamydiae to the success of primary endosymbiosis?

Huang and Gogarten reported in 2007 that some genes of putative chlamydial origin found in green plants were also shared among all Archaeplastida and that their phylogeny recovered the monophyly of the whole supergroup [46]. They concluded that HGT from Chlamydiae occurred in the common ancestor of Archaeplastida, before or after the recruitment of the cyanobiont. Moreover, they argued that the relatively high amount of HGT from Chlamydiae in genomes of Archaeplastida compared to other bacterial sources (excluding Cyanobacteria) was a testimony of a long period of close relationship between these two partners, probably of endosymbiotic nature considering the intracellular lifestyle of Chlamydiae. Additionally, they could not detect any of these genes in heterotrophic eukaryotes, indicating that this relationship did not or only shortly predated primary endosymbiosis. Using these observations, they hypothesized that primary endosymbiosis happened with a host that was already in an endosymbiotic relationship with an ancient Chlamydiae that resembled a modern *Protochlamydia*. The presence of a Chlamydiae symbiont would have allowed the recruitment of functions crucial to the long-term establishment of the cyanobiont, rather than its digestion. In the hypothesis of Huang and Gogarten, their view of the Chlamydiae symbiont is that of a genuine long lasting endosymbiont, providing a large amount of EGT, more than what is detectable in modern Archaeplastida genomes. For them, the loss of the symbiont provoked the loss of most of the related EGT, like what is observed in eukaryotes with a reduced mitochondria or reverting from phototrophy to heterotrophy.

Moustafa et al. contributed in 2008 to this issue with a re-evaluation of the number of HGT from Chlamydiae into Archaeplastida genomes using updated genome data [47]. Though they agreed with Huang and Gogarten on the pivotal involvement of Chlamydiae gene transfers in the success of primary endosymbiosis, they proposed an alternative interpretation of the phylogenies regarding the timing of these gene transfers. First, they observed that some HGTs from Chlamydiae were detected not in all Archaeplastida but only in red alga or in glaucophytes. This can be explained with almost the same likelihood, either by common gains in the common ancestor of Archaeplastida followed by lineage specific gene losses or by independent acquisitions in a single lineage. Moreover, they pointed out that many bacterivorous eukaryotes harbor prokaryotic genes of (probable) HGT origin in their genomes [53]. With that in mind, they considered that a scenario based on multiple HGT from Chlamydiae in a common ancestor of Archaeplastida couldn't be completely excluded. In their model, the phagotrophic ancestor of Archaeplastida captured a cyanobacterium and settled a "metabolic connection" allowing to switch to a mixotrophic lifestyle. After this event, and for a certain period, this new mixotrophic cell "continuously" acquired genes from various preys or parasites including Chlamydiae. At a specific point, the set of genes that were available in the nucleus became sufficient to start a new form of endosymbiotic relationship with the cyanobiont, leading to the evolution of the plastid.

The same year, Becker et al. published their own evaluation of Chlamydiae contribution to Archaeplastida genomes and discussed how gene phylogenies can help inferring if the presence of a Chlamydiae symbiont predated, co-occured of was subsequent to the engulfment of the cyanobiont during primary endosymbiosis [45]. Part of their discussion focused on genes having different origins between Archaeplastida lineages, for instance functions for which red algae and glaucophytes use a gene of cyanobacterial origin whereas green algae and plants use a homolog of Chlamydiae origin. They noted that there seems to be an exclusion rule applying on these HGT paralogs, because no cases where both paralogs were kept were observed. For these authors, the existence of such paralogy cases of HGT/EGT is a clue that the cyanobiont and Chlamydiae symbiont co-existed for a long period during the evolution of Archaeplastida.

In 2013, Ball et al. inspired by these previous publications, also provided their own evaluation of the contribution of Chlamydiae genes to Archaeplastida genomes and proposed a functional hypothesis for the interaction of the Chlamydiae symbiont, the cyanobiont and the host in a tripartite model of primary endosymbiosis [51]. This tripartite model has recently spread around several groups working on the subject and is used as a framework to infer additional evolutionary hypotheses on primary endosymbiosis.

Metabolic symbiosis and primary endosymbiosis

Given the high number of heterotrophic eukaryotes feeding by phagotrophy on cyanobacteria or using them as temporary symbionts, one can argue that opportunities to settle obligate symbiotic relationships are frequent. Shifting to an obligate relationship probably starts with a strong compromise where each partner "subcontracts" a metabolic service in exchange of one other to the point of being unable to process it and become definitely dependent. Thus, the limitation to start a dynamics of primary endosymbiosis resides on the kind of mutual services one cyanobacterium and one eukaryote can provide to each other, but also on the ability to settle an exchange of the specific metabolites upon which the symbiosis resides. The actual structure and localization of metabolic pathways common to all Archaeplastida, as well as the nature and phylogenetic origin of their components can provide hints to understand the nature of the metabolic symbiosis that could have definitely settled the primary endosymbiosis.

In 2006, Deschamps et al. proposed a hypothesis on the nature of the metabolic symbiosis that originally associated the cyanobacterial ancestor of the plastid and its host [54]. The model resides on the fact that one of the properties that emerged from primary endosymbiosis is the substitution of glycogen by starch to store glucose. Glycogen and starch share a similar molecular structure but have different physical properties. The chemical advantage of starch is that it can store more glucose in a single cell because, as opposed to soluble glycogen, insoluble starch granules do not account to osmotic pressure anymore. Every eukaryotic species known for synthesizing starch is either photosynthetic or derive from a photosynthetic ancestor, meaning that starch metabolism is probably an innovation that occurred together with the acquisition of photosynthesis. All three lineages of Archaeplastida synthesize starch, either in the cytoplasm (Glaucophyta and Rhodophyta) or in the chloroplast (green algae and plants) using components found separately in cyanobacteria and eukaryotes [54]. The survey of all enzymes involved in starch metabolism in all three Archaeplastida lineages allows to infer the most probable composition of this pathway in the common ancestor of all Archaeplastida. Considering the actual cellular localization of the starch pathway in extant Archaeplastida, it was most probably located in the cytoplasm in their common ancestor. Phylogenetic analyses of the protein sequences of these enzymes showed that they were recruited from the glycogen synthesis pathways of both the cyanobiont and the host. It seems that primary endosymbiosis, and the accompanying EGT flow, have for the first time gathered all genes of both pathways in the same genome. These genes, expressed into the cytoplasm of the host, arranged a new hybrid pathway from which starch emerged, offering a high storage capacity for the carbon fixed by the new photosynthetic symbiont.

A second observation inspiring the hypothesis of Deschamps et al. [54] is the discovery of starch-like structures in unicellular group V diazotrophic cyanobacteria [55]. These starch-like molecules are synthesized using enzymes that are phylogenetically related to the ones involved in starch metabolism in Archaeplastida, and have not been found in any other group of Cyanobacteria so far. For this reason, Deschamps el al. proposed that unicellular diazotrophic cyanobacteria were possibly the ancestors of primary plastids. The fixation of nitrogen in diazotrophic cyanobacteria involves an oxygen-intolerant nitrogenase [56]. Filamentous cyanobacteria use specialized cells called heterocysts to physically isolate nitrogen fixation from oxygenic photosynthesis [57]. In unicellular diazotrophic cyanobacteria nitrogen fixation only occurs in the dark when photosynthesis is deactivated. For these unicellular cyanobacteria, becoming an endosymbiont offered an opportunity to acquire nitrogen directly from the host, thus eliminating diazothrophy and breaking the incompatibility with oxygenic photosynthesis. Thus, the first hypothesis of Deschamps and coworkers was that primary endosymbiosis was settled by a metabolic symbiosis based on the exchange of carbon against nitrogen [54]. The nature of the metabolite used for exporting glucose from the cyanobiont was proposed to be ADP-glucose, a specific precursor restricted to starch biosynthesis in the chloroplast of green algae and land plants and to glycogen synthesis in bacteria. Red algae and glaucophytes build starch from UDP-glucose, a widely spread substrate of many other enzymatic pathways requiring glucose. Starch synthesis in the cytoplasm of the common ancestor of Archaeplastida probably used both precursors, because if not, ADP-glucose specific enzymes acquired by EGT from the cyanobiont, encoded in the nucleus of green algae and plants and active in their chloroplast, would have never been positively selected until their diversification. Thus, the model of Deschamps et al. [54] is that ADP-glucose was synthesized in the cyanobiont using a cyanobacterial ADP-Glc pyrophosphorylase and exported toward the cytoplasm by an unknown transporter and used for starch

accumulation. Glucose could then be mobilized again from this stock for other uses. The hypothesis on the reasons that could have subsequently selected for a relocation of starch metabolism in the chloroplast of Viridiplantae is described in references [58,59].

In 2010, Colleoni and coworkers proposed that a nucleotide sugar transporter (NST) already present in the eukaryotic ancestor of Archaeplastida, could have been involved in the export of ADP-glucose from the plastid [60]. This transporter is a putative ancestor of a modern protein, shared by all Archaeplastida, specific for GDP-mannose, but also able to transport ADP-glucose at lower rates and with a lower affinity. This observation provides additional likelihood for a metabolic symbiosis based on the export of ADP-glucose.

Ménage à trois

One weakness of the metabolic symbiosis hypothesis presented above is that the described metabolic link cannot be available immediately after the engulfment of the cyanobiont. First, it requires the efficient integration of a protein able to transport ADP-glucose through the double membrane of the cyanobiont. This assumes that the phagotrophic membrane that was probably originally present around the symbiont had already vanished. Secondly, the biochemical link requires at least one EGT event that would transfer a gene coding for an ADP-glucose specific glycogen synthase to the nucleus of the host and activate its expression. Given the stochasticity of the EGT process, this would require a certain time before such a gene gets indeed expressed into the cytoplasm of the host. Thus, how were the symbiont and the host interacting together before that? What would have prevented the digestion of the cyanobiont if a direct benefit for both partners was not instantly available?

In 2013, Ball et al. published a revision of the metabolic symbiosis model of primary endosymbiosis called the "ménage à trois" [51]. This new version assumes the simultaneous infection by a Chlamydiae symbiont together with the engulfment of the cyanobiont in the host of primary endosymbiosis. The advantage of this new model is that there is no need to wait for an EGT event to start consuming ADP-glucose in the host cytoplasm. Indeed, Ball et al. proposed that some effectors secreted by a Chlamydiae symbiont could have provided essential functions in the host cytoplasm to help priming the metabolic symbiosis with the cyanobiont. For instance, it has been recently shown that *Chlamydia trachomatis* exports enzymes of the glycogen synthesis pathway [61] provoking a change in the levels of glucose and glycogen production in the host. Ball et al. provide additional experimental evidence that many enzymes involved in glycogen metabolism in Chlamydiae are secreted in the host cytoplasm [51]. They also observe that two enzymatic components of starch metabolism in plants (an ADP-glucose specific starch synthase, GlgA, and an isoamylase, GlgX) are in fact derived from HGT of Chlamydiae genes. Given these new information, Ball et al. propose that Chlamydiae helped the settlement of metabolic symbiosis between the cyanobiont and its host by providing, directly through the secretion system, the necessary enzymes to reorganize the host glycogen synthesis pathway, allowing the use of ADP-glucose and motivating its export from the cyanobiont. Later on, genes of the glycogen metabolism of Chlamydiae and cyanobacteria would have been horizontally transferred to the nucleus and used to start accumulating starch.

At the exact same period, Facchinelli et al. published a proteomic analysis of the muroplast of the glaucophyte algae *Cyanophora paradoxa* [62]. This analysis unveils a potential new route for exporting glucose from the cyanelle to the cytoplasm using a transporter of glucose-6-phosphate (G6P). In bacteria, the uhp operon encodes four genes involved in the "uptake of hexose phosphates" [63]. The UhpC protein usually serves as a sensor, except in Chlamydiae where it became able to translocate G6P [43]. A homolog of *uhpC* was found by Price et al. in the genome of *C. paradoxa* [64], and Facchinelli et al. describe the presence of the UhpC protein in the muroplast of this species [62]. The phylogeny of this gene, available in reference [65], shows that it is present in the genome of glaucophytes, red algae and green algae and that it was probably acquired by HGT from Chlamydiae in the common ancestor of Archaeplastida. All these observations, together with the absence of phosphate transporters of the NST family in *Cyanophora*, prompted Facchinelli et al. [62] to speculate that UphC may have been the first protein used to export carbon from the cyanobiont, making it directly available for glycogen synthesis in the host. Thus, the recruitment of an ADP-glucose transporter related to NST and the selection for an ADP-glucose based carbon storage metabolism would have only happened after the split of Glaucophyta from the other Archaeplastida lineages.

Proposing a scenario where there is no selection on ADP-glucose specific enzyme in the common ancestor of extant Archaeplastida questions the entire validity of the metabolic symbiosis hypothesis of Deschamps et al. [54]. This is also incompatible with the apparent presence of an ADP-glucose specific starch synthase in *Cyanophora paradoxa* [64] that must have been positively selected before the divergence of glaucophytes. Facchinelli and Ball published a joined paper to formulate a revision of the "ménage à trois" scenario compatible with all the recent data and restoring a putative selection on ADP-glucose based carbon storage metabolism [66]. This latest version assumes that the cyanobiont would have been integrated in the inclusion vesicle of the reticulate body of the Chlamydiae co-symbiont. According to Facchinelli et al. [66] G6P is exported out of the cyanobiont to the intravesicular space thanks to the UhpC protein encoded in the cyanobiont genome after a single HGT event from the Chlamydiae symbiont, possibly via the type IV secretion system (another kind of secretion system able to export proteins of protein-DNA complexes). G6P is then converted to ADP-glucose and stored into glycogen particles in the intravesicular space. Extra ADP-glucose is exported from the inclusion vesicle thanks to the NST-like transporter that would be installed there using the host Golgi secretion pathway. ADP-glucose is finally used by TTSS secreted enzymes of the chlamydial glycogen pathway, originally there for hijacking the host metabolism for parasitism purposes. Later on, after several EGT events, components of starch metabolism would have replaced the TTSS secreted chlamydial enzymes.

All these hypothetical models are based on very strong assumptions that will not be discussed in detail here. The main supposition is that Chlamydiae and cyanobacteria did interact in the same host at the same time during primary endosymbiosis. Are there any clues indicating that?

What is the actual genomic contribution of Chlamydiae to Archaeplastida?

At least four independent genome wide detections of potential HGT from Chlamydiae to Archaeplastida have been made available in the literature: Huang and Gogarten listed 21 putative HGTs [46], Becker et al. listed 39 putative HGTs [45], Moustafa et al. listed 55 putative HGTs [47] and Ball et al. listed 48 putative HGTs [51]. Even if the latest analysis dates back to 2013, a significant amount of new genomes/transcriptomes were released since then, for instance in reference [67]. To have an updated view of the contribution of Chlamydiae to Archaeplastida, I decided to compile all genes reported in these 4 publications and to reanalyze their phylogeny using the most complete available sample dataset. Merging the four lists ended up with a non-redundant set of 86 candidate genes possibly acquired by HGT. The procedure for this analysis was as follows: (*i*) the reference protein indicated for each reported HGT was used as a query for similarity search using BlastP [68] against a local custom protein database; (*ii*) the top Chlamydiae and Archaeplastida BlastP hits were extracted from the database and used as queries for Net-BlastP searches against the RefSeq protein database [69] on NCBI servers; (*iii*) for each HGT candidate, the 300 first hit sequences were extracted from the local BlastP as well as the 800 first hit sequences from the two Net-BlastP searches. All sequences were merged in a single file and duplicates automatically discarded; (*iv*) a first round of multiple alignments (MAFFT default parameters [70]), trimming (BMGE, default parameters [71]) and phylogenetic tree reconstructions (FastTree [72]) was processed. Every tree was manually inspected to extract the subset of sequences corresponding to the part of the tree centered on Chlamydiae; (*v*) a second round of tree reconstruction was done on this reduced dataset using the same software tools except for maximum likelihood (ML) tree inferences that were computed with TreeFinder [73] using the WAG+Γ substitution model. Final trees are available in Newick and PDF formats in the supporting material.

All 86 ML trees were inspected to determine if their topology was compatible with a HGT acquired from Chlamydiae and if the topology was consistent with a transfer toward the common ancestor of Archaeplastida. A summary of this analysis is available in Tab. 1. Ten gene phylogenies showing a clear vertical descent in eukaryotes were discarded because not compatible with a HGT event. Moreover, 7 trees were insufficiently sampled to allow any interpretation, and in 18 other trees the topology was compatible with a HGT but the closest sister group to Archaeplastida was not Chlamydiae but other bacterial groups. This later observation is in accordance with previous estimations of the composition of the protein repertoire of plastids [33–35].

Fig. 1 depicts a time-line of the evolution of Archaeplastida, from their common heterotrophic ancestor to their diversification in three lineages. On this figure are also indicated the kind of gene tree topologies that would support a HGT from Chlamydiae at certain points of this time-line. All topology compatible with a HGT event from Chlamydiae toward the common ancestor of Archaeplastida must recover the monophyly of Archaeplastida and show Chlamydiae as its closest bacterial sister group. If any non-photosynthetic eukaryotes (i.e. eukaryotes known for having no photosynthetic ancestors) branch between the Chlamydiae and the Archaeplastida, then one can infer that this gene was acquired before primary endosymbiosis (Fig. 1a). Eleven trees showed this topology, and in almost all cases the non-photosynthetic eukaryotes include Amoebozoa, a phylum known for often carrying Chlamydiae parasites. These trees can be interpreted in three different ways: either Amoebozoa and Archaeplastida acquired independently the same gene by HGT from Chlamydiae; or these genes are ancestral in eukaryotes and were transferred to the common ancestor of Chlamydiae; or, finally, the common ancestor of Amoebozoa and Archaeplastida acquired this gene from Chlamydiae, and it was subsequently lost in the whole descent except in these two groups. There is no way to clearly favor one these three scenarios, and thus these genes should not be counted as cases of direct chlamydial contribution to Archaeplastida genomes. As a matter of fact, proteins encoded by all these genes are reported to be located in the cytoplasm in *Arabidopsis thaliana* [74]. This is evidence that HGT of Chlamydiae origin can be positively selected in eukaryotes in the absence of a plastid. This fact was already mentioned by Thiergart et al. in 2012 [27].

Gene trees where Chlamydiae were closely related with primary photosynthetic eukaryotes but where Archaeplastida were not monophyletic were all compatible with a replacement event of the same function by genes of different origins (6 cases). There is no way to determine in what order these consecutive HGT occurred.

Trees compatible with HGT from Chlamydiae and where species of 2 or 3 Archaeplastida lineages were present (Fig. 1b) are the most consistent with an HGT during the time of endosymbiosis (24 cases). Nonetheless, these trees give no additional clues on when exactly each HGT happened during the possibly long lasting period between the initial engulfment of the cyanobiont and the diversification of Archaeplastida.

Finally, trees compatible with HGT from Chlamydiae and where only one Archaeplastida lineage was present can lead to two interpretations (10 cases). Indeed, there is an equal probability that these genes were acquired before the diversification of Archaeplastida and differentially lost (Fig. 1b), or that they were acquired independently in one lineage after diversification (Fig. 1c).

The above phylogenetic analyses suggest that a continuous flow of HGT from Chlamydiae took place during the evolution of eukaryotes. Some occurred before primary endosymbiosis, others happened in the time between the engulfment of the cyanobiont and the diversification of Archaeplastida, and some others probably took place in only one lineage after their diversification. This demonstrates that

HGT from bacteria, including the ones from Chlamydiae should be seen as a "strafe" process, rather than a "one shot" event. Considering the lifestyle of extant Chlamydiae, we can speculate that most of these transfers are linked to periods of parasitism or symbiosis of the Chlamydiae donor with the respective eukaryotic recipient. This is nonetheless not mandatory, as HGT could have happened without internal contact. Indeed, many cases of HGT from diverse bacterial sources toward eukaryotes exist, in photosynthetic species [37] but also in heterotrophic ones [75,76], and do not necessarily imply endosymbiosis or parasitism. If we focus only on HGT events that seem to have happened during primary endosymbiosis, as we stated above, molecular phylogenies (if interpreted sensu stricto) cannot help us inferring when exactly these HGT occurred.

At this point, alternative and more mechanistic arguments can be invoked. For instance, one can argue that genes acquired by HGT are only kept if their function is positively selected at the time they start being expressed by the recipient organism. Because many of the described Chlamydiae HGTs encode plastidial functions, it was stated that their acquisition cannot predate primary endosymbiosis, but must be concomitant or subsequent to it. This is true only for genes that do not have any functional equivalent already encoded in the recipient genome. HGT replacing or duplicating existing functions can be positively selected during the evolution of a lineage, at least for a certain time. Many of the genes listed in Tab. 1, are involved in central or informational metabolisms that exist both in the cytoplasm and in the plastid. Thus, they do not necessarily need the co-occurrence of a cyanobiont/plastid to be positively selected but could have been selected for a cytoplasmic role in the non-photosynthetic ancestor of Archaeplastida and then converted to the same role in the plastid after primary endosymbiosis. Functional or structural constraints are also often invoked when trying to justify why a gene of a specific donor was selected after HGT in place of another one from a different origin. This is a particularly interesting question in a context where chlamydial and cyanobacterial endosymbionts could have provided a lot of homologous gene at the same time, or asynchronously, toward the same recipient genome. In my analysis, I detected 6 cases of gene replacement indicating that not every Archaeplastida encodes the same function with the same gene of a unique HGT origin. This observation illustrates that the selection applying on a specific function can be relaxed with respect to the origin of the gene encoding it. Thus arguments of functional constraints must always be used with caution.

Chlamydiae as helpers of the cyanobiont during primary endosymbiosis

There is no doubt that a set of genes acquired by HGT from Chlamydiae is mandatory to the functioning of plastids in extant Archaeplastida. There is also a strong indication that some of these HGTs probably occurred from a Chlamydiae symbiont in an ancestor of Archaeplastida. These two points do not imply that a Chlamydiae symbiont was necessarily present during the very first stage of primary endosymbiosis

Tab. 1 Summary of the phylogenetic reanalysis of genes reported as acquired by horizontal gene transfer from Chlamydiae to Archaeplastida by Huang and Gogarten [46], Ball et al. [51], Moustafa et al. [47] and Becker et al. [45].

	Huang 2007	Moustafa 2008	Becker 2008	Ball 2013	Ch	Ro	Gl	Topology interpretation	Putative function	Chlamydial ref. gene
1	•	•	•	•	•	•	-	**HGT from Chlamydiales**	**3-Oxoacyl-(acyl-carrier-protein) synthase**	**WP_006340838**
2	-	•	-	•	•	•	•	Eukaryotic gene	3-Phosphoadenosine 5-phosphosulfate 3- phosphatase	WP_021757506
3	-	•	-	•	•	•	-	Chlamydiales not closest	Mg^{2+}-dependent DNA exonuclease	WP_006340671.1
4	•	•	-	-	•	•	•	Chlamydiales not closest	Phosphoribosylanthranilate isomerase	WP_013943364.1
5	-	•	-	•	•	•	•	**HGT from Chlamydiales**	**Cytosine/adenosine deaminases**	**WP_011175198.1**
6	-	•	•	•	•	•	•	**HGT from Chlamydiales**	**Predicted sulfur transferase**	**WP_011174928.1**
7	-	•	-	•	•	•	•	**HGT from Chlamydiales**	**Sodium:hydrogen antiporter 1**	**WP_006340811.1**
8	-	•	-	•	•	•	•	N.P. Euka are present	tRNA d(2)-isopentenylpyrophosphate transferase	WP_011175793.1
9	•	•	•	•	•	•	-	**HGT from Chlamydiales**	**Malate/lactate dehydrogenases**	**WP_011176317.1**
10	-	•	•	-	•	•	-	HGT from Chlamydiales	F-box and associated interaction domains-containing protein	WP_011175702.1
11	•	•	•	•	-	•	•	Archaeplastida polyphyletic	Chloroplast polynucleotide phosphorylase	WP_011175193.1
12	-	•	•	•	•	•	-	**HGT from Chlamydiales**	**S-adenosylmethionine-dependent methyltransferase**	**WP_011174861.1**
13	-	•	•	•	•	•	-	Archaeplastida polyphyletic	Involved deoxyxylulose pathway of isoprenoid biosynthesis	WP_011175290.1
14	•	•	•	•	•	•	•	**HGT from Chlamydiales**	**ATP/ADP antiporter**	**YP_007249.1**
15	-	•	•	•	•	•	•	**HGT from Chlamydiales**	**Glycosyltransferase family A (GT-A)**	**WP_011174874.1**
16	•	•	•	•	•	•	•	**HGT from Chlamydiales**	**L,L-diaminopimelate aminotransferase**	**WP_011175235.1**
17	•	•	•	•	•	•	•	**HGT from Chlamydiales**	**isoamylase**	**WP_011175656.1**
18	•	•	-	•	•	•	•	**HGT from Chlamydiales**	**Cation transport ATPase**	**WP_006341401.1**
19	•	•	•	•	•	•	•	**HGT from Chlamydiales**	**4-Diphosphocytidyl-2C-methyl-D-erythritol 2-P synthase**	**WP_011176135.1**
20	-	•	•	•	•	•	•	**HGT from Chlamydiales**	**SAM-dependent methyltransferases**	**WP_011176541.1**
21	-	•	-	•	•	•	•	Eukaryotic gene	tRNA-dihydrouridine synthase	YP_008986.1
22	-	•	-	•	•	•	-	Chlamydiales not closest	Putative carbonic anhydrase	WP_011175608.1
23	-	•	-	•	•	•	•	**HGT from Chlamydiales**	**Anthranilate phosphoribosyltransferase**	**WP_013943362.1**
24	-	•	-	•	•	•	•	Eukaryotic gene	Inosine triphosphate pyrophosphatase family protein	WP_011175160.1
25	-	•	-	•	•	•	•	Eukaryotic gene	Glycine-rich RNA-binding protein 4	WP_011175443.1
26	-	•	-	-	-	•	-	Very poor sampling	Glycosyl transferase GT2	WP_011176127.1
27	-	•	-	•	•	•	•	Eukaryotic gene	tRNA guanine N7 methyltransferase	WP_011175834.1
28	•	•	-	•	•	•	-	HGT from Chlamydiales	Uncharacterized conserved protein	WP_011176253.1
29	•	•	•	•	•	•	•	**HGT from Chlamydiales**	**Oligoendopeptidase F**	**WP_011175728.1**
30	-	•	•	•	•	•	•	N.P. Euka are present	CMP-2-keto-3-deoxyoctulosonic acid synthetase	WP_011175375.1
31	-	•	-	•	•	•	-	Chlamydiales not closest	Glycosyl transferase	WP_011175669.1
32	•	•	•	•	•	•	•	**HGT from Chlamydiales**	**glycerol-3-phosphate acyltransferase**	**WP_011175867.1**
33	•	•	•	•	•	•	-	HGT from Chlamydiales	Phosphoglycerate mutase 1	WP_011174711.1
34	-	•	-	-	•	•	-	Chlamydiales not closest	Putative dimethyladenosine transferase	WP_011174945.1
35	-	•	-	•	•	•	•	Chlamydiales not closest	S-adenosyl-L-methionine-dependent methyltransferase	WP_011176166.1
36	-	•	-	-	•	•	-	Chlamydiales not closest	Tonoplast intrinsic protein	WP_011175346.1
37	-	•	•	•	•	•	•	**HGT from Chlamydiales**	**Starch synthase**	**WP_011176142.1**
38	-	•	-	•	•	•	•	Eukaryotic gene	Superoxide dismutase [Cu-Zn]	WP_011176304.1
39	-	•	-	-	•	•	•	Chlamydiales not closest	Leucine-rich repeat-containing protein	WP_011174886.1
40	-	•	•	•	•	•	•	N.P. Euka are present	Pyrophosphate-fructose 6-phosphate 1-phosphotransferase	WP_011175430.1
41	•	•	-	•	•	•	•	**HGT from Chlamydiales**	**2-C-Methyl-D-erythritol 4-phosphate cytidylyltransferase**	**WP_011174877.1**
42	-	•	-	•	•	•	•	N.P. Euka are present	UDP-glucuronate 4-epimerase 2	WP_011174629.1
43	-	•	•	•	•	•	•	**HGT from Chlamydiales**	**Uncharacterized conserved protein**	**WP_011175877.1**
44	•	•	-	•	•	-	-	HGT from Chlamydiales	16S rRNA uridine-516 pseudouridylate synthase	WP_011174710.1
45	-	•	•	•	•	•	-	HGT from Chlamydiales	FOG: CBS domain	WP_011176327.1
46	-	•	-	•	•	•	•	N.P. Euka are present	Biotin/lipoate A/B protein ligase family protein	WP_011176383.1
47	-	•	-	•	•	•	•	Eukaryotic gene	Superoxide dismutase [Cu-Zn]	WP_011174820.1
48	•	•	-	•	•	•	-	Archaeplastida polyphyletic	Tyrosyl-tRNA synthetase	WP_011175719.1
49	-	•	•	•	•	•	•	N.P. Euka are present	Queuine tRNA-ribosyltransferase	WP_011175344.1
50	-	•	-	•	•	•	-	Chlamydiales not closest	Ribosomal 5S rRNA binding domain-containing protein	WP_011176140.1
51	•	•	•	•	•	•	-	Archaeplastida polyphyletic	Phosphate/sulfate permeases	WP_011174649.1
52	-	•	-	•	•	•	-	HGT from Chlamydiales	D-alanine-D-alanine ligase family protein	WP_011174948.1
53	-	•	-	•	•	•	•	Eukaryotic gene	rRNA methylases	WP_011174691.1
54	-	•	-	-	-	•	-	Very poor sampling	Predicted ATPase of the PP-loop superfamily	WP_011175189.1
55	-	•	-	•	•	•	•	Archaeplastida polyphyletic	tRNA and rRNA cytosine-C5-methylases	WP_006341520.1
56	-	•	-	•	•	•	-	N.P. Euka are present	Uncharacterized conserved protein	WP_013181531.1
57	-	-	-	•	•	•	•	HGT from Chlamydiales	Aspartokinases	WP_011175315.1
58	-	-	-	•	•	•	•	HGT from Chlamydiales	FKBP-type peptidyl-prolyl cis-trans isomerase 1	WP_011174933.1
59	-	•	•	•	•	•	•	Chlamydiales not closest	Pseudouridylate synthases, 23S RNA-specific	WP_011174905.1
60	-	-	-	•	•	•	•	Archaeplastida polyphyletic	Phosphoenolpyruvate synthase/pyruvate phosphate dikinase	WP_013924500.1
61	-	-	-	•	•	•	•	**HGT from Chlamydiales**	**6-Pyruvoyl-tetrahydropterin synthase**	**WP_013181942.1**
62	-	-	-	•	•	•	-	HGT from Chlamydiales	UDP-N--acetylmuramate dehydrogenase	WP_011176170.1
63	-	-	-	•	•	•	-	Chlamydiales not closest	Predicted SAM-dependent methyltransferases	WP_011175593.1
64	-	-	-	•	•	•	-	Chlamydiales not closest	2-Succinyl-6-hydroxy-2,4-cyclohexadiene-1-carboxylate synthase	WP_013943765.1

Tab. 1 *(continued)*

	Huang 2007	Moustafa 2008	Becker 2008	Ball 2013	Ch	Ro	Gl	Topology interpretation	Putative function	Chlamydial ref. gene
65	-	-	-	•	•	•	-	Archaeplastida polyphyletic	Isochorismate synthase	WP_013943764.1
66	-	-	-	•	-	-	-	Very poor sampling	D-Ala-D-Ala ligase and related ATP-grasp enzymes	WP_013943388.1
67	-	-	-	•	-	•	•	N.P. Euka are present	Membrane-associated lipoprotein involved in thiamine biosynthesis	WP_011176247.1
68	-	-	-	•	-	-	•	HGT from Chlamydiales	Uncharacterized conserved protein	WP_011176413.1
69	-	-	-	-	•	•	•	**HGT from Chlamydiales**	**UphC, Glucose-6-phosphate transporter**	**WP_011174937.1**
70	-	-	-	•	-	-	•	N.P. Euka are present	Predicted Rossmann fold nucleotide binding protein	WP_011175131.1
71	-	-	-	•	•	•	-	**HGT from Chlamydiales**	**Predicted metal-dependent hydrolase**	**WP_013944685.1**
72	-	-	-	•	•	•	-	Very poor sampling	Predicted O-linked N-acetylglucosamine transferase	WP_013925630.1
73	-	-	-	•	•	•	•	N.P. Euka are present	FOG: PPR repeat	WP_013943667.1
74	-	-	-	•	•	•	-	Very poor sampling	Putative uncharacterized protein	WP_011175444.1
75	-	-	-	•	•	•	•	Chlamydiales not closest	Protein involved in glycerolipid metabolism	WP_020966676.1
76	-	-	•	-	•	•	•	N.P. Euka are present	Asparaginyl-tRNA synthetase	WP_011174896
77	-	-	•	-	•	•	-	Eukaryotic gene	7-dehydrocholesterol reductase	WP_011175770
78	-	-	•	-	•	•	•	**HGT from Chlamydiales**	**tRNA-pseudouridine synthase I**	**YP_007681.1**
79	-	-	•	-	•	•	•	Eukaryotic gene	Lon protease 1	WP_011175012.1
80	-	-	•	-	•	•	-	Very poor sampling	DNA mismatch repair protein MutS	WP_011175771.1
81	-	-	•	-	•	•	•	Chlamydiales not closest	Ribosome recycling factor	WP_011176421.1
82	-	-	•	-	•	•	•	Chlamydiales not closest	Tyrosine transporter	WP_011174718.1
83	-	-	•	-	•	•	•	Chlamydiales not closest	50S rRNA methyltransferase	WP_011174889.1
84	-	-	•	-	•	•	•	Chlamydiales not closest	Cysteinyl-tRNA synthetase	WP_011175785.1
85	-	-	•	-	-	-	-	No Archaeplastida	Folylpolyglutamate synthase	WP_011176172.1
86	-	-	•	-	•	•	•	Chlamydiales not closest	Transketolase	WP_011176060.1

The citation of each gene in the three publications is indicated by "•". The presence of Chloroplastida (Ch), Rhodophyceae (Ro) and Glaucophyta (Gl) in each tree is indicated by "•". For each gene, an interpretation of the tree topology is provided, as well as its putative function and a chlamydial protein reference accession number. Bold lines are trees compatible with a HGT from Chlamydiae to the common ancestor of Archaeplastida. N.P. – non photosynthetic.

to help its settlement. There is no solid evidence for that but only speculation. In fact, the settlement of the metabolic relationship between the host and the cyanobiont could have predated the arrival of a Chlamydiae symbiont, and the subsequent HGTs would have only helped to convert the cyanobiont into its actual plastid form by providing a set of important or accessory functions. This alternative point of view is supported by the fact that HGT from Chlamydiae that were selected to function in the plastid can, for the most, only have occurred or been selected after the development of the plastid protein targeting system. This innovation took probably a certain time during which the cyanobiont and its host had to find a way to communicate without the help of chlamydial genes.

This timing problem affects the metabolic symbiosis hypothesis of primary endosymbiosis presented above, because it requires EGT events, either from the cyanobiont or from the Chlamydiae symbiont, to start exporting carbon from the cyanobiont in the form of ADP-Glc or G6P. The last version of the "ménage à trois" hypothesis was designed to prevent this flaw [66]. Proposing that the cyanobiont was included in the same vesicle as the Chlamydiae symbiont brings more opportunity for gene transfers between them. A single transfer of the *uhpC* gene toward the cyanobiont would be enough to make it export G6P in the intravesicular space. G6P will then be converted to ADP-glucose by the chlamydial ADP-glucose pyrophosphorylase and used internally for glycogen synthesis or exported into the cytoplasm. In this version of the model, neither EGT nor HGT toward

the host nucleus are needed to settle a metabolic exchange of carbon, and the positive pressure for keeping ADP-glucose specific enzymes is maintained. Still, this very last version of the "ménage à trois" remains highly hypothetical. We should keep in mind that all is based on the assumptions made from the first metabolic symbiosis model, speculating that an ADP-glucose based metabolism was necessarily selected immediately in the common ancestor of extant Archaeplastida. This is compatible with phylogenetic data, but not at all demonstrated. There is no reason to exclude a scenario where the selection for ADP-glucose utilizing enzymes occurred relatively late after the engulfment of the cyanobiont, which is also compatible with phylogenetic data. The "finite" set of EGT that we detect emerged probably from a long and iterative process, with several rounds of transfers in an evolving selective environment. Even if it was repeatedly observed that genes related to carbon metabolism tend to be easily lost from genomes of parasitic species [77], nobody can evaluate the amount of time required to purge these genes definitely, preventing their transfer to the nucleus. With that in mind, why not considering the possibility that a metabolite different from ADP-glucose was first used to export carbon from the cyanobiont? Are we sure that the primitive exchange between cyanobiont and host has been kept intact in extant Archaeplastida? The time before their diversification was a succession of tunings and adjustments, what we infer today is the last state of their common ancestor, which is not necessarily the same as the first step of primary endosymbiosis.

In a recent review, Zimorski et al. [37] stress out, using data from reference [32], that Chlamydiae are not the only and major contributors to essential functions of the plastid by HGT. However, a lesser interest is put into explaining how and when these other genes of various bacterial origins were acquired by ancestors of Archaeplastida. What could have been their influence in helping the success of primary endosymbiosis? Zimorski et al., like Dagan et al. and Thiergart et al. before them [27,32], also point out that it is probably an error to infer the genome of the ancestor of the cyanobiont only from extant cyanobacteria. HGTs between prokaryotes are frequent, and were certainly already frequent in the past. The ancestor of the cyanobiont was a cyanobacterium with a core genome typical of cyanobacteria but also with a singular accessory genome of his own. If some of the genes constituting this accessory genome were subject to EGT toward the host nucleus, they cannot be detected today as cyanobacterial genes, but only as bacterial HGT. A scenario where a Chlamydiae symbiont would have transfered a gene to the cyanobiont while they were sharing the same host (like the one proposed for *uhpC* in

the "Ménage à trois" hypothesis) can also be explained with equal likelihood by the presence of this gene in the pan genome of the future cyanobiont before endosymbiosis. What if the cyanobacteria about to become the cyanobiont contained many genes acquired from other bacterias, including Chlamydiae. This was proposed by Martin et al. [29] to explain why *A. thaliana* seems to have acquired EGT from so many bacterial sources without having actually shared an endosymbiotic relationship with all of them. This idea was refused by Brinkman et al. [48] with the argument that extant Cyanobacteria do not possess chlamydial genes in their genomes, at least not in comparable amount with what was transferred to Archaeplastida. As Zimorski et al. explain [37], cyanobacterial genomes have evolved an additional billion years since primary endosymbiosis, this is more than enough to add or eliminate a lot of genes from their accessory genome. Can't we imagine that the peculiarity of primary endosymbiosis is the precise content of the genome of the cyanobiont? A genome that may have contained the key to its special relationship with the host during endosymbiosis. A key that vanished since then.

Acknowledgments

The author would like to thank David Moreira for the numerous discussions on the topic, as well as the six anonymous reviewers of this manuscript for the time they put into writing extensive, precise and very constructive reviews. This work was not supported by any specific funding.

Competing interests

No competing interests have been declared.

Supplementary material

The supplementary material for this article is available online at http://pbsociety.org.pl/journals/index.php/asbp/rt/suppFiles/asbp.2014.048/0:

References

1. Mereschkowsky C. Über Natur und Ursprung der Chromatophoren im Pflanzenreiche. Biol Cent. 1905;25(18):593–604.

2. Margulis L. Symbiosis and evolution. Sci Am. 1971;225(2):48–57.

3. Martin W, Kowallik K. Annotated English translation of Mereschkowsky's 1905 paper "Über Natur und Ursprung der Chromatophoren imPflanzenreiche." Eur J Phycol. 1999;34(3):287–295.

4. Adl SM, Simpson AGB, Lane CE, Lukeš J, Bass D, Bowser SS, et al. The revised classification of eukaryotes. J Eukaryot Microbiol. 2012;59(5):429–514.

5. Rodríguez-Ezpeleta N, Brinkmann H, Burey SC, Roure B, Burger G, Löffelhardt W, et al. Monophyly of primary photosynthetic eukaryotes: green plants, red algae, and glaucophytes. Curr Biol. 2005;15(14):1325–1330.

6. Moreira D, Le Guyader H, Philippe H. The origin of red algae and the evolution of chloroplasts. Nature. 2000;405(6782):69–72.

7. Martin W, Brinkmann H, Savonna C, Cerff R. Evidence for a chimeric nature of nuclear genomes: eubacterial origin of eukaryotic glyceraldehyde-3-phosphate dehydrogenase genes. Proc Natl Acad Sci USA. 1993;90(18):8692–8696.

8. Martin W, Stoebe B, Goremykin V, Hapsmann S, Hasegawa M, Kowallik KV. Gene transfer to the nucleus and the evolution of chloroplasts. Nature. 1998;393(6681):162–165.

9. Gutensohn M, Fan E, Frielingsdorf S, Hanner P, Hou B, Hust B, et al. Toc, Tic, Tat et al.: structure and function of protein transport machineries in chloroplasts. J Plant Physiol. 2006;163(3):333–347.

10. Steiner JM, Löffelhardt W. Protein import into cyanelles. Trends Plant Sci. 2002;7(2):72–77.

11. Koussevitzky S, Nott A, Mockler TC, Hong F, Sachetto-Martins G, Surpin M, et al. Signals from chloroplasts converge to regulate nuclear gene expression. Science. 2007;316(5825):715–719.

12. Chan KX, Crisp PA, Estavillo GM, Pogson BJ. Chloroplast-to-nucleus communication: current knowledge, experimental strategies and relationship to drought stress signaling. Plant Signal Behav. 2010;5(12):1575–1582.

13. Keeling PJ. The endosymbiotic origin, diversification and fate of plastids. Phil Trans R Soc Lond B. 2010;365(1541):729–748.

14. Marin B, M. Nowack EC, Melkonian M. A plastid in the making: evidence for a second primary endosymbiosis. Protist. 2005;156(4):425–432.

15. Yoon HS, Nakayama T, Reyes-Prieto A, Andersen RA, Boo SM, Ishida K, et al. A single origin of the photosynthetic organelle in different Paulinella lineages. BMC Evol Biol. 2009;9(1):98.

16. Nowack ECM, Vogel H, Groth M, Grossman AR, Melkonian M, Glockner G. Endosymbiotic gene transfer and transcriptional regulation of transferred genes in *Paulinella chromatophora*. Mol Biol Evol. 2011;28(1):407–422.

17. Mackiewicz P, Bodył A, Gagat P. Possible import routes of proteins into the cyanobacterial endosymbionts/plastids of *Paulinella chromatophora*. Theory Biosci. 2012;131(1):1–18.

18. Nowack ECM, Grossman AR. Trafficking of protein into the recently established photosynthetic organelles of *Paulinella chromatophora*. Proc Natl Acad Sci USA. 2012;109(14):5340–5345.

19. Simonson AB, Servin JA, Skophammer RG, Herbold CW, Rivera MC, Lake JA. Decoding the genomic tree of life. Proc Natl Acad Sci USA. 2005;102(1 suppl):6608–6613.

20. Ribeiro S, Golding GB. The mosaic nature of the eukaryotic nucleus. Mol Biol Evol. 1998;15(7):779–788.

21. Anantharaman V, Koonin EV, Aravind L. Comparative genomics and evolution of proteins involved in RNA metabolism. Nucl Acids Res. 2002;30(7):1427–1464.

22. Langer D, Hain J, Thuriaux P, Zillig W. Transcription in archaea: similarity to that in eucarya. Proc Natl Acad Sci USA. 1995;92(13):5768–5772.

23. Esser C. A genome phylogeny for mitochondria among alpha-proteobacteria and a predominantly eubacterial ancestry of yeast nuclear genes. Mol Biol Evol. 2004;21(9):1643–1660.

24. López-Garćia P, Moreira D. Metabolic symbiosis at the origin of eukaryotes. Trends Biochem Sci. 1999;24(3):88–93.

25. Martin W, Müller M. The hydrogen hypothesis for the first eukaryote. Nature. 1998;392(6671):37–41.

26. Martin W. Archaebacteria (Archaea) and the origin of the eukaryotic nucleus. Curr Opin Microbiol. 2005;8(6):630–637.

27. Thiergart T, Landan G, Schenk M, Dagan T, Martin WF. An evolutionary network of genes present in the eukaryote common ancestor polls genomes on eukaryotic and mitochondrial origin. Genome Biol Evol. 2012;4(4):466–485.

28. Rochette NC, Brochier-Armanet C, Gouy M. Phylogenomic test of the hypotheses for the evolutionary origin of eukaryotes. Mol Biol Evol. 2014;31(4):832–845.

29. Martin W, Rujan T, Richly E, Hansen A, Cornelsen S, Lins T, et al. Evolutionary analysis of *Arabidopsis*, cyanobacterial, and chloroplast genomes reveals plastid phylogeny and thousands of cyanobacterial genes in the nucleus. Proc Natl Acad Sci USA. 2002;99(19):12246–12251.

30. Reyes-Prieto A, Hackett JD, Soares MB, Bonaldo MF, Bhattacharya D. Cyanobacterial contribution to algal nuclear genomes is primarily limited to plastid functions. Curr Biol. 2006;16(23):2320–2325.

31. Deusch O, Landan G, Roettger M, Gruenheit N, Kowallik KV, Allen JF, et al. Genes of cyanobacterial origin in plant nuclear genomes point to a heterocyst-forming plastid ancestor. Mol Biol Evol. 2008;25(4):748–761.

32. Dagan T, Roettger M, Stucken K, Landan G, Koch R, Major P, et al. Genomes of Stigonematalean cyanobacteria (subsection V) and the evolution of oxygenic photosynthesis from prokaryotes to plastids. Genome Biol Evol. 2013;5(1):31–44.

33. Reyes-Prieto A, Moustafa A. Plastid-localized amino acid biosynthetic pathways of Plantae are predominantly composed of non-cyanobacterial enzymes. Sci Rep. 2012;2:955.

34. Qiu H, Price DC, Weber APM, Facchinelli F, Yoon HS, Bhattacharya D. Assessing the bacterial contribution to the plastid proteome. Trends Plant Sci. 2013;18(12):680–687.

35. Suzuki K, Miyagishima SY. Eukaryotic and eubacterial contributions to the establishment of plastid proteome estimated by large-scale phylogenetic analyses. Mol Biol Evol. 2010;27(3):581–590.

36. Stegemann S, Bock R. Experimental reconstruction of functional gene transfer from the tobacco plastid genome to the nucleus. Plant Cell. 2006;18(11):2869–2878.

37. Zimorski V, Ku C, Martin WF, Gould SB. Endosymbiotic theory for organelle origins. Curr Opin Microbiol. 2014;22:38–48.

38. Kamneva OK, Knight SJ, Liberles DA, Ward NL. Analysis of genome content evolution in pvc bacterial super-phylum: assessment of candidate genes associated with cellular organization and lifestyle. Genome Biol Evol. 2012;4(12):1375–1390.

39. Horn M, Collingro A, Schmitz-Esser S, Beier CL, Purkhold U, Fartmann B, et al. Illuminating the evolutionary history of

chlamydiae. Science. 2004;304(5671):728–730.

40. Horn M. Chlamydiae as symbionts in eukaryotes. Annu Rev Microbiol. 2008;62(1):113–131.

41. Wolf YI, Aravind L, Koonin EV. Rickettsiae and Chlamydiae: evidence of horizontal gene transfer and gene exchange. Trends Genet. 1999;15(5):173–175.

42. Haferkamp I, Schmitz-Esser S, Wagner M, Neigel N, Horn M, Neuhaus HE. Tapping the nucleotide pool of the host: novel nucleotide carrier proteins of *Protochlamydia amoebophila*. Mol Microbiol. 2006;60(6):1534–1545.

43. Schwöppe C, Winkler HH, Neuhaus HE. Properties of the glucose-6-phosphate transporter from *Chlamydia pneumoniae* (HPTcp) and the glucose-6-phosphate sensor from *Escherichia coli* (UhpC). J Bacteriol. 2002;184(8):2108–2115.

44. Subtil A, Collingro A, Horn M. Tracing the primordial Chlamydiae: extinct parasites of plants? Trends Plant Sci. 2014;19(1):36–43.

45. Becker B, Hoef-Emden K, Melkonian M. Chlamydial genes shed light on the evolution of photoautotrophic eukaryotes. BMC Evol Biol. 2008;8(1):203.

46. Huang J, Gogarten J. Did an ancient chlamydial endosymbiosis facilitate the establishment of primary plastids? Genome Biol. 2007;8(6):R99.

47. Moustafa A, Reyes-Prieto A, Bhattacharya D. Chlamydiae has contributed at least 55 genes to Plantae with predominantly plastid functions. PLoS ONE. 2008;3(5):e2205.

48. Brinkman FSL, Blanchard JL, Cherkasov A, Av-Gay Y, Brunham RC, Fernandez RC, et al. Evidence that plant-like genes in *Chlamydia* species reflect an ancestral relationship between Chlamydiaceae, cyanobacteria, and the chloroplast. Genome Res. 2002;12(8):1159–1167.

49. Stephens RS. Genome sequence of an obligate intracellular pathogen of humans: *Chlamydia trachomatis*. Science. 1998;282(5389):754–759.

50. Collingro A, Tischler P, Weinmaier T, Penz T, Heinz E, Brunham RC, et al. Unity in variety – the pan-genome of the Chlamydiae. Mol Biol Evol. 2011;28(12):3253–3270.

51. Ball SG, Subtil A, Bhattacharya D, Moustafa A, Weber APM, Gehre L, et al. Metabolic effectors secreted by bacterial pathogens: essential facilitators of plastid endosymbiosis? Plant Cell. 2013;25(1):7–21.

52. Reinhold T, Alawady A, Grimm B, Beran KC, Jahns P, Conrath U, et al. Limitation of nocturnal import of ATP into *Arabidopsis* chloroplasts leads to photooxidative damage. Plant J. 2007;50(2):293–304.

53. Watkins RF, Gray MW. The frequency of eubacterium-to-eukaryote lateral gene transfers shows significant cross-taxa variation within amoebozoa. J Mol Evol. 2006;63(6):801–814.

54. Deschamps P, Colleoni C, Nakamura Y, Suzuki E, Putaux JL, Buléon A, et al. Metabolic symbiosis and the birth of the plant kingdom. Mol Biol Evol. 2008;25(3):536–548.

55. Nakamura Y, Takahashi J, Sakurai A, Inaba Y, Suzuki E, Nihei S, et al. Some cyanobacteria synthesize semi-amylopectin type alpha-polyglucans instead of glycogen. Plant Cell Physiol. 2005;46(3):539–545.

56. Gallon JR. The oxygen sensitivity of nitrogenase: a problem for biochemists and micro-organisms. Trends Biochem Sci. 1981;6:19–23.

57. Wolk CP, Ernst A, Elhai J. Heterocyst metabolism and development. In: Bryant DA, editor. The molecular biology of cyanobacteria. Dordrecht: Springer; 2004. p. 769–823. (Advances in photosynthesis and respiration; vol 1).

58. Deschamps P, Moreau H, Worden AZ, Dauvillée D, Ball SG. Early

gene duplication within chloroplastida and its correspondence with relocation of starch metabolism to chloroplasts. Genetics. 2008;178(4):2373–2387.

59. Deschamps P, Haferkamp I, d' Hulst C, Neuhaus HE, Ball SG. The relocation of starch metabolism to chloroplasts: when, why and how. Trends Plant Sci. 2008;13(11):574–582.

60. Colleoni C, Linka M, Deschamps P, Handford MG, Dupree P, Weber APM, et al. Phylogenetic and biochemical evidence supports the recruitment of an ADP-glucose translocator for the export of photosynthate during plastid endosymbiosis. Mol Biol Evol. 2010;27(12):2691–2701.

61. Lu C, Lei L, Peng B, Tang L, Ding H, Gong S, et al. *Chlamydia trachomatis* GlgA is secreted into host cell cytoplasm. PLoS ONE. 2013;8(7):e68764.

62. Facchinelli F, Pribil M, Oster U, Ebert NJ, Bhattacharya D, Leister D, et al. Proteomic analysis of the *Cyanophora paradoxa* muroplast provides clues on early events in plastid endosymbiosis. Planta. 2013;237(2):637–651.

63. Shattuck-Eidens DM, Kadner RJ. Molecular cloning of the uhp region and evidence for a positive activator for expression of the hexose phosphate transport system of *Escherichia coli*. J Bacteriol. 1983;155(3):1062–1070.

64. Price DC, Chan CX, Yoon HS, Yang EC, Qiu H, Weber APM, et al. *Cyanophora paradoxa* genome elucidates origin of photosynthesis in algae and plants. Science. 2012;335(6070):843–847.

65. Rockwell NC, Lagarias JC, Bhattacharya D. Primary endosymbiosis and the evolution of light and oxygen sensing in photosynthetic eukaryotes. Front Ecol Evol. 2014;2:66.

66. Facchinelli F, Colleoni C, Ball SG, Weber APM. *Chlamydia*, cyanobiont, or host: who was on top in the ménage à trois? Trends Plant Sci. 2013;18(12):673–679.

67. Keeling PJ, Burki F, Wilcox HM, Allam B, Allen EE, Amaral-Zettler LA, et al. The marine microbial eukaryote transcriptome sequencing project (MMETSP): illuminating the functional diversity of eukaryotic

68. Altschul SF, Madden TL, Schäffer AA, Zhang J, Zhang Z, Miller W, et al. Gapped BLAST and PSI-BLAST: a new generation of protein database search programs. Nucl Acids Res. 1997;25(17):3389–3402.

69. Pruitt KD, Tatusova T, Maglott DR. NCBI reference sequences (RefSeq): a curated non-redundant sequence database of genomes, transcripts and proteins. Nucl Acids Res. 2007;35(database):D61–D65.

70. Katoh K, Toh H. Recent developments in the MAFFT multiple sequence alignment program. Brief Bioinform. 2008;9(4):286–298.

71. Criscuolo A, Gribaldo S. BMGE (block mapping and gathering with entropy): a new software for selection of phylogenetic informative regions from multiple sequence alignments. BMC Evol Biol. 2010;10(1):210.

72. Price MN, Dehal PS, Arkin AP. FastTree 2 – approximately maximum-likelihood trees for large alignments. PLoS ONE. 2010;5(3):e9490.

73. Jobb G, von Haeseler A, Strimmer K. TREEFINDER: a powerful graphical analysis environment for molecular phylogenetics. BMC Evol Biol. 2004;4(1):18.

74. Lamesch P, Berardini TZ, Li D, Swarbreck D, Wilks C, Sasidharan R, et al. The *Arabidopsis* information resource (TAIR): improved gene annotation and new tools. Nucl Acids Res. 2012;40(D1):D1202–D1210.

75. Keeling PJ, Palmer JD. Horizontal gene transfer in eukaryotic evolution. Nat Rev Genet. 2008;9(8):605–618.

76. Marcet-Houben M, Gabaldón T. Acquisition of prokaryotic genes by fungal genomes. Trends Genet. 2010;26(1):5–8.

77. Henrissat B, Deleury E, Coutinho PM. Glycogen metabolism loss: a common marker of parasitic behaviour in bacteria? Trends Genet. 2002;18(9):437–440.

life in the oceans through transcriptome sequencing. PLoS Biol. 2014;12(6):e1001889.

19

Evolution of the cell wall components during terrestrialization

Alicja Banasiak*

Department of Developmental Plant Biology, Institute of Experimental Biology, University of Wrocław, Kanonia 6/8, 50-328 Wrocław, Poland

Abstract

Colonization of terrestrial ecosystems by the first land plants, and their subsequent expansion and diversification, were crucial for the life on the Earth. However, our understanding of these processes is still relatively poor. Recent intensification of studies on various plant organisms have identified the plant cell walls are those structures, which played a key role in adaptive processes during the evolution of land plants. Cell wall as a structure protecting protoplasts and showing a high structural plasticity was one of the primary subjects to changes, giving plants the new properties and capabilities, which undoubtedly contributed to the evolutionary success of land plants.

In this paper, the current state of knowledge about some main components of the cell walls (cellulose, hemicelluloses, pectins and lignins) and their evolutionary alterations, as preadaptive features for the land colonization and the plant taxa diversification, is summarized. Some aspects related to the biosynthesis and modification of the cell wall components, with particular emphasis on the mechanism of transglycosylation, are also discussed. In addition, new surprising discoveries related to the composition of various cell walls, which change how we perceive their evolution, are presented, such as the presence of lignin in red algae or MLG $(1\rightarrow3),(1\rightarrow4)$-$\beta$-D-glucan in horsetails. Currently, several new and promising projects, regarding the cell wall, have started, deciphering its structure, composition and metabolism in the evolutionary context. That additional information will allow us to better understand the processes leading to the terrestrialization and the evolution of extant land plants.

Keywords: cell wall; evolution; land plants; terrestrialization; ancestral genes; polysaccharide; lignin

Introduction

Emergence of the land plants was one of the most important events in the history of life on the Earth, which happened about 480–470 million years ago [1–5]. It is generally accepted that land plants, including bryophytes and vascular plants, are a monophyletic group derived from ancient freshwater charophycean green algae (CGA) [1,6,7]. Their rapid expansion and diversification led to the changes in the atmosphere and soil composition, significantly affecting the development and evolution of all living organisms [8–10].

Currently, the vascular plants are a dominant group in terrestrial ecosystems, showing great diversity of structural and physiological adaptations [11]. The cell walls and their modifications undoubtedly played a key role in that evolutionary success. As a structure protecting the protoplasts and responding to extracellular conditions it contributed significantly to the plant adaptation to different environments, which changed during million years of evolution [6–8]. As a consequence, the plant cell walls are extremely variable in their structure, composition and ongoing metabolic processes, depending on the plant phylogenetic position, developmental stage, cell type, or even the region within a single cell [12]. Regardless of this diversity, the essential components of the cell walls are carbohydrates. Cellulose is usually a major structural element forming supporting rigid network, while other polysaccharides, glycoproteins, enzymes and phenolic compounds form a matrix, in which this network is immersed [11,13–16]. Due to such a composition the cell wall is a tensegral structure, in which tensile and compressive forces are balanced giving its exceptional strength while maintaining the flexibility. Relatively well understood are the structure and the composition of the cell walls in seed plants, whereas in evolutionarily less advanced plant groups this knowledge is rather scarce. However in recent decades, the representatives of these plant groups have become the subject of exhaustive studies, in appreciation of their significance for understanding the origin of land plants, the mechanisms responsible for their radiation and the relationships between structure and function of various cell walls.

In this review, the main focus is on the polysaccharide components of the cell wall and on lignin, in the context of evolutionary changes (Tab. 1) associated with their significance for the land plant emergence and diversification.

* Email: alicja.banasiak@uni.wroc.pl

Handling Editor: Beata Zagórska-Marek

Tab. 1 Cell wall components, genes involved in their biosynthesis and some metabolic activity in different groups of organisms.

Group	Groups	Cellulose in the cell wall	CesA family genes	Rosette terminal complexes	Xylans	Mannans	CslA family genes	Xyloglucans	CslC family genes	XET activity	MLG	CslJ/H/F group of genes	MXE activity	Glucurono-arabinoxylans	HG	RG-I	RG-II	XGA	Lignin monomers (G, H, S)	Lignin biosynthetic pathway
Bacteria	Cyanophyta	+	+																	
Algae	Glaucophyta	+	+																	
	Rhodophyta	+	+		+	+													G, H, S	+
	Chlorophycean green algae	+	+		+	+				+	+		+		+					
	Charophycean green algae (CGA)	+	+	+	+	+			+	+	+		+		+	+				
Bryophytes	Bryophyta (mosses)	+	+	+	+	+	+	+	+	+					+		+			+
	Anthocerotophyta (hornworts)	+	+	+	+	+	+	+	+	+						+	+			
	Marchantiophyta (liverworts)	+	+	+	+	+	+	+	+	+					+	+	+			
Tracheophtes	Lycopodiophyta	+	+	+	+	+	+	(+)	+	+	+				+	+	+		G, H, some species – S	+
	Pteridophyta (ferns)	+	+	+	+	+	+	(+)	+	+		H			+	+	+		G, H, some species – S	+
	Equisetopsida (horsetails)	+	+	+	+	+	+	(+)	+	+	+		+		+	+	+	+	G, H	+
	Gymnosperms	+	+	+	+	+	+	(+)	+	+					+	+	+	+	G, H, some species – S	+
	Monocots without Poales	+	+	+	+	+	+	(+)	+	+					+	+	+	+	G, H, S	+
	Poales	+	+	+	+	+	+	+	+	+	+	J, H, F		(+)	+	+	+	+	G, H, S, p-coumaric acids	+
	Eudicots	+	+	+	+	+	+	(+)	+	+		J			+	+	+	+	G, H, S	+

+ – identified component; (+) – hemicellulose dominating in the primary cell wall.

The origin of the cell walls of land plants

Polysaccharide-rich cell wall, in addition to the acquisition by eukaryotes the ability to conduct the photosynthesis, was a key innovation in plant development and evolution. It is suggested that both these events are the outcome of primary endosymbiosis [12,17], when a cyanobacteria penetrated the host cell, giving rise to the common ancestor of the Archaeplastida group [18], which includes glaucophytes, red and green algae, and land plants [19–21]. Although cyanobacteria cell walls are composed of peptidoglycan and differ from the polysaccharide-rich plant cell walls, it is hypothesized that endosymbiont genes, involved in carbohydrates biosynthesis, due to horizontal gene transfer, provided the molecular background for the cell wall formation in the land plant lineage [22].

New potentials achieved due to the primary endosymbiosis were later tested by the algae of Archaeplastida group in order to better protect sensitive protoplasts, resulting in high diversity of cell wall structures and compositions [10–23]. It seems that some of the new solutions gained became the pivotal point for terrestrialization, although most of the compounds and structural diversity of cell wall in extant terrestrial plants were established later, as innovations associated with their adaptive radiation and diversification [11].

Cellulose – the main component of plant cell wall

The fundamental component of the plant cell walls is cellulose, which is deposited outside the plasma membrane in the form of microfibrils, forming the extracellular network. This polysaccharide is synthesized by the complexes of membrane proteins called cellulose synthase, also named terminal complexes (TCs), which include catalytic subunits encoded by the *CesA* family genes [24–26]. In seed plants, TCs form hetero-oligomeric rosettes, supposedly consist of 36 catalytic subunits that synthesize 36 cellulose chains forming a single microfibril [27], although recent studies suggest that plant microfibril can contain only 18–24 cellulose chains [28].

Genes of the *CesA* family are supposed to be of the ancient cyanobacteria origin [17,26,29,30]. Results of the phylogenetic analyses showed that in CGA the *CesA* was a single gene, whereas already in basal land plants *CesA* formed a single monophyletic group of genes. Following, due to the duplication and diversification events, many different *CesA* subfamilies have evolved in the vascular plants before the divergence of ferns [24,26]. The functions of *CesA* genes were conserved during the evolution, from the cyanobacterial endosymbiont, by algae ancestors of land plants, to extant groups of Tracheophyta [26,31], indicating the significance of cellulose for development and functioning of all plants.

Despite the ability of diverse plant organisms to synthesize cellulose, only one lineage colonized the land. In the ancestors of this lineage (CGA), for further successful terrestrialization seems to be crucial the structural change, which occurred in the terminal complexes. In CGA and all land plants the terminal complexes have a rosette shape [32]. In addition, the CesA in these TCs show high amino acid similarity and the presence of highly conserved domains [32].

In contrast, more ancestral chlorophycean green algae possess the linear, ancestral type of TCs [33]. This strongly suggests that the rosette complex was formed in CGA, as direct ancestors of land plants [32]. The reorganization of the TC structure from linear to rosette was possibly one of the evolutionary innovations which enabled the expansion to the terrestrial habitats but how this structural change has happened still remains unknown. As the domains that characterize the CesA family of seed plants, evolved before the emergence of land plants and were conserved from CGA to angiosperms, it is assumed that they play a pivotal role in maintaining the structure of the rosette complexes [32].

During evolution, *CesA* genes encoding the catalytic subunits of TCs were subjected to specialization [26], which in seed plants had occurred at least at two levels: first – specialization of entire complexes to cellulose biosynthesis for primary and secondary walls; second – diversification of CesA catalytic subunits within a single rosette and formation of hetero-oligomeric complexes [34]. The sequence of specialization events is a crucial aspect necessary for understanding the evolution of cellulose biosynthesis. A recent study on seed plants and a representative of mosses *Physcomitrella patens*, showed specialization of entire complexes in both these groups [32,35]. In seed plants, terminal complexes are specialized to form cellulose of primary and secondary cell walls, whereas in *P. patens* to deposit cellulose during apical and diffusive growth. Phylogenetic analyses showed that *CesAs* in *P. patens* and in seed plants are not orthologs. Thus, specialization of entire terminal complexes in these two groups had to occur independently. Their common ancestor had only one *CesA* gene, and unspecialized homo-oligomeric complexes, but these already forming the rosette structure of TCs. Therefore, functional specialization of entire terminal complexes occurred after the formation of rosettes [31]. Complexes specialized to produce primary and secondary walls evolved to hetero-oligomeric TCs due to the diversification of *CesAs* [31,34,36–39]. This second specialization event took place probably after the divergence of lycophytes, in a common ancestor of ferns and seed plants. *CesA* genes, already in ferns, form many subfamilies, which contain genes orthologous to that of extant seed plants, including orthologs of *AtCesA4*, *AtCesA7* and *AtCesA8*, involved in secondary cell wall formation in *Arabidopsis* [26,40]. The emergence of hetero-oligomeric complexes in a common ancestor of seed plants [27] suggests their importance for the origin and diversification of this group of plants and for functional specialization at the cellular and tissue level.

Polysaccharide components of the cell wall matrix

The cell wall matrix is composed of different compounds, which determine cell wall properties [41]. In extant seed plants, significant differences in the matrix composition became the reason for the cell walls classification to two types [15]. Type I cell wall occurs in dicots, non-commelinoid monocots and in gymnosperms, and consists of cellulose microfibrils surrounded by the xyloglucan (XyG), pectins and structural proteins. Type II cell wall is characteristic only of commelinoid monocots; it contains cellulose fibrils coated with glucuronoarabinoxylan (GAX), very low concentrations of pectins and structural proteins, and also high level of hydroxycinnamic acids [42].

Hemicelluloses – cellulose-binding glycans

Hemicelluloses are especially important matrix components of the cell wall. Due to their complex structure and composition, which are specific for the taxonomic groups of plants, tissues, cells and layers of the wall, they are considered as the most diverse and variable components of the cell walls. According to the current models, they form connections between cellulose microfibrils and are a subject of various modifications affecting the properties of the cell walls [11,12,22,43,44]. In extant seed plants, the most important for the cell walls functioning are mannans, xyloglucans, $(1{\rightarrow}3),(1{\rightarrow}4)$-β-D-glucans (mixed-linkage glucans; MLG) and xylans [45]. These hemicelluloses were probably involved in the evolution of land plants affecting their adaptive abilities, beneficial during colonization of new habitats, as well as contributing to taxa diversification and functional specialization of the cells [17,41,46,47].

MANNANS. Mannans are present in all land plants and probably were one of the first hemicelluloses to appear. Large amounts of mannan are found in the cell walls of all CGA and basal land plants [48–50]. However, in angiosperms mannans are much less common, suggesting that during the evolution of terrestrial plants, in parallel to increasing diversity and abundance of hemicelluloses, there was a tendency to reduce mannan level [49]. Mannans are present in all green algae and only in some red algae, which both however resulted from the same endosymbiotic events, but are not found in brown algae, differing in their origin. It is therefore speculated that the presence of mannans is responsible for the green algae diversity and evolutionary success [22].

Mannans can be of pure or more structurally complicated form, with a wide range of physicochemical properties, and therefore complex functions, which are poorly understood [51]. In some algae, mannans occur in crystalline form [52], forming mannan microfibrils, which replace the cellulose skeleton as the main structural component of the cell wall [53]. Mannans can be also enabling the formation of polysaccharide network by linking cellulose microfibrils [54]. In primary cell walls of seed plants, however, xyloglucan takes over this function as binding glycan, causing probably a decrease in mannans amount [55]. However, mannans structural role is still preserved in the secondary cell walls.

XYLOGLUCANS. Xyloglucan (XyG) is the dominant hemicellulose in primary cell walls of all seed plants [56], with the exception of commelinoid monocots, where the main hemicellulose is glucuronoarabinoxylan [15,42]. Since the presence of xyloglucan seems to be limited to land plants (embryophytes) it is believed that XyG may constitute an important evolutionary innovation [11]. It cannot be ruled out that this polysaccharide was present already in the CGA ancestors of land plants, however, the experimental results on xyloglucan detection in *Chara* are disputable [22,46,57].

The structural and biochemical analyses of XyGs revealed differences in their side chains and fucosylated subunit in different groups of plants [58–60]. Most vascular plants and hornworts produce the structurally homologous XXXG xyloglucan type, with the conservative pattern of branching and fucosylated subunit, although Poales and Solanales have low-fucose xyloglucan [59]. In non-seeds plants the xyloglucan structure exhibits greater variety. For example, basal bryophytes have a XXGGG- and XXGG xyloglucans containing β-D-galactosyluronic acid and branched xylose residues [58], while *Equisetum* and *Selaginella* have xyloglucan containing α-L-arabinopyranose in place of many of the β-D-galactopyranose residues [39,58] (Tuomivaara et al. [61] xyloglucan oligosaccharide nomenclature). The presence of XXXG xyloglucan in only one bryophyte group – hornworts and in tracheophytes proves their sister relationship [58]. In addition, it suggests that this form of xyloglucan evolved in the direct ancestor of all vascular plants and therefore could be important for Tracheophyta emergence.

Commelinoid monocots, which include grasses, are the only group of angiosperms with small amount or even lack of xyloglucan. Instead they have the type II cell wall and glucuronoarabinoxylan as the dominating hemicellulose in their primary walls [15,62]. It is supposed that the formation of the type II cell wall in commelinoids has been a relatively recent event in plant evolution, related to some changes in the strategy of survival and adaptation to new habitats [63]. Such simultaneous rapid and extensive changes in the cell wall composition only in one group of plants and in relatively short time, leading to the type II cell wall formation, are difficult to explain. Furthermore, the major components and structure of the cell walls were rather conserved in seed plant evolution. It is speculated that even so complex changes could result from a small modification of one component, forcing other elements to adjust. This hypothesis can be supported by the experiments, showing that a small change in the chemistry of cell walls can drastically affect their structure and functioning due to the mutual relationship between the individual components [17].

MIXED LINKAGE GLUCANS. Cell walls of grasses (Poaceae), belonging to commelinoids, contain a rare hemicellulose called mixed-linkage glucan (MLG), which for a long time was believed to be unique for these plants [42,62,64]. However, with development of chemical, molecular and bioinformatic methods for cell wall analysis, it became evident that MLG is much more common in the plant kingdom than previously considered [64]. The most surprising was however the discovery of mixed glucan in horsetail *Equisetum arvense* [47,65] and *Selaginella moellendorffii* since Poaceae diverged several millions years after these plants [40,66]. These results were obtained by few independent research teams, with the use of different experimental approaches: detection of anti-MLG antibodies, as well as by biochemical analyzes [40,47,65]. Interestingly, that MLG was found in CGA and polysaccharide similar to MLG in algae *Ulva lactuca* and *Ulva rigida* belonging to Chlorophyta lineage [46]. The occurrence of MLG in so many different and evolutionary distant groups of plants is difficult to explain. It is suggested that the MLG presence might be related to the silica accumulations in the cell walls [22,47]. Poales and horsetails produce MLG and synthesize abundance of silica [67]. However, this hypothesis requires further research.

XYLANS. Xylans are ubiquitous hemicelluloses of vascular plants [68], present mainly in their secondary cell walls [69,70]. Therefore, it was suggested that the acquisition of xylan synthesis capabilities could have been an evolutionary innovation of this group of plants [49,68]. However, xylans were also found in hornworts [68], mosses [48], CGA [57], chlorophycean green algae and red algae [71], suggesting that they are much more common. Nevertheless, it is still not known whether xylans present in these species are structurally similar to those of vascular plants [70]. Xylans have the (1→3) or (1→4) glycosidic linkages [72]. Some red algae, including relatively basal *Bangia fuscopurpurea*, have only (1→3) glycosidic bonds in xylans, while more evolutionary advanced, as e.g. *Palmaria palmata*, probably contain both types, (1→3) and (1→4) glycosidic linkages in the same molecule [72,73]. In contrast, more evolutionarily advanced CGA, including Charales, Coleochaetales and Zygnematales, as well as all land plants contain only (1,4)-ß-D-xylan in their cell walls [57,68]. These findings suggest that CGA and terrestrial plants gained the ability of (1,4)-ß-D-xylan biosynthesis from red algae, in which diverse xylan biosynthetic pathways were present [72].

The origins of the secondary cell walls with xylans, as main hemicelluloses, are not fully elucidated. Nevertheless, their evolution was probably an important preadaptive feature of vascular plants, that enabled the emergence of conducting and mechanical tissues, which allowed the plants to accelerate their growth and increase the biomass production [11,49,68].

Hemicellulose biosynthesis

Matrix polysaccharides are encoded by *CesA*-like *Csl* genes that together with *CesA* form a superfamily *CesA/Csl* [74]. In seed plants nine *Csl* gene groups are known [26]. According to recent phylogenetic studies, *Csls* derived from two ancestral genes, gained due to the primary endosymbiotic event [26]. The first of them – ancestral *Csl*, was the precursor of *CslA/C/K* families, while the second one – ancestral *CesA*, gave rise to both extant *CesAs* and to the remaining *Csl* families [26]. *Csl* genes are present in all land plant lineage genomes [75], and their diversification led to the evolution of hemicelluloses and diversity of terrestrial plants [12].

It is believed that duplication of the first ancestral *Csl* gene could have resulted in the separation of direct ancestors of monophyletic group of land plants – CGA from chlorophycean green algae [12], and thus it could be a key event for successful expansion to the land. In CGA, probably as a result of subsequent duplication, ancestral *Csl* gene evolved to *CslA* and *CslC* genes [24]. In land plants, *CslAs* were reported as those involved in the mannan synthesis [76,77], while *CslC* in the xyloglucan synthesis [78]. Extant CGA contain mannans in their cell walls, but are devoid of the *CslA* genes, which were probably lost during evolution [26]. Therefore, the question remains unresolved, which genes in these organisms are responsible for the mannan synthesis. Mannans are present also in cell walls of chlorophycean green algae, where the ancestral *Csl* is probably responsible for their biosynthesis [24]. It is therefore likely that because the *CslA* gene was lost, its function in CGA is executed by *CslC*, especially that xyloglucan, which is synthesized in seed plants by *CslC*, was not found in CGA [12]. Alternatively,

in the synthesis of mannans in CGA, the *CslD* gene can be involved, which derived from a second ancestral gene – *CesA* [26]. The exact role of *CslD* in CGA is not specified, but recent studies suggest that in seed plants *CslD* can be responsible for glucomannan synthesis [45,79,80]. Both these alternative assumptions indicate that in CGA, the ability to synthesis of mannans was emerged independently of *CslA* and is a result of convergent evolution.

CGA were the first group of plants, in which the *CslC* gene appeared and it is considered to be involved in the xyloglucan biosynthesis in seed plants [78]. Xyloglucan is crucial for the cell wall expansion [41], thus it is necessary for the plant growth and for the increase of plant size, which was an important tendency during land plant evolution. Because the presence of xyloglucan in CGA has not been clearly confirmed so far, it seems likely that in these algae, acquisition of *CslC* gene could be an important preadaptive innovation for further land expansion.

Duplication and diversification of the second ancestral gene – *CesA*, probably resulted in two genes, one of which was the last common ancestor of *CesA/CslD/F* group and the second one of *CslB/H/E/J/G* group [26]. The ancestor of the first group subsequently duplicated and diverged to the *CesA* and *CslD* clades. Both of these genes appeared in CGA and were highly conserved during the evolution of land plants, indicating that their divergence could have been another important preadaptation to the life on the land, and that it can play a pivotal role in terrestrialization. Interestingly, probably as a result of next, relatively recent *CslD* duplication and further rapid divergence and diversification, a *CslF* gene family emerged [26]. Genes of this family are present only in some Poales [81].

The diversification of the ancestor gene for the second *CslB/H/E/J/G* group happened before the evolution of ferns. For this reason, it is believed that these genes are important not as much for the origin of land plants as rather for their radiation and diversification [26]. Interestingly, gymnosperms have no obvious homologues or orthologs of this gene group but have genes relatively basal to *CslE/G* and *CslB/H*, which are specific to gymnosperms. Because in ferns, which are phylogenetically older than gymnosperms, the *CslB/H/E/J/G* group of genes is present, it is postulated that gymnosperms lost these genes during evolution, although it is still not clear how [26]. Alternatively, it is suggested that gymnosperms have a lower rate of substitution in their genome and therefore a lower rate of evolution [82].

The function of the *CesA/CslD/F* and *CslB/H/E/J/G* group of genes remains largely unclear [26]. However, *CslF*, *CslH* and *CslJ* were shown to be involved in the biosynthesis of MLG [74,81–84]. These three gene families occur in Poales, which synthesize MLG. The *CslF* gene is unique for this taxon, while two other families (*CslH* and *CslJ*) are also present in other groups of vascular plants [26], where MLG is not found. Therefore, it is possible that these genes may have some additional functions. Because in recent years, the presence of MLG was reported also for plant groups different than Poales, including some liverworts, *Selaginella* [2,40], horsetails [47,65,85,86] and CGA [57], it seems that the evolution of MLG biosynthesis is more complicated. In CGA, the presence of two *Csl* genes was noted: *CslC* and

CslD, which exact functions have not been determined yet, but theoretically they may be responsible for the MLG synthesis. The *CslD* gene is the ancestor of the *CslF* family, which encodes known MLG synthases in grasses [26], therefore *CslD* are likely to perform this function in CGA. The other possibility is that MLG in CGA is synthesized by a different, convergent biosynthetic pathway comparing to grasses, which can be independent of *Csls* [11]. The presence of MLG in other phylogenetically distant organisms [12], which emerged before the divergence of known MLG synthases: CslF, CslH, CslJ, is an additional support for the hypothesis of multiple origins of this polymer [85].

Lack of conservation of MLG synthesis during evolution, and the relatively rare occurrence of this polysaccharide in different groups of plants, indicates that although the MLG biosynthetic pathway existed in a direct algal ancestor of land plants, the presence of this glucan was not necessary for terrestrialization. It seems more likely, that the MLG biosynthesis have a specific, not fully understood yet, significance for adaptation to certain environmental conditions or to specific functions [12]. In addition, not only the biosynthetic pathway, but also the MLG function could have changed during evolution and differ between ancient CGA, horsetails and most modern monocots – Poales.

Many hemicelluloses of the cell walls of land plants are synthesized by Csls [24], but so far there is no evidence for their contribution to the xylan biosynthesis [87]. Possibly, the xylan biosynthesis can be carried out by a pathway independent of Csl, involving other glycosyltransferases (GTs). Recent studies show that in the biosynthesis of the xylan backbone are involved mainly IRX10/IRX10L [88–90], belonging to the glycosyltransferases family 47 (GT47), which have the xylan xylosyltransferase activity and can function together with other proteins, such as IRX9/IRX9L, IRX14/IRX14L [58,69,89,91–93] belonging to the GT43 family. Phylogenetic analyzes have shown that both *Selaginella* and *Physcomitrella* have orthologs of IRX10, as well as IRX14/IRX14L and IRX9L, suggesting that the mechanism of 1,4-ß-D-xylan biosynthesis could be conserved in evolution. Interestingly, *Ostreococcus* (the smallest known eukaryote) contains several genes of GT43 families, which are known as those involved in the synthesis of 1,4-ß-D-xylan [91,94,95]. It is therefore possible, that the process of 1,4-ß-D-xylan biosynthesis preceded the evolution of land plant lineage [22].

Hemicelluloses rearrangement in cell wall

In the land plant lineage, changes in the hemicelluloses composition are associated with the alterations of their metabolism. Particularly important for the terrestrial plants evolution seems to be the mechanisms, that modifies the cell wall structure enabling its expansion what in turn increase the overall plant size. Experimental studies on seed plants showed that in the cell wall expansion xyloglucan endo-transglycosylases/hydrolases (XTH) are involved, for which the xyloglucan is a substrate [96,97]. XTH belong to the group of glycoside hydrolases GH16, and have two different enzymatic activities: the activity of xyloglucan endohydrolase (XEH), which enables the hydrolysis of xyloglucan backbone, and the activity of xyloglucan endotransglycosylase

(XET), resulting in the interpolymer grafting between two xyloglucan molecules [98]. Structural and phylogenetic analyses showed that XEH evolved secondarily with XET being the ancestral one [99].

The XET activity is necessary for the cell wall expansion, when the sliding of cellulose microfibrils occur due to the relaxation of an integral cellulose-xyloglucan network [41]. XTHs cut and try to connect the associated xyloglucans with other available xyloglucans (XET activity), causing a reorganization of the cell wall, which enabling its growth [41]. XTH and xyloglucan are present in all land plants and the XET activity was shown to be significantly correlated with the intensively growing cells and/or tissues [8,100]. Interestingly, in CGA, the algal ancestors of land plants, despite the absence of confirmed xyloglucan, the presence of XET activity was detected [41]. This means, that XTH probably emerged in the evolution before its known substrate – xyloglucan. It is assumed that the ancestral XTH had lower substrate specificity and therefore initially was able to act on other hemicelluloses [100]. Because phylogenetic and structural relations between XTHs, xylan endohydrolases and (1,3-1,4)-ß-D-endoglucanases [41,101] have been showed in many studies, it seems likely that xylans [101] or MLGs [41] can be the substrates for ancestral XET/XTH. In light of these assumptions, results of the research on *Ulva linza*, belonging to chlorophycean algae, which is more evolutionarily distant from terrestrial plants than CGA, are extremely interesting. In this species, the cell wall has cellulose microfibrils associated with polysaccharides similar to XyG [102], which have mixed bonds resembling MLG (mixed-linkage xyloglucan-like polysaccharide) [46,103]. Possibility, that this polysaccharide had a common ancestor with MLG and xyloglucan from the land plants lineage, is intriguing. The most interesting however is the XET activity found in this species [104]. The polysaccharide with mixed features of XyG and MLG is a putative substrate for this XET activity in *Ulva*. It is also a substrate for (1,3-1,4)-ß-D-glucan endohydrolases, what confirms the evolutionary relationships between XTHs and (1,3-1,4)-ß-D-glucan endohydrolases [41]. An additional confirmation of these associations is the recent discovery in *Populus trichocarpa* of the unique endohydrolases with the broad substrate specificity, able to cleave different linear glucans such as MLGs, cello-oligosaccharides and highly branched oligo- and polysaccharides of xyloglucan. These hydrolases have features of both XTH and (1,3-1,4)-β-D-glucan endohydrolases [105]).

Many studies showed that XTH enzyme and hemicelluloses changed in order to adjust the XET activity and its substrate for the optimal growth of cell walls. The ultimate result, a XTH/xyloglucan system, was probably a key event for successful expansion to the land, as confirmed by the fact that during further evolution and diversification of land plants, this system was highly conserved [41,101].

In this context, extremely puzzling seems to be the possibility that the XTH/xyloglucan system lost its importance in commelinoid monocots, where the XET activity lost its strong substrate specificity [101]. In this group of plants, xyloglucan is present only in a small amount, what is surprising regarding the great abundance of XTH genes and the high level of XET activity. Recent studies indicate that the substrates for XET activity in these plants may be not only xyloglucans, but also dominating in their cell walls glucuronoarabinoxylan and/or MLG [101].

In green algae and horsetails, which contain MLG, another transglycosylation activity the MXE (MLG: xyloglucan endotransglycosylase) has recently been found, which is responsible for the graft of MLG to xyloglucan [106]. This activity is distinct from the XET activity. MXE localization in horsetails showed that, in contrast to the XET activity detected in young growing tissues, MXE is present only in mature organs, possibly reinforcing the cell walls in senescent tissues [106,107]. In grasses, in which MLG is an abundant component of cell walls, the MXE activity is either absent or very low, and probably connected with XTH [108], because the active MXE proteins were not found in these plants [107]. In seed plants that do not have MLG, both functions, i.e. the loosening as well as the strengthening of cell wall structure, seem to be conducted by XTHs [44,96,109]. Perhaps, XTHs perform the same functions in grasses, as in these plants the MXE proteins were not detected.

Many recent studies show, that many other endohydrolases occur in plants, which like XTH and MXE have the hydrolase and endotransglycosylase activities [39]. Relatively recent discovery provided information that among others, for example endo-β-mannanases belonging to the family 5 of glycoside hydrolases (GH5) have both of these activities [51]. Their activity as mannan endotranglycosylases was detected in many species of seed plants, and the first such an enzyme was isolated from the tomato fruit which was found to belong to the multigene family. For the genes encoding the endo-beta-mannanases having the hydrolase and transglycosylase activity the name MTH (mannan transglycosylase/hydrolase) was proposed [51]. Phylogenetic analysis of endo-β-mannanases showed that the plant MTHs are related with fungal, bacterial and animal enzymes, suggesting their ancient evolutionary origin. In contrast, the diversifications of MTHs have probably occurred already after the separation of the plants from the other groups of organisms [92].

Another group of recently discovered and still poorly known enzymes, which possess hydrolase and endotransglycosylase activity are endo-β-1,4-xylanases, belonging to the family 10 of glycoside hydrolases (GH10). The xylan endotransglycosylation activity was found in both, the primary cell walls of the many seeds plants and representatives of basal land plants [39], as well as in the secondary walls of poplar [95]. Endo-β-1,4-xylanases probably are involved in the modification of the xylan in primary and secondary cell walls. Their exact function had not been explained yet, although recent studies have shown that in the secondary wood they are likely to be responsible for the releases tension in the secondary cell walls [95]. Because xylan endotransglycosylation activity is a relatively recent discovery, currently little is known about its emergence in the evolutionary aspect.

Pectins – important polysaccharides of the primary cell wall

Pectins are acidic polysaccharides of primary cell walls of all land plants, where they play different functions. The most important seems to be the regulation of cell adhesion [110,111], which is essential for the development and functioning of multicellular organisms, as land plants. Pectins participate also in the apical and diffusive growth of cells [112], are involved in the defense response against pathogens [113] and regulate the cell wall strengthening [114]. In seed plants, three main groups of pectins are distinguished: homogalacturonan (HG), rhamnogalacturonan I (RG-I) and rhamnogalacturonan II (RG-II) [113]. Sometimes, as the fourth separate group, the xylogalacturonans (XGA) are also given [112,115].

Pectins are abundant already in CGA, occurring even in single-celled ones, as *Penium margaritaceum* [49,112]. The presence of pectin in the ancestors of land plants suggests they importance for the expansion of land or at least for the development of first non-vascular land plants, which contain much more pectins than more evolutionarily advanced vascular plants. In the latter, the amount of pectins, in relation to the bryophytes, is markedly reduced [49], but they still remain one of the major components of the cell wall, essential for the proper cell wall functioning. Pectic polysaccharides, as components of cell walls, are generally conserved during the evolution of land plants, although minor structural differences in the side chains occur.

The main pectic component of the cell wall of land plants is HG, which may be also the polysaccharide backbone for branched pectin types, such as XGA and RG-II [112]. HG is an important structural component of the primary cell walls, since the pair of HG molecules interconnected by Ca^{2+} bridges stiffens the walls [114]. Homogalacturonan seems to be the primary pectin, because it occurs not only in land plants, but also in CGA [11,116,117], where as in embryophytes it may be bridged by calcium. In addition, the distribution of HG epitopes with different degree of esterification with methanol suggests, that in tested algae species, similarly to embryophytes, methyl-esterification is associated with processes of cell growth and differentiation [46,117,118].

RG-Is are pectins widely conserved in land plants and present only in some more evolutionary advanced charophycean green algae [57], suggesting that they originated later than HG. Their branched structure enables modifications of the side chains, resulting in the structural diversity of this group of pectins [113].

RG-II molecules can be cross-linked through the boron ester bonds between the side chains [119]. Reduced ability to form the RG-II dimmers leads to dwarfism, indicating the importance of RG-II cross-linking for plant growth [120]. RG-II occurs almost exclusively in vascular plants [49,121,122]. In bryophytes, if found it occurs in a very small amount [122]. The RGII level in the cell walls appears to increase with the ongoing evolution of vascular plants, and this trend is likely to correlate with an increase in upward growth and the ability to form lignified secondary cell walls [122]. Therefore, RG II is considered to be an evolutionary innovation of land plants, which enabled the development and expansion of vascular plants.

The RG-II structure is highly conserved [112,123], although some seed tracheophytes have RG-II where the L-rhamnose residue is replaced by 3-O-metylrhamnose [11]. Despite the fact that RG-II is characteristic of the terrestrial plants [112], some rare sugar residues in RG-II, like 3-deoxy-D-manno-2-octulosonic acid (KDO) [124], were also found in scales of prasinophycean algae [124,125]. Furthermore, the sequence of CMP-KDO synthase, essential for the synthesis of KDO-containing polymers, was detected in both land plants and in gram-negative bacteria [126]. These results suggest that some sugar residues and some genes involved in the RG-II biosynthesis can be of more ancient origin than this polymer [12].

Pectin metabolism

Many of pectin specific features and the evolution of genes responsible for their biosynthesis and modifications are not fully elucidated [112]). The complicated structure and function of pectins require numerous genes, associated with their production and metabolism [115,127]. Many of pectin related genes have already been identified [112], but in most cases their exact function is not clear. Recent research on *P. patens*, as a species representing the early stage of the transition from water to land, have provided new data on the evolution of pectin metabolism and its importance for the development of land plants [112]. It has been found that five of the 16 analyzed pectin related gene families (homogalacturonan galacturonosyltransferases, polygalacturonases, pectin methylesterases, homogalacturonan methyltransferases, and pectate lyase-like proteins), form a multigene family already in the early terrestrial plants. This indicates that the pectin related genes diversification, occurred probably before the development of land plants [112]. These genes are associated mainly with HG – the most primary pectin, and their early diversification suggests complex function of this polysaccharide before the expansion of land [128]. Seven different gene families (UDP-rhamnose synthases, UDP-glucuronic acid epimerases, homogalacturonan galacturonosyltransferase-like proteins, β-1,4-galactan and β-1,4-galactosyltransferases, rhamnogalacturonan II xylosyltransferases, and pectin acetylesterases) were shown to have a single member in the common ancestor of land plants, and their subsequent diversification suggests its relation to the land expansion [112]. The last four families of 16 analyzed genes (xylogalacturonan xylosyltransferase, rhamnogalacturonan I arabinosyltransferase, pectin methylesterase inhibitor, or polygalacturonase inhibitor protein families) were not detected in *P. patens* [112]. These genes probably have emerged in the evolution later and therefore they could not be relevant for the early stages of the land colonization. They could have however contributed to the diversification of land plants, or more likely to cell specialization. This is indicated by the fact that the XGA (pectin) as well as XGA xylosyltransferase, responsible for its biosynthesis, are not present in *P. patens* [48], while in evolutionarily younger taxa they have a distinct, tissue-specific location.

Lignin – phenol component important for terrestrialization

The key evolutionary innovation that enabled the development of vascular plants that currently dominate in the terrestrial environment was the ability to produce lignified cell walls [8,129,130]. During land colonization, plants were exposed to various stresses, associated with the changes in the environment, such as UV radiation, mechanical stress, drying, and also to the co-evolution of herbivores and pathogens [131]. Lignin and the cell metabolism associated with lignin production were important adaptive factors, which helped to overcome most of these problems and resulted in the rapid development and evolutionary success of land plants. The first land plants that did not form lignin yet, had the primitive phenylpropanoids metabolism, which allowed them to accumulate simple phenylpropanoids, playing a role in the protection from UV radiation. The cell wall lignification, which appeared later in the evolution, gave plants the mechanical strength enabling the vertical growth, thereby improving access to the light and efficient photosynthesis, which in turn led to a significant increase in the plant body size [132]. Lignin, through strengthening the water-conducting cells, contributed to the formation of the efficient long-distance conducting system [12,133]. Furthermore, lignin is one of the most difficult to degrade biopolymers, so that it constitutes an effective protection against pathogens and herbivores, which have been evolving together with vascular plants [134,135].

Lignin is a three-dimensional phenolic hetero-polymer which by crosslinking the cellulose and hemicellulose impregnates cell walls [15,109,136] to provide them with mechanical strength [137]. It is polymerized by oxidative coupling mainly of three p-hydroxycinnamyl alcohols as a result of enzymatic reaction catalyzed by laccases and class III of plant peroxidases [138–141]. As a result, hydrophobic heteropolymers are produced, consisting of H (p-hydroxyphenyl), G (guaiacyl) and S (syringyl) subunits originating from p-coumaryl, coniferyl and sinapyl alcohols, respectively [141,142].

Lignins are present in all Tracheophyta. However, the different groups of extant vascular plants differ in the composition and arrangement of lignin monomers, deposited in their cell walls [132]. In addition, various cell types may also differ in lignin composition. In most gymnosperm plants studied, lignins are typically composed of G subunits with a small amount of H subunits. In angiosperms, besides G and H subunits, additionally a large ratio of S lignin monomers is present [143,144], that were previously regarded as the evolutionary innovation in this group of plants. However, the presence of S lignin was reported also in representatives of other Tracheophyta groups, like all living gnetophytes, some lycophytes, ferns and gymnosperms [141]. The presence of S lignin in so phylogenetically distant groups of plants, and sometimes only in some species within them, can hypothetically be explained by the loss of the ability of S lignin biosynthesis in some taxa or as the example of convergent biosynthetic pathways [132]. Biochemical studies indicate the latter possibility, arguing that the S lignin biosynthetic pathways differ in various groups of plants [145]. Interestingly even more complex composition of lignin is present in grasses [142], because besides of the G, S, and H subunits of the p-hydroxycinnamyl alcohol they contain additionally a significant amount of ester related subunits of p-coumaric acids [141].

Lignin is considered as the characteristic feature of tracheophytes, although there are some early and highly controversial studies [146], reporting the presence of lignin or compounds similar to lignin, in non-vascular plants such as brown algae, charophycean green algae and mosses [147–149]. The presence of lignin in brown algae and green algae has never been confirmed though. Despite the recent detection of epitopes associated with lignin in *Nitella* [150], charophycean green algae, now it is assumed that true lignin is not present in CGA [151]. However, the presence of these epitopes indicates that the primitive form of phenylpropanoids metabolism may already occur in chlorophycean, before the colonization of land by plants [132]. Epitopes associated with lignin were also detected in certain mosses, including *Sphagnum cuspidatum*, where the intensity of the fluorescent labeling of antibodies is similar to the control tracheophytes [150]. Bryophytes have already been known as those producing the phenylpropanoid compounds associated with precursors of lignin [49], but the presence of the true lignin in mosses is still rather controversial. Contemporary results obtained for mosses suggest, however, that the ability to transport and accumulate polymerized phenolic compounds in the cell walls could evolve even in non-vascular terrestrial plants. These phenolic compounds, which in some cases are similar to the true lignin, can exist as polyphenols associated with the cell wall [132].

In light of these assumptions the recent finding of the secondary wall and lignin in red algae *Calliarthron cheilosporioides* was very surprising, because the last common ancestor of red algae and land plants is dated for more than 1.3 billion years ago [152]. Moreover, lignins present in *Calliarthron* are highly specialized, as they have all three G, H and S subunits and a composition typical of angiosperms [152]. The homology of all three types of lignin monomers in *Calliarthron* and vascular plants could indicate their deeply conserved evolution, what is difficult to imagine given the phylogenetic distance between these groups and more than a billion years of independent evolution. More likely, the presence of lignin in *Calliarthron* could be explained by the convergent evolution [152]. Such convergent evolution of S lignin biosynthesis probably occurred in the case of lycopods and angiosperms [22], as S lignins in these two groups are not homologous and they are synthesized by distinctly different biosynthetic pathways [132,145]. However, in the case of *Calliarthron*, where all three types of lignin can be found in the cell wall, convergent evolution would imply the simultaneous evolution of all lignin biosynthetic pathways, which seems unlikely. Therefore, it is not determined yet whether the presence of lignin in *Callianthron* is a result of inheritance or convergent evolution [12].

Lignin biosynthesis
Lignin is synthesized in two main steps: monoligol biosynthesis and their assembly in lignin polymers, and next their subsequent attachment to the hemicellulose and cellulose [151]. Peroxidases and laccases are enzymes considered as those involved in dimerization of monolignols and their incorporation to lignin, although these mechanisms are not fully elucidated [153]. Biochemical pathways of monolignols biosynthesis are highly conserved in vascular plants [151].

Based on the recent analyses of genes involved in the lignin biosynthesis it has been suggested that at last two distinct turning points in their evolution existed [151]. The first one was probably the appearance of the complete lignin biosynthetic pathway, with all nine gene families involved in this process, in mosses. Furthermore, the sudden expansion in number of these gene families' members occurred in this group of plants [151]. These results indicate that mosses are not only the basal line of terrestrial plants, but also a key point in the evolution of tracheophytes [154].

The second turning point for the lignin biosynthesis was the emergence of the *F5H* gene family, encoding P450 enzymes that convert G monolignol to S monolignol in angiosperms, increasing the number of different monomers and therefore increasing lignin diversity and complexity in the cell walls [155]. The *F5H* genes occurrence could lead to the formation of specialized cell phenotypes and morphological diversity in angiosperms [151]. However, reports on the S lignin presence in lycophytes and *Ginkgo* are contradictory to the assumption that *F5H* occurs exclusively in angiosperms. Furthermore, in a representative of lycophyte, *Selaginella moellendorffii*, a novel *SmF5H* gene was identified, that is involved in the S lignin biosynthesis [132,145]. It has been shown that SmF5H has the phenylpropanoid 3-hydroxylase activity, the catalytic property absent in angiosperms, proving a difference in the strategy of S lignin biosynthesis in these two plant groups [132]. These results, as well as the phylogenetic analysis, suggest that *SmF5H* evolved independently of *F5H* in angiosperms, and probably comes from the independent radiation of P450 enzymes, unique to lycophytes [145]. This finding supports the hypothesis that the S lignin may has evolved many times in different lines of vascular plants [151].

The expansion of most gene families, involved in lignin biosynthesis was correlated with the substrate diversity and occurred mostly already after the divergence of monocots and dicots [151], significantly increasing the diversity of cell walls. This somehow correlates with the mechanisms of defense against pathogens [151], because more varied and unpredictable lignin structure in the cell walls increases the chance of protection against its degradation.

Conclusions

The cell wall of the extant land plants is a structure with very diverse and complex composition. Although it is still not clear, whether the changes in the cell walls were the primary factors enabling adaptation to the changing environmental conditions or they were a secondary effect of the adaptation to new habitats, undoubtedly changes in the chemistry of the cell wall accompanied the main steps in the evolution of land [156]. Entirely new possibilities opened up, causing the acceleration of the evolution, and contributing to a multi-level diversification among extant land plants. One of the first key innovations was indisputably the emergence of polysaccharide-rich cell walls. Subsequent evolutionary events, such as e.g. the emergence of multicellularity and the cell elongation mechanism, terrestrialization process, vascularization and lignification, were closely related to changes in these polysaccharide-rich cell walls.

Currently there are several projects aimed at understanding the evolution of a variety of cell walls, including the ancestral groups of plants, which deliver new and surprising discoveries verifying various hypotheses concerning the cell wall evolution. However, the full explanation, how plant cell walls originated and how they changed during the evolution of extant land plants, or how specialization of cells and cellular systems have arisen to perform certain functions, are still ahead of us.

Acknowledgments
This work has not been funded from additional sources.

Competing interests
No competing interests have been declared.

References

1. Becker B, Marin B. Streptophyte algae and the origin of embryophytes. Ann Bot. 2009;103(7):999–1004.

2. Sørensen I, Pettolino FA, Bacic A, Ralph J, Lu F, O'Neill MA, et al. The charophycean green algae provide insights into the early origins of plant cell walls. Plant J. 2011;68(2):201–211.

3. Rubinstein CV, Gerrienne P, de la Puente GS, Astini RA, Steemans P. Early Middle Ordovician evidence for land plants in Argentina (eastern Gondwana). New Phytol. 2010;188(2):365–369.

4. Sanderson MJ, Thorne JL, Wikstrom N, Bremer K. Molecular evidence on plant divergence times. Am J Bot. 2004;91(10):1656–1665.

5. Wodniok S, Brinkmann H, Glöckner G, Heidel AJ, Philippe H, Melkonian M, et al. Origin of land plants: do conjugating green algae hold the key? BMC Evol Biol. 2011;11(1):104.

6. Karol KG, McCourt RM, Cimino MT, Delwiche CF. The closest living relatives of land plants. Science. 2001;294(5550):2351–2353.

7. McCourt RM, Delwiche CF, Karol KG. Charophyte algae and land plant origins. Trends Ecol Evol. 2004;19(12):661–666.

8. Kenrick P, Crane PR. The origin and early evolution of plants on land. Nature. 1997;389(6646):33–39.

9. Niklas KJ, Kutschera U. The evolution of the land plant life cycle. New Phytol. 2010;185(1):27–41.

10. Domozych DS, Ciancia M, Fangel JU, Mikkelsen MD, Ulvskov P, Willats WGT. The cell walls of green algae: a journey through evolution and diversity. Front Plant Sci. 2012;3:82.

11. Sørensen I, Domozych D, Willats WGT. How have plant cell walls evolved? Plant Physiol. 2010;153(2):366–372.

12. Popper ZA, Michel G, Hervé C, Domozych DS, Willats WGT, Tuohy MG, et al. Evolution and diversity of plant cell walls: from algae to flowering plants. Annu Rev Plant Biol. 2011;62(1):567–590.

13. Bacic A, Harris PJ, Stone BA. Structure and function of plant cell walls. In: Preiss J, editor. The biochemistry of plants. New York, NY: Academic Press; 1988. p. 297–371. (vol 14).

14. O'Neill M, Albersheim P, Darvill A. The pectic polysaccharides of primary cell wall. In: Dey PM, editor. Methods in plant biochemistry. London: Academic Press; 1990. p. 415–441. (vol 2).

15. Carpita NC, Gibeaut DM. Structural models of primary cell walls in flowering plants: consistency of molecular structure with the physical properties of the walls during growth. Plant J. 1993;3(1):1–30.

16. Ridley BL, O'Neill MA, Mohnen D. Pectins: structure, biosynthesis, and oligogalacturonide-related signaling. Phytochemistry. 2001;57(6):929–967.

17. Niklas KJ. The cell walls that bind the tree of life. Bioscience. 2004;54(9):831–841.

18. Adl SM, Simpson AGB, Farmer MA, Andersen RA, Anderson OR, Barta JR, et al. The new higher level classification of eukaryotes with emphasis on the taxonomy of protists. J Eukaryot Microbiol. 2005;52(5):399–451.

19. Bhattacharya D, Yoon HS, Hackett JD. Photosynthetic eukaryotes unite: endosymbiosis connects the dots. Bioessays. 2004;26(1):50–60.

20. Palmer JD, Soltis DE, Chase MW. The plant tree of life: an overview and some points of view. Am J Bot. 2004;91(10):1437–1445.

21. Baldauf SL. An overview of the phylogeny and diversity of eukaryotes. J Syst Evol. 2008;46(3):263–273.

22. Popper ZA, Tuohy MG. Beyond the green: understanding the evolutionary puzzle of plant and algal cell walls. Plant Physiol. 2010;153(2):373–383.

23. Baldan B, Andolfo P, Navazio L, Tolomio C, Mariani P. Cellulose in algal cell wall: an "in situ" localization. Eur J Histochem. 2001;45(1):51–56.

24. Yin Y, Huang J, Xu Y. The cellulose synthase superfamily in fully sequenced plants and algae. BMC Plant Biol. 2009;9(1):99.

25. Nobles DR Jr, Brown RM Jr. Many paths up the mountain: tracking the evolution of cellulose biosynthesis. In: Brown RM Jr, Saxena IM, editors. Cellulose: molecular and structural biology. Dordrecht: Springer; 2007. p. 1–15.

26. Yin Y, Johns MA, Cao H, Rupani M. A survey of plant and algal genomes and transcriptomes reveals new insights into the evolution and function of the cellulose synthase superfamily. BMC Genomics. 2014;15(1):260.

27. Carroll A, Specht CD. Understanding plant cellulose synthases through a comprehensive investigation of the cellulose synthase family sequences. Front Plant Sci. 2011;2:5.

28. Newman RH, Hill SJ, Harris PJ. Wide-angle x-ray scattering and solid-state nuclear magnetic resonance data combined to test models for cellulose microfibrils in mung bean cell walls. Plant Physiol. 2013;163(4):1558–1567.

29. Tsekos I. The sites of cellulose synthesis in algae: diversity and evolution of cellulose-synthesizing enzyme complexes. J Phycol. 1999;35(4):635–655.

30. Nobles DR Jr, Brown RM Jr. The pivotal role of cyanobacteria in the evolution of cellulose synthases and cellulose synthase-like proteins. Cellulose. 2004;11(3–4):437–448.

31. Roberts AW, Bushoven JT. The cellulose synthase (CESA) gene superfamily of the moss *Physcomitrella patens*. Plant Mol Biol. 2006;63(2):207–219.

32. Roberts AW, Roberts EM, Delmer DP. Cellulose synthase (CesA) genes in the green alga *Mesotaenium caldariorum*. Eukaryot Cell. 2002;1(6):847–855.

33. Lewis LA, McCourt RM. Green algae and the origin of land plants. Am J Bot. 2004;91(10):1535–1556.

34. Roberts AW, Roberts EM, Haigler CH. Moss cell walls: structure and

biosynthesis. Front Plant Sci. 2012;3:166.

35. Wise HZ, Saxena IM, Brown RM. Isolation and characterization of the cellulose synthase genes *PpCesA6* and *PpCesA7* in *Physcomitrella patens*. Cellulose. 2011;18(2):371–384.

36. Tanaka K. Three distinct rice cellulose synthase catalytic subunit genes required for cellulose synthesis in the secondary wall. Plant Physiol. 2003;133(1):73–83.

37. Djerbi S, Lindskog M, Arvestad L, Sterky F, Teeri TT. The genome sequence of black cottonwood (*Populus trichocarpa*) reveals 18 conserved cellulose synthase (*CesA*) genes. Planta. 2005;221(5):739–746.

38. Nairn CJ, Haselkorn T. Three loblolly pine *CesA* genes expressed in developing xylem are orthologous to secondary cell wall *CesA* genes of angiosperms. New Phytol. 2005;166(3):907–915.

39. Franková L, Fry SC. Phylogenetic variation in glycosidases and glycanases acting on plant cell wall polysaccharides, and the detection of transglycosidase and trans-β-xylanase activities. Plant J. 2011;67(4):662–681.

40. Harholt J, Sørensen I, Fangel J, Roberts A, Willats WGT, Scheller HV, et al. The glycosyltransferase repertoire of the spikemoss *Selaginella moellendorffii* and a comparative study of its cell wall. PLoS ONE. 2012;7(5):e35846.

41. van Sandt VST, Stieperaere H, Guisez Y, Verbelen JP, Vissenberg K. XET activity is found near sites of growth and cell elongation in bryophytes and some green algae: new insights into the evolution of primary cell wall elongation. Ann Bot. 2007;99(1):39–51.

42. Vogel J. Unique aspects of the grass cell wall. Curr Opin Plant Biol. 2008;11(3):301–307.

43. Mellerowicz EJ, Sundberg B. Wood cell walls: biosynthesis, developmental dynamics and their implications for wood properties. Curr Opin Plant Biol. 2008;11(3):293–300.

44. Nishikubo N, Takahashi J, Roos AA, Derba-Maceluch M, Piens K, Brumer H, et al. XET-mediated xyloglucan rearrangements in developing wood of hybrid aspen. Plant Physiol. 2011;155:399–413.

45. Scheller HV, Ulvskov P. Hemicelluloses. Annu Rev Plant Biol. 2010;61(1):263–289.

46. Popper ZA. Primary cell wall composition of bryophytes and charophytes. Ann Bot. 2003;91(1):1–12.

47. Fry SC, Nesselrode BHWA, Miller JG, Mewburn BR. Mixed-linkage (1→3,1→4)-β-D-glucan is a major hemicellulose of *Equisetum* (horsetail) cell walls. New Phytol. 2008;179(1):104–115.

48. Moller I, Sørensen I, Bernal AJ, Blaukopf C, Lee K, Øbro J, et al. High-throughput mapping of cell-wall polymers within and between plants using novel microarrays: glycan microarrays for plant cell-wall analysis. Plant J. 2007;50(6):1118–1128.

49. Popper Z. Evolution and diversity of green plant cell walls. Curr Opin Plant Biol. 2008;11(3):286–292.

50. Estevez JM, Fernandez PV, Kasulin L, Dupree P, Ciancia M. Chemical and in situ characterization of macromolecular components of the cell walls from the green seaweed *Codium fragile*. Glycobiology. 2008;19(3):212–228.

51. Schröder R, Atkinson RG, Redgwell RJ. Re-interpreting the role of endo-β-mannanases as mannan endotransglycosylase/hydrolases in the plant cell wall. Ann Bot. 2009;104(2):197–204.

52. Chanzy HD, Grosrenaud A, Vuong R, Mackie W. The crystalline polymorphism of mannan in plant cell walls and after recrystallisation. Planta. 1984;161(4):320–329.

53. Mackie W, Preston RD. The occurrence of mannan microfibrils in the green algae *Codium fragile* and *Acetabularia crenulata*. Planta. 1968;79(3):249–253.

54. Whitney SEC, Brigham JE, Darke AH, Reid JSG, Gidley MJ. Structural aspects of the interaction of mannan-based polysaccharides with bacterial cellulose. Carbohydr Res. 1998;307(3–4):299–309.

55. Hosoo Y, Imai T, Yoshida M. Diurnal differences in the supply of glucomannans and xylans to innermost surface of cell walls at various developmental stages from cambium to mature xylem in *Cryptomeria japonica*. Protoplasma. 2006;229(1):11–19.

56. Popper ZA, Fry SC. Primary cell wall composition of pteridophytes and spermatophytes. New Phytol. 2004;164(1):165–174.

57. Domozych DS, Sorensen I, Willats WGT. The distribution of cell wall polymers during antheridium development and spermatogenesis in the Charophycean green alga, *Chara corallina*. Ann Bot. 2009;104(6):1045–1056.

58. Pena MJ, Darvill AG, Eberhard S, York WS, O'Neill MA. Moss and liverwort xyloglucans contain galacturonic acid and are structurally distinct from the xyloglucans synthesized by hornworts and vascular plants. Glycobiology. 2008;18(11):891–904.

59. Hoffman M, Jia Z, Peña MJ, Cash M, Harper A, Blackburn AR, et al. Structural analysis of xyloglucans in the primary cell walls of plants in the subclass Asteridae. Carbohydr Res. 2005;340(11):1826–1840.

60. Hsieh YSY, Harris PJ. Xyloglucans of monocotyledons have diverse structures. Mol Plant. 2009;2(5):943–965.

61. Tuomivaara ST, Yaoi K, O'Neill MA, York WS. Generation and structural validation of a library of diverse xyloglucan-derived oligosaccharides, including an update on xyloglucan nomenclature. Carbohydr Res. 2015;402:56–66.

62. Smith BG, Harris PJ. The polysaccharide composition of Poales cell walls. Biochem Syst Ecol. 1999;27(1):33–53.

63. Sarkar P, Bosneaga E, Auer M. Plant cell walls throughout evolution: towards a molecular understanding of their design principles. J Exp Bot. 2009;60(13):3615–3635.

64. Trethewey JAK, Campbell LM, Harris PJ. (1→3),(1→4)-β-D-glucans in the cell walls of the Poales (sensu lato): an immunogold labeling study using a monoclonal antibody. Am J Bot. 2005;92(10):1660–1674.

65. Sørensen I, Pettolino FA, Wilson SM, Doblin MS, Johansen B, Bacic A, et al. Mixed-linkage (1→3),(1→4)-β-D-glucan is not unique to the Poales and is an abundant component of *Equisetum arvense* cell walls. Plant J. 2008;54(3):510–521.

66. Bell PR. Green plants: their origin and diversity. 2nd ed. Cambridge: Cambridge University Press; 2000.

67. Hodson MJ, White PJ, Mead A, Broadley MR. Phylogenetic variation in the silicon composition of plants. Ann Bot. 2005;96(6):1027–1046.

68. Carafa A, Duckett JG, Knox JP, Ligrone R. Distribution of cell-wall xylans in bryophytes and tracheophytes: new insights into basal interrelationships of land plants. New Phytol. 2005;168(1):231–240.

69. York W, Oneill M. Biochemical control of xylan biosynthesis – which end is up? Curr Opin Plant Biol. 2008;11(3):258–265.

70. Kulkarni AR, Peña MJ, Avci U, Mazumder K, Urbanowicz BR, Pattathil S, et al. The ability of land plants to synthesize glucuronoxylans predates the evolution of tracheophytes. Glycobiology. 2012;22(3):439–451.

71. Lahaye M, Robic A. Structure and functional properties of ulvan, a polysaccharide from green seaweeds. Biomacromolecules. 2007;8(6):1765–1774.

72. Painter TJ, Aspinall GO. Algal polysaccharides. In: The polysaccharides. New York, NY: Academic Press; 1983. p. 195–285. (vol 2).

73. Turvey JR, Williams EL. The structures of some xylans from red algae. Phytochemistry. 1970;9(11):2383–2388.

74. Richmond TA. The cellulose synthase superfamily. Plant Physiol. 2000;124(2):495–498.

75. Keegstra K, Walton J. Plant science. Beta-glucans – brewer's bane, dietician's delight. Science. 2006;311(5769):1872–1873.

76. Liepman AH, Wilkerson CG, Keegstra K. Expression of cellulose synthase-like (Csl) genes in insect cells reveals that CslA family members encode mannan synthases. Proc Natl Acad Sci USA. 2005;102(6):2221–2226.

77. Lerouxel O, Cavalier DM, Liepman AH, Keegstra K. Biosynthesis of plant cell wall polysaccharides – a complex process. Curr Opin Plant Biol. 2006;9(6):621–630.

78. Cocuron JC, Lerouxel O, Drakakaki G, Alonso AP, Liepman AH, Keegstra K, et al. A gene from the cellulose synthase-like C family encodes a beta-1,4 glucan synthase. Proc Natl Acad Sci USA. 2007;104(20):8550–8555.

79. Verhertbruggen Y, Yin L, Oikawa A, Scheller HV. Mannan synthase activity in the CSLD family. Plant Signal Behav. 2011;6(10):1620–1623.

80. Yin L, Verhertbruggen Y, Oikawa A, Manisseri C, Knierim B, Prak L, et al. The cooperative activities of CSLD2, CSLD3, and CSLD5 are required for normal Arabidopsis development. Mol Plant. 2011;4(6):1024–1037.

81. Burton RA, Wilson SM, Hrmova M, Harvey AJ, Shirley NJ, Medhurst A, et al. Cellulose synthase-like CslF genes mediate the synthesis of cell wall (1,3;1,4)-β-D-glucans. Science. 2006;311(5769):1940–1942.

82. Buschiazzo E, Ritland C, Bohlmann J, Ritland K. Slow but not low: genomic comparisons reveal slower evolutionary rate and higher dN/dS in conifers compared to angiosperms. BMC Evol Biol. 2012;12(1):8.

83. Burton RA, Jobling SA, Harvey AJ, Shirley NJ, Mather DE, Bacic A, et al. The genetics and transcriptional profiles of the cellulose synthase-like HvCslF gene family in barley (Hordeum vulgare L.). Plant Physiol. 2008;146(4):1821–1833.

84. Doblin MS, Pettolino FA, Wilson SM, Campbell R, Burton RA, Fincher GB, et al. A barley cellulose synthase-like CSLH gene mediates (1,3;1,4)-β-D-glucan synthesis in transgenic Arabidopsis. Proc Natl Acad Sci USA. 2009;106(14):5996–6001.

85. Burton RA, Fincher GB. (1,3;1,4)-β-D-glucans in cell walls of the Poaceae, lower plants, and fungi: a tale of two linkages. Mol Plant. 2009;2(5):873–882.

86. Fincher GB. Exploring the evolution of (1,3;1,4)-β-D-glucans in plant cell walls: comparative genomics can help! Curr Opin Plant Biol. 2009;12(2):140–147.

87. Zhou HL, He SJ, Cao YR, Chen T, Du BX, Chu CC, et al. OsGLU1, a putative membrane-bound endo-1,4-β-D-glucanase from rice, affects plant internode elongation. Plant Mol Biol. 2006;60(1):137–151.

88. Ren Y, Hansen SF, Ebert B, Lau J, Scheller HV. Site-directed mutagenesis of IRX9, IRX9L and IRX14 proteins involved in xylan biosynthesis: glycosyltransferase activity is not required for IRX9 function in Arabidopsis. PLoS ONE. 2014;9(8):e105014.

89. Jensen JK, Johnson NR, Wilkerson CG. Arabidopsis thaliana IRX10 and two related proteins from psyllium and Physcomitrella patens are xylan xylosyltransferases. Plant J. 2014;80(2):207–215.

90. Urbanowicz BR, Peña MJ, Moniz HA, Moremen KW, York WS. Two Arabidopsis proteins synthesize acetylated xylan in vitro. Plant J. 2014;80(2):197–206.

91. Cantarel BL, Coutinho PM, Rancurel C, Bernard T, Lombard V, Henrissat B. The carbohydrate-active enzymes database (CAZy): an expert resource for glycogenomics. Nucl Acids Res. 2009;37(database):D233–D238.

92. Yuan JS, Yang X, Lai J, Lin H, Cheng ZM, Nonogaki H, et al. The endo-β-mannanase gene families in Arabidopsis, rice, and poplar. Funct Integr Genomics. 2006;7(1):1–16.

93. Brown DM, Goubet F, Wong VW, Goodacre R, Stephens E, Dupree P, et al. Comparison of five xylan synthesis mutants reveals new insight into the mechanisms of xylan synthesis. Plant J. 2007;52(6):1154–1168.

94. Lee C, Zhong R, Richardson EA, Himmelsbach DS, McPhail BT, Ye ZH. The PARVUS gene is expressed in cells undergoing secondary wall thickening and is essential for glucuronoxylan biosynthesis. Plant Cell Physiol. 2007;48(12):1659–1672.

95. Derba-Maceluch M, Awano T, Takahashi J, Lucenius J, Ratke C, Kontro I, et al. Suppression of xylan endotransglycosylase PtxtXyn10A affects cellulose microfibril angle in secondary wall in aspen wood. New Phytol. 2015;205(2):666–681.

96. Fry SC, Smith RC, Renwick KF, Martin DJ, Hodge SK, Matthews KJ. Xyloglucan endotransglycosylase, a new wall-loosening enzyme activity from plants. Biochem J. 1992;282(pt 3):821–828.

97. Nishitani K, Tominaga R. Endo-xyloglucan transferase, a novel class of glycosyltransferase that catalyzes transfer of a segment of xyloglucan molecule to another xyloglucan molecule. J Biol Chem. 1992;267(29):21058–21064.

98. Nishitani K, Vissenberg K. Roles of the XTH protein family in the expanding cell. In: Verbelen JP, Vissenberg K, editors. The expanding cell. Berlin: Springer; 2006. p. 89–116. (Plant cell monographs).

99. Baumann MJ, Eklof JM, Michel G, Kallas AM, Teeri TT, Czjzek M, et al. Structural evidence for the evolution of xyloglucanase activity from xyloglucan endo-transglycosylases: biological implications for cell wall metabolism. Plant Cell. 2007;19(6):1947–1963.

100. Bateman RM, Crane PR, DiMichele WA, Kenrick PR, Rowe NP, Speck T, et al. Early evolution of land plants: phylogeny, physiology, and ecology of the primary terrestrial radiation. Annu Rev Ecol Syst. 1998;29(1):263–292.

101. Strohmeier M, Hrmova M, Fischer M, Harvey AJ, Fincher GB, Pleiss J. Molecular modeling of family GH16 glycoside hydrolases: potential roles for xyloglucan transglucosylases/hydrolases in cell wall modification in the Poaceae. Protein Sci. 2009;13(12):3200–3213.

102. Lahaye M, Jegou D, Buleon A. Chemical characteristics of insoluble glucans from the cell wall of the marine green alga Ulva lactuca (L.) Thuret. Carbohydr Res. 1994;262(1):115–125.

103. Ray B, Lahaye M. Cell-wall polysaccharides from the marine green alga Ulva "rigida" (Ulvales, Chlorophyta). Chemical structure of ulvan. Carbohydr Res. 1995;274:313–318.

104. Kim YH, Kim CY, Song WK, Park DS, Kwon SY, Lee HS, et al. Overexpression of sweetpotato swpa4 peroxidase results in increased hydrogen peroxide production and enhances stress tolerance in tobacco. Planta. 2008;227(4):867–881.

105. Eklof JM, Shojania S, Okon M, McIntosh LP, Brumer H. Structure-function analysis of a broad specificity Populus trichocarpa endo-glucanase reveals an evolutionary link between bacterial licheninases and plant XTH gene products. J Biol Chem. 2013;288(22):15786–15799.

106. Fry SC, Mohler KE, Nesselrode BHWA, Frankov L. Mixed-linkage β-glucan: xyloglucan endotransglucosylase, a novel wall-remodeling enzyme from Equisetum (horsetails) and charophytic algae. Plant J. 2008;55(2):240–252.

107. Mohler KE, Simmons TJ, Fry SC. Mixed-linkage glucan:xyloglucan endotransglucosylase (MXE) re-models hemicelluloses in Equisetum shoots but not in barley shoots or

Equisetum callus. New Phytol. 2013;197(1):111–122.

108. Hrmova M, Farkas V, Lahnstein J, Fincher GB. A barley xyloglucan xyloglucosyl transferase covalently links xyloglucan, cellulosic substrates, and (1,3;1,4)-β-D-glucans. J Biol Chem. 2007;282(17):12951–12962.

109. Cosgrove DJ. Growth of the plant cell wall. Nat Rev Mol Cell Biol. 2005;6(11):850–861.

110. Mouille G, Ralet MC, Cavelier C, Eland C, Effroy D, Hématy K, et al. Homogalacturonan synthesis in *Arabidopsis thaliana* requires a Golgi-localized protein with a putative methyltransferase domain. Plant J. 2007;50(4):605–614.

111. Krupková E, Immerzeel P, Pauly M, Schmülling T. The *TUMOROUS SHOOT DEVELOPMENT2* gene of *Arabidopsis* encoding a putative methyltransferase is required for cell adhesion and coordinated plant development. Plant J. 2007;50(4):735–750.

112. McCarthy TW, Der JP, Honaas LA, dePamphilis CW, Anderson CT. Phylogenetic analysis of pectin-related gene families in *Physcomitrella patens* and nine other plant species yields evolutionary insights into cell walls. BMC Plant Biol. 2014;14(1):79.

113. Atmodjo MA, Hao Z, Mohnen D. Evolving views of pectin biosynthesis. Annu Rev Plant Biol. 2013;64(1):747–779.

114. Braccini I, Pérez S. Molecular basis of Ca²⁺-induced gelation in alginates and pectins: the egg-box model revisited. Biomacromolecules. 2001;2(4):1089–1096.

115. Mohnen D. Pectin structure and biosynthesis. Curr Opin Plant Biol. 2008;11(3):266–277.

116. Proseus TE, Boyer JS. Calcium pectate chemistry controls growth rate of *Chara corallina*. J Exp Bot. 2006;57(15):3989–4002.

117. Domozych DS, Serfis A, Kiemle SN, Gretz MR. The structure and biochemistry of charophycean cell walls: I. Pectins of *Penium margaritaceum*. Protoplasma. 2007;230(1-2):99–115.

118. Eder M, Lütz-Meindl U. Analyses and localization of pectin-like carbohydrates in cell wall and mucilage of the green alga *Netrium digitus*. Protoplasma. 2010;243(1–4):25–38.

119. O'Neill MA, Warrenfeltz D, Kates K, Pellerin P, Doco T, Darvill AG, et al. Rhamnogalacturonan-II, a pectic polysaccharide in the walls of growing plant cell, forms a dimer that is covalently cross-linked by a borate ester. J Biol Chem. 1996;271(37):22923–22930.

120. O'Neill MA, Eberhard S, Albersheim P, Darvill AG. Requirement of borate cross-linking of cell wall rhamnogalacturonan II for *Arabidopsis* growth. Science. 2001;294(5543):846–849.

121. Perez S, Rodríguez-Carvajal MA, Doco T. A complex plant cell wall polysaccharide: rhamnogalacturonan II. A structure in quest of a function. Biochimie. 2003;85(1–2):109–121.

122. Matsunaga T, Ishii T, Matsumoto S, Higuchi M, Darvill A, Albersheim P, et al. Occurrence of the primary cell wall polysaccharide rhamnogalacturonan II in pteridophytes, lycophytes, and bryophytes. Implications for the evolution of vascular plants. Plant Physiol. 2004;134(1):339–351.

123. Pabst M, Fischl RM, Brecker L, Morelle W, Fauland A, Köfeler H, et al. Rhamnogalacturonan II structure shows variation in the side chains monosaccharide composition and methylation status within and across different plant species. Plant J. 2013;76:61–72.

124. York WS, Darvill AG, McNeil M, Albersheim P. 3-deoxy-d-manno-2-octulosonic acid (KDO) is a component of rhamnogalacturonan II, a pectic polysaccharide in the primary cell walls of plants. Carbohydr Res. 1985;138(1):109–126.

125. Becker B, Becker D, Kamerling JP, Melkonian M. 2-keto-sugar acids in green flagellates: a chemical marker for prasinophycean scales. J Phycol. 1991;27(4):498–504.

126. Royo J, Gımez E, Hueros G. CMP–KDO synthetase: a plant gene borrowed from gram-negative eubacteria. Trends Genet. 2000;16(10):432–433.

127. Harholt J, Suttangkakul A, Vibe Scheller H. Biosynthesis of pectin. Plant Physiol. 2010;153(2):384–395.

128. Willats WG, Orfila C, Limberg G, Buchholt HC, van Alebeek GJ, Voragen AG, et al. Modulation of the degree and pattern of methyl-esterification of pectic homogalacturonan in plant cell walls. Implications for pectin methyl esterase action, matrix properties, and cell adhesion. J Biol Chem. 2001;276(22):19404–19413.

129. Peter G, Neale D. Molecular basis for the evolution of xylem lignification. Curr Opin Plant Biol. 2004;7(6):737–742.

130. Boyce CK, Zwieniecki MA, Cody GD, Jacobsen C, Wirick S, Knoll AH, et al. Evolution of xylem lignification and hydrogel transport regulation. Proc Natl Acad Sci USA. 2004;101(50):17555–17558.

131. Raven JA. Physiological correlates of the morphology of early vascular plants. Biol J Linn Soc. 1984;88(1–2):105–126.

132. Weng JK, Chapple C. The origin and evolution of lignin biosynthesis. New Phytol. 2010;187(2):273–285.

133. Coleman HD, Samuels AL, Guy RD, Mansfield SD. Perturbed lignification impacts tree growth in hybrid poplar – a function of sink strength, vascular integrity, and photosynthetic assimilation. Plant Physiol. 2008;148(3):1229–1237.

134. Quentin M, Allasia V, Pegard A, Allais F, Ducrot PH, Favery B, et al. Imbalanced lignin biosynthesis promotes the sexual reproduction of homothallic oomycete pathogens. PLoS Pathog. 2009;5(1):e1000264.

135. Yuan JS, Köllner TG, Wiggins G, Grant J, Degenhardt J, Chen F. Molecular and genomic basis of volatile-mediated indirect defense against insects in rice. Plant J. 2008;55(3):491–503.

136. Fry SC. Primary cell wall metabolism: tracking the careers of wall polymers in living plant cells. New Phytol. 2004;161(3):641–675.

137. Somerville C, Bauer S, Brininstool G, Facette M, Hamann T, Milne J, et al. Toward a systems approach to understanding plant cell walls. Science. 2004;306(5705):2206–2211.

138. Boerjan W, Ralph J, Baucher M. Lignin biosynthesis. Annu Rev Plant Biol. 2003;54:519–546.

139. Boudet AM, Lapierre C, Grima-Pettenati J. Biochemistry and molecular biology of lignification. New Phytol. 1995;129(2):203–236.

140. Barceló AR, Ros LVG, Gabaldón C, López-Serrano M, Pomar F, Carrión JS, et al. Basic peroxidases: the gateway for lignin evolution? Phytochem Rev. 2004;3(1–2):61–78.

141. Espiñeira JM, Novo Uzal E, Gómez Ros LV, Carrión JS, Merino F, Ros Barceló A, et al. Distribution of lignin monomers and the evolution of lignification among lower plants. Plant Biol. 2011;13(1):59–68.

142. Ralph J, Bunzel M, Marita JM, Hatfield RD, Lu F, Kim H, et al. Peroxidase-dependent cross-linking reactions of p-hydroxycinnamates in plant cell walls. Phytochem Rev. 2004;3(1–2):79–96.

143. Jin Z, Matsumoto Y, Tange T, Akiyama T, Higuchi M, Ishii T, et al. Proof of the presence of guaiacyl–syringyl lignin in

Selaginella tamariscina. J Wood Sci. 2005;51(4):424–426.

144. Gómez Ros LV, Gabaldón C, Pomar F, Merino F, Pedreño MA, Barceló AR. Structural motifs of syringyl peroxidases predate not only the gymnosperm-angiosperm divergence but also the radiation of tracheophytes. New Phytol. 2007;173(1):63–78.

145. Weng JK, Li X, Stout J, Chapple C. Independent origins of syringyl lignin in vascular plants. Proc Natl Acad Sci USA. 2008;105(22):7887–7892.

146. Lewis NG, Yamamoto E. Lignin: occurrence, biogenesis and biodegradation. Annu Rev Plant Physiol Plant Mol Biol. 1990;41(1):455–496.

147. Siegel SM. Evidence for the presence of lignin in moss gametophytes. Am J Bot. 1969;56(2):175.

148. Reznikov VM, Mikhaseva M, Zil'bergleit M. The lignin of the alga Fucus vesiculosus. Chem Nat Comp. 1978;14:554–556.

149. Delwiche CF, Graham LE, Thomson N. Lignin-like compounds and sporopollenin *Coleochaete*, an algal model for land plant ancestry. Science. 1989;245(4916):399–401.

150. Ligrone R, Carafa A, Duckett JG, Renzaglia KS, Ruel K. Immunocytochemical detection of lignin-related epitopes in cell walls in bryophytes and the charalean alga *Nitella*. Plant Syst Evol. 2008;270(3–4):257–272.

151. Xu Z, Zhang D, Hu J, Zhou X, Ye X, Reichel KL, et al. Comparative genome analysis of lignin biosynthesis gene families across the plant kingdom. BMC Bioinformatics. 2009;10(11 suppl):S3.

152. Martone PT, Estevez JM, Lu F, Ruel K, Denny MW, Somerville C, et al. Discovery of lignin in seaweed reveals convergent evolution of cell-wall architecture. Curr Biol. 2009;19(2):169–175.

153. Boudet AM, Kajita S, Grima-Pettenati J, Goffner D. Lignins and lignocellulosics: a better control of synthesis for new and improved uses. Trends Plant Sci. 2003;8(12):576–581.

154. Nishiyama T, Fujita T, Shin-I T, Seki M, Nishide H, Uchiyama I, et al. Comparative genomics of *Physcomitrella patens* gametophytic transcriptome and *Arabidopsis thaliana*: implication for land plant evolution. Proc Natl Acad Sci USA. 2003;100(13):8007–8012.

155. Meyer K, Shirley AM, Cusumano JC, Bell-Lelong DA, Chapple C. Lignin monomer composition is determined by the expression of a cytochrome P450-dependent monooxygenase in *Arabidopsis*. Proc Natl Acad Sci USA. 1998;95(12):6619–6623.

156. Xue X, Fry SC. Evolution of mixed-linkage (1→3,1→4)-β-D-glucan (MLG) and xyloglucan in *Equisetum* (horsetails) and other monilophytes. Ann Bot. 2012;109(5):873–886.

20

Unique genome evolution in an intracellular N$_2$-fixing symbiont of a rhopalodiacean diatom

Takuro Nakayama[1]*, Yuji Inagaki[1,2]

[1] Center for Computational Sciences, University of Tsukuba, 1-1-1 Tennoudai, Tsukuba, Ibaraki 305-8577, Japan

[2] Graduate School of Life and Environmental Sciences, University of Tsukuba, 1-1-1 Tennoudai, Tsukuba, Ibaraki 305-8572, Japan

Abstract

Cyanobacteria, the major photosynthetic prokaryotic lineage, are also known as a major nitrogen fixer in nature. N$_2$-fixing cyanobacteria are frequently found in symbioses with various types of eukaryotes and supply fixed nitrogen compounds to their eukaryotic hosts, which congenitally lack N$_2$-fixing abilities. Diatom species belonging to the family Rhopalodiaceae also possess cyanobacterial symbionts called spheroid bodies. Unlike other cyanobacterial N$_2$-fixing symbionts, the spheroid bodies reside in the cytoplasm of the diatoms and are inseparable from their hosts. Recently, the first spheroid body genome from a rhopalodiacean diatom has been completely sequenced. Overall features of the genome sequence showed significant reductive genome evolution resulting in a diminution of metabolic capacity. Notably, despite its cyanobacterial origin, the spheroid body was shown to be truly incapable of photosynthesis implying that the symbiont energetically depends on the host diatom. The comparative genome analysis between the spheroid body and another N$_2$-fixing symbiotic cyanobacterial group corresponding to the UCYN-A phylotypes – both were derived from cyanobacteria closely related to genus *Cyanothece* – revealed that the two symbionts are on similar, but explicitly distinct tracks of reductive evolution. Intimate symbiotic relationships linked by nitrogen fixation as seen in rhopalodiacean diatoms may help us better understand the evolution and mechanisms of bacterium-eukaryote endosymbioses.

Keywords: nitrogen fixation; endosymbiosis; genome reduction; spheroid body; rhopalodiacean diatom

Spheroid bodies in rhopalodiacean diatoms

Nitrogen is one of the most important and fundamental elements for all living cells. However, only prokaryotic species are able to fix and utilize the dinitrogen that abundantly exists in the atmosphere [1]. As no eukaryotic cell possesses a N$_2$-fixing capacity, multiple eukaryotic lineages separately developed symbiotic relationships with N$_2$-fixing prokaryotes to secure a nitrogen source (e.g., the rhizobia-legume symbiosis [1]).

Cyanobacteria, one of the major contributors to aquatic primary production, contain many species that are able to fix molecular nitrogen and to perform photosynthesis. These N$_2$-fixing cyanobacteria are frequently found to form various symbiotic relationships with diverse eukaryotes (as seen in a water fern *Azolla*, hornworts, cycads, *Gunnera*; [1,2]). In particular, cyanobacterial symbioses have been documented often in phylogenetically diverse diatoms [3,4], suggesting that separate diatom lineages established symbiotic partnerships with cyanobacteria.

Rhopalodiacean diatoms, a taxonomically small group of pennate diatoms, are one of those symbiont-possessing lineages [5,6]. The family Rhopalodiaceae comprises only three genera, *Rhopalodia*, *Epithemia* and *Protokeelia* [7]. With exception of marine *Protokeelia* species, rhopalodiacean diatoms can be seen widely in freshwater habitats. The cyanobacterial symbionts, so-called "spheroid bodies", has been found in species of *Rhopalodia* and *Epithemia*, and previous observations of the *R. gibba* spheroid body (4–6 mm in width, 5–7 mm in length) showed that the symbiont resides in the host cytoplasm and has two envelope membranes with putatively distinct origins – the outer one was derived from the eukaryotic host cell, while the inner one is the plasmamembrane of the symbiont [8–10]. It has been shown that the number of the spheroid bodies per diatom cell can vary depending on the availability of nitrogen compounds in culture media [11]. Phylogenetic analyses clearly showed that the spheroid bodies in rhopalodiacean diatoms were derived from a cyanobacterium closely related to a N$_2$-fixing genus, *Cyanothece* [5,9]. The host-symbiont association in rhopalodiacean diatoms is seemingly more intimate than those in other diatoms: (*i*) the spheroid bodies reside inside of the host plasmamembrane, while some other cyanobacterial symbionts are extracellularly attached to the diatom hosts

---

* Corresponding author. Email: tak.ae100@gmail.com

Handling Editor: Andrzej Bodył

or found in the periplasmic space between the silicated cell wall and the plasmamembrane [8,12]; (*ii*) the spheroid bodies are believed to be inseparable from the diatoms, since these "cyanobacteria" have never been successfully cultivated independently from the hosts [9]; (*iii*) the most distinctive characteristic that can separate the spheroid bodies from other cyanobacterial symbionts is the lack of chlorophyll autofluorescence [13], implying that the spheroid bodies do not or barely possess photosynthetic activity.

Investigations of rhopalodiacean diatoms bearing unique cyanobacterial symbionts provide new insights into eukaryote-prokaryote symbioses linked by nitrogen fixation. However, only a handful of molecular studies have been done on the spheroid bodies and their host diatoms to date, and the biological, evolutionary, and/or environmental backgrounds, which facilitated this unique symbiosis, remain uncertain. Recently our research group successfully determined the first whole genome sequence of a spheroid body in the rhopalodiacean diatom *Epithemia turgida* [14]. The detailed metabolic functions deduced from the spheroid body genome indicated that the cyanobacterial symbionts reduced its metabolic capacity including photosynthesis, suggesting that the symbiont has abandoned a photoautotrophic lifestyle and energetically depends on its host.

The complete spheroid body genome sequence

To our knowledge, there is a single pioneering study on the genome of a spheroid body. Kneip et al. [15] carried out shotgun sequencing of the spheroid body of *Rhopalodia gibba*, and provided the first clue for the evolutionary status of the cyanobacterial symbionts in rhopalodiacean diatoms. They generated over 140 Kbp of non-contiguous DNA sequence along with a contiguous 51 Kbp fragment. As anticipated from the N$_2$-fixing ability of *R. gibba*, an almost complete *nif* gene cluster, which encodes a set of the proteins required for nitrogen fixation, was found in the 51 Kbp genome fragment. In addition, the first genome sequencing effort revealed the signatures of genome reduction such as pseudogenizations, losses, truncations, and fusions of genes. However, the partial nature of this genome data impedes comparisons to the genomes of closely related cyanobacteria (free-living or symbiotic), and consequently it remained unclear how the intracellular lifestyle altered the spheroid body genome.

Against this background, we determined a complete genome sequence of the spheroid body in the rhopalodiacean diatom *E. turgida* [14]. A 16S rDNA phylogenetic tree of the symbionts indicated that the spheroid bodies of genera *Rhopalodia* and *Epithemia* have a single origin, implying that the spheroid bodies diverged along with the speciation of rhopalodiacean diatoms [5]. The genome of the *E. turgida* spheroid body (*Et*SB) consists of a single circular chromosome with a size of 2.79 Mbp, which is slightly larger than the size predicted by Kneip et al. [15]. As universally observed in comparisons between obligate bacterial symbionts and their free-living close relatives [16–19], the *Et*SB genome was found to be greatly reduced compared to the genome of *Cyanothece* sp. PCC 8801 (4.68 Mbp in size [20]), a free-living

close relative of the spheroid bodies. The difference in genome size between *Et*SB and *Cyanothece* sp. PCC 8801 coincides with the number of protein-coding genes: The *Et*SB genome possesses 1720 protein-coding genes, which is only 39% of the number of protein-coding genes in the genome of *Cyanothece* sp. PPC 8801. In addition, the G+C content of the *Et*SB genome (33.4%) is lower than that of *Cyanothece* sp. PCC 8801 (39.8%).

Gene retentions and losses in the spheroid body genome

The *Et*SB genome successfully provided the first comprehensive picture of the metabolic activities in the cyanobacterial endosymbiont. Confirming the result from the partial sequence of the *R. gibba* spheroid body genome [15], the *Et*SB also contains the complete *nif* gene cluster, with the exception of *fdxN* and *nifU* (discussed in the following section). Importantly, incorporation of gaseous nitrogen into chlorophyll *a* of the host diatom plastids was clearly confirmed by a ^{15}N-isotope tracing analysis [14], indicating that the host diatoms indeed utilize nitrogen fixed by the spheroid bodies.

Fig. 1 shows gene status in the *Et*SB genome compared against a consensus set of protein-coding genes from three free-living cyanobacterial relatives (*Cyanothece* spp. PCC 8801, PCC 8802 and ATCC 51142). A phylogenomic analysis suggested that the three cyanobacteria have a close evolutionary affinity to the spheroid bodies [14], and consequently the ancestral spheroid body likely possessed genes similar to the consensus gene set for the three species. The hypothesized ancestral gene repertoire is represented by KEGG Orthology IDs (KO IDs [21]) in Fig. 1. In comparison with this gene set (1174 KO IDs in total), the spheroid body was found to possess 69% of the "ancestral" genes (Fig. 1a). As seen in Fig. 1b, genes in major functional categories related to basic cellular functions (i.e., categories for "translation", "transcription", "nucleotide metabolism", and "replication and repair"; Fig. 1b) mostly remained intact. In addition to these "housekeeping" genes, the *Et*SB genome retains genes for all amino acid biosynthetic pathways, which are often discarded in obligate bacterial symbionts in various symbiotic systems [17,19,22,23], implying that the spheroid bodies do not require an external amino acid supply.

The greatest gene loss was found in the category for "energy metabolism" (Fig. 1b). The reduction of gene numbers in this category is related to the lack of photosynthesis, which had been suspected from previous works [9,15]. Indeed the *Et*SB genome was found to possess none of the functional genes for photosystem I/II, phycobilisome (cyanobacterial light harvesting complex), or chlorophyll biosynthesis. This situation clearly indicates that *Et*SB is unable to carry out photosynthesis, and consequently this intracellular symbiont energetically depends on its diatom host. The *Et*SB lacks a functional RuBisCO, the fundamental enzyme for the Calvin cycle, further supporting the above idea. To our knowledge, none of cyanobacteria, free-living or symbiotic, are found to have completely abandoned a photoautotrophic lifestyle other than the spheroid bodies.

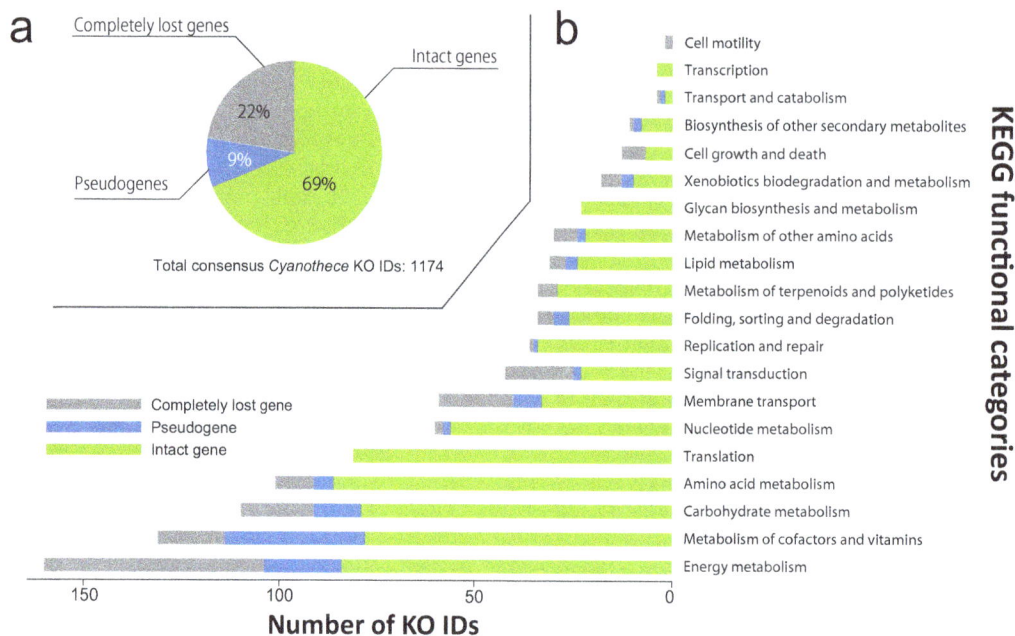

Fig. 1 Pseudogenization and gene loss during the spheroid body symbiosis predicted by an analysis based on KEGG-Orthology (KO). KO IDs shared among the genomes of free-living *Cyanothece* spp. PCC 8801, PCC 8802, and ATCC 51142 were regarded as that of the ancestral spheroid body genome. The number of the "ancestral" KO IDs is 1174. **a** A pie chart representing the overall trend in pseudogenization/gene loss. The proportion of the ancestral KO IDs predicted to be still functional in *Et*SB (i.e. the corresponding open reading frames are intact in the genome) is colored in green. The proportion of the ancestral KO IDs that correspond to pseudogenes in the *Et*SB genome is colored in blue. The proportion of the ancestral KO IDs not found in the *Et*SB genome, which were most likely lost during the symbiosis, is colored in grey. **b** A bar graph representing the functional category-wise trend in pseudogenization/gene loss. The color scheme is the same as described in (**a**).

Another important feature of the *Et*SB genome is that a number of pseudogenes still remain detectable based on sequence similarity. In total, the *Et*SB genome was found to retain 225 pseudogenes. Amongst the missing genes in comparison with the ancestral gene set (Fig. 1a), nearly one third of those missing genes have been detected as pseudogenes (9% of total ancestral gene set). From the point of view of functional categories, pseudogenes were found to be the most abundant in "metabolism of cofactors and vitamins" (Fig. 1b). Within this category, biosynthetic pathways for chlorophyll *a* and vitamin B_{12} are intriguing in terms of pseudogenization. In both pathways, nearly all of the genes were identified as pseudogenes (Fig. 2). These observations imply that not enough evolutionary time to eliminate pseudogenes from the genome has passed since the two pathways were inactivated. This assumption is consistent with an estimation that the rhopalodiacean diatoms can be traced back to approximately 12 Mya [5,24,25], while the 40–60 Myr is likely required to have pseudogenes completely disintegrated [26].

Another interesting pseudogene in the *Et*SB genome is that encoding NifU, a scaffold protein for the assembly of [4Fe-4S] clusters, which are fundamental compounds for the nitrogenase. Kneip et al. [15] identified the *nifU* gene in the genome fragment of the *R. gibba* spheroid body, albeit the putative protein appeared to be severely truncated at the N-terminus and lacked four out of the five catalytically

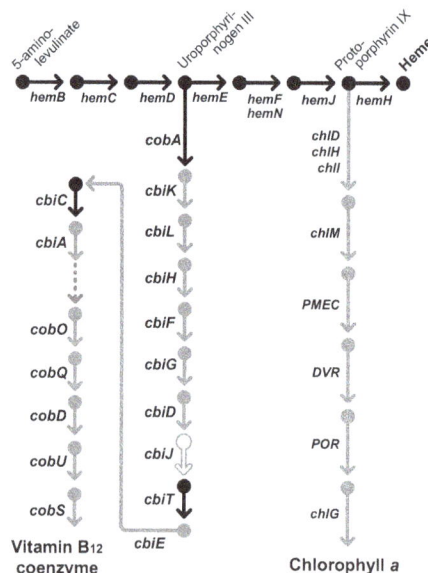

Fig. 2 Heme, vitamin B_{12}, and chlorophyll *a* biosynthetic pathways in the spheroid body of *Epithemia turgida*. Enzymatic reactions are displayed as arrows with the corresponding gene names. Grey arrows indicate inactive reactions due to disruptions of the genes encoding the corresponding enzymes. Black arrows indicate the reactions that can be catalyzed by functional enzymes. An open arrow represents the reaction by *cbiJ*, which is absent in the spheroid body genome. A dashed grey arrow indicates an uncertain pathway.

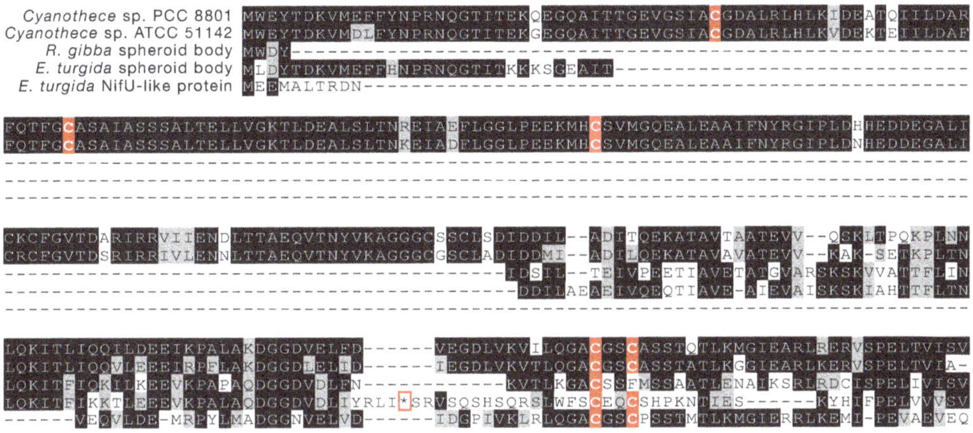

Fig. 3 Alignment of NifU proteins encoded in the genomes of the spheroid bodies of *Rhopalodia gibba* and *Epithemia turgida*, *Cyanothece* strains (PCC 8801 and ATCC 51142), and an *E. turgida* NifU-like protein. Conserved cysteine residues predicted to be important for the function of NifU, and an in-frame stop codon encoded in the *E. turgida* spheroid body genome are highlighted by filled and open red colored boxes, respectively.

important cysteine residues (Fig. 3). In the *Et*SB genome, we detected the homologous sequence to the *R. gibba nifU* gene, but its coding region was interrupted by a stop codon. Thus, we concluded that the *Et*SB *nifU* gene is dysfunctional (Fig. 3). The *Et*SB should possess a different protein for [4Fe-4S] cluster assembly instead of NifU, as nitrogenase activity in this symbiont was confirmed by a nitrogen isotope tracing analysis [14]. We hypothesize that the NifU function in Fe-S cluster synthesis is fulfilled by a small protein with a considerable sequence similarity to NifU C-terminal region (Fig. 3; NifU-like protein).

UCYN-A – another symbiotic N$_2$-fixer related to the spheroid bodies

A unicellular cyanobacterial phylotype called UCYN-A was identified by environmental DNA analyses of ocean water samples [27,28], and was recently nominated as a major nitrogen-fixer in oceans, second to a filamentous cyanobacterium, *Trichodesmium* [29]. A detailed cellular identity related to the UCYN-A phylotype has not been entirely revealed, as no culture strain has been established. Nevertheless, there are two complete genome sequences corresponding to two closely related UCYN-A phylotypes, UCYN-A1 and UCYN-A2 [23,30]. Interestingly, the UCYN-A genomes (~1.5 Mbp) were found to be much smaller than the *Et*SB genome (~2.8 Mbp), suggesting a severe diminution of metabolic capacity. According to the genomic information, the UCYN-A cyanobacterium lacks the entire tricarboxylic acid (TCA) cycle and biosynthetic pathways for several amino acids and purine nucleotides, which are partially or completely retained in *Et*SB.

Based on the genome size and metabolic capacity deduced from the genome data, the UCYN-A cyanobacteria most likely experienced a more severe genome reduction than *Et*SB. The most prominent difference between the UCYN-A cyanobacteria and *Et*SB is photosynthetic ability: the former

is most likely photosynthetic, as the genomes still retain the complete gene set for photosystem I and that for chlorophyll *a* biosynthesis. In sharp contrast, *Et*SB has entirely discarded photosynthetic ability, as described in the previous section. The magnitude of genome reduction and gene repertoires is different between the two cyanobacterial genomes, as the spheroid bodies of rhopalodiacean diatoms and the UCYN-A cyanobacteria have independent progenitors, which are phylogenetically closely related, but explicitly distinct from each other [14].

The incomplete nature of the metabolic capacity deduced from the UCYN-A genomes suggests that the corresponding cyanobacteria depend on external nutrient supplies. Consistent with the above prediction, there is some evidence for a symbiotic relationship between the UCYN-A cyanobacteria and a prymnesiophyte unicellular alga [10,31]. Currently, it remains unclear how intimate the host-symbiont relationship is: Thompson et al. [31] suggested the cyanobacteria reside in the epicellular space of the host alga, while electron microscopic images identified the UCYN-A cyanobacteria as an endosymbiont [10]. To reveal the host dependency of the UCYN-A cyanobacteria, transcriptomic and genome data from the host are indispensable for the future. As the spheroid bodies and UCYN-A cyanobacteria have independently established deep symbiotic associations with photosynthetic algae, comparisons between the two symbiotic systems may provide key insights into bacterium-eukaryote partnerships built on a nitrogen supply in the hydrosphere.

Concluding remarks

The complete spheroid body genome greatly advanced our understanding of the functions and metabolism of the cyanobacterial symbiont. Nevertheless, we currently know little about how the host system controls the division of the symbiont and distribution of the divided symbionts to the daughter cells, and enables the trafficking of metabolites from and to the symbiont. To address the above issue, future investigations should focus on the host diatom. We anticipate that the host (diatom) system controlling the spheroid bodies

is useful for understanding the processes that gave birth to two major bacterium-derived organelles, mitochondria and plastids, as well as the early evolution of eukaryotes.

Acknowledgments
We would like to thank Bruce A. Curtis (Dalhousie University, Canada) for his kind help with proofreading the manuscript. Work in our laboratory is currently supported by a grant from Japanese Society for the Promotion of Sciences to YI (No. 23117006).

Authors' contributions
The following declarations about authors' contributions to the research have been made: survey of literature and writing the manuscript: TN, YI.

Competing interests
No competing interests have been declared.

References
1. Kneip C, Lockhart P, Voß C, Maier UG. Nitrogen fixation in eukaryotes – new models for symbiosis. BMC Evol Biol. 2007;7(1):55.

2. Rai AN, Bergman B, Rasmussen U, editors. Cyanobacteria in symbiosis. Dordrecht: Kluwer Academic Publishers; 2002.

3. Janson S. Cyanobacteria in symbiosis with diatoms. In: Rai AN, Bergman B, Rasmussen U, editors. Cyanobacteria in symbiosis. Dordrecht: Springer; 2002. p. 1–10.

4. Foster RA, Kuypers MMM, Vagner T, Paerl RW, Musat N, Zehr JP. Nitrogen fixation and transfer in open ocean diatom–cyanobacterial symbioses. ISME J. 2011;5(9):1484–1493.

5. Nakayama T, Ikegami Y, Nakayama T, Ishida K, Inagaki Y, Inouye I. Spheroid bodies in rhopalodiacean diatoms were derived from a single endosymbiotic cyanobacterium. J Plant Res. 2011;124(1):93–97.

6. Adler S, Trapp EM, Dede C, Maier UG, Zauner S. Rhopalodia gibba : the first steps in the birth of a novel organelle? In: Löffelhardt W, editor. Endosymbiosis. Vienna: Springer; 2014. p. 167–179.

7. Round FE, Crawford RM, Mann DG. The diatoms: biology & morphology of the genera. Cambridge: Cambridge University Press; 1990.

8. Drum RW, Pankratz S. Fine structure of an unusual cytoplasmic inclusion in the diatom genus, Rhopalodia. Protoplasma. 1965;60(1):141–149.

9. Prechtl J. Intracellular spheroid bodies of Rhopalodia gibba have nitrogen-fixing apparatus of cyanobacterial origin. Mol Biol Evol. 2004;21(8):1477–1481.

10. Hagino K, Onuma R, Kawachi M, Horiguchi T. Discovery of an endosymbiotic nitrogen-fixing cyanobacterium UCYN-A in Braarudosphaera bigelowii (Prymnesiophyceae). PLoS ONE. 2013;8(12):e81749.

11. de Yoe HR, Lowe RL, Marks JC. Effects of nitrogen and phosphorus on the endosymbiont load of Rhopalodia gibba and Epithemia turgida (Bacillariophyceae). J Phycol. 1992;28(6):773–777.

12. Hilton JA, Foster RA, James Tripp H, Carter BJ, Zehr JP, Villareal TA. Genomic deletions disrupt nitrogen metabolism pathways of a cyanobacterial diatom symbiont. Nat Commun. 2013;4:1767.

13. Kies L. Glaucocystophyceae and other protists harbouring prokaryotic endocytobionts. In: Reisser W, editor. Algae and symbioses. Bristol: Biopress; 1992. p. 353–377.

14. Nakayama T, Kamikawa R, Tanifuji G, Kashiyama Y, Ohkouchi N, Archibald JM, et al. Complete genome of a nonphotosynthetic cyanobacterium in a diatom reveals recent adaptations to an intracellular lifestyle. Proc Natl Acad Sci USA. 2014;111(31):11407–11412.

15. Kneip C, Voß C, Lockhart PJ, Maier UG. The cyanobacterial endosymbiont of the unicellular algae Rhopalodia gibba shows reductive genome evolution. BMC Evol Biol. 2008;8(1):30.

16. Wernegreen JJ. Genome evolution in bacterial endosymbionts of insects. Nat Rev Genet. 2002;3(11):850–861.

17. Kuwahara H, Yoshida T, Takaki Y, Shimamura S, Nishi S, Harada M, et al. Reduced genome of the thioautotrophic intracellular symbiont in a deep-sea clam, Calyptogena okutanii. Curr Biol. 2007;17(10):881–886.

18. Moran NA, McCutcheon JP, Nakabachi A. Genomics and evolution of heritable bacterial symbionts. Annu Rev Genet. 2008;42(1):165–190.

19. Nowack ECM, Melkonian M, Glöckner G. Chromatophore genome sequence of Paulinella sheds light on acquisition of photosynthesis by eukaryotes. Curr Biol. 2008;18(6):410–418.

20. Bandyopadhyay A, Elvitigala T, Welsh E, Stockel J, Liberton M, Min H, et al. Novel metabolic attributes of the genus Cyanothece, comprising a group of unicellular nitrogen-fixing cyanobacteria. mBio. 2011;2(5):e00214–11.

21. Kanehisa M. The KEGG resource for deciphering the genome. Nucl Acids Res. 2004;32(90001):277D–280.

22. Shigenobu S, Watanabe H, Hattori M, Sakaki Y, Ishikawa H. Genome sequence of the endocellular bacterial symbiont of aphids Buchnera sp. APS. Nature. 2000;407(6800):81–86.

23. Tripp HJ, Bench SR, Turk KA, Foster RA, Desany BA, Niazi F, et al. Metabolic streamlining in an open-ocean nitrogen-fixing cyanobacterium. Nature. 2010;464(7285):90–94.

24. Hajós M. Stratigraphy of Hungary's Miocene diatomaceous Earth deposits. Budapest: Institutum Geologicum Hungaricum; 1986. (Geologica Hungarica; vol 49).

25. Simonsen R. The diatom system: ideas on phylogeny. Bacillaria. 1979;2:9–71.

26. Gomez-Valero L. The evolutionary fate of nonfunctional DNA in the bacterial endosymbiont Buchnera aphidicola. Mol Biol Evol. 2004;21(11):2172–2181.

27. Zehr JP, Mellon MT, Zani S. New nitrogen-fixing microorganisms detected in oligotrophic oceans by amplification of nitrogenase (nifH) genes. Appl Env. Microbiol. 1998;64(9):3444–3450.

28. Zehr JP, Bench SR, Carter BJ, Hewson I, Niazi F, Shi T, et al. Globally distributed uncultivated oceanic N_2-fixing cyanobacteria lack oxygenic photosystem II. Science. 2008;322(5904):1110–1112.

29. Goebel NL, Turk KA, Achilles KM, Paerl R, Hewson I, Morrison AE, et al. Abundance and distribution of major groups of diazotrophic cyanobacteria and their potential contribution to N_2 fixation in the tropical Atlantic Ocean. Environ Microbiol. 2010;12(12):3272–3289.

30. Bombar D, Heller P, Sanchez-Baracaldo P, Carter BJ, Zehr JP. Comparative genomics reveals surprising divergence of two closely related strains of uncultivated UCYN-A cyanobacteria. ISME J. 2014;8(12):2530–2542.

31. Thompson AW, Foster RA, Krupke A, Carter BJ, Musat N, Vaulot D, et al. Unicellular cyanobacterium symbiotic with a single-celled eukaryotic alga. Science. 2012;337(6101):1546–1550.

21

Protein translocons in photosynthetic organelles of *Paulinella chromatophora*

Przemysław Gagat, Paweł Mackiewicz*

Department of Genomics, Faculty of Biotechnology, University of Wrocław, Fryderyka Joliot-Curie 14a, 50-383 Wrocław, Poland

Abstract

The rhizarian amoeba *Paulinella chromatophora* harbors two photosynthetic cyanobacterial endosymbionts (chromatophores), acquired independently of primary plastids of glaucophytes, red algae and green plants. These endosymbionts have lost many essential genes, and transferred substantial number of genes to the host nuclear genome via endosymbiotic gene transfer (EGT), including those involved in photosynthesis. This indicates that, similar to primary plastids, *Paulinella* endosymbionts must have evolved a transport system to import their EGT-derived proteins. This system involves vesicular trafficking to the outer chromatophore membrane and presumably a simplified Tic-like complex at the inner chromatophore membrane. Since both sequenced *Paulinella* strains have been shown to undergo differential plastid gene losses, they do not have to possess the same set of Toc and Tic homologs. We searched the genome of *Paulinella* FK01 strain for potential Toc and Tic homologs, and compared the results with the data obtained for *Paulinella* CCAC 0185 strain, and 72 cyanobacteria, eight Archaeplastida as well as some other bacteria. Our studies revealed that chromatophore genomes from both *Paulinella* strains encode the same set of translocons that could potentially create a simplified but fully-functional Tic-like complex at the inner chromatophore membranes. The common maintenance of the same set of translocon proteins in two *Paulinella* strains suggests a similar import mechanism and/or supports the proposed model of protein import. Moreover, we have discovered a new putative Tic component, Tic62, a redox sensor protein not identified in previous comparative studies of *Paulinella* translocons.

Keywords: *Paulinella chromatophora*; chromatophores; endosymbiosis; primary plastids; protein import; translocons

Introduction

The primary plastid endosymbiosis is one of the most important transitions in the evolution of life on our planet. It took place sometime before 1.5 billion years ago, when a phagotrophic eukaryote, the ancestor of glaucophytes, red algae and green plants, enslaved a β-cyanobacterium, which was transformed into a two-membrane photosynthetic primary plastid [1,2]. In order to become a true cell organelle, the prokaryote underwent a tremendous transformation that involved two key processes: (*i*) endosymbiotic gene transfer (EGT), i.e. gene transfer from the endosymbiont to the host nuclear genome, and (*ii*) origin of an import machinery in its envelope membranes for host genome-encoded proteins, including the products of EGT-derived genes [3–5].

Interestingly, primary plastid endosymbiosis was not a singular event in the history of our biosphere, but has happened at least twice, questioning thereby a paradigm that endosymbiont-to-organelle transformation is an exceptionally

rare evolutionary phenomenon [6]. For the second time, it took place in the case of *Paulinella chromatophora*, a testate filose amoeba, belonging to the supergroup Rhizaria, a lineage that is very distantly related to the primary plastid bearing lineages, i.e. glaucophytes, red algae and green plants, united in the supergroup Archaeplastida [7,8].

Paulinella chromatophora harbors two photosynthetic cyanobacterial endosymbionts (chromatophores), acquired independently of Archaeplastida primary plastids ~60 million years ago [8–11]. Similarly, to primary plastids, chromatophores are surrounded by a two-membrane envelope and are deeply integrated with their host cell. They divide synchronously with *Paulinella* cell (after being distributed to daughter cells), exchange metabolites with it and are incapable of independent life [9,12,13]. They are especially similar in their structure to glaucophyte primary plastids (cyanelles) because they still retain the peptidoglycan wall in the intermembrane space, a clear sign of their bacterial inheritance [12]. The intimacy between the *Paulienlla* host and its endosymbiotic bodies is well emphasized by substantial reduction of the latter genome, which was sequenced for two *Paulinella* strains CCAC 0185 [11] and FK01 [14]. They were reduced in size and coding capacity by a factor

* Corresponding author. Email: pamac@smorfland.uni.wroc.pl

Handling Editor: Andrzej Bodył

of three (to 1 Mb and ~900 genes), compared with their closest free-living relative, the cyanobacterium *Cyanobium gracile* PCC 6307 (~3 Mb and ~3300 genes) [11,14] (Fig. 1). The genome-size reduction involved loss of many genes, including those engaged in essential biosynthetic pathways and endosymbiotic gene transfer [11,14]. Transcriptome analyses have identified more than 30 EGT-derived genes in the *Paulinella* nuclear genome [14–16]; however, many more are expected; Nowack et al. [16] estimated their number between 40 and 125. Because the lost genes are of the utmost importance for cell functioning, and at least some of the transferred ones are transcriptionally regulated by the host and involved in photosynthesis or photosynthesis-related processes, *Paulinella* chromatophores are expected to have evolved a protein import route, or routes.

To check if chromatophore protein import system is similar to the primary plastid translocons at the outer and inner envelope membrane, Toc and Tic, respectively, Bodył et al. [17] searched for homologs of *Toc* and *Tic* genes in the completely sequenced chromatophore genome of *Paulinella* CCAC 0185 strain. They found that the *Paulinella* chromatophore genome encodes homologs to Toc12, Toc64, Tic21 and Tic32, but lost those of Toc75, Tic20, Tic55 and Tic62. They suggested that the missing genes, which are still present in the genomes of closely related cyanobacteria, were most probably relocated via the EGT to the host nuclear genome, and that their products are now imported into the photosynthetic bodies' membranes to create a Toc-Tic-like import machinery [17]. Mackiewicz and Bodył [18] also found that an EGT-derived gene encoding the subunit IV of PS I reaction center (PsaE) form *Paulinella* FK01 strain has a clear signal peptide predicted from an alternative initiation site [18]. Since signal peptides are involved in protein import into the endoplasmic reticulum, such a presequence in the case of PsaE indicates the possibility of endomembrane-dependent targeting to the outer chromatophore membrane. Subsequent studies confirmed the presence of a signal peptide-like presequence in another four of nine investigated EGT-derived proteins that evolved by modification of their N-terminal mature parts, but not the missing Toc and Tic components in the nuclear genome [16,19].

On the basis of their bioinformatics analyses, Mackiewicz et al. [20] formulated a model for protein import into *Paulinella* chromatophores. According to the model, *Paulinella* EGT-derived proteins established endomembrane-dependent mechanism, vesicular trafficking, to pass the outer chromatophore membrane, while the inner membrane is crossed by a simplified Tic-like complex. All investigated EGT-derived proteins, presumed to follow this pathway, characterize low molecular weight and nearly neutral charge, which constitute good adaptations to their passage through the peptidoglycan wall still present in the chromatophores. Mackiewicz et al. [20] also identified additional putative elements of the system, for example chaperons in the intermembrane space and components of the molecular motor responsible for pulling the imported proteins into the chromatophore stroma. It should be noted that part of the model, concerning the vesicular trafficking to the outer chromatophore membrane, has recently been confirmed by Nowack et al. [21]. They used immunogold labeling electron microscopy

to show that PsaE, PsaK1 and PsaK2 from *Paulinella* CCAC 0185 strain are localized both in chromatophores and the endomembrane system, including the Golgi apparatus. They also indicated that these proteins are indeed PS I subunits as they associate with PS I components. Interestingly, Nowack et al. [21] revealed that the three proteins do not have typical signal peptides at their N-termini. In the case of PsaE some internal targeting information was suggested by Mackiewicz et al. [20], whereas, for PsaK proteins algorithms recognized uncleavable signal peptides in their N-terminal hydrophobic domains, suggesting that their N-termini might fulfil a signal peptide role in the protein import after all [19].

Both *Paulinella* strains have undergone differential plastid gene losses [14,22] and have even been suggested to represent two different species on the basis of some morphological and phylogenetic analyses [8]. Therefore it is interesting to check if they possess the same set of potential Toc and Tic homologs. The common maintenance of the same set of proteins would indicate a similar import mechanism and/or support the model proposed by Mackiewicz et al. [20]. Therefore, we searched *Paulinella* FK01 strain genome for Toc and Tic homologs, and compared the results with the data obtained for *Paulinella* CCAC 0185 strain, 72 cyanobacteria, eight Archaeplastida, as well as some other bacteria. We carried out more sensitive homology and conserved domain searches than in previous approaches, which is justified by a short length and relatively high divergence of these proteins [17].

Material and methods

We downloaded 867 and 841 sequences of chromatophore-encoded proteins for *Paulinella* CCAC 0185 strain from Genbank [23] and for *Paulinella* FK01 strain from Debashish Bhattacharya's Laboratory website (http://dblab.rutgers.edu/home/index.php), respectively. We also acquired proteomes of 72 completely sequenced cyanobacteria, eight Archaeplastida and two reference bacteria for comparative studies from Genbank [23]. On the basis of the obtained sequences a local protein database was created. Sequences of 16S and 23S rRNA for phylogenetic studies were extracted from gbk files downloaded from GenBank (genome database) for appropriate organisms. This set included sequences, which were derived from two *Paulinella chromatophora* genomes, 72 completely sequenced cyanobacterial genomes, eight primary plastid genomes and five genomes representing different bacterial lineages as an outgroup.

To find potential Toc-Tic components, we searched the database using *Pisum sativum* translocons as query sequences with PsiBlast (*E*-value set to 0.01, word size to two, filtering for low complexity regions on, five iterations) to achieve better sensitivity than with standard BLAST or FASTA [24]. In the case of Tic21 a homolog from *Arabidopsis thaliana* was used as Tic21 is absent from *P. sativum* [25]. The obtained candidates were verified in terms of domain content by searching CDD database with *E*-value < 0.01 (Tab. 1) [26]. Alignments of Toc12, Tic21, Tic32 and Tic62 from *P. sativum* or *A. thaliana* and two strains of *P. chromatophora*, CCAC0185 and FK01, were performed in PSI-Coffee [27], slow and accurate algorithm dedicated to distantly related

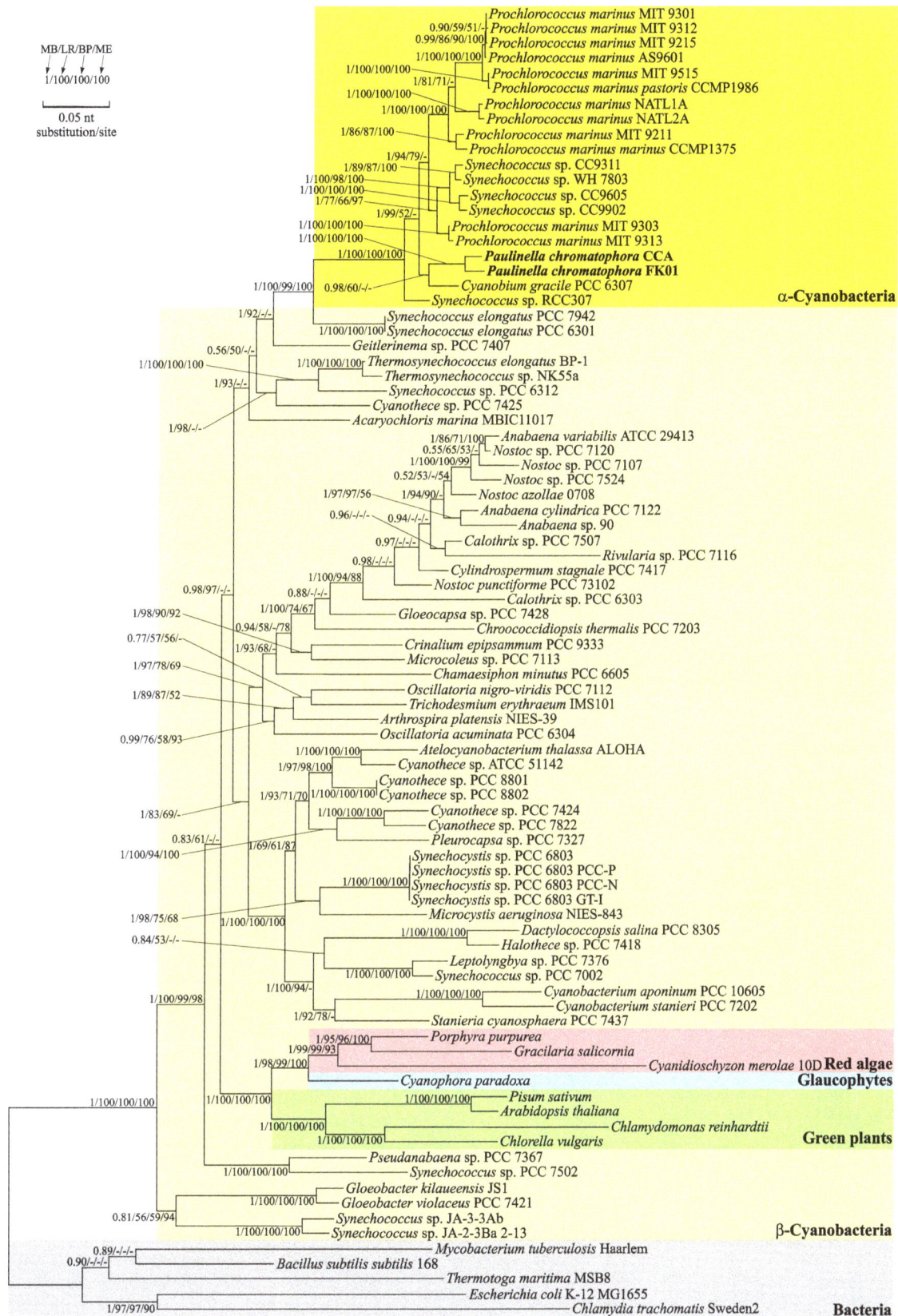

MB/LR/BP/ME
1/100/100/100

0.05 nt
substitution/site

Prochlorococcus marinus MIT 9301
Prochlorococcus marinus MIT 9312
Prochlorococcus marinus MIT 9215
Prochlorococcus marinus AS9601
Prochlorococcus marinus MIT 9515
Prochlorococcus marinus pastoris CCMP1986
Prochlorococcus marinus NATL1A
Prochlorococcus marinus NATL2A
Prochlorococcus marinus MIT 9211
Prochlorococcus marinus marinus CCMP1375
Synechococcus sp. CC9311
Synechococcus sp. WH 7803
Synechococcus sp. CC9605
Synechococcus sp. CC9902
Prochlorococcus marinus MIT 9303
Prochlorococcus marinus MIT 9313
Paulinella chromatophora CCA
Paulinella chromatophora FK01
Cyanobium gracile PCC 6307
Synechococcus sp. RCC307
α-Cyanobacteria

Synechococcus elongatus PCC 7942
Synechococcus elongatus PCC 6301
Geitlerinema sp. PCC 7407
Thermosynechococcus elongatus BP-1
Thermosynechococcus sp. NK55a
Synechococcus sp. PCC 6312
Cyanothece sp. PCC 7425
Acaryochloris marina MBIC11017
Anabaena variabilis ATCC 29413
Nostoc sp. PCC 7120
Nostoc sp. PCC 7107
Nostoc sp. PCC 7524
Nostoc azollae 0708
Anabaena cylindrica PCC 7122
Anabaena sp. 90
Calothrix sp. PCC 7507
Rivularia sp. PCC 7116
Cylindrospermum stagnale PCC 7417
Nostoc punctiforme PCC 73102
Calothrix sp. PCC 6303
Gloeocapsa sp. PCC 7428
Chroococcidiopsis thermalis PCC 7203
Crinalium epipsammum PCC 9333
Microcoleus sp. PCC 7113
Chamaesiphon minutus PCC 6605
Oscillatoria nigro-viridis PCC 7112
Trichodesmium erythraeum IMS101
Arthrospira platensis NIES-39
Oscillatoria acuminata PCC 6304
Atelocyanobacterium thalassa ALOHA
Cyanothece sp. ATCC 51142
Cyanothece sp. PCC 8801
Cyanothece sp. PCC 8802
Cyanothece sp. PCC 7424
Cyanothece sp. PCC 7822
Pleurocapsa sp. PCC 7327
Synechocystis sp. PCC 6803
Synechocystis sp. PCC 6803 PCC-P
Synechocystis sp. PCC 6803 PCC-N
Synechocystis sp. PCC 6803 GT-I
Microcystis aeruginosa NIES-843
Dactylococcopsis salina PCC 8305
Halothece sp. PCC 7418
Leptolyngbya sp. PCC 7376
Synechococcus sp. PCC 7002
Cyanobacterium aponinum PCC 10605
Cyanobacterium stanieri PCC 7202
Stanieria cyanosphaera PCC 7437
Porphyra purpurea
Gracilaria salicornia
Cyanidioschyzon merolae 10D **Red algae**
Cyanophora paradoxa **Glaucophytes**
Pisum sativum
Arabidopsis thaliana
Chlamydomonas reinhardtii
Chlorella vulgaris
Green plants
Pseudanabaena sp. PCC 7367
Synechococcus sp. PCC 7502
Gloeobacter kilaueensis JS1
Gloeobacter violaceus PCC 7421
Synechococcus sp. JA-3-3Ab
Synechococcus sp. JA-2-3Ba 2-13
β-Cyanobacteria
Mycobacterium tuberculosis Haarlem
Bacillus subtilis subtilis 168
Thermotoga maritima MSB8
Escherichia coli K-12 MG1655
Chlamydia trachomatis Sweden2 **Bacteria**

Fig. 1 MrBayes tree based on 16S and 23S rRNA alignment. An independent origin of *Paulinella* endosymbionts from primary plastids of red algae, glaucophytes and green plants is clearly visible. The former are descendants of α-cyanobacteria whereas the latter of β-cyanobacteria. Numbers at nodes, in the presented order, correspond to posterior probabilities estimated in MrBayes (MB), support values calculated by local rearrangements–expected likelihood weights method (LR), as well as bootstrap percentages calculated in TreeFinder (BP) and PAUP using minimum evolution method (ME). Values of the posterior probabilities and bootstrap percentages lower than 0.50 and 50%, respectively, were omitted or indicated by a dash "–". The length of branches leading to Bacteria taxa was shortened by 50% in comparison to their original length because of very high substitution rate.

proteins using profile information. The alignments were visualized and edited in Jalview 2.4.0.b2 [28]. The N-terminal transmembrane β-barrel regions were predicted with BOC-TOPUS [29], whereas α-helical transmembrane regions with TopPred [30] and TMpred [31].

Phylogenetic trees for the concatenated alignment of 16S and 23S rRNA were inferred by maximum likelihood (ML) method in TreeFinder [32], Bayesian approach in MrBayes 3.2.1 [33] and minimum evolution (ME) method based on logDet/paralinear distance in PAUP [34]. In TreeFinder and MrBayes approaches, we used separate models of nucleotide substitutions for each type of rRNA. Two different GTR+Γ(5) models for each data partition were applied in the ML method as suggested by the Propose Model module in Treefinder considering all criteria (–lnL, AIC, AICc,

BIC, HQ). In MrBayes analyses, we assumed two separate mixed+I+Γ(5) models for each rRNA partition to sample appropriate models across the substitution model space in the Bayesian MCMC analysis itself avoiding the need for a priori model testing [35]. In TreeFinder detailed search depth set to 2 was applied whereas in PAUP the final tree was searched from 10 starting trees obtained by stepwise and random sequence addition followed by the tree-bisection-reconnection (TBR) branch-swapping algorithm. To assess significance of particular branches, non-parametric bootstrap analyses were performed on 1000 replicates in these two methods. Additionally, we applied the local rearrangements-expected likelihood weights (LR-ELW) method in TreeFinder. In MrBayes analyses, two independent runs starting from random trees were applied, each using eight Markov chains. Trees were sampled every 100 generations for 10 000 000 generations. In the final analysis, we selected trees from the last 2 984 000 generations that reached the stationary phase and convergence (i.e. the standard deviation of split frequencies stabilized and was lower than the proposed threshold of 0.01). The temperature parameter for heating the chains was suitably adjusted to keep the proportions of successful state exchanges between chains close to the suggested range from 0.10 to 0.70.

Results and discussion

We found that both *Paulinella* chromatophore genomes encode the same set of sequence homologs to Toc12, Toc34, Toc64, Toc159, Tic21, Tic32 and Tic62 but lost those of Toc75, Tic20, Tic22 and Tic55 (Fig. 2). Interestingly, the missing genes are still present in the α-cyanobacterial genomes (*Cyanobium gracile*, *Prochlorococcus marinus* and *Synechococccus* CC9311, CC9605, CC9902 and WH 7803) closely related to *Paulinella* (Fig. 1). This indicates that they might have been lost during chromatophore genome reduction or transferred to the nucleus. An interesting case is Tic22 homolog, which is absent not only from *Paulinella* but also from all α-cyanobacteria grouped with *Paulinella* except for *Synechococcus* RCC307 (Fig. 1). The latter α-cyanobacterium is placed at the basal position to the clade indicating that the gene encoding Tic22 must have been lost after the divergence of this basal lineage from the other α-cyanobacteria. It should be mentioned that Tic22 homolog is present in all other, more than 50 studied cyanobacterial taxa (Fig. 2). Independent losses concern also the gene for Tic62 homolog in the cyanobacterium *Atelocyanobacterium thalassa*, Tic21 in the land plant *Pisum sativum* as well as Tic55 in the cyanobacterium *Synechococcus* sp. JA-3-3Ab and

Tab. 1 Domains characteristic of Toc and Tic proteins from *Pisum sativum* and *Arabidopsis thaliana*, which were used to predicted their homologs in *Paulinella*, cyanobacteria and bacteria proteomes.

Protein name	CDD domain	Additional information
Tic20	233225	Tic20, chloroplast protein import component, length = 267
Tic21	256942	DUF3611, protein of unknown function, length = 183
Tic22	233226	TIC22, chloroplast protein import component, length = 270
	252499	Tic22-like family, length = 268
Tic32	212492	Retinol dehydrogenase, length = 269
	235736	Oxidoreductase, length = 315
Tic55	239830	Tic55, a 55 kDa LLS1-related non-heme iron oxygenase, length = 134
Tic62	178748	Translocon at the inner envelope of chloroplast, length = 576
Toc12	197617	DnaJ molecular chaperone homology domain, length = 60
	249696	DnaJ domain, length = 63
Toc34	130064	GTP-binding protein (chloroplast envelope protein translocase, length = 313
	206652	Toc34 like, length = 248
Toc75	215598	Protein TOC75, length = 796
	130065	Chloroplast envelope protein translocase, IAP75 family, length = 718
Toc64	166363	Indole-3-acetamide amidohydrolase, length = 422
	181375	Amidase, length = 395
	238112	TPR, Tetratricopeptide repeat domain, length = 100
Toc159	233224	Chloroplast protein import component Toc86/159, length = 763

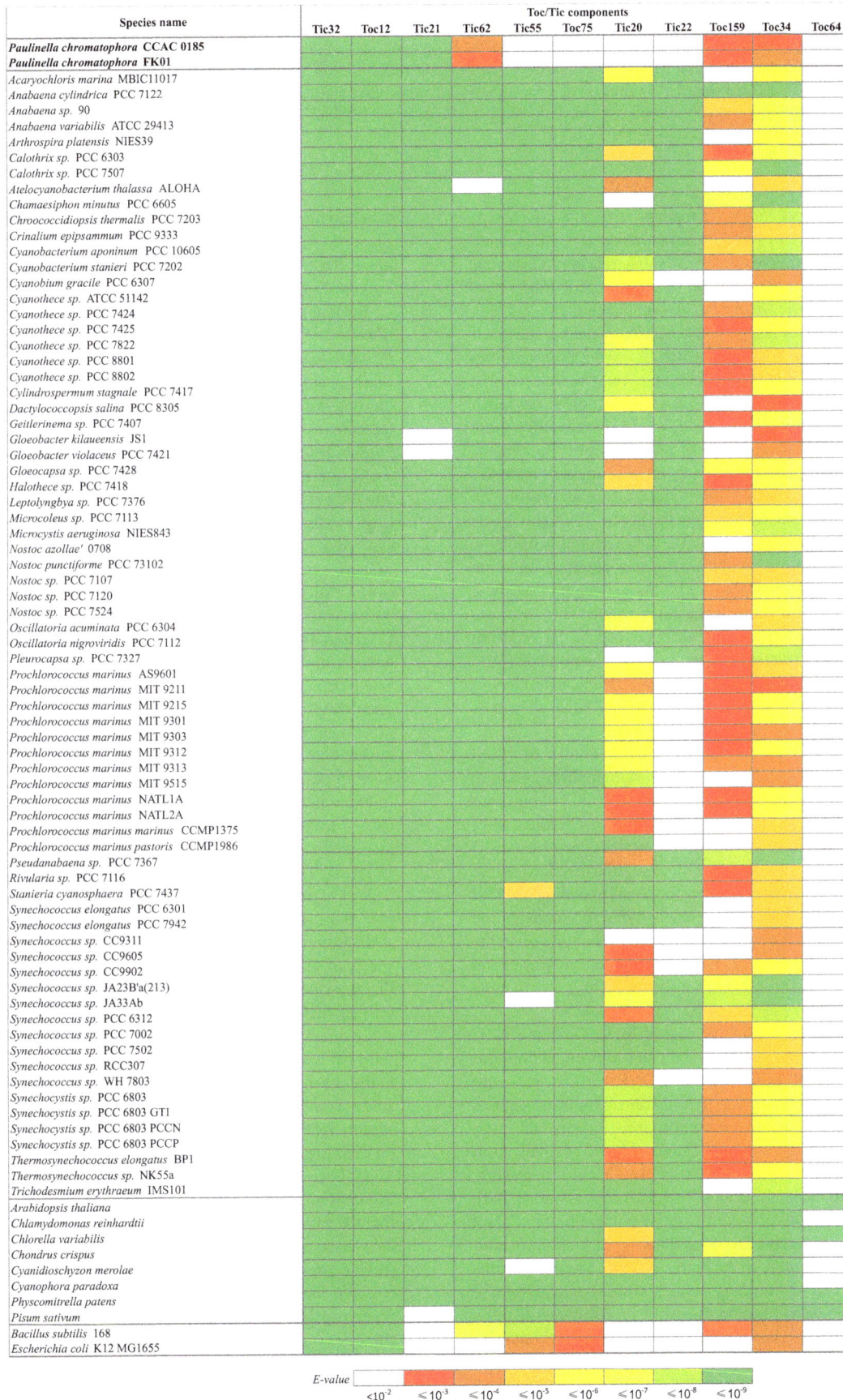

Species name	Tic32	Toc12	Tic21	Tic62	Tic55	Toc75	Tic20	Tic22	Toc159	Toc34	Toc64

Toc/Tic components

Paulinella chromatophora CCAC 0185
Paulinella chromatophora FK01
Acaryochloris marina MBIC11017
Anabaena cylindrica PCC 7122
Anabaena sp. 90
Anabaena variabilis ATCC 29413
Arthrospira platensis NIES39
Calothrix sp. PCC 6303
Calothrix sp. PCC 7507
Atelocyanobacterium thalassa ALOHA
Chamaesiphon minutus PCC 6605
Chroococcidiopsis thermalis PCC 7203
Crinalium epipsammum PCC 9333
Cyanobacterium aponinum PCC 10605
Cyanobacterium stanieri PCC 7202
Cyanobium gracile PCC 6307
Cyanothece sp. ATCC 51142
Cyanothece sp. PCC 7424
Cyanothece sp. PCC 7425
Cyanothece sp. PCC 7822
Cyanothece sp. PCC 8801
Cyanothece sp. PCC 8802
Cylindrospermum stagnale PCC 7417
Dactylococcopsis salina PCC 8305
Geitlerinema sp. PCC 7407
Gloeobacter kilaueensis JS1
Gloeobacter violaceus PCC 7421
Gloeocapsa sp. PCC 7428
Halothece sp. PCC 7418
Leptolyngbya sp. PCC 7376
Microcoleus sp. PCC 7113
Microcystis aeruginosa NIES843
Nostoc azollae' 0708
Nostoc punctiforme PCC 73102
Nostoc sp. PCC 7107
Nostoc sp. PCC 7120
Nostoc sp. PCC 7524
Oscillatoria acuminata PCC 6304
Oscillatoria nigroviridis PCC 7112
Pleurocapsa sp. PCC 7327
Prochlorococcus marinus AS9601
Prochlorococcus marinus MIT 9211
Prochlorococcus marinus MIT 9215
Prochlorococcus marinus MIT 9301
Prochlorococcus marinus MIT 9303
Prochlorococcus marinus MIT 9312
Prochlorococcus marinus MIT 9313
Prochlorococcus marinus MIT 9515
Prochlorococcus marinus NATL1A
Prochlorococcus marinus NATL2A
Prochlorococcus marinus marinus CCMP1375
Prochlorococcus marinus pastoris CCMP1986
Pseudanabaena sp. PCC 7367
Rivularia sp. PCC 7116
Stanieria cyanosphaera PCC 7437
Synechococcus elongatus PCC 6301
Synechococcus elongatus PCC 7942
Synechococcus sp. CC9311
Synechococcus sp. CC9605
Synechococcus sp. CC9902
Synechococcus sp. JA23B'a(213)
Synechococcus sp. JA33Ab
Synechococcus sp. PCC 6312
Synechococcus sp. PCC 7002
Synechococcus sp. PCC 7502
Synechococcus sp. RCC307
Synechococcus sp. WH 7803
Synechocystis sp. PCC 6803
Synechocystis sp. PCC 6803 GT1
Synechocystis sp. PCC 6803 PCCN
Synechocystis sp. PCC 6803 PCCP
Thermosynechococcus elongatus BP1
Thermosynechococcus sp. NK55a
Trichodesmium erythraeum IMS101
Arabidopsis thaliana
Chlamydomonas reinhardtii
Chlorella variabilis
Chondrus crispus
Cyanidioschyzon merolae
Cyanophora paradoxa
Physcomitrella patens
Pisum sativum
Bacillus subtilis 168
Escherichia coli K12 MG1655

E-value: $<10^{-2}$ $\leq 10^{-3}$ $\leq 10^{-4}$ $\leq 10^{-5}$ $\leq 10^{-6}$ $\leq 10^{-7}$ $\leq 10^{-8}$ $\leq 10^{-9}$

Fig. 2 Distribution of Toc and Tic homologs in *Paulinella chromatophora*, cyanobacteria, Archaeplastida and two bacteria. The homologs were found by PsiBlast searches and verified for domain content by searching conserved domain database (CDD). Only best hits with *E*-values obtained from CDD equal or lower than 0.01 are indicated. The color scheme corresponds to *E*-value range. Please note that the genomes of *Paulinella* chromatophores encode homologs of Toc12,Toc34, Toc159, Tic21, Tic32 and Tic62 but those of, Toc75, Tic20, Tic22, Tic55 and Toc64 were lost. In the case of Toc64 significant homologous sequences were found; however, they did not contain a tetratrico-peptide domain, responsible for Hsp90 docking, and therefore we left this column blank.

---

the red alga *Cyanidioshyzon merolae*. The gene for Tic21 is also absent from two studied *Gleobacter* species, which often take basal position to the rest cyanobacteria in phylogenetic trees [36–38], suggesting that lack of this gene might have characterised the ancestor of all cyanobacteria.

Because no homolog to the crucial outer membrane trans-locon Toc75 was found in the two chromatophore genomes and the *Paulinella* nuclear genome [16], the discovery of homologs to its two receptors, Toc34 and Toc159, was much unexpected (Fig. 2). However, we do not consider these proteins to be components of protein import machinery without Toc75, especially taking into account their relatively low *E*-values, in both PsiBlast and CDD analyses (Fig. 2). They may be distantly related to the plastid Toc components but they do not seem to be their functional homologs. Interestingly, quite significant *E*-values obtained for some cyanobacterial Toc34/Toc159 homologs suggest that these proteins might be of cyanobacterial origin. This contrasts with the common prevailing idea that they are derived from an ancient eukaryotic GTPase, Toc159 as the result of Toc34 duplication [39–41]. A thorough phylogenetic investigation is, however, necessary to clarify the issue, which should be at present considered only a hypothesis.

We also found sequences significantly similar to *Pisum sativum* Toc64 in both *Paulinella* chromatophore genomes and bacterial genomes (Fig. 2). However, they should not be considered its true functional homologs because all of them did not have an important tetratrico-peptide (TPR) domain, responsible for Hsp90 docking, and therefore cannot fulfil Toc64 function. The gene for Toc64 encoding the TPR domain is also absent from glaucophytes, rhodophytes and the green alga *Chlamydomonas reinhardtii* but present in the other green alga *Chlorella variabilis*, some prasinophytes and land plants (Fig. 2 and data not shown). It indicates that the gene is a late acquisition in the green lineage.

In contrast to Toc64, homologs of Toc12, Tic21 and Tic32 in both *Paulinella* strains contain all the essential domains present in plant counterparts (Fig. 3). *Paulinella* Tic21 with its four well predicted α-helical transmembrane regions is suggested to create a protein conducting channel at the inner chromatophore membrane and can be involved in protein import. Such function of plant Tic21 was recently reported by Kikuchi et al. [42] and Hirabayashi et al. [43], and Lv et al. [44] proved functional compatibility of plant and cyanobacterial Tic21 by showing that the knockout mutant of *Arabidopsis tic21* was rescued by *Synechocystis* ortholog. Tic32 homologs, which possess dehydrogenase and calmodulin-binding domains, probably regulate the import via redox sensing [20]. Interestingly, Toc12 homologs, instead of N-terminal transmembrane β-barrel domains typical of plant proteins, comprise C-terminal transmembrane

α-helical regions. Toc12, is one of the four chromatophore-encoded Hsp40 (DnaJ) proteins and a putative component of molecular motor responsible for pulling imported proteins into the chromatophore stroma [20].

The new discovery of the undertaken analyses was identification of potential homologs to Tic62 in *Paulinella*. In *Pisum sativum* Tic62 consists of two characteristic regions: the N-terminal part with a highly conserved NAD(P)-binding domain and C-terminal part with biding sites for ferredoxin-NAD(P)-oxido-reductase (FNR; Fig. 3) [45]. Similar N-terminal organization was shown for *Paulinella* Tic62 homologs, but the C-terminal region with FNR interacting repeats is missing from both *Paulinella* proteins (Fig. 3). However, the C-terminal region, which apparently evolved only in vascular plants, is not necessary for binding to the Tic translocon/inner membrane. The region responsible for mediating binding to the Tic complex is localized in the central part of *Pisum sativum* Tic62 [46], whereas the C-terminal region allows for specific and strong binding of ferredoxin-NAD(P)-oxido-reductase molecules, especially when Tic62 is present at the stroma lamellae of the thylakoid membrane [47]. The reversible interaction with Tic complex/inner membrane is supposed to be mediated by some hydrophobic contacts [48]. One or two transmembrane domains were proposed in *Pisum sativum* Tic62 by Küchler et al. [45]. In agreement with that, we predicted two such domains using TopPred [30] and TMpred [31]. Interestingly, we also identified one potential transmembrane domains in each of *Paulinella* Tic62 proteins using these two programs (Fig. 3).

We suggest that *Paulinella* Tic62 homolog is a new component of the simplified but fully-functional Tic-like translocon at the inner chromatophore membrane, composed, beside Tic62, of Tic21, Tic32 and Toc12 (Hsp40; Fig. 4). The fact that these proteins possess all the essential domains to fulfil the task further strengthens the reliability of a Tic complex at the inner chromatophore membrane. Because Tic55, the third redox sensing protein [49–51], has not been discovered in the chromatophore genome in previous [17] and present study, as well as in the *Paulinella* nuclear genome [16], we suggest Tic32 and Tic62 to perform the task of redox sensing on their own or with some unknown partner.

An interesting argument for functioning of a simplified Tic system provided Kikuchi et al. [52]. They suggested that the primordial Archaeplastida Tic complex was probably based on Tic20. Similarly to Tic21 and mitochondrial translocon at the inner membrane Tim23/Tim17, Tic20 contains four well predicted α-helical transmembrane domains capable of forming a protein conducting channel, and therefore together with Tic21 and Tim23/Tim17 represent a good example of parallel evolution of import functions [42]. Since a Tic20-based system might have functioned

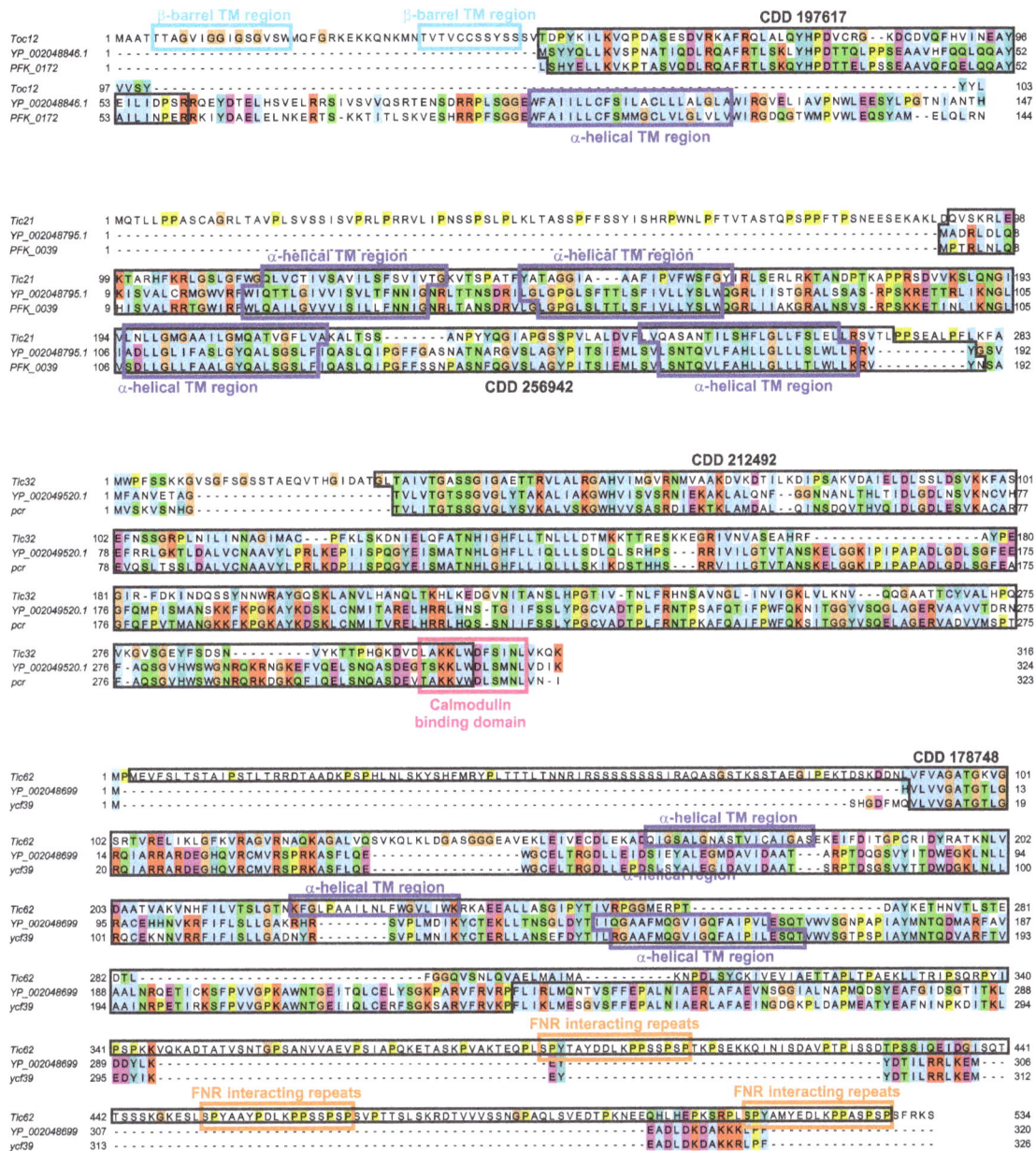

Fig. 3 Alignment of Toc12, Tic21, Tic32 and Tic62 from *Pisum sativum* and two strains of *Paulinella chromatophora* CCAC0185 and FK01. The range of domains characteristic of a given protein was indicated according to CDD database prediction. A characteristic motif of calmodulin binding domain in Tic32 was also marked [20], as well as NADH(P) binding domain (CDD 257784) and FNR-interacting repeats in Tic62 [46]. The N-terminal transmembrane β-barrel regions were predicted with BOCTOPUS [29], α-helical transmembrane regions with TopPred [30] and TMpred [31]. The average range of the regions was calculated based on the two latter predictions.

in the ancestor of Archaeplastida, it is easy to imagine the possibility of analogous system based on Tic21 in the case of *Paulinella*. In support of this protein import via Tic21 was experimentally proved in plants [42,43]. Although additional core translocon subunits of eukaryotic origin, such as Tic214, Tic100 and Tic56 were added to Tic20 in the 'green' lineage, they are missing from basal Archaeplastida, glaucophytes and red algae [52]. This suggests that a simple Tic apparatus is responsible for protein import into primary plastids of glaucophytes and red algae and probably *Paulinella chromatophora* as well. Kikuchi et al. [52] also showed that eukaryote-derived Tic110, previously considered the main translocation pore, is not a component of the complex consisting of Tic20, Tic214, Tic100 and Tic56, but plays a role of a scaffold for stromal molecular chaperons at a later stage during protein import [49,50].

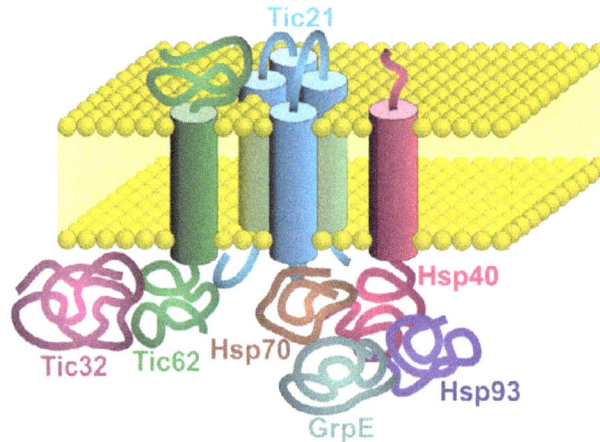

Fig. 4 Model for a protein translocon in inner membrane of *Paulinella* photosynthetic chromatophores. The translocon is a simplified Tic-like apparatus and is responsible for final import step of nuclear-encoded proteins across the inner chromatophore membrane. It is composed of Tic21 (protein-conducting channel), Tic32 and Tic62 (calcium and redox-sensing regulatory proteins) as well as a molecular motor responsible for pulling imported proteins into the organelle stroma. The latter could consist of Hsp93, Hsp70, Hsp40 (Toc12) and GrpE.

Conclusion

Our studies reveal that chromatophore genomes from both *Paulinella* strains encode the same set of translocons that could potentially create a simplified but fully-functional Tic-like translocon at the inner chromatophore membranes. Moreover, we have discovered a new putative Tic component, namely Tic62, a redox sensor protein not identified in previous comparative studies of *Paulinella* translocons [17]. The common maintenance of the same set of Toc/Tic proteins indicates a similar import mechanism in the two investigated *Paulinella* strains and supports the proposed model. We also suggest a possibility that Toc34 and Toc159, GTPases of presumed eukaryotic origin [39–41], might in fact be derived from cyanobacterial proteins.

Acknowledgments
We are very grateful to the Reviewers and the Editor for their excellent comments and insightful remarks which significantly improved the paper. This work was supported by Polish National Science Centre grant No. 2011/01/N/NZ8/00150 and Wrocław Centre for Networking and Supercomputing grant No. 307 to PG and PM.

Authors' contributions
The following declarations about authors' contributions to the research have been made: conducting analyses and writing the manuscript: PG, PM.

Competing interests
No competing interests have been declared.

References

1. Yoon HS, Hackett JD, Ciniglia C, Pinto G, Bhattacharya D. A molecular timeline for the origin of photosynthetic eukaryotes. Mol Biol Evol. 2004;21:809–818.

2. Deusch O, Landan G, Roettger M, Gruenheit N, Kowallik KV, Allen JF, et al. Genes of cyanobacterial origin in plant nuclear genomes point to a heterocyst-forming plastid ancestor. Mol Biol Evol. 2008;25:748–761.

3. Cavalier-Smith T, Lee JJ. Protozoa as hosts for endosymbioses and the conversion of symbionts into organelles. J Eukaryot Microbiol. 1985;32:376–379.

4. Bodył A, Mackiewicz P, Stiller JW. Early steps in plastid evolution: current ideas and controversies. Bioessays. 2009;31:1219–1232.

5. Gross J, Bhattacharya D. Mitochondrial and plastid evolution in eukaryotes: an outsiders' perspective. Nat Rev Genet. 2009;10:495–505.

6. Bodył A, Mackiewicz P, Gagat P. Organelle evolution: *Paulinella* breaks a paradigm. Curr Biol. 2012;22:R304–306.

7. Bhattacharya D, Helmchen T, Melkonian M. Molecular evolutionary analyses of nuclear-encoded small subunit ribosomal RNA identify an independent rhizopod lineage containing the Euglyphina and the Chlorarachniophyta. J Eukaryot Microbiol. 1995;42:65–69.

8. Yoon HS, Nakayama T, Reyes-Prieto A, Andersen RA, Boo SM, Ishida K, et al. A single origin of the photosynthetic organelle in different *Paulinella* lineages. BMC Evol Biol. 2009;9:98.

9. Marin B, Nowack EC, Melkonian M. A plastid in the making: evidence for a second primary endosymbiosis. Protist. 2005;156:425–432.

10. Yoon HS, Reyes-Prieto A, Melkonian M, Bhattacharya D. Minimal plastid genome evolution in the *Paulinella* endosymbiont. Curr Biol. 2006;16:R670–672.

11. Nowack EC, Melkonian M, Glockner G. Chromatophore genome sequence of *Paulinella* sheds light on acquisition of photosynthesis by eukaryotes. Curr Biol. 2008;18:410–418.

12. Kies L. Electron microscopical investigations on *Paulinella chromatophora* Lauterborn, a thecamoeba containing blue-green endosymbionts (Cyanelles). Protoplasma. 1974;80:69–89.

13. Kies L, Kremer BP. Function of cyanelles in the tecamoeba *Paulinella chromatophora*. Naturewissenschaften. 1979;66:578–579.

14. Reyes-Prieto A, Yoon HS, Moustafa A, Yang EC, Andersen RA, Boo SM, et al. Differential gene retention in plastids of common recent origin. Mol Biol Evol. 2010;27:1530–1537.

15. Nakayama T, Ishida KI. Another acquisition of a primary photosynthetic organelle is underway in *Paulinella chromatophora*. Curr Biol. 2009;19:R284–R285.

16. Nowack ECM, Vogel H, Groth M, Grossman AR, Melkonian M, Glöckner G. Endosymbiotic gene transfer and transcriptional regulation of transferred genes in *Paulinella chromatophora*. Mol Biol Evol. 2011;28:407–422.

17. Bodył A, Mackiewicz P, Stiller JW. Comparative genomic studies suggest that the cyanobacterial endosymbionts of the amoeba *Paulinella chromatophora* possess an import apparatus for nuclear-encoded proteins. Plant Biol. 2010;12:639–649.

18. Mackiewicz P, Bodył A. A hypothesis for import of the nuclear-encoded PsaE protein of *Paulinella chromatophora* (Cercozoa, Rhizaria) into its cyanobacterial endosymbionts/plastids via the endomembrane system. J Phycol 2010;46:847–859.

19. Mackiewicz P, Bodył A, Gagat P. Possible import routes of proteins into the cyanobacterial endosymbionts/plastids of *Paulinella chromatophora*. Theory Biosci. 2012;131:1–18.

20. Mackiewicz P, Bodył A, Gagat P. Protein import into the photosynthetic organelles of and its implications for primary plastid endosymbiosis. Symbiosis. 2012;58:99–107.

21. Nowack EC, Grossman AR. Trafficking of protein into the recently established photosynthetic organelles of *Paulinella chromatophora*. Proc Natl Acad Sci USA. 2012;109:5340–5345.

22. Qiu H, Yang EC, Bhattacharya D, Yoon HS. Ancient gene paralogy may mislead inference of plastid phylogeny. Mol Biol Evol. 2012;29:3333–3343.

23. Benson DA, Cavanaugh M, Clark K, Karsch-Mizrachi I, Lipman DJ, Ostell J, et al. GenBank. Nucleic Acids Res. 2013;41:D36–D42.

24. Altschul SF, Madden TL, Schäffer AA, Zhang J, Zhang Z, Miller W, et al. Gapped BLAST and PSI-BLAST: a new generation of protein database search programs. Nucleic Acids Res. 1997;25:3389–3402.

25. Kalanon M, McFadden GI. The chloroplast protein translocation complexes of *Chlamydomonas reinhardtii*: a bioinformatic comparison of Toc and Tic components in plants, green algae and red algae. Genetics. 2008;179:95–112.

26. Marchler-Bauer A, Anderson JB, Cherukuri PF, DeWeese-Scott C, Geer LY, Gwadz M, et al. CDD: a conserved domain database for protein classification. Nucleic Acids Res. 2005;33:D192–196.

27. Di Tommaso P, Moretti S, Xenarios I, Orobitg M, Montanyola A, Chang JM, et al. T-Coffee: a web server for the multiple sequence alignment of protein and RNA sequences using structural information and homology extension. Nucleic Acids Res. 2011;39:W13–17.

28. Waterhouse AM, Procter JB, Martin DM, Clamp M, Barton GJ. Jalview version 2 – a multiple sequence alignment editor and analysis workbench. Bioinformatics. 2009;25:1189–1191.

29. Hayat S, Elofsson A. BOCTOPUS: improved topology prediction of transmembrane β barrel proteins. Bioinformatics. 2012;28:516–522.

30. Claros MG, von Heijne G. TopPred II: an improved software for membrane protein structure predictions. Comput Appl Biosci. 1994;10:685-686.

31. Hofmann. TMbase – a database of membrane spanning proteins segments. Biol Chem Hoppe Seyler. 1993;374.

32. Jobb G, von Haeseler A, Strimmer K. TREEFINDER: a powerful graphical analysis environment for molecular phylogenetics. BMC Evol Biol. 2004;4:18.

33. Ronquist F, Teslenko M, van der Mark P, Ayres DL, Darling A, Hohna S, et al. MrBayes 3.2: efficient Bayesian phylogenetic inference and model choice across a large model space. Syst Biol. 2012;61:539–542.

34. Rogers JS, Swofford DL. A fast method for approximating maximum likelihoods of phylogenetic trees from nucleotide sequences. Syst Biol. 1998;47:77–89.

35. Huelsenbeck JP, Larget B, Alfaro ME. Bayesian phylogenetic model selection using reversible jump Markov chain Monte Carlo. Mol Biol Evol. 2004;21:1123–1133.

36. Shih PM, Wu D, Latifi A, Axen SD, Fewer DP, Talla E, et al. Improving the coverage of the cyanobacterial phylum using diversity-driven genome sequencing. Proc Natl Acad Sci USA. 2013;110:1053–1058.

37. Criscuolo A, Gribaldo S. Large-scale phylogenomic analyses indicate a deep origin of primary plastids within cyanobacteria. Mol Biol Evol. 2011;28:3019–3032.

38. Ochoa de Alda JAG, Esteban R, Diago ML, Houmard J. The plastid ancestor originated among one of the major cyanobacterial lineages. Nat Commun. 2014;5.

39. Reumann S, Inoue K, Keegstra K. Evolution of the general protein import pathway of plastids (review). Mol Membr Biol. 2005;22:73–86.

40. Gross J, Bhattacharya D. Revaluating the evolution of the Toc and Tic protein translocons. Trends Plant Sci. 2009;14:13–20.

41. Hernandez Torres J, Maldonado MA, Chomilier J. Tandem duplications of a degenerated GTP-binding domain at the origin of GTPase receptors Toc159 and thylakoidal SRP. Biochem Biophys Res Commun. 2007;364:325–331.

42. Kikuchi S, Oishi M, Hirabayashi Y, Lee DW, Hwang I, Nakai M. A 1-megadalton translocation complex containing Tic20 and Tic21 mediates chloroplast protein import at the inner envelope membrane. Plant Cell. 2009;21:1781–1797.

43. Hirabayashi Y, Kikuchi S, Oishi M, Nakai M. In vivo studies on the roles of two closely related *Arabidopsis* Tic20 proteins, AtTic20-I and AtTic20-IV. Plant Cell Physiol. 2011;52:469–478.

44. Lv HX, Guo GQ, Yang ZN. Translocons on the inner and outer envelopes of chloroplasts share similar evolutionary origin in *Arabidopsis thaliana*. J Evol Biol. 2009;22:1418–1428.

45. Kuchler M, Decker S, Hormann F, Soll J, Heins L. Protein import into chloroplasts involves redox-regulated proteins. EMBO J. 2002;21:6136–6145.

46. Stengel A, Benz P, Balsera M, Soll J, Bolter B. TIC62 redox-regulated translocon composition and dynamics. J Biol Chem. 2008;283:6656–6667.

47. Benz JP, Stengel A, Lintala M, Lee YH, Weber A, Philippar K, et al. Arabidopsis Tic62 and ferredoxin-NADP(H) oxidoreductase form light-regulated complexes that are integrated into the chloroplast redox poise. Plant Cell. 2009;21:3965–3983.

48. Balsera M, Stengel A, Soll J, Bölter B. Tic62: a protein family from metabolism to protein translocation. BMC Evol Biol. 2007;7:43.

49. Li HM, Chiu CC. Protein transport into chloroplasts. Annu Rev Plant Biol. 2010;61:157–180.

50. Shi LX, Theg SM. The chloroplast protein import system: from algae to trees. Biochim Biophys Acta. 2013;1833:314–331.

51. Stengel A, Benz JP, Soll J, Bolter B. Redox-regulation of protein import into chloroplasts and mitochondria: similarities and differences. Plant Signal Behav. 2010;5:105–109.

52. Kikuchi S, Bedard J, Hirano M, Hirabayashi Y, Oishi M, Imai M, et al. Uncovering the protein translocon at the chloroplast inner envelope membrane. Science. 2013;339:571–574.

22

Contrasting patterns in the evolution of the Rab GTPase family in Archaeplastida

Romana Petrželková, Marek Eliáš*

Department of Biology and Ecology, Faculty of Science, University of Ostrava, Chittussiho 10, 710 00 Ostrava, Czech Republic

Abstract

Rab GTPases are a vast group of proteins serving a role of master regulators in membrane trafficking in eukaryotes. Previous studies delineated some 23 Rab and Rab-like paralogs ancestral for eukaryotes and mapped their current phylogenetic distribution, but the analyses relied on a limited sampling of the eukaryotic diversity. Taking advantage of the recent growth of genome and transcriptome resources for phylogenetically diverse plants and algae, we reanalyzed the evolution of the Rab family in eukaryotes with the primary plastid, collectively constituting the presumably monophyletic supergroup Archaeplastida. Our most important novel findings are as follows: (*i*) the ancestral set of Rabs in Archaeplastida included not only the paralogs Rab1, Rab2, Rab5, Rab6, Rab7, Rab8, Rab11, Rab18, Rab23, Rab24, Rab28, IFT27, and RTW (=Rabl2), as suggested previously, but also Rab14 and Rab34, because Rab14 exists in glaucophytes and Rab34 is present in glaucophytes and some green algae; (*ii*) except in embryophytes, Rab gene duplications have been rare in Archaeplastida. Most notable is the independent emergence of divergent, possibly functionally novel, in-paralogs of Rab1 and Rab11 in several archaeplastidial lineages; (*iii*) recurrent gene losses have been a significant factor shaping Rab gene complements in archaeplastidial species; for example, the Rab21 paralog was lost at least six times independently within Archaeplastida, once in the lineage leading to the "core" eudicots; (*iv*) while the glaucophyte *Cyanophora paradoxa* has retained the highest number of ancestral Rab paralogs among all archaeplastidial species studied so far, rhodophytes underwent an extreme reduction of the Rab gene set along their stem lineage, resulting in only six paralogs (Rab1, Rab2, Rab6, Rab7, Rab11, and Rab18) present in modern red algae. Especially notable is the absence of Rab5, a virtually universal paralog essential for the endocytic pathway, suggesting that endocytosis has been highly reduced or rewired in rhodophytes.

Keywords: Archaeplastida; Chloroplastida; endocytosis; evolution; Glaucophyta; Rab GTPases; Rhodophyta

Introduction

The family of Rab GTPases, constituting the largest subgroup of the Ras GTPase superfamily, is one of the hallmarks of the eukaryotic cell. Rabs serve as central regulators of membrane trafficking and are involved in maintaining identity of the various compartments of the membrane system and in ensuring specificity of the transport events between the compartments [1]. Our understanding of the Rab function derives primarily from studies on a few selected models systems, primarily mammalian cells and the yeast *Saccharomyces cerevisiae*, but more limited knowledge exists also for other species representing different distantly related eukaryotic lineages, for example the kinetoplastid *Trypanosoma brucei*, the ciliate *Tetrahymena thermophila*, and the plant *Arabidopsis thaliana* [2–4]. One of the puzzling aspects of the biology of the Rab family is the fact that the

total number of Rab genes may differ profoundly between different species: whereas some eukaryotic cells are able to secure their proper functioning with less than ten Rabs, other species exhibit tens or even hundreds of Rab paralogs [3,5].

Comparative genomic and phylogenetic studies revealed that a large number of distinct Rab paralogs have been established early in the evolution of eukaryotes [3,5,6]. Reconstructions of the Rab complement in the deepest point of the phylogeny of extant eukaryotes, i.e. the last eukaryotic common ancestor (LECA), suggest the existence of over 20 paralogs [5,6]. This is consistent with the presence of a highly elaborate endomembrane system in the LECA, in line with the emerging view of the LECA as a fully fledged and surprisingly complex eukaryotic cell [7,8].

However, the exact number of ancestral eukaryotic Rab paralogs still remains uncertain due to three factors. Firstly, classification of some Rab-like proteins lacking the C-terminal tail with a prenylation motif, such as RTW (=Rabl2; [6,9]) or IFT27 (=Rabl4; [6,10]), as bona fide Rab family members is controversial due to the poor resolution of Ras superfamily phylogenies. Second, the inference on

---

* Corresponding author. Email: marek.elias@osu.cz

Handling Editor: Andrzej Bodył

the ancestral Rab complement depends on the position of the root of the eukaryotic phylogeny; several competing hypothesis have been recently discussed in the literature, but no consensus currently exists on where the root actually lies (see [11]). Third, the accuracy of the reconstruction of the Rab complement in the LECA significantly depends on the sampling of the eukaryotic phylogenetic diversity. Indeed, the analyses published so far [5,6,12] relied on only a limited number of genome sequences or only transcriptomic (expressed sequence tag – EST) data for many crucial eukaryotic lineages, while other lineages have not been studied at all. Recent progress in DNA sequencing technologies has enabled to dramatically improve the sampling of the eukaryotic phylogenetic diversity by full genome or deep transcriptome sequencing, although important gaps still persist [13,14].

Archaeplastida, often called Plantae, are a major eukaryotic supergroup defined by the synapomorphic presence of a primary plastid, i.e. a direct product of the original endosymbiotic acquisition of a cyanobacterial ancestor of eukaryotic plastids [15]. Such a plastid is found in three living eukaryotic lineages – glaucophytes (Glaucophyta or Glaucocystophyta), rhodophytes (Rhodophyta, Rhodophyceae, or Rhodoplantae), and the "green lineage" comprising green algae and their descendants land plants (Chloroplastida or Viridiplantae) [16,17]. In the most parsimonious scenario, these three lineages constitute a monophyletic grouping to the exclusion of other eukaryotes. However, phylogenomic analyses have so far failed to provide conclusive evidence for this hypothesis, because some other lineages, specifically haptophytes and/or cryptists, tend to disrupt the monophyly of the three archaeplastidial groups in some analyses (see, e.g., [18,19]). This would suggest a secondary loss of the primary plastid from some eukaryotes. Regardless these controversies, the monophyly of the Archaeplastida sensu Adl et al. [15] remains the preferred working hypothesis that will also be assumed in this study.

Our knowledge about the cell biology of the different archaeplastidial lineages is extremely uneven and biased towards Chloroplastida, particularly towards land plants (embryophytes) and the model green alga *Chlamydomonas reinhardtii*. In rhodophytes, *Cyanidioschyzon merolae* representing the basal lineage of red algae (Cyanidiophyceae) has been established as a highly useful model system for addressing diverse cell biological questions and it happened to become the first alga with a sequenced nuclear genome [20]. However, it is questionable to what extent this unicellular extremophilic species may be representative for red algae as a whole. *Cyanophora paradoxa* is a glaucophyte that has been used as a model organism for the whole group [21], but the knowledge on this species lags far behind the model systems of the other two archaeplastidial lineages.

However, a lot of key insights into the cell biology, biochemistry or physiology of any organisms can be obtained by computational analyses of their genetic blueprints. Fortunately, recent years have witnessed a rapid accumulation of genomic or transcriptomic data from both red algae and glaucophytes. These include draft genome sequences of the phylogenetically diverse rhodophytes *Chondrus crispus* [22], *Porphyridium purpureum* [23], *Galdieria sulphuraria* [24], and *Pyropia yezoensis* [25], and of the glaucophyte

C. paradoxa [26]. Transcriptomes of an even broader set of red algal and glaucophyte species have been deeply sequenced thanks to the Marine Microbial Eukaryote Transcriptome Sequencing Project (MMETSP) [14]. Hence, we now have a unique opportunity to quickly improve our knowledge about the molecular underpinnings of red algal and glaucophyte cells by exploring the wealth of these data by computational analyses.

Substantial effort has been put into studying Rab GTPases in model plant species, primarily *A. thaliana*, which enabled to demonstrate that the plant Rab complement exhibits at the functional level both features shared with other eukaryotes as well as novel, plant specific features [2,27]. Comparative genomic and phylogenetic studies have additionally shown that land plants have retained only a subset of the presumed ancestral eukaryotic Rab paralogs, but secondarily expanded the Rab family by extensive gene duplications [6,28,29]. Much less is known about Rabs in algae. Except occasional early studies (e.g. [30]), attempts to functionally characterize algal Rab proteins are virtually lacking. Phylogenetic studies on algal Rabs have been also limited. Concerning Archaeplastida, the most comprehensive analysis published to date [6] included data from only four green algal genomes (*C. reinhardtii*, *Chlorella variabilis*, *Ostreococcus lucimarinus* and *Micromonas pusilla* CCMP1545) and two red algal genomes (*C. merolae*, *G. sulphuraria*), while only partial transcriptomic data from classical (Sanger sequencing-based) EST surveys were available for glaucophytes (*C. paradoxa* and *Glaucocystis nostochinearum*).

The aim of this study is to improve our knowledge about the diversity and evolution of Rab GTPases in Archaeplastida, focusing specifically on algal lineages. A significantly expanded sampling of the archaeplastidial diversity offered many new important insights helping us to refine the view of the cellular evolution in this highly significant assemblage of eukaryotic organisms.

Material and methods

Species and sequence data analyzed

The sequence dataset used in this study resulted as an expansion of the dataset of Rab sequences analyzed previously by Elias et al. [6], which included, in addition to the sequences from algal species mentioned above, Rabs from the eudicot *A. thaliana* and the lycophyte *Selaginella moellendorffii*. This dataset was revised and expanded using newly available data. We replaced the incomplete representation of Rab sequences from *C. paradoxa* as derived from an EST survey by a (presumably) complete set of sequences deduced from a recently reported draft genome assembly. We added sequences from a deeply sequenced transcriptome of the glaucophyte *Cyanoptyche gloeocystis* and we instead removed the extremely fragmentary set of sequences from the glaucophyte *G. nostochinearum* (which was anyway not informative beyond what was implied by the *C. paradoxa* and *C. gloeocystis* datasets). We also tested two independent transcriptome assemblies from the glaucophyte *Gloeochaete wittrockiana* made available by the MMESTP project (MMETSP0308 and MMETSP1089;

http://camera.calit2.net/mmetsp/list.php), but it turned out that both assemblies are heavily contaminated by another organism, most likely an amoebozoan (data not shown), perhaps due to contamination of the original culture (SAG 46.84). We therefore omitted *G. wittrockiana* from our analyses. We expanded the sampling of the rhodophyte diversity (only the class Cyanidiophyceae had been represented in the original dataset) by extracting Rab sequences from three newly released genome sequences representing three additional rhodophyte classes (Florideophyceae, Bangiophyceae, Porphyridiophyceae). Transcriptome assemblies additionally allowed us to include one representative of each of the remaining three classes (Compsopogonophyceae, Stylonematophyceae, Rhodellophyceae). Although transcriptome assemblies have recently become available for a number of phylogenetically diverse green algae, we decided not to include them in our analysis due to frequent contamination issues and to keep the size of the dataset within reasonable limits (we nevertheless used the assemblies for certain targeted analyses, see below). However, to improve the coverage of the Chloroplastida group in our analysis, we added Rab sequences from four green algal genomes and two land plant genomes. All species systematically analyzed are listed in Tab. 1. Links to the sources of the sequence data are available in Tab. S1.

Extraction and curation of Rab sequences

We used the program BLAST and its appropriate variants (blastp, tblastn, blastn) [31] to identify sequences of candidate Rab genes and proteins in the genome or transcriptome assemblies or the corresponding protein sequence predictions. The identified sequences were BLASTed against our local database of annotated Ras superfamily GTPases to discriminate genuine Rabs from other subgroups of the superfamily. As mentioned above, the delimitation of the Rab family is not completely settled; to be consistent with our previous study [6] we included in the analysis two Rab-like paralogs (RTW and IFT27) and the GTPase RAN. The latter is a universal eukaryotic gene found in every species investigated so far, so it provided an internal control of the completeness of the genome or transcriptome resource for a given species. Genome sequences were checked by tblastn for the possible presence of genes missing in the respective predicted protein sequence sets. Transcriptome data for the same species (EST databases or TSA – transcript shotgun assemblies) were also checked to identify possible genes missing in the draft genome assemblies. If needed, partial gene sequences due to gaps in the genome assembly were combined with the corresponding transcript sequence to obtain a complete coding sequence of the gene. Some Rab genes in the red alga *C. crispus* and in *Oryza sativa* could be identified in contigs or scaffolds not included in the most recent genome release, but their authenticity was indisputable. Existing protein sequence predictions were carefully checked by inspecting alignments to related sequences and all suspicious cases were reevaluated by investigating the respective nucleotide sequence, in many cases leading to a revision of the gene model and the resulting protein sequence. In species with only TSA and no genome sequence available, some Rab genes were represented by incomplete

transcript sequences, but for many of them a complete or at least longer coding sequence could be obtained by iterative addition of matching raw Illumina reads in the Short sequence archive (http://www.ncbi.nlm.nih.gov/sra/). Two transcripts from the red alga *Rhodella maculata* remained too incomplete to be included in phylogenetic analyses, but their identity as Rab2 and Rab7 was indisputable from BLAST comparisons with Rabs from other rhodophytes. A list of all sequences analyzed in this study, together with the corresponding accession numbers or sequence identifiers, is available in Tab. S1. Revised or newly predicted protein sequences are provided in a separate supplementary file.

Alignment and phylogenetic analyses

The newly identified Rab protein sequences were divided into groups each representing a different ancestral paralog (the assignment of individual sequences to the paralogs was in virtually all cases straightforward based on BLAST comparisons) and for each group multiple alignment was created using MAFFT (version 7, default parameters; http://mafft.cbrc.jp/alignment/server/ [32]). Each aligned group was then added manually to a large master alignment built for our previous study [6], using previously aligned sequences of the same paralog as a guide. From the expanded master alignment subsets of sequences were selected to create desired smaller alignments for phylogenetic analyses. We applied the same mask as before [6] to remove columns where the alignment was too uncertain. Phylogenetic tress were inferred using the maximum likelihood (ML) method as implemented in the program RAxML-HPC BlackBox (8.0.24) [33] accessible at the CIPRES Science Gateway (https://www.phylo.org/portal2 [34]). The substitution model employed was LG+Γ, branch support was assessed by the rapid bootstrapping algorithm that is an inherent part of the best tree search strategy of RAxML. To test the robustness of the tree topologies we also employed ML inference using the program PhyML-CAT and the empirical profile mixture model C20 [35] with gamma correction (four categories) of the among-site rate heterogeneity; Chi2-based parametric branch support was calculated using the approximate likelihood ratio test implemented in PhyML (-b -2 option). Trees were visualized using iTOL (http://itol.embl.de/ [36]) and rendered for publication using a graphical editor.

Results and discussion

Virtually all Rab genes in Archaeplastida can be readily assigned to known ancestral Rab paralogs

We relied on complete or high-quality draft genome sequences and/or deeply sequenced transcriptomes to build a manually curated set of Rab family protein sequences from 22 species of the Archaeplastida supergroup: two representatives of Glaucophyta, eight members of Rhodophyta, and twelve members of Chloroplastida (Tab. 1). While the sampling for glaucophytes, here represented only by one species with a draft genome sequence and one species with a deeply sequenced transcriptome, remains rather limited, the phylogenetic diversity of Rhodophyta and Chloroplastida is covered more comprehensively, which gives us an

Tab. 1 Rab GTPases in Archaeplastida.

		Rab1	Rab2	Rab5	Rab6	Rab7	Rab8	Rab11	Rab14	Rab18	Rab21	Rab23	Rab24	Rab28	Rab34	IFT27	RTW
Glaucophyta	*Cyanophora paradoxa*	4	1	1	1	1	1	1	1	1		1	1	1	1	1	1
	Cyanoptyche gloeocystis (T)	5	1	?	1	1	1	4	?	?		1(2?)	?	1(2?)	1	?	1
Rhodophyta	*Galdieria sulphuraria*	1	1		1	1	1	2									
	Cyanidioschyzon merolae	1	1		1	1	1	1		1							
	Chondrus crispus	1	1		1	1	1	2		1							
	Pyropia yezoensis	1			1	1	1	1		1							
	Rhodella maculata (T)	1	1		1	1	1	1		?							
	Porphyridium purpureum	2	1		1	1	1	1		1							
	Rhodosorus marinus (T)	2	1		1	1	1	2		1							
	Compsopogon coeruleum (T)	2	1		1	1	1	1		1							
Chloroplastida	*Chlamydomonas reinhardtii*	1	1	1	1	1	1	1		3		1		1		1	1
	Volvox carteri	1	1	?	1	1	1	1		3		1		1		1	1
	Coccomyxa subellipsoidea	1	1	2 (1xF1)	1	1	1	1		1	1	1	1				
	Chlorella variabilis	1	1	2 (1xF1)	1	2	1	2		1			1				
	Ostreococcus lucimarinus	1	1	2 (1xF1)	1	1	1	1		1	1						
	Micromonas pusilla CCMP1545	1	1	1	1	1	1	1		1		1		1		1	
	Micromonas pusilla NOUM17	1	1	2 (1xF1)	1	1	1	1		1		1				1	1
	Klebsormidium flaccidum	2	1	2 (1xF1)	1	1	1	1		1	1	1					
	Physcomitrella patens	3	4(+1)	5 (3xF1)	7	4(+1)	5	12		5(+1)	3(+2)	1					
	Selaginella moellendorffii	2	1	2 (1xF1)	1	1	3	7			1	1					
	Arabidopsis thaliana	4	3	3 (1xF1)	5	8	5	25(+1)		3							
	Oryza sativa	5	3(+2)	5 (2xF1)	2	4	4	17		3	1						

Distribution of Rab GTPases, including the Rab-like proteins IFT27 (=Rabl4) and RTW (=Rabl2) in 22 selected archaeplastidial species. For species marked with "(T)" only a deeply sequenced transcriptome (no genome sequence) is available. For each ancestral paralog the number of in-paralogs in each species is given. The problematic Rab gene from *Cyanoptyche gloeocystis*, Rab1L (see the text) is here counted as one of the Rab1 in-paralogs. For Rab23 and Rab28 in *C. gloeocystis*, two highly similar variants are recorded in the transcriptome and it is unclear whether they represent two different genes or allelic variants. In *Physcomitrella patens*, *Arabidopsis thaliana*, and *Oryza sativa* apparent pseudogenes are present; their number for the respective ancestral paralog is indicated in parentheses. In species represented only by a transcriptome assembly question mark indicates a suspicious absence of a gene present in related species. For the Rab5 paralog in Chloroplastida the table indicates the total number of in-paralogs and in parentheses how many of them correspond to the RabF1 form.

opportunity to infer conclusions with a potentially general validity for the respective archaeplastidial lineages.

Each Rab sequence was initially assigned to one of the ancestral paralogs (as defined by Elias et al. [6]) based on BLAST similarity scores to previously annotated sequences in our in-house database of Ras superfamily GTPases. The assignment was unambiguous for virtually all sequences, except one gene from the glaucophyte *C. gloeocystis* (eventually labelled Rab1L; Tab. S1, Fig. S1), which gave similar scores to members of the closely related Rab1 and Rab8 paralogs. One additional sequence from *C. variabilis* (Rab7L; Tab. S1) was most similar in BLAST-based comparisons to various Rab7 proteins, but was so divergent (maximal identity was only around 30%) that it could not be included in the phylogenetic analysis of Rab sequences. However, the fact that the protein is so far specific for *C. variabilis* and is consistently most similar to Rab7 indicates that it is most likely an extremely divergent lineage-specific offshoot of the Rab7 paralog (*C. variabilis* additionally harbours a canonical Rab7 gene; Tab. 1). The ability to readily annotate on the basis of sequence similarity most of the archaeplastidial Rab sequences indicates that most of them have been evolving rather slowly. This contrasts with the situation in some other eukaryotic groups, e.g. Amoebozoa or Ciliata, which tend to accumulate large numbers of highly divergent Rab paralogs that are often difficult to classify even by using phylogenetic analyses [6].

To corroborate the initial annotation we performed a phylogenetic analysis of a multiple sequence alignment comprising not only the archaeplastidial sequences, but also Rab sequences from other eukaryotic groups representing ancestral eukaryotic paralogs that were not found in Archaeplastida in our previous study [6]. The ML tree inferred using the program RAxML is displayed in Fig. S1; a tree obtained using the program PhyML and a different substitution model (see "Material and methods") was topologically different in many regions, but agreed with the RAxML tree in all branches relevant for defining the main ancestral Rab paralogs (data not shown). As observed in previous phylogenetic reconstructions of the Rab family, the topology of the tree has many poorly supported branches at all levels of the phylogenetic depth and is not free from obvious topological artefacts, including the paraphyly of the Rab1 paralog due to the Rab8 paralog nested within (encountered also in a previous analysis [6]) and the paraphyly of the Rab2 paralog due to a clade of Rab4 and Rab14 nested within. Nevertheless, the assignment of virtually all archaeplastidial Rab sequences to different ancestral paralogs can be easily deduced from the tree and is compatible with the BLAST-based assignment. The only somewhat problematic sequence is the *C. gloeocystis* Rab1L gene mentioned above. In our tree its affiliation to the group comprising Rab1 and Rab8 paralogs indicated by BLAST is confirmed, suggesting that it is most likely a lineage-specific divergent in-paralog of Rab1 and Rab8. We consider the origin from Rab1 as more probable, since Rab1 shows a general tendency to recurrently duplicate in Arachaeplastida (see below) while no Rab8 duplications were found for algal taxa in Archaeplastida, and because a focused phylogenetic analysis of Rab1 and Rab8 sequences using PhyML-CAT and the empirical profile mixture model

C20 suggested that it may actually be an extremely divergent relative of some glaucophyte Rab1 sequences (see below).

The expanded sampling for the first time documents the presence of Rab14 and Rab34 in Archaeplastida

The main novel finding of the analyses described above is the revision of the number of Rab paralogs inferred to have been present in the last common ancestor (LCA) of Archaeplastida. This is primarily thanks to the improved representation of the glaucophyte gene complement, which revealed the presence of two paralogs so far unknown from any archaeplastidial species – Rab14 and Rab34. Hence, the set of Rab paralogs that were apparently present in the archaeplastidial LCA now includes the following items: Rab1, Rab2, Rab5, Rab6, Rab7, Rab8, Rab11, Rab14, Rab18, Rab21, Rab23, Rab24, Rab28, Rab34, IFT27 and RTW. Identification of Rab14 and Rab34 in Archaeplastida is significant also for the general understanding of the evolution of the Rab family in eukaryotes, since it strengthens the notion that these two paralogs were present already in the LECA. This was previously uncertain, since with the hypotheses on the position of the root of the eukaryote phylogeny placed between Archaeplastida and remaining eukaryotes (see [37]), it was still possible that the absence of Rab14 and Rab34 in Archaeplastida is a primitive state.

For the remaining Rab paralogs thought to be present in the LECA [6], i.e. Rab4, Rab20, Rab22, Rab32A, Rab32B, Rab50, and RabTitan, no candidate orthologs were found in any archaeplastidial species systematically analyzed here. The most parsimonious scenario concerning the fate of these paralogs is that they had been lost before the archaeplastidial LCA (Fig. 1).

However, caution must be taken when inferences on early gene losses are made from a limited sampling of extant taxa. As a test case we decided to probe a possible presence of Rab14 and Rab34 orthologs in archaeplastidial species not systematically analyzed in this study. Using BLAST we screen transcriptomic data at NCBI and in the MMETSP database with the glaucophyte sequences as queries. Obvious Rab14 orthologs were found in a few transcriptome assemblies (from the charophyte *Nitella hyalina* and some angiosperms), but the sequences most likely represent contaminations from metazoan sources (data not shown; contaminations in the *Nitella* transcriptome assembly released by Finet et al. [38] were noticed also by others [39]). On the other hand, apparently authentic sequences similar to Rab34 were encountered in several "prasinophytes" and in the basally branching streptophyte alga *Chlorokybus atmophyticus*, and their assignment as Rab34 orthologs was confirmed by a phylogenetic analysis (Fig. S2, Tab. S1). Hence, despite the fact that Chloroplastida are represented in our set of systematically analyzed species by the highest number of genome sequences, the sampling was still insufficient to capture the actual repertoire of ancestral Rab paralogs retained in Chloroplastida.

Returning to the question of the set of Rab paralogs retained in Archaeplastida as a whole, we thus cannot exclude the possibility that future sampling of their phylogenetic diversity eventually reveals that at least some of the paralogs still unreported from this group (Rab4,

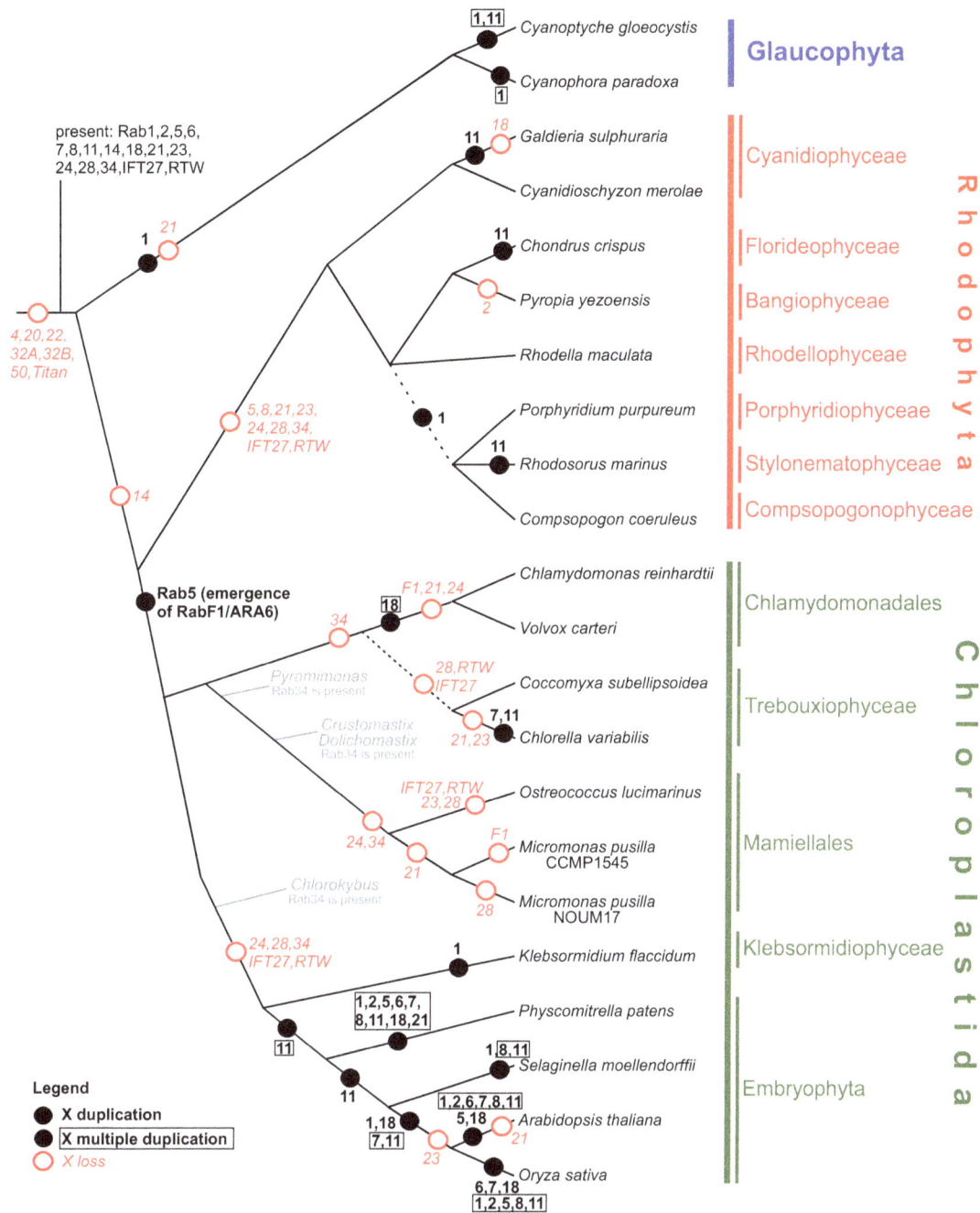

Fig. 1 Major events in the evolution of the Rab GTPase family in Archaeplastida. The events inferred from phylogenetic analyses of Rab sequences are mapped onto a schematic tree depicting phylogenetic relationships among the species analyzed in this study. The tree reflects the current understanding of the archaeplastidial phylogeny. Dashed lines indicate two branches that are uncertain. The clade comprising the red algal classes Porphyridiophyceae, Stylonematophyceae, and Compsopogonophyceae was suggested by some phylogenetic analyses [45], but statistical support was lacking. The monophyly of the green algal class Trebouxiophyceae, including *Coccomyxa subellipsoidea* and *Chlorella variabilis*, is generally accepted in the literature, but recent phylogenetic analyses based on plastid genome data cast doubt on this assumption [62].

Rab20, Rab22, Rab32A, Rab32B, Rab50, or RabTitan) do exist in some members. In addition, it was recently suggested that the group of Rab5-like sequences known from some Chloroplastida and typified by the *A. thaliana* gene RabF1 (also called ARA6 [40]) may represent a paralog that originated before the divergence of Archaeplastida [41].

This group is characterized by the absence of a C-terminal geranylgeranylated tail, which is functionally replaced by an N-terminal extension modified by myristoylation and palmitoylation [40,41]. Hoepflinger et al. noticed the existence of Rab5-like proteins with the same modification in some members of the Alveolata group (e.g. the apicomplexan *Plasmodium falciparum*) [41]. However, their phylogenetic analysis failed to provide evidence for the common origin of the myristoylated/palmitolyated Rab5-like paralogs in Chloroplastida and Alveolata, and this negative result is consistent with our own, even broader phylogenies of the Rab5-related group (unpublished data). In fact, analyses of Rab sequences from diverse protist lineages indicate that such a replacement of the C-terminal geranylgeranylation by an N-terminal acylation has occurred convergently many times in the evolution of several different ancestral Rab paralogs (not only Rab5; unpublished data). Hence, we prefer a scenario in which the RabF1 (ARA6) group emerged by Rab5 duplication and modification in the stem lineage of Chloroplastida (Fig. 1).

Rab gene duplications have been very unevenly distributed across different Rab paralogs and organismal lineages in Archaeplastida

It is well established that gene duplication has been a very potent factor shaping the Rab family during the eukaryote evolution [5,6,12]. We used the expanded set of archaeplastidial Rab genes to make a fresh reconstruction of individual gene duplication events in the Rab family in this group (Fig. 1). Note that in case of organisms represented only by transcriptome data, we ignored highly similar variants of some Rab genes for which it is difficult to decide whether they represent allelic variants or recently duplicated genes.

The pattern of Rab duplications in Archaeplastida exhibits several notable features. The most conspicuous one is that duplication events have been very frequent in embryophytes, whereas their occurrence in other archaeplastidial taxa is much rarer. For some paralogs, including Rab2, Rab6, Rab8, and Rab21, no duplications have been recorded outside embryophytes. For Rab7, the only duplication outside embryophytes seem to be the one thought to give rise to the highly divergent Rab7-like gene in *C. variabilis* (see above). For Rab5, only the duplication at the base of Chloroplasida that resulted in the RabF1 group (see above) has been inferred outside embryophytes. Rab18, in addition to having multiplied in embryophytes, has triplicated in the lineage leading to chlamydomonadalean algae (*C. reinhardtii*, *V. carteri*; Fig. S1, Tab. 1). One of the chlamydomonadalean Rab18 in-paralogs (Rab18c; Fig. S1) is rather divergent and we speculate that it may have acquired a novel cellular function to regulate a membrane trafficking process unique for the respective group of algae.

The raison d'etre for the expanded Rab gene sets in embryophytes may be the need of a multicellular body to finely regulate various, possibly tissue-specific, membrane trafficking pathways. At the same time it is clear that a substantial portion of the duplication events can be accounted for by the frequent occurrence of whole genome duplications (WGD) in the embryophyte evolution [42]. This is also obvious from the fact that the lycophyte *S. moellendorffii*, which does not seem to have an ancestor with a duplicated

genome [43], has a much smaller set of Rab genes than the moss *Physcomitrella patens*, *A. thaliana* and rice (Tab. 1), which all underwent WGD in their past (the angiosperms even multiple times [42]). However, it is beyond the scope of this paper to associate individual Rab gene duplications with the various inferred WGD events in plant evolution.

Two Rab paralogs – Rab1 and Rab11, stand out for their tendency to duplicate in a recurrent manner in Archaeplastida. To better understand the evolutionary history of these two paralogs, we conducted separate phylogenetic analyses of archaeplastidial Rab1 and Rab11 sequences using related paralogs (Rab8 and Rab2, respectively) as outgroups. Annotated ML trees are displayed as Fig. 2 and Fig. 3. Although the resolution of many relationships within the trees is poor, both trees enabled us to derive interesting conclusions.

In case of Rab1, independent duplications events appear to have occurred early in the evolution of three different archaeplastidial groups – glaucophytes, rhodophytes, and embryophytes, in all cases leading to an asymmetric evolution of the resulting in-paralogs (Fig. 2). While one in-paralog has stayed conservative in its sequence (note relatively short branches of the sequences denoted as "prototypical" in-paralogs in Fig. 2), the other is more divergent, suggesting that it has undergone neofunctionalization ("novel" in-paralogs in Fig. 2). In glaucophytes this duplication most likely occurred before the divergence of all known glaucophyte genera: although we included only sequences from *C. paradoxa* and *C. gloeocystis* in our trees, candidate orthologs of both the "prototypical" and the "novel" Rab1 in-paralogs are discernible in the transcriptome assembly of *G. wittrockiana* and among EST sequences from *G. nostochninearum* (data not shown). Interestingly, a PhyML-CAT tree inferred from the Rab1 and Rab8 sequences using the empirical profile mixture model C20 suggested that the somewhat problematic Rab1L gene from *C. gloeocystis* (see above) may actually be an extremely divergent additional member of the "novel" glaucophyte Rab1 in-paralog (data not shown).

In rhodophytes the Rab1 duplication is manifested in only three lineages, recently defined as classes separated from the traditionally circumscribed paraphyletic class Bangiophyceae [44]: Porphyridophyceae, Stylonematophyceae, and Compsopogonophyceae (Fig. 2). The simplest explanation is that these three classes constitute a clade to the exclusion of other rhodophyte classes, and that the Rab1 duplication is exclusive for this putative clade. The interrelationships at the base of the rhodophyte phylogeny (except the firmly established basal position of Cyanidiophyceae and the sisterhood of Bangiophyceae sensu stricto and Florideophyceae) have not yet been resolved with confidence, but at least some phylogenies (e.g., [45]) do support the existence of the putative clade mentioned above.

Finally, in embryophytes an early Rab1 duplication (in addition to numerous ones specific for different terminal branches) can be traced back before the radiation of monocots and eudicots, but perhaps after the radiation of lycophytes (as there is no corresponding duplication in *S. moellendorffii*) and euphyllophytes (Fig. 2). Following the nomenclature of Rab genes in *A. thaliana* [46], the two in-paralogs are denoted RabD1 and RabD2, with the

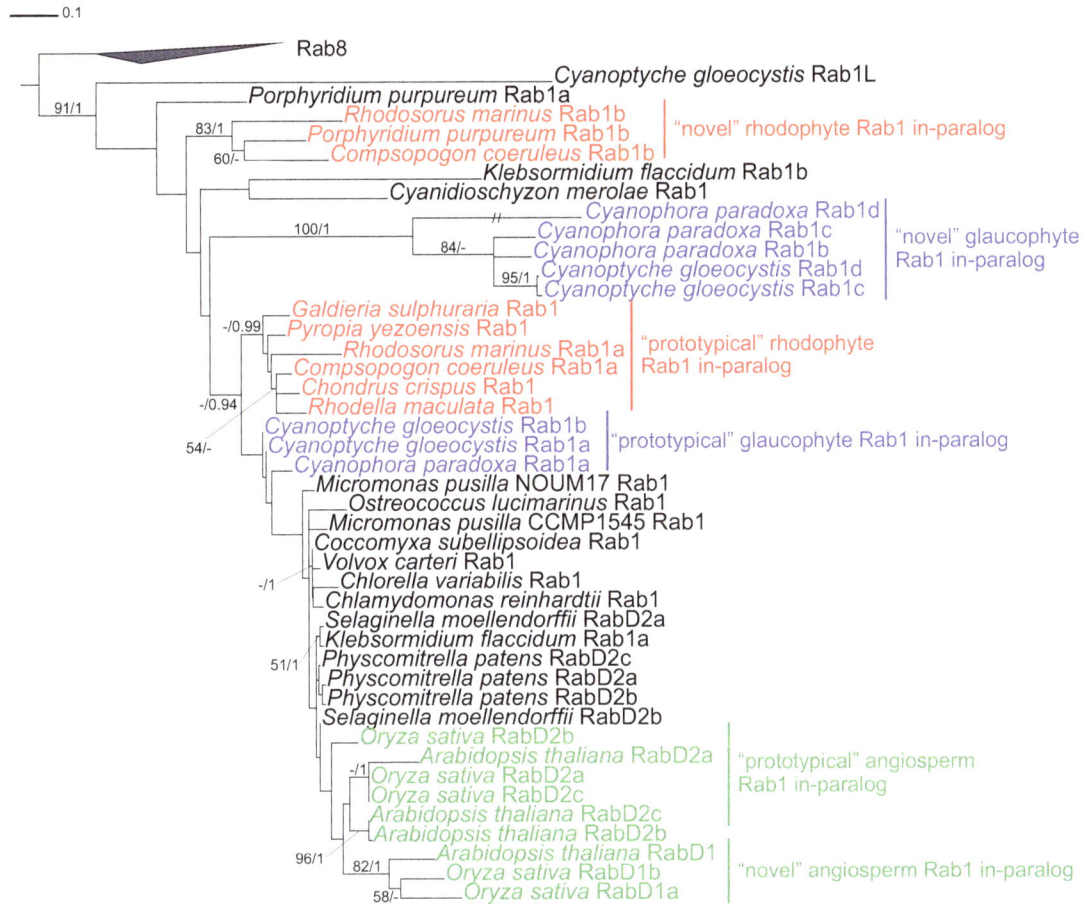

Fig. 2 Phylogenetic analysis of Rab1 genes in Archaeplastida. Portrayed is a maximum likelihood tree (RAxML, LG+Γ model) inferred from a multiple alignment of protein sequences representing the ancestral Rab1 paralog in 22 archaeplastidial species. Sequences representing the Rab8 paralog were used as an outgroup. Numbers at branches correspond to bootstrap support values (higher than 50) calculated using the rapid bootstrapping algorithm of the RAxML program. The bar on the top corresponds to the estimated number of substitutions per site. Groups of sequences of special interest are shown in color; their annotation is discussed in the text. Note that the sole Rab1 gene from *Cyanidioschyzon merolae* and the Rab1a gene from *Porphyridium purpureum* probably represent the "prototypical" rhodophyte Rab1 paralog and their failure to branch with the other rhodophyte sequences may be an artifact caused by their higher divergence. Sequences IDs are available in Tab. S1.

former being the more divergent ("novel"). Although a more comprehensive analysis of Rab1 genes in embryophytes is needed to more precisely pinpoint the emergence of the two in-paralogs, a previously published phylogeny of the Rab1 subfamily including also data from the conifer *Pinus taeda* indicates that RabD1 and RabD2 split before the divergence of gymnosperms and angiosperms [47].

Similar to Rab1, duplications of Rab11 genes frequently exhibit a patter suggesting neofunctionalization of one of the duplicated versions (Fig. 3). Thus, the glaucophyte *C. gloeocystis*, some rhodophytes (*G. sulphuraria*, *Rhodosorus marinus*, and *C. crispus*) and the green alga *C. variabilis* each harbour markedly divergent Rab11 in-paralogs in addition to more canonical ones. One of the *C. gloeocystis* paralogs is even so divergent that we denoted it Rab11L (Rab11-like). The more divergent Rab11 in-paralogs from *G. sulphuraria*

and *C. variabilis* (Rab11b in Fig. 3) cluster together, which might suggest that they arose from the same duplication event. However, statistical support for this cluster is low and the scenario mentioned above would be quite complex, as a number of independent losses of this in-paralog within rhodophytes and Chloroplastida would have to be assumed, so we prefer independent duplications as a more likely explanation.

Our analysis finally provided hints for an interpretation of the evolutionary history of Rab11 genes in embryophytes. A model compatible with the phylogenetic tree of Rab11 sequences and with the phylogenetic distribution of the different in-paralogs assumes the following successive events: (*i*) a triplication of Rab11 at the base of embryophytes (or at least before the divergence of mosses and vascular plants); (*ii*) an additional duplication at the base

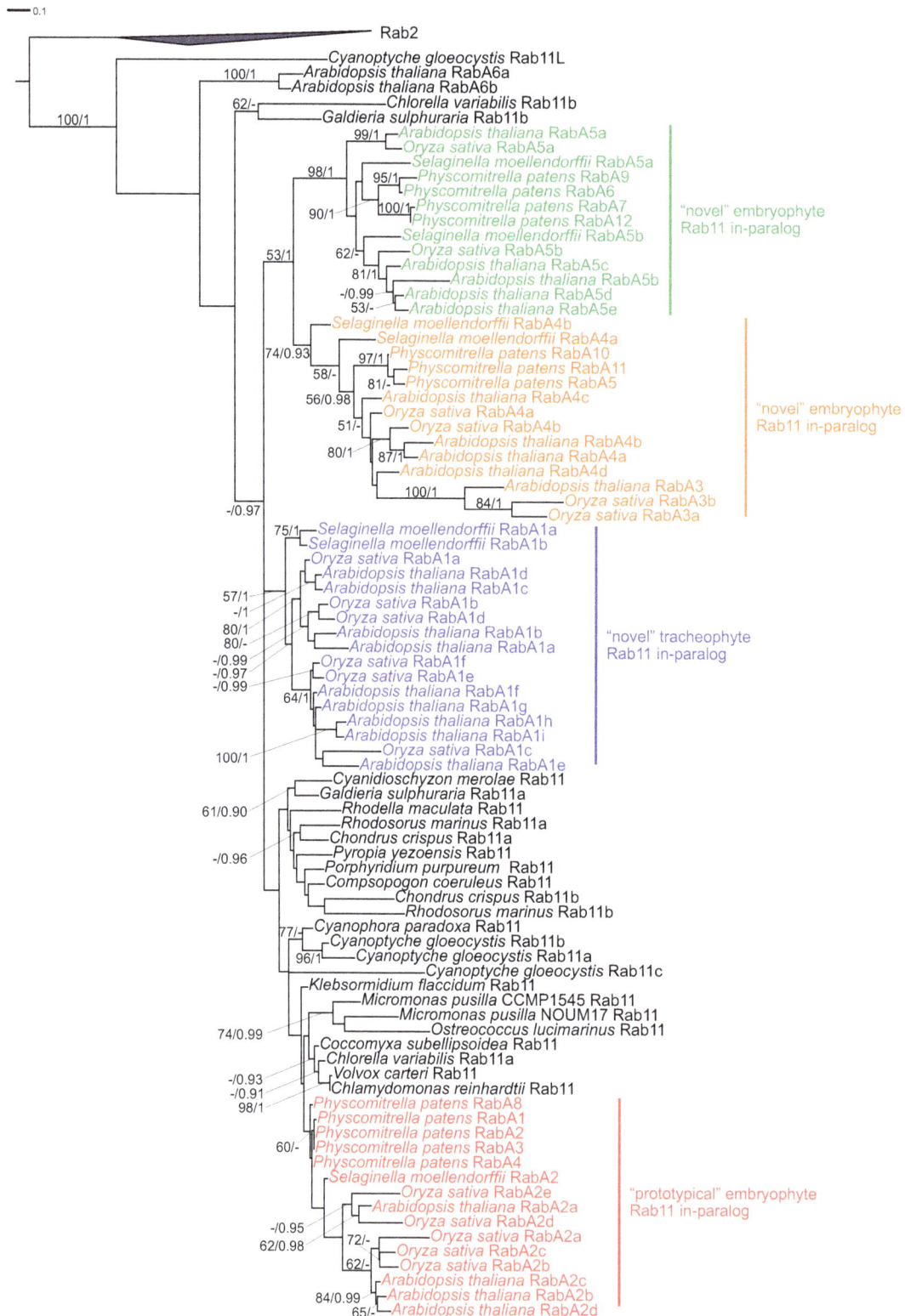

Fig. 3 Phylogenetic analysis of Rab11 genes in Archaeplastida. Portrayed is a maximum likelihood tree (RAxML, LG+Γ model) inferred from a multiple alignment of protein sequences representing the ancestral Rab11 paralog in 22 archaeplastidial species. Sequences representing the Rab2 paralog were used as an outgroup. Numbers at branches correspond to bootstrap support values (higher than 50) calculated using the rapid bootstrapping algorithm of the RAxML program. The bar on the top corresponds to the estimated number of substitutions per site. Groups of sequences of special interest are shown in color; their annotation is discussed in the text. Sequences IDs are available in Tab. S1.

of vascular plants (before the divergence of lycophytes and euphyllophytes); (***iii***) multiple additional duplications at the base of angiosperms and within the eudicot and monocot lineages (Fig. 1 and Fig. 3). From the perspective of Rab11 subgroups defined for *A. thaliana* (see [46]), the angiosperm RabA2 group represents the least derived ("prototypical") in-paralog, the RabA1 group (apparently having specific orthologs in *S. moellendorffii* but not in *P. patens*), is a novelty of tracheophytes, whereas the RabA5 group and a combined RabA4 plus RabA3 group represent two separate in-paralogs that emerged from the "prototypical" paralog by duplications before the moss-tracheophyte divergence. The RabA6 group defined in *A. thaliana* probably represents a very divergent in-paralog specific for the *A. thaliana* lineage, but its exact origin and phylogenetic distribution need to be investigated further using a much broader sampling of angiosperm Rab11 diversity.

Gene loss has significantly sculpted the Rab family in Archaeplastida

Although gene duplications as a means of increasing the complexity of an organism may be viewed as the more interesting and significant events in the evolutionary history of a gene family, the relevance of an opposite process – simplification due to gene loss – may be even higher in particular cases. Concerning Rabs, there are ancestral paralogs that have been rarely duplicated, yet they were lost many times independently from different eukaryotic lineages. One such example is Rab24, whose scattered phylogenetic distribution in eukaryotes implies a high number of independent losses [48]. The same pattern is seen also within Archaeplastida, where Rab24 was lost at least four times independently, specifically in the rhodophyte stem lineage, in chlamydomonadalean green algae, in the Mamiellales lineage, and in streptophytes (Fig. 1).

The glaucophyte *C. paradoxa*, is an extant archaeplastidial species with the most complete set of ancestral Rab paralogs known so far. However, the genomic and transcriptomic data currently available for glaucophytes suggest that this lineage may have lost one Rab paralog certainly present in the last common archaeplastidial ancestor – Rab21 (note that a Rab21 sequence could be found in the MMETSP transcriptome assembly for *G. wittrockiana*, but it most likely represents a contamination, see "Materials and methods"). The loss of Rab21 from glaucophytes needs to be confirmed by further genome and transcriptome sampling, but it would not be at all unprecedented, as Rab21 appears to have been lost many times independently in eukaryotes [6]. Within Archaeplastida, Rab21 is missing not only from glaucophytes, but also from rhodophytes and some lineages of Chloroplastida, in our taxon sampling represented by chlamydomonadalean algae (both *C. reinhardtii* and *Volvox carteri*), *Chlorella variabilis*, both *Micromonas* strains, and *A. thaliana* (Tab. 1). Hence, at least six independent losses of Rab21 need to be assumed to explain the current distribution of Rab21 genes in Archaeplastida (Fig. 1).

One of the earliest reductions of the Rab complement within Archaeplastida was probably that affecting Rab14. In the most parsimonious scenario of archaeplastidial phylogeny, which assumes that glaucophytes are sister to a clade comprising Rhodophyta and Chloroplastida [49], a single

loss of the Rab14 gene in the exclusive ancestor of the latter two lineages would explain its phylogenetic distribution. However, the branching order of the three archaeplastidial lineages has not yet been resolved with confidence (see [50]), so independent losses of Rab14 in Rhodophyta and Chloroplastida cannot be excluded.

A branch with the highest number of inferred Rab losses within Archaeplastida is the stem lineage of Rhodophyta (Fig. 1). Specifically, Rab5, Rab8, Rab21, Rab23, Rab24, Rab28, Rab34, IFT27, and RTW appear to have been lost from rhodophytes before the radiation of modern forms. In combination with losses that occurred earlier in the archaeplastidial evolution, modern rhodophytes have retained only six out of the 23 paralogs of Rab and Rab-like proteins inferred as ancestral for eukaryotes [6]. Most rhodophytes thus possess genes representing Rab1, Rab2, Rab6, Rab7, Rab11, and Rab18 paralogs (Tab. 1), but some red algal species have reduced this set even further. *Galdieria sulphuraria* apparently lacks Rab18 (this paralog is also missing from a related species, *Galdieria phlegrea*; data not shown) and *Pyropia yezoensis* seems to lack Rab2 (no Rab2 could be identified also in the transcriptome data for a related species, *Pyropia haitanensis*). With only five Rab genes, *P. yezoensis*, a multicellular red alga, exhibits the minimal number of Rab paralogs ever recorded for a eukaryotic cell. According to our knowledge, the same number is found only in the extremely reduced and divergent parasitic group of Microsporidia (unpublished data).

The configuration of Rab paralogs currently inferred for the LCA of Chloroplastida was this one: Rab1, Rab2, Rab5 (the "prototypical" version, in *A. thaliana* called RabF2), RabF1 (=ARA6), Rab6, Rab7, Rab8, Rab11, Rab18, Rab21, Rab23, Rab24, Rab28, Rab34, IFT27 and RTW. While Rab1, Rab2, Rab5, Rab6, Rab7, Rab8, Rab11, and Rab18 are always represented by at least one gene in every chloroplastidial species studied so far, the remaining ancestrally present paralogs are completely missing from some extant taxa. Losses affecting Rab21 and Rab24 were already mentioned above. The unexpected finding of Rab34 in "prasinophytes" (*Dolichomastix*, *Crustomastix*, *Pyramimonas*) specifically related to the Mamiellales [51,52], which are devoid of Rab34 genes, and in *C. atmophyticus*, which represents a streptophyte lineage basal to Rab34-lacking streptophytes covered by this study (Klebsormidiophyceae, embryophytes), necessitates at least three independent losses of Rab34 in Chloroplastida (Fig. 1). The "green" lineage also seems to easily dispense with Rab28 (lost at least four times) and with Rab23, IFT27, and RTW (each lost at least three times).

Finally, we could infer at least two losses of the RabF1 paralog, specifically from *M. pusilla* CCMP1545 and from chlamydomonadalean algae (*C. reinhardtii* and *V. carteri*). It was previously claimed that RabF1 is missing also from *O. lucimarinus* [41], but we found in this species an obvious ortholog that also possess the characteristic N-terminal extension (Tab. S1 and Fig. S3). The failure of Hoepflinger et al. [41] to detect this ortholog may relate to the fact that the respective gene model is incomplete in the official genome annotation release for this species. We were also able to correct the gene model for the protein sequence EFN55859 from *Chlorella variabilis* that was noticed by Hoepflinger et al. [41] as lacking the N-terminal region; the revised sequence has the N-terminal extension typical for all RabF1 proteins (Fig. S3).

Cell biological implications of the varying Rab complement in different archaeplastidial taxa

The pattern of Rab gene duplications and losses in Archaeplastida described above is important per se, but its main value is primarily in that it can serve as a framework for understanding the evolution of archaeplastidial cells and their functionalities. Below we discuss some implications that we consider most significant.

One of the dominant patterns seems to be the recurrent simplification of cellular complexity due to the loss of particular Rab paralogs. A massive series of losses appears to have occurred already during the emergence of Archaeplastida as a group (Fig. 1), which may imply substantial reduction of the complexity of membrane trafficking processes of the original eukaryotic host cell concomitant with the evolution of a plastid from a cyanobacterial endosymbiont. The loss of Rab14, inferred to have happened in an exclusive ancestor of Rhodophyta and Chloroplastida, may perhaps be viewed as a continuation of this trend, although it is not at all clear why it has been preserved in glaucophytes, as little specific information on membrane trafficking and the endomembrane system in general is available for this group. Indeed, no obvious functional explanation is available to account for the evolutionary younger loss events that affected many other paralogs. This is not only due to very poor knowledge of the cell biology of most archaeplastidial lineages, but also because for many Rab paralogs (e.g. Rab21, Rab24, Rab28, or Rab34) only restricted functional information is available in general.

The lack of a significant correlation between differences or similarities in the configuration of Rab genes and differences or similarities at the organismal level in different organisms can be documented by several notable examples. One is provided by the green algae *C. reinhardtii* and *V. carteri*. The former is a single-cell organism with isogamous mating, while the latter is a multicellular organism with differentiated somatic and germ lines and with oogamous sexual reproduction [53]. However, their sets of Rab genes are exactly the same (Tab. 1), suggesting that expansions of the Rab family is not a sine qua non for achieving a higher morphological complexity. On the other hand, Rab21 has been kept in monocots and some basal "dicots", but is missing from *A. thaliana* and perhaps all other "core" eudicots (Tab. 1 and data not shown). What is so fundamentally different between cells of the different angiosperms that this Rab paralog, and presumably a specific cellular process regulated by Rab21 (most likely in the endocytic pathway [54]), could have been lost from the *A. thaliana* lineage? An even more extreme case is represented by the pair of organisms nominally representing the same species, *M. pusilla* CCMP1545 and *M. pusilla* NOUM17 (sometimes referred to as *Micromonas* sp. RCC299, e.g. in [41]). Global analyses of the gene complements of these two morphologically indistinguishable strains revealed hundreds of genes specific for one or the other strain [55], and this difference pertains also to the Rab family, as RabF1 is missing from *M. pusilla*

CCMP1545 and Rab28 is missing from *M. pusilla* NOUM17 (Tab. 1).

However, for a group of paralogs there is an emerging cellular correlate for their presence/absence pattern. At least three Rab and Rab-like proteins – Rab23, IFT27, and RTW (Rabl2) – are functionally associated with the flagellar apparatus, so they are generally missing from species unable to make flagella or cilia [9,10,56]. Hence, their absence from red algae, angiosperms, *O. lucimarinus*, and *C. variabilis*, which all lack flagella (or where a typical flagellum is at least unknown, see [57]), is not at all surprising. However, IFT27 and RTW may get lost independently of a flagellum loss, since both paralogs are absent from the streptophytes *Klebsormidium flaccidum*, *P. patens*, and *S. moellendorffii*, although they do exhibit flagellated reproductive cells [58]. It is possible that since these flagellated stages are only transient, some original flagellum-associated functions may have been lost in these groups, which is consistent with the fact that many other conserved flagellar proteins are missing in land plants with flagellated stages [59]. The trebouxiophyte alga *C. subellipsoidea* is notable for possessing Rab23 (but not IFT27 and RTW; Tab. 1), although to our knowledge, no flagellated stages have been described for the genus *Coccoymyxa*. One possible explanation is that a cryptic, rarely occurring stage with a (possibly reduced) flagellum does exist in the life cycle of *C. subellipsoidea*, but has not been documented yet (as seems to be the case for another trebouxiophyte, *C. variabilis*, see [57]). Alternatively, Rab23 might have kept or evolved a new, flagellum-independent, function in *Coccomyxa*.

In light of the previous discussion, the massive reduction of the Rab gene complement in the stem rhodophyte lineage (Fig. 1) is partially explained as a consequence of the flagellum loss in red algae. However, most of the remaining paralogs lost in rhodophytes perhaps lack a specific functional connection to the flagellum, so their absence from red algal cells is indicative of a general simplification of their endomembrane system compared to other eukaryotes. Quite unusual is the absence of Rab8, which occurs in most eukaryotes, perhaps functioning as a general exocytotic factor [6]. But even more striking is the lack of Rab5, an essential component of the endocytic machinery associated with early (recycling) endosomes [60]. Rab5 is nearly universal in eukaryotes and the only known group outside rhodophytes devoid of an apparent Rab5 ortholog are the extremely divergent diplomonads (*Giardia intestinalis*, *Spironucleus salmonicida*; [6] and unpublished data). Interestingly, other components of the canonical endocytic machinery are missing from at least some rhodophytes. For example, *C. crispus* (a multicellular red alga) lacks the AP-2 adaptor complex and endocytic Qc-SNARE proteins [22]. Future systematic comparative analyses of rhodophyte genomes combined with cell biological experiments will hopefully clarify whether and how endocytosis takes place in red algal cells.

Conclusions

Thanks to a considerably improved sampling of the phylogenetic diversity of Archaeplastida we were able to draw a more comprehensive picture of the intricate evolutionary history of the Rab family in this significant segment of the eukaryotic tree of life. However, a recurrent theme of the previous discussion was that the data from a much higher number of archaeplastidial lineages are needed to provide more robust answers to many important questions, such as the actual composition of Rab gene complements at different nodes of the archaeplastidial phylogeny. Some

lineages may be especially informative and should become the prime targets for next genome sequencing projects. For example, some "prasinophytes" have retained the ability to phagocytose prey [61] and it is possible that their membrane trafficking machinery exhibits some primitive features lost in other Archaeplastida. We are sure that there still are exciting unexpected aspects of the Rab family (not only) in Archaeplastida to be revealed.

Acknowledgments
We thank Aleš Horák and Vladimír Klimeš for their technical help with phylogenetic analyses and Vyacheslav Yurchenko for critical comments on the manuscript. This study was supported by the Czech Science Foundation (grant No. 13-24983S), the project CZ.1.05/2.1.00/03.0100 (IET) financed by the structural funds of the EU, the project LO1208 of the National Feasibility Programme I of the Czech Republic, and the internal student grant SGS26/PřF/2014-2015.

Authors' contributions
The following declarations about authors' contributions to the research have been made: carried out the bioinformatic analyses, prepared figures and contributed to the final text: RP; conceived the study, contributed to the bioinformatic analyses and drafted the manuscript: ME.

Competing interests
No competing interests have been declared.

Supplementary material
The following supplementary material for this article is available online at http://pbsociety.org.pl/journals/index.php/asbp/rt/suppFiles/asbp.2014.052/0:
1. Tab. S1: list of sequences analyzed.
2. Figs. S1–S3.
3. A text file with revised or newly predicted protein sequences.

References
1. Stenmark H. Rab GTPases as coordinators of vesicle traffic. Nat Rev Mol Cell Biol. 2009;10(8):513–525.

2. Woollard AA, Moore I. The functions of Rab GTPases in plant membrane traffic. Curr Opin Plant Biol. 2008;11(6):610–619.

3. Brighouse A, Dacks JB, Field MC. Rab protein evolution and the history of the eukaryotic endomembrane system. Cell Mol Life Sci. 2010;67(20):3449–3465.

4. Briguglio JS, Turkewitz AP. *Tetrahymena thermophila*: a divergent perspective on membrane traffic. J Exp Zool B Mol Dev Evol. 2014;322(7):500–516.

5. Klöpper TH, Kienle N, Fasshauer D, Munro S. Untangling the evolution of Rab G proteins: implications of a comprehensive genomic analysis. BMC Biol. 2012;10:71.

6. Elias M, Brighouse A, Gabernet-Castello C, Field MC, Dacks JB. Sculpting the endomembrane system in deep time: high resolution phylogenetics of Rab GTPases. J Cell Sci. 2012;125(pt 10):2500–2508.

7. Koonin EV. The origin and early evolution of eukaryotes in the light of phylogenomics. Genome Biol. 2010;11(5):209.

8. Koumandou VL, Wickstead B, Ginger ML, van der Giezen M, Dacks JB, Field MC. Molecular paleontology and complexity in the last eukaryotic common ancestor. Crit Rev Biochem Mol Biol. 2013;48(4):373–396.

9. Lo JC, Jamsai D, O'Connor AE, Borg C, Clark BJ, Whisstock JC, et al. RAB-like 2 has an essential role in male fertility, sperm intra-flagellar transport, and tail assembly. PLoS Genet. 2012;8(10):e1002969.

10. Huet D, Blisnick T, Perrot S, Bastin P. The GTPase IFT27 is involved

in both anterograde and retrograde intraflagellar transport. Elife. 2014;3:e02419.

11. Burki F. The eukaryotic tree of life from a global phylogenomic perspective. Cold Spring Harb Perspect Biol. 2014;6(5):a016147.

12. Diekmann Y, Seixas E, Gouw M, Tavares-Cadete F, Seabra MC, Pereira-Leal JB. Thousands of rab GTPases for the cell biologist. PLoS Comput Biol. 2011;7(10):e1002217.

13. del Campo J, Sieracki ME, Molestina R, Keeling P, Massana R, Ruiz-Trillo I. The others: our biased perspective of eukaryotic genomes. Trends Ecol Evol. 2014;29(5):252–259.

14. Keeling PJ, Burki F, Wilcox HM, Allam B, Allen EE, Amaral-Zettler LA, et al. The Marine Microbial Eukaryote Transcriptome Sequencing Project (MMETSP): illuminating the functional diversity of eukaryotic life in the oceans through transcriptome sequencing. PLoS Biol. 2014;12(6):e1001889.

15. Adl SM, Simpson AG, Farmer MA, Andersen RA, Anderson OR, Barta JR, et al. The new higher level classification of eukaryotes with emphasis on the taxonomy of protists. J Eukaryot Microbiol. 2005;52(5):399–451.

16. Adl SM, Simpson AG, Lane CE, Lukeš J, Bass D, Bowser SS, et al. The revised classification of eukaryotes. J Eukaryot Microbiol. 2012;59(5):429–493.

17. de Clerck O, Bogaert KA, Leliaert F. Diversity and evolution of algae: primary endosymbiosis. Adv Bot Res. 2012;64:55–86.

18. Hampl V, Hug L, Leigh JW, Dacks JB, Lang BF, Simpson AG, et al. Phylogenomic analyses support the monophyly of Excavata and resolve relationships among eukaryotic "supergroups". Proc Natl Acad Sci USA. 2009;106(10):3859–3864.

19. Yabuki A, Kamikawa R, Ishikawa SA, Kolisko M, Kim E, Tanabe AS, et al. *Palpitomonas bilix* represents a basal cryptist lineage: insight into the character evolution in Cryptista. Sci Rep. 2014;4:4641.

20. Matsuzaki M, Misumi O, Shin-I T, Maruyama S, Takahara M, Miyagishima SY, et al. Genome sequence of the ultrasmall unicellular red alga *Cyanidioschyzon merolae* 10D. Nature. 2004;428(6983):653–657.

21. Bhattacharya D, Price DC, Chan CX, Gross J, Steiner JM, Löffelhardt. Analysis of the genome of *Cyanophora paradoxa*: an algal model for understanding primary endosymbiosis. In: Löffelhardt W, editor. Endosymbiosis. Vienna: Springer; 2014. p. 135–148.

22. Collén J, Porcel B, Carré W, Ball SG, Chaparro C, Tonon T, et al. Genome structure and metabolic features in the red seaweed *Chondrus crispus* shed light on evolution of the Archaeplastida. Proc Natl Acad Sci USA. 2013;110(13):5247–5252.

23. Bhattacharya D, Price DC, Chan CX, Qiu H, Rose N, Ball S, et al. Genome of the red alga *Porphyridium purpureum*. Nat Commun. 2013;4:1941.

24. Schönknecht G, Chen WH, Ternes CM, Barbier GG, Shrestha RP, Stanke M, et al. Gene transfer from bacteria and archaea facilitated evolution of an extremophilic eukaryote. Science. 2013;339(6124):1207–1210.

25. Nakamura Y, Sasaki N, Kobayashi M, Ojima N, Yasuike M, Shigenobu Y, et al. The first symbiont-free genome sequence of marine red alga, Susabi-nori (*Pyropia yezoensis*). PLoS One. 2013;8(3):e57122.

26. Price DC, Chan CX, Yoon HS, Yang EC, Qiu H, Weber AP, et al. *Cyanophora paradoxa* genome elucidates origin of photosynthesis in algae and plants. Science. 2012;335(6070):843–847.

27. Saito C, Ueda T. Functions of RAB and SNARE proteins in plant life. Int Rev Cell Mol Biol. 2009;274:183–233.

28. Zhang Z, Hill DR, Sylvester AW. Diversification of the RAB guanosine triphosphatase family in dicots and monocots. J Integr Plant Biol. 2007;49(8):1129–1141.

29. Rensing SA, Lang D, Zimmer AD, Terry A, Salamov A, Shapiro H, et al. The *Physcomitrella* genome reveals evolutionary insights into the conquest of land by plants. Science. 2008;319(5859):64–69.

30. Huber H, Beyser K, Fabry S. Small G proteins of two green algae are localized to exocytic compartments and to flagella. Plant Mol Biol. 1996;31(2):279–293.

31. Altschul SF, Madden TL, Schäffer AA, Zhang J, Zhang Z, Miller W, et al. Gapped BLAST and PSI-BLAST: a new generation of protein database search programs. Nucleic Acids Res. 1997;25(17):3389–3402.

32. Katoh K, Standley DM. MAFFT multiple sequence alignment software version 7: improvements in performance and usability. Mol Biol Evol. 2013;30(4):772–780.

33. Stamatakis A. RAxML version 8: a tool for phylogenetic analysis and post-analysis of large phylogenies. Bioinformatics. 2014;30(9):1312–1313.

34. Miller MA, Pfeiffer W, Schwartz T. Creating the CIPRES science gateway for inference of large phylogenetic trees. In: Proceedings of the gateway computing environments workshop (GCE). New Orleans, LA; 2010. p. 1–8.

35. Quang LS, Gascuel O, Lartillot N. Empirical profile mixture models for phylogenetic reconstruction. Bioinformatics. 2008;24(20):2317–2323.

36. Letunic I, Bork P. Interactive tree of life v2: online annotation and display of phylogenetic trees made easy. Nucleic Acids Res. 2011;39(web server issue):W475–W478.

37. Rogozin IB, Basu MK, Csürös M, Koonin EV. Analysis of rare genomic changes does not support the unikont-bikont phylogeny and suggests cyanobacterial symbiosis as the point of primary radiation of eukaryotes. Genome Biol Evol. 2009;1:99–113.

38. Finet C, Timme RE, Delwiche CF, Marlétaz F. Multigene phylogeny of the green lineage reveals the origin and diversification of land plants. Curr Biol. 2010;20(24):2217–2222.

39. Laurin-Lemay S, Brinkmann H, Philippe H. Origin of land plants revisited in the light of sequence contamination and missing data. Curr Biol. 2012;22(15):R593–R594.

40. Ueda T, Yamaguchi M, Uchimiya H, Nakano A. Ara6, a plant-unique novel type Rab GTPase, functions in the endocytic pathway of *Arabidopsis thaliana*. EMBO J. 2001;20(17):4730–4741.

41. Hoepflinger MC, Geretschlaeger A, Sommer A, Hoeftberger M, Nishiyama T, Sakayama H, et al. Molecular and biochemical analysis of the first ARA6 homologue, a RAB5 GTPase, from green algae. J Exp Bot. 2013;64(18):5553–5568.

42. Vanneste K, Maere S, van de Peer Y. Tangled up in two: a burst of genome duplications at the end of the Cretaceous and the consequences for plant evolution. Philos Trans R Soc Lond B Biol Sci. 2014;369(1648).

43. Banks JA, Nishiyama T, Hasebe M, Bowman JL, Gribskov M, dePamphilis C, et al. The *Selaginella* genome identifies genetic changes associated with the evolution of vascular plants. Science. 2011;332(6032):960–963.

44. Yoon HS, Müller KM, Sheath RG, Ott FD, Bhattacharya D. Defining the major lineages of red algae (Rhodophyta). J Phycol. 2006;42(2):482–492.

45. Verbruggen H, Maggs CA, Saunders GW, Le Gall L, Yoon HS, de Clerck O. Data mining approach identifies research priorities and data requirements for resolving the red algal tree of life. BMC Evol Biol. 2010;10:16.

46. Rutherford S, Moore I. The *Arabidopsis* Rab GTPase family: another enigma variation. Curr Opin Plant Biol. 2002;5(6):518–528.

47. Elias M, Patron NJ, Keeling PJ. The RAB family GTPase Rab1A from *Plasmodium falciparum* defines a unique paralog shared by Chromalveolates and Rhizaria. J Eukaryot Microbiol. 2009;56(4):348–356.

48. Elias M. Patterns and processes in the evolution of the eukaryotic endomembrane system. Mol Membr Biol. 2010;27(8):469–489.

49. Reyes-Prieto A, Bhattacharya D. Phylogeny of nuclear-encoded plastid-targeted proteins supports an early divergence of glaucophytes within Plantae. Mol Biol Evol. 2007;24(11):2358–2361.

50. Deschamps P, Moreira D. Signal conflicts in the phylogeny of the primary photosynthetic eukaryotes. Mol Biol Evol. 2009;26(12):2745–2753.

51. Marin B, Melkonian M. Molecular phylogeny and classification of the Mamiellophyceae class. nov. (Chlorophyta) based on sequence comparisons of the nuclear- and plastid-encoded rRNA operons. Protist. 2010;161(2):304–336.

52. Lemieux C, Otis C, Turmel M. Six newly sequenced chloroplast genomes from prasinophyte green algae provide insights into the relationships among prasinophyte lineages and the diversity of streamlined genome architecture in picoplanktonic species. BMC Genomics. 2014;15:857.

53. Umen JG. Green algae and the origins of multicellularity in the plant kingdom. Cold Spring Harb Perspect Biol. 2014;6(11).

54. Ali M, Leung KF, Field MC. The ancient small GTPase Rab21 functions in intermediate endocytic steps in trypanosomes. Eukaryot Cell. 2014;13(2):304–319.

55. Worden AZ, Lee JH, Mock T, Rouzé P, Simmons MP, Aerts AL, et al. Green evolution and dynamic adaptations revealed by genomes of the marine picoeukaryotes *Micromonas*. Science. 2009;324(5924):268–272.

56. Lumb JH, Field MC. Rab23 is a flagellar protein in *Trypanosoma brucei*. BMC Res Notes. 2011;4:190.

57. Blanc G, Duncan G, Agarkova I, Borodovsky M, Gurnon J, Kuo A, et al. The *Chlorella variabilis* NC64A genome reveals adaptation to photosymbiosis, coevolution with viruses, and cryptic sex. Plant Cell. 2010;22(9):2943–2955.

58. Lewis LA, McCourt RM. Green algae and the origin of land plants. Am J Bot. 2004;91(10):1535–1556.

59. Hodges ME, Wickstead B, Gull K, Langdale JA. The evolution of land plant cilia. New Phytol. 2012;195(3):526–540.

60. Woodman PG. Biogenesis of the sorting endosome: the role of Rab5. Traffic. 2000;1(9):695–701.

61. Maruyama S, Kim E. A modern descendant of early green algal phagotrophs. Curr Biol. 2013;23(12):1081–1084.

62. Lemieux C, Otis C, Turmel M. Chloroplast phylogenomic analysis resolves deep-level relationships within the green algal class Trebouxiophyceae. BMC Evol Biol. 2014;14:211.

23

Stromuling when stressed!

Björn Krenz[1]*, Tai Wei Guo[2], Tatjana Kleinow[3]

[1] Lehrstuhl für Biochemie, Department Biologie, Staudtstr. 5, 91058 Erlangen, Germany

[2] Department of Plant Pathology and Plant-Microbe Biology, Cornell University, Ithaca NY, USA

[3] Institut für Biomaterialien und biomolekulare Systeme, Abteilung für Molekularbiologie und Virologie der Pflanzen, Universität Stuttgart, Pfaffenwaldring 57, 70550 Stuttgart, Germany

Abstract

Stromules are stroma-filled tubules, extruding from the plastid and surrounded by both envelope membranes, but so far, stromules remain enigmatic structures and their function unknown. Stromules can interconnect plastids and have been found to associate with the nucleus, endoplasmic reticulum, Golgi complex, plasma membrane, mitochondria and peroxisomes. This minireview briefly summarizes markers to visualize stromules, inducers of stromules and provides new data about plant virus induced stromules.

Keywords: stromule; geminivirus; plant virus; chloroplast

Introduction

Plastids are major and important organelles of the plant cell, as they perform essential biosynthetic and metabolic functions. These include photosynthetic carbon fixation, synthesis of amino acids, fatty acids, starch and secondary metabolites such as pigments [1]. There are several forms of differentiated plastids, i.e. chromoplast, leucoplast or chloroplast; and they are classified on the basis of their structure, pigment composition, metabolism and function.

Chloroplasts are characterized by their high concentration of chlorophyll and carry out a number of other functions besides photosynthesis, including fatty acid and amino acid synthesis [1], as well as immune response [2]. These organelles are not static, as they can circulate and move in plant cells when influenced by environmental factors, especially light. Cytoskeletal elements, such as actin, are involved in moving and positioning of chloroplasts [3].

Chloroplasts have two envelope membranes, the outer and inner chloroplast membranes, and are generally lens-shaped [4]. Protrusions emanating from chloroplasts are sometimes visible by light, fluorescence and electron microscopy. These thin, stroma-filled tubules, extruding from the plastid, and surrounded by both envelope membranes, are called stromules and have been observed for all plastid types examined so far [5–9]. Other organelles can have protrusions too, and corresponding terms were given to describe those extensions, according to the stromules for plastids convention:

namely matrixules for mitochondria [10] and peroxules for peroxisomes [11]. Stromules are highly dynamic, branching and elongating across the plant cell, usually less than 1 μm in diameter and of variable length. They can appear as short beak-like projections, to linear or branched structures up to several hundred μm long.

Stromules occur in all cell types, but stromule morphology and the proportion of plastids with stromules, vary from tissue to tissue and depend on plant developmental stages as well as biotic and abiotic stress conditions [12,13]. Although a lot of work has been done to understand stromules (reviewed in [8,14]), there are still a lot of unanswered questions. Here, we briefly summarize the current status on markers and inducers of stromules, with emphasis on plant pathogens as inducers.

Stromule markers

In addition to direct stromule observation by light and electron microscopy, or a fluorescence dye [15] in untransformed plants, a common method to visualize stromules in vivo is by the use of a marker protein fused with a fluorescent tag like the green fluorescent protein (GFP). One of the first markers applied was the CT-GFP construct [16,17]. This marker protein localizes to the chloroplast stroma and thereby also visualizes stromules.

Meanwhile, various constructs for stroma-targeted fluorescent marker proteins and also transgenic plant lines expressing these markers were established for plastid and stromule observation [18]. A transgenic *Nicotiana benthamiana* line, which is available to highlight the plastid

---

* Corresponding author. Email: bjoern.krenz@fau.de

Handling Editor: Andrzej Bodył

stroma and stromules, expresses a GFP fusion protein with ferredoxin NADP(H) oxidoreductase (FNR), FNR::eGFP [19–21]. Other constructs have also been reported to visualize stromules, e.g. chloroplast-localized heat shock cognate 70-kDa protein (cpHSC70-1) [22,23], outer envelope protein 16 -2.1, -2.2 and -1.3 [24] and mutants of the *Arabidopsis* resistance protein RPW8.2 [25].

Markers based on chloroplast outer envelope proteins have to be considered carefully. Breuers et al. [26] showed that stromule formation can simply be induced by over-expression of an outer envelope protein, e.g. outer envelope proteins AtLACS9 and AtTOC64-III. They showed that the formation is independent of the function or structure of the protein and posited that the proliferation of the membrane was a direct effect of the protein content in the membrane. So it must be taken into consideration that visualizing stromules usually requires protein overexpression, which might lead to artifacts. Therefore controls are necessary and essential. In general, expression of a stroma-targeted protein or a fluorescence dye in fluorescence microscopy analysis of stromules is more advisable than expression of a chloroplast membrane-localized protein.

Stromule formation upon plant virus infection

Kwok and Hanson [27] suggested that stromules may serve as pathways between nuclei, the cell periphery and possibly even other cells. They observed a concentration of plastids around nuclei with stromules reaching to the plasma membrane and through nuclear grooves. This close contact between plastids and the nuclei and/or the plasma membrane implied a function for the exchange of molecules between plastids and other organelles or diverse regions of the plant cell. Furthermore, Kwok and Hanson [27] found that stromules from two neighboring cells appeared to meet at either side of an adjoining cell wall. The cpHSC70-1-containing stromules detected upon an Abutilon mosaic virus (AbMV, plant ssDNA virus, *Geminiviridae*)-infection that was also in close association with the nucleus, appeared to interconnect plastids and extended from plastids outward to the cell periphery. This led to speculations of intra- and intercellular geminiviral transport in association with stromules [22,23]. A geminiviral nucleoprotein complex moves out of the plant nucleus via plasmodesmata into the adjacent cell to systemically infect the plant. Therefore we wanted to know, if stromules form structures spanning the nucleus, chloroplasts and plasmodesmata and if we can visualize an association of stromules and plasmodesmata. This association then could serve as an intracellular virus transport highway to the neighboring cell.

In our hands, stromule formation was never visualized by the overexpression of cpHSC70-1 alone, but only together with a plant DNA virus infection. Similar results were obtained with inoculation and co-expression experiments with AbMV and the *Arabidopsis thaliana* outer envelope protein 7 (OEP7; Fig. 1a). However, it was shown elsewhere that only the overexpression of OEP7 can lead to stromule-like structures formation [28]. OEP7 was fused to GFP and transiently expressed in *N. benthamiana* epidermal leaves.

OEP7::GFP signals were exclusively found in association with chloroplasts, but no stromules or stromule-like structures could be detected, except induced upon AbMV infection. Leaf material was also co-infiltrated with agrobacteria harboring an expression construct for the AbMV nuclear shuttle protein (NSP) fused to the red fluorescent protein (RFP). NSP is located in the plant nucleus, but additionally mobilized by the AbMV movement protein (MP) to the plasma membrane ([23] and references therein). NSP and MP together mediate cell-to-cell transfer and long distance transport of viral DNA replicated within the nucleus throughout the whole plant body. Thus, an AbMV-infected plant cell displays NSP::RFP signals in the nucleus and at the cell periphery.

In agreement with previous data, the OEP7::GFP marker also highlighted the induction of what appeared to be a stromule network upon an AbMV infection. What remains to be demonstrated is whether, under the condition of viral infection, the membranes of different plastids fuse with each other or if they just closely associate to each other. Chloroplasts tended to surround the nucleus and one could observe stromule formation connecting chloroplasts, along the nucleus or reaching towards the plasma membrane. Interestingly, stromules associated with the plasma membrane seemed to follow the membrane (Fig. 1c,d). Occasionally bulges along the stromules were observed. The OEP7::GFP labeled stromules induced by AbMV-infection exhibited similar appearance to those visualized by cpHSC70-1 [22,23].

In an additional experiment we wanted to test stromule association with plasmodesmata – microscopic channels that traverse the cell walls of plant cells [29]. Plasmodesmata enable direct, regulated, symplastic intercellular transport of substances between cells. The 3a movement protein (MP) of cucumber mosaic virus (CMV) targets plasmodesmata and accumulates in the central cavity of the pore [30] and was used in this assay as a marker protein for plasmodesmata. CMV MP was fused with RFP and co-expressed with OEP7::GFP and AbMV. CMV MP::RFP highlighted plasmodesmata, and again, AbMV infection was sufficient to induce stromules visualized by OEP7::GFP, but we could never observe an association of CMV MP::RFP and OEP7::GFP in this experimental set-up. Also the bulges or packet structures along the stromules were never associated with the plasmodesmata marker (data not shown).

Inducers of stromules

Schattat et al. [14,31,32] do not support a function of stromules in trafficking of macromolecules between plastids, despite the strong microscopic impression of interplastid connectivity via stromules. They could not observe an exchange of a stroma marker protein between independent plastids. Hanson and Sattarzadeh [33] then commented and showed evidence for the flow of proteins between interconnected plastids. They claimed that Schattat et al. [14] were unable to observe movement because of stromule breakage or disruption of the fluorescent protein by the high-intensity laser power they employed. Mathur and Barton [34] disagreed with this explanation, stating that their procedure was very mild in comparison to procedures using intense lasers. While the groups disagree about whether proteins can flow between plastids already connected with stromules, they both agree that the debate surrounding the formation of new connections between separate plastids will continue until a stromule in the act of attaching to another one is found during a photoconversion experiment.

Beside possible interplastid connectivity via stromules, it is striking that stromules or stromule-like structures also extend to the plasma membrane, as well as the nucleus. It may indicate a specific response to viral infection resulting in increased communication between chloroplasts and other cell compartments, or a more general stress response of the cell. Various abiotic and biotic stress conditions including heat [9], subcellular redox stress [35], application of extracellular sucrose or glucose [21], colonization by an arbuscular mycorrhizal fungus [36] and infiltration of agrobacteria [31] were described as inducers of stromules. Gray et al. [37] showed that various stress treatments, including drought and salt stress, are able to induce stromule formation in tobacco epidermal cells. Application of abscisic acid (ABA) to tobacco

and wheat seedlings also induced stromule formation very effectively. Stromules were more abundant in dark-grown seedlings than in light-grown seedlings, and stromule formation was sensitive to red and far-red light. Stromules were induced by treatment with the ethylene precursor ACC (1-aminocyclopropane-1-carboxylic acid) and by treatment with methyl jasmonate.

Undoubtedly, stromule formation can be induced by plant viruses [22,23,38–40] (and this study). RNA-virus infected sugar beets showed structures resembling stromules [40]. Shalla [38] described stromule occurrence in TMV-infected tomatoes. Caplan and colleagues described the same in TMV-infected tobacco plants [39], just as for *N. benthamiana* plants locally infected with the AbMV [22,23]. Several geminiviruses were found to interact with diverse plant hormone pathways, such as the salicylic acid, ethylene, jasmonic acid pathways and to the brassinosteroid pathway. They activate the salicylic acid and ethylene pathways, which both participate in the host defense response [41]. Transcriptomic analysis has recently shown that geminiviruses

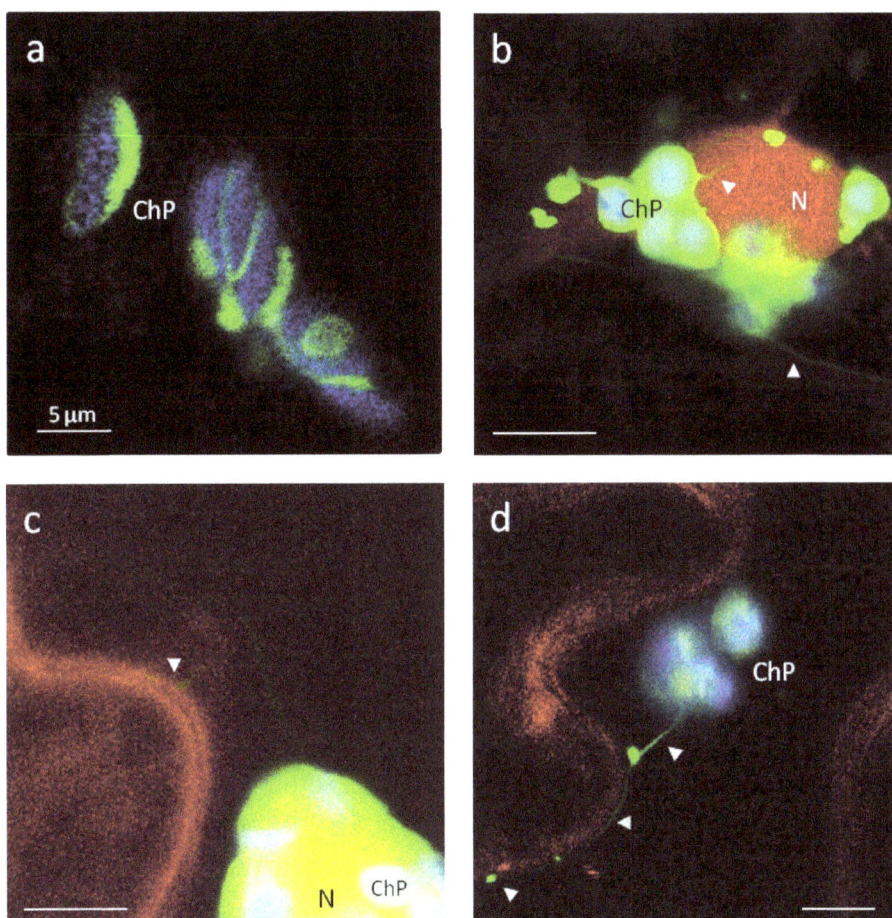

Fig. 1 Abutilon mosaic virus-induced OEP7-containing stromules extending from plastids to the nucleaus and cell periphery. Transient co-expression of test constructs in leaf tissues of non-infected and locally AbMV-infected *N. benthamiana* and confocal fluorescence microscopy. AbMV infection was established by simultaneous agro-infiltration of infectious DNA A and DNA B clones with the fluorescent protein expression constructs. **a** Merged image of chloroplast (ChP) with full length OEP7:green fluorescent protein (GFP) driven under the CaMV 35S promotor in non-infected plants, chloroplast autofluorescence is shown in blue, 2dpi. **b–d** Merged image of chloroplasts (ChP) and nucleus (N) with OEP7:green fluorescent protein (GFP) and NSP:red fluorescent protein (RFP) in AbMV-infected cells, NSP::RFP marks nucleus and plasma membrane, chloroplast autofluorescence is shown in blue, 2dpi. Arrowheads highlight stromules. Bar: 5 μm.

also induce ABA- and ethylene responsive genes in tomato [42]. So there is the possibility that geminiviruses induce stromules by interfering with the plant hormone signaling. But if stromule formation is part of an antiviral immune response then it either comes too late or the virus has the capability to neutralize the stress response, irrespective of stromule formation.

Signaling molecules and stromule function?

Although the function of stromules is unknown, a putative correlation between plastid metabolic activity and stromule biogenesis has been proposed, enabling increased molecular exchange between plastids and other organelles or adjacent cells to process environmental stimuli [17,18,27,33]. The identity of proteins, metabolites and other signaling components, which may be transported are unknown, but there are data indicating that some molecules are excluded. Thylakoid membranes, plastid DNA, and ribosomes have not been detected within stromules [43], while Rubisco and aspartate aminotransferase have been observed [44].

Newell et al. [43] tried to determine if plastid DNA or plastid ribosomes are able to enter stromules, potentially permitting the transfer of genetic information between plastids. Plastid DNA and ribosomes were marked with green fluorescent protein (GFP) fusions to LacI, the *lac* repressor, and their observations indicated that plastid DNA and plastid ribosomes do not routinely move into stromules in tobacco and *Arabidopsis*. Nevertheless, the possibility of a rare movement of plastid DNA and ribosomes or the transfer of much smaller DNA molecules, e.g., plastid transformation vectors, via stromules cannot be completely excluded. We observed stromules associated along a stretch of plasma membrane (Fig. 1c,d) and a possible explanation for this phenomenon is that the intimate association of plasma membrane and stromule provides a connection for rapid exchange of membrane lipids or other molecules. But this is rather speculative, so far.

Stromules are also discussed as structures supporting the autophagy of plastid material [45]. Reactive oxygen species lead to the fragmentation and oxidation of stroma proteins, which then bud to form Rubisco-containing bodies (RCB). Autophagosomes drive RCBs then to the vacuole for degradation [46].

Future directions

The biggest challenge in the future will be to decipher the function of stromules and the molecular mechanisms in stromule formation. Although numerous stromule markers and inducers are known, there is no stromule mutant available, which would be a huge step in understanding stromules. It is still unclear if stromules act as interaction hubs between subcellular compartments or cells and if molecules are trafficking. There is reasonable doubt that these are stroma-localized proteins, but maybe outer membrane envelope proteins, Ca^{2+} or lipids? We are curious about the first answers in the stromule riddle.

Acknowledgments

We would like to acknowledge the Optical Imaging Centre Erlangen (OICE). We are also grateful to Dr. Christopher E. Lane for critically reading of the manuscript. This work was partially financed by the CRC796, Erlangen.

Authors' contributions

The following declarations about authors' contributions to the research have been made: performed the experiments: TWG; designed the experiments and wrote the manuscript: BK, TK.

Competing interests

No competing interests have been declared.

References

1. Rolland N, Curien G, Finazzi G, Kuntz M, Maréchal E, Matringe M, et al. The biosynthetic capacities of the plastids and integration between cytoplasmic and chloroplast processes. Annu Rev Genet. 2012;46(1):233–264.

2. Stael S, Kmiecik P, Willems P, van der Kelen K, Coll NS, Teige M, et al. Plant innate immunity – sunny side up? Trends Plant Sci. 2014 (in press).

3. Oikawa K, Kasahara M, Kiyosue T, Kagawa T, Suetsugu N, Takahashi F, et al. CHLOROPLAST UNUSUAL POSITIONING1 is essential for proper chloroplast positioning. Plant Cell. 2003;15(12):2805–2815.

4. Cavalier-Smith T. Membrane heredity and early chloroplast evolution. Trends Plant Sci. 2000;5(4):174–182.

5. Köhler RH, Hanson MR. Plastid tubules of higher plants are tissue-specific and developmentally regulated. J Cell Sci. 2000;113 (pt 1):81–89.

6. Hanson MR, Köhler RH. GFP imaging: methodology and application to investigate cellular compartmentation in plants. J Exp Bot. 2001;52(356):529–539.

7. Sage TL, Sage RF. The functional anatomy of rice leaves: implications for refixation of photorespiratory CO_2 and efforts to engineer C_4 photosynthesis into rice. Plant Cell Physiol. 2009;50(4):756–772.

8. Mathur J, Mammone A, Barton KA. Organelle extensions in plant cells. J Integr Plant Biol. 2012;54:851–867.

9. Holzinger A, Buchner O, Lütz C, Hanson MR. Temperature-sensitive formation of chloroplast protrusions and stromules in mesophyll cells of *Arabidopsis thaliana*. Protoplasma. 2007;230(1–2):23–30.

10. Logan DC. The mitochondrial compartment. J Exp Bot. 2006;57(6):1225–1243.

11. Bishop GJ. Refining the plant steroid hormone biosynthesis pathway. Trends Plant Sci. 2007;12(9):377–380.

12. Gunning BES. Plastid stromules: video microscopy of their outgrowth, retraction, tensioning, anchoring, branching, bridging, and tip-shedding. Protoplasma. 2005;225(1–2):33–42.

13. Waters MT, Fray RG, Pyke KA. Stromule formation is dependent upon plastid size, plastid differentiation status and the density of plastids within the cell. Plant J. 2004;39(4):655–667.

14. Schattat MH, Barton KA, Mathur J. The myth of interconnected plastids and related phenomena. Protoplasma. 2014 (in press).

15. Menzel D. An interconnected plastidom in *Acetabularia*: implications

for the mechanism of chloroplast motility. Protoplasma. 1994;179(3–4):166–171.

16. Köhler RH, Zipfel WR, Webb WW, Hanson MR. The green fluorescent protein as a marker to visualize plant mitochondria in vivo. Plant J. 1997;11(3):613–621.

17. Köhler RH, Cao J, Zipfel WR, Webb WW, Hanson MR. Exchange of protein molecules through connections between higher plant plastids. Science. 1997;276(5321):2039–2042.

18. Natesan SKA, Sullivan JA, Gray JC. Stromules: a characteristic cell-specific feature of plastid morphology. J Exp Bot. 2005;56(413):787–797.

19. Erickson JL, Ziegler J, Guevara D, Abel S, Klösgen RB, Mathur J, et al. *Agrobacterium*-derived cytokinin influences plastid morphology and starch accumulation in *Nicotiana benthamiana* during transient assays. BMC Plant Biol. 2014;14(1):127.

20. Marques JP. In vivo transport of folded EGFP by the pH/TAT-dependent pathway in chloroplasts of *Arabidopsis thaliana*. J Exp Bot. 2004;55(403):1697–1706.

21. Schattat M, Klösgen R. Induction of stromule formation by extracellular sucrose and glucose in epidermal leaf tissue of *Arabidopsis thaliana*. BMC Plant Biol. 2011;11(1):115.

22. Krenz B, Windeisen V, Wege C, Jeske H, Kleinow T. A plastid-targeted heat shock cognate 70 kDa protein interacts with the *Abutilon* mosaic virus movement protein. Virology. 2010;401(1):6–17.

23. Krenz B, Jeske H, Kleinow T. The induction of stromule formation by a plant DNA-virus in epidermal leaf tissues suggests a novel intra- and intercellular macromolecular trafficking route. Front Plant Sci. 2012;3:291.

24. Mueller SJ, Lang D, Hoernstein SNW, Lang EGE, Schuessele C, Schmidt A, et al. Quantitative analysis of the mitochondrial and plastid proteomes of the moss *Physcomitrella patens* reveals protein macrocompartmentation and microcompartmentation. Plant Physiol. 2014;164(4):2081–2095.

25. Wang W, Zhang Y, Wen Y, Berkey R, Ma X, Pan Z, et al. A comprehensive mutational analysis of the *Arabidopsis* resistance protein RPW8.2 reveals key amino acids for defense activation and protein targeting. Plant Cell. 2013;25(10):4242–4261.

26. Breuers FKH, Bräutigam A, Geimer S, Welzel UY, Stefano G, Renna L, et al. Dynamic remodeling of the plastid envelope membranes – a tool for chloroplast envelope in vivo localizations. Front Plant Sci. 2012;3:7.

27. Kwok EY, Hanson MR. Plastids and stromules interact with the nucleus and cell membrane in vascular plants. Plant Cell Rep. 2004;23(4):188–195.

28. Machettira AB, Groß LE, Tillmann B, Weis BL, Englich G, Sommer MS, et al. Protein-induced modulation of chloroplast membrane morphology. Front Plant Sci. 2012;2:118.

29. Oparka KJ. Getting the message across: how do plant cells exchange macromolecular complexes? Trends Plant Sci. 2004;9(1):33–41.

30. Blackman LM, Boevink P, Cruz SS, Palukaitis P, Oparka KJ. The movement protein of cucumber mosaic virus traffics into sieve elements in minor veins of *Nicotiana clevelandii*. Plant Cell. 1998;10(4):525–538.

31. Schattat MH, Griffiths S, Mathur N, Barton K, Wozny MR, Dunn N, et al. Differential coloring reveals that plastids do not form networks for exchanging macromolecules. Plant Cell. 2012;24(4):1465–1477.

32. Schattat MH, Klösgen RB, Mathur J. New insights on stromules: stroma filled tubules extended by independent plastids. Plant Signal Behav. 2012;7(9):1132–1137.

33. Hanson MR, Sattarzadeh A. Trafficking of proteins through plastid stromules. Plant Cell. 2013;25(8):2774–2782.

34. Mathur J, Barton KA, Schattat MH. Fluorescent protein flow within stromules. Plant Cell. 2013;25(8):2771–2772.

35. Itoh RD, Yamasaki H, Septiana A, Yoshida S, Fujiwara MT. Chemical induction of rapid and reversible plastid filamentation in *Arabidopsis thaliana* roots. Physiol Plant. 2010;139(2):144–158.

36. Lohse S. Organization and metabolism of plastids and mitochondria in arbuscular mycorrhizal roots of *Medicago truncatula*. Plant Physiol. 2005;139(1):329–340.

37. Gray JC, Hansen MR, Shaw DJ, Graham K, Dale R, Smallman P, et al. Plastid stromules are induced by stress treatments acting through abscisic acid: stress induction of plastid stromules. Plant J. 2012;69(3):387–398.

38. Shalla TA. Assembly and aggregation of tobacco mosaic virus in tomato leaflets. J Cell Biol. 1964;21:253–264.

39. Caplan JL, Mamillapalli P, Burch-Smith TM, Czymmek K, Dinesh-Kumar SP. Chloroplastic protein NRIP1 mediates innate immune receptor recognition of a viral effector. Cell. 2008;132(3):449–462.

40. Esau K. Anatomical and cytological studies on beet mosaic. J Agric Res. 1944;69:95–117.

41. Ascencio-Ibanez JT, Sozzani R, Lee TJ, Chu TM, Wolfinger RD, Cella R, et al. Global analysis of *Arabidopsis* gene expression uncovers a complex array of changes impacting pathogen response and cell cycle during geminivirus infection. Plant Physiol. 2008;148(1):436–454.

42. Miozzi L, Napoli C, Sardo L, Accotto GP. Transcriptomics of the interaction between the monopartite phloem-limited geminivirus tomato yellow leaf curl Sardinia virus and *Solanum lycopersicum* highlights a role for plant hormones, autophagy and plant immune system fine tuning during infection. PLoS ONE. 2014;9(2):e89951.

43. Newell CA, Natesan SKA, Sullivan JA, Jouhet J, Kavanagh TA, Gray JC. Exclusion of plastid nucleoids and ribosomes from stromules in tobacco and *Arabidopsis*: absence of nucleoids and ribosomes from stromules. Plant J. 2012;69(3):399–410.

44. Kwok EY. GFP-labelled Rubisco and aspartate aminotransferase are present in plastid stromules and traffic between plastids. J Exp Bot. 2004;55(397):595–604.

45. Ishida H, Yoshimoto K, Izumi M, Reisen D, Yano Y, Makino A, et al. Mobilization of rubisco and stroma-localized fluorescent proteins of chloroplasts to the vacuole by an ATG gene-dependent autophagic process. Plant Physiol. 2008;148(1):142–155.

46. Avila-Ospina L, Moison M, Yoshimoto K, Masclaux-Daubresse C. Autophagy, plant senescence, and nutrient recycling. J Exp Bot. 2014;65(14):3799–3811.

24

Astroecology, cosmo-ecology, and the future of life

Michael N. Mautner*

Department of Chemistry, Virginia Commonwealth University, 1001 West Main Street, Richmond, VA 23284-2006, USA

Abstract

Astroecology concerns the relations between life and space resources, and cosmo-ecology extrapolates these relations to cosmological scales. Experimental astroecology can quantify the amounts of life that can be derived from space resources. For this purpose, soluble carbon and electrolyte nutrients were measured in asteroid/meteorite materials. Microorganisms and plant cultures were observed to grow on these materials, whose fertilities are similar to productive agricultural soils. Based on measured nutrient contents, the 10^{22} kg carbonaceous asteroids can yield 10^{18} kg biomass with N and P as limiting nutrients (compared with the estimated 10^{15} kg biomass on Earth). These data quantify the amounts of life that can be derived from asteroids in terms of time-integrated biomass [$BIOTA_{int}$ = biomass (kg) × lifetime (years)], as 10^{27} kg-years during the next billion years of the Solar System (a thousand times the 10^{24} kg-years to date). The 10^{26} kg cometary materials can yield biota 10 000 times still larger. In the galaxy, potential future life can be estimated based on stellar luminosities. For example, the Sun will develop into a white dwarf star whose 10^{15} W luminosity can sustain a $BIOTA_{int}$ of 10^{34} kg-years over 10^{20} years. The 10^{12} main sequence and white and red dwarf stars can sustain 10^{46} kg-years of $BIOTA_{int}$ in the galaxy and 10^{57} kg-years in the universe. Life has great potentials in space, but the probability of present extraterrestrial life may be incomputable because of biological and ecological complexities. However, we can establish and expand life in space with present technology, by seeding new young solar systems. Microbial representatives of our life-form can be launched by solar sails to new planetary systems, including extremophiles suited to diverse new environments, autotrophs and heterotrophs to continually form and recycle biomolecules, and simple multicellulars to jump-start higher evolution. These programs can be motivated by life-centered biotic ethics that seek to secure and propagate life. In space, life can develop immense populations and diverse new branches. Some may develop into intelligent species that can expand life further in the galaxy, giving our human endeavors a cosmic purpose.

Keywords: asteroids; astrobiology; astroecology; cosmo-ecology; life in space; nutrients; biotic ethics; in situ resources

Introduction

It is the basic human identity that we belong to life. Life is unique in nature, and for us it is precious, as we are all fundamentally united with our family of organic DNA/protein life. Life also stands out in its intricately complex structures, and because the laws of physics precisely allow life to exist.

Further, all life are united in the active pursuit of self-propagation. Belonging to life then implies a human purpose to secure, expand and propagate life. Indeed, we seek a higher purpose to our existence, and filling the universe with life can give our human existence a cosmic purpose. This purpose is best achieved in space, where life can have an immense future. While life is fragile on Earth, it can be secure in multiple worlds in space [1–4].

Astrobiology addresses the place of life in nature: its origins, prevalence and future, all of which depend on the interactions of life with its environment. Astroecology addresses these interactions, to answer and quantify some basic questions:

(*i*) Were plausible past environments conducive to the origins and early sustenance of microorganisms?
(*ii*) Can we quantify the probability that life arises in favorable environments?
(*iii*) What is then the probability that extraterrestrial life exists?
(*iv*) Can space resources support life, and if so, how much?
(*v*) Can life migrate in space, and what roles may humans play?
(*vi*) What is the future of life in the Solar System on astronomical time-scales, and in the galaxy on cosmological scales?

These questions have been speculated on since antiquity. However, they can be addressed now scientifically, and in some aspects quantitatively.

* Corresponding author. Email: mmautner@vcu.edu

Handling Editor: Beata Zagórska-Marek

For example, experimental astroecology tests of meteorites showed that similar materials in asteroids can support bacteria, algae, plant cultures and even shrimp hatchlings, with fertilities comparable to agricultural soils (Fig. 1). Measured bioavailable nutrients in these materials allow estimating the potential amounts of life (time-integrated biomass) that can be constructed in this and similar solar systems. Further, predicted future energy sources in space allow estimating the potential amounts of life in the galaxy on cosmological time-scales. These subjects will be reviewed in the present paper.

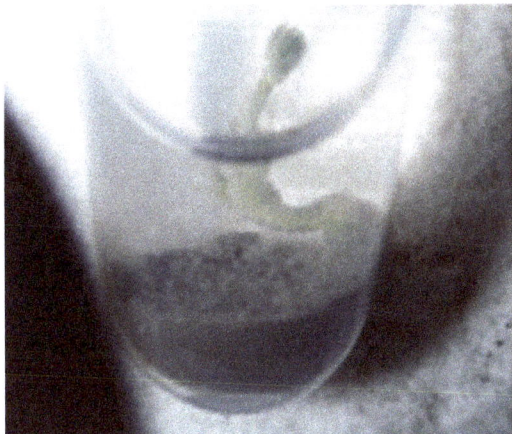

Fig. 1 Space is fertile: asparagus culture growing on meteorite soil [4].

Astroecology

The definition, and a quantitative measure, of life

Astroecology concerns the interactions of life with the space environment. For scientific purposes, it is necessary to define life and to measure or estimate its amounts quantitatively.

To estimate the amounts of life in ecosystems, "what is life?" must be first defined. In fact, defining life becomes a practical actionable question when we can control and alter life. If we aim to propagate life, we must define: what do we accept as fellow life that we seek to propagate?

Life can be defined broadly in terms of entropy and information [5]. However, our family of organic DNA/protein life focuses on propagating the species, which continues the biological, genetic, chemical patterns of life. At the heart of this process is the genetic code, which is used in translating genetic information encoded in nucleic acids into proteins. These proteins include enzymes that directly or indirectly help to reproduce the genetic code [6].

By this definition, life is a process. This process requires the flow of materials and energy, and uses information in molecular structures. Self-propagation is achieved by cycles in which DNA sequences are transcribed to RNA and then translated into proteins, including enzymes which help reproducing the DNA code, for example, by catalizing the biosynthesis of nucleic bases. All known biological cellular life, and only life, share these features.

We can therefore define life rationally: life is a process of active self-propagation by complex molecular structures through DNA/protein cycles.

Quantifying life: biomass integrated over time active ($BIOTA_{int}$)

For a scientific study of astroecology, life has to be quantified. A quantitative measure of life in an ecosystem can be formulated using the total amount of active biomass and its duration. This can be expressed in Equation (1) in terms of time-integrated biomass (biomass integrated over times available, $BIOTA_{int}$) measured in kg-years (similar to labor measured in men-years).

$$BIOTA_{int} = \int m_{biomass,t}\, dt \qquad (1)$$

Here $m_{biomass,t}$ is the amount of biomass at time t and integration is from time when life starts in the ecosystem to any given time t. Integration to the final inhabited time of the ecosystem, t_f, yields the total amount of life in the ecosystem. This $BIOTA_{int}$ may be measured in kg-years [7].

For a constant steady-state biomass $BIOTA_{steady-state}$ (i.e., a constant biomass maintained by a balance of formation and destruction) lasting for time t in an ecosystem, the time-integrated $BIOTA_{int}$ is then given simply by Equation (2).

$$BIOTA_{int} = m_{biomass,steady-state}\, t \qquad (2)$$

For example, assuming that the present amount of life on Earth, on the order of 10^{15} kg [5] has been constant for the last billion years then $BIOTA_{int}$ on Earth has been 10^{15} kg $\times 10^9$ years = 10^{24} kg-years. The potential life in the galaxy is immensely greater, on the order of 10^{48} kg-years, and the potential life in the universe can be 10^{59} kg-years [7].

Equation (2) yields $BIOTA_{int}$ that can be derived from the resource materials. Since this depends on the duration of the biomass, it may be limited if wastage removes biomass irreversibly, such as by mineralization or leakage to space, as discussed below.

Relations between resources and biomass

The maximum possible amount of life in a finite ecosystem is defined by the amount of resource materials, their nutrient contents that can be used to construct biomass and the elemental requirements of the biomass.

The biomass that can be constructed from element x in a resource material is given by Equation (3) [8,9].

$$m_{x,biomass} = m_{resource} c_{x,resource} / c_{x,biomass} \qquad (3)$$

Here $m_{resource}$ is the mass of the resource material, $c_{x,resource}$ and $c_{x,biomass}$ are the concentrations of element x in the resource materials and in biomass, respectively. Accordingly, $m_{x,biomass}$ (kg) of biomass could be constructed from $m_{resource}$ (kg) of resource material if x was the limiting element, i.e., the element in the resource materials that gives the smallest biomass. The limiting plant nutrients in nature, including meteorite soils, are usually bioavailable nitrogen (N) or phosphorus (P).

For example, Tab. 1 shows the amounts of biomass that can be constructed based on several biologically key elements in carbonaceous chondrite meteorites. The yields vary among the various meteorite classes, with phosphate or nitrate as the limiting factors, while carbon and K, Ca, Mg and sulfate would allow larger biomass yields [7].

The previous sections quantify life in terms of active biomass. This does not account for the quality of life. However, a fraction of the biomass can support intelligent life. For example, presently about 10^{15} kg biomass on Earth [10] supports 7×10^9 people, requiring about 10^5 kg supporting biomass per person. We assume below that in an efficient designed ecosystem 10^4 kg biomass per person will be required.

Alternative to materials, energy may be the limiting factor. Equation (4) gives the biomass that a power source can sustain.

$$m_{biomass} = P_{source}/P_{biomass} \qquad (4)$$

Here $m_{biomass}$ is the biomass sustained by the power source that outputs P_{source} power (energy/time), while the biomass requires $P_{biomass}$ (power/kg) to function, including an efficiency factor. We consider below a requirement of 100 W per kilogram human biomass, i.e., 10 kW for a 100 kg human [4].

Experimental astroecology

Meteorite models of asteroid soils

Human settlement of the Solar System will require large scale in situ resource utilization. Settlements may be established in space colonies [3], on asteroids [11], on planets and their moons, including Mars [12,13]. These environments need to provide bioresources of organic carbon, inorganic plant nutrients, and water. The most accessible sources in the Solar System are found in carbonaceous C type asteroids that contain these materials.

Can these materials really support life, and if so, how much? Fortunately, samples of the asteroids are available in carbonaceous chondrite meteorites, and samples of planetary resources in Martian meteorites. We measured their soluble, bioavalable contents of the nutrients shown in Tab. 1, and tested if soil microorganisms, colonizing cyanobacteria, algae, edible plants (asparagus and potato), and even small animals (brine shrimp) can grow on these space materials.

First, we tested the growth of microorganisms on meteorites. The results showed that a mixture of autotrophs (algae) and heterotrophs (fungi) can grow on meteorite and planetary materials, as required for a sustainable ecosystem that recycles nutrients (Fig. 2). The algal populations in the extracts were substantially larger than in the control deionized water, and approached populations reached in optimized BG11 nutrient medium (Fig. 3).

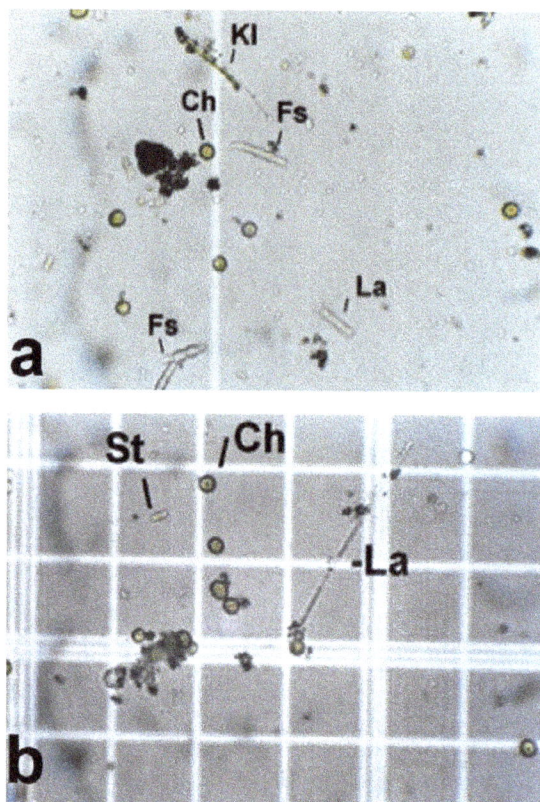

Fig. 2 Algal populations (colony forming units, CFU/ml) in extracts of Murchison carbonaceous chondrite (**a**) and Dar al Ghani 476 Martian meteorites (supplemented by 0.5 millimolar NH_4NO_3; **b**) after 32 days growth. Ch – *Chlorella*; Kl – *Klebsormidium*; La – *Leptolyngbya*; St – *Stichococcus*; Fs – fungal spores [9].

Tab. 1 Biomass yields from water-soluble and total elemental contents in carbonaceous chondrite materials[a].

Meteorite	Type	Biomass yield from element (g/kg)					
		C	N	K+	Nitrate-N	Phosphate-P	Sulfate- S
Murchison	CM2	10.3	2.2	7.0	0.1	0.06	823
GRA 95229	CR2	4.0	3.8	3.8	0.1	0.03	279
Allende	CV3	2.1	1.9	0.8	0.1	0.19	24
ALH 85002	CK4	0.4	0.7	0.1	0.7	0.09	51
Average biomass (from soluble elements)[b]		2.4	1.5	2.5	0.3	0.08	301
Average biomass (from total elements)[c]		15.9	5.9	17.8	----	74	2343

[a] Units of g biomass/kg meteorite. Maximum biomass (g) of average composition that could be constructed from a given soluble element x in 1 kg of each meteorite, if x was the limiting nutrient. [b] Calculated using soluble contents from eight meteorites and average elemental concentrations in dry biomass. [c] As in [b], but total elemental contents [26].

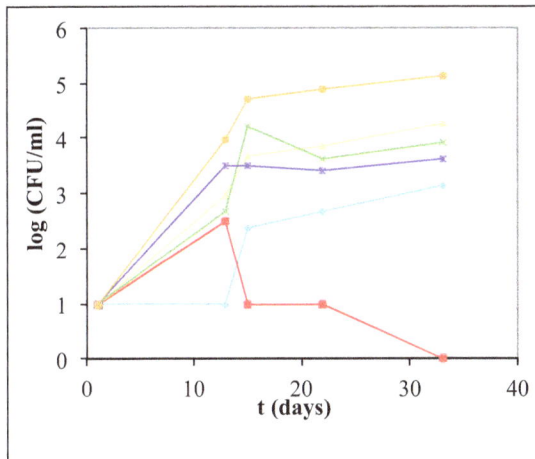

Fig. 3 Algal population growth in meteorite and simulant extracts and BG-11 nutrient medium. Allende (yellow line), Murchison (violet), DaG 476 (blue), Hawaii lava Mars simulant (green), BG-11 medium (orange), deionized water (red) [9].

Further, asparagus cultures showed that nutrients from carbonaceous chondrite and Martian meteorites enhanced plant growth, especially the Martian meteorite because of its larger phosphate content (Fig. 4) [8,9,14]. We also found that brine shrimp eggs can uptake meteorite materials and can hatch in meteorite extracts, which shows that the extracts are not significantly toxic.

These experimental astroecology studies were followed by measuring the concentrations of soluble bioavailable carbon and nutrient electrolytes in carbonaceous chondrite meteorites. With these data, we calculated the potential biomass yields from several types of carbonaceous chondrites according to Equation (2) with P as the limiting nutrient [9]. Tab. 1 shows the calculated results for a composite soil with the average elemental contents of eight carbonaceous chondrite meteorites of various classes.

The measured bioavailable nutrient contents and the biological yields from asteroid/meteorite materials can be combined for rating their soil fertilities as compared with productive agricultural soils (Tab. 2).

The results showed that the carbonaceous chondrite meteorites have soil fertilities comparable to agricultural soils. Martian meteorites had the highest fertilities because of their high bioavailable phosphorus contents.

Further, these results allowed an experiment-based estimate of potential biomass in the Solar System that can be derived from asteroid resources, as discussed below.

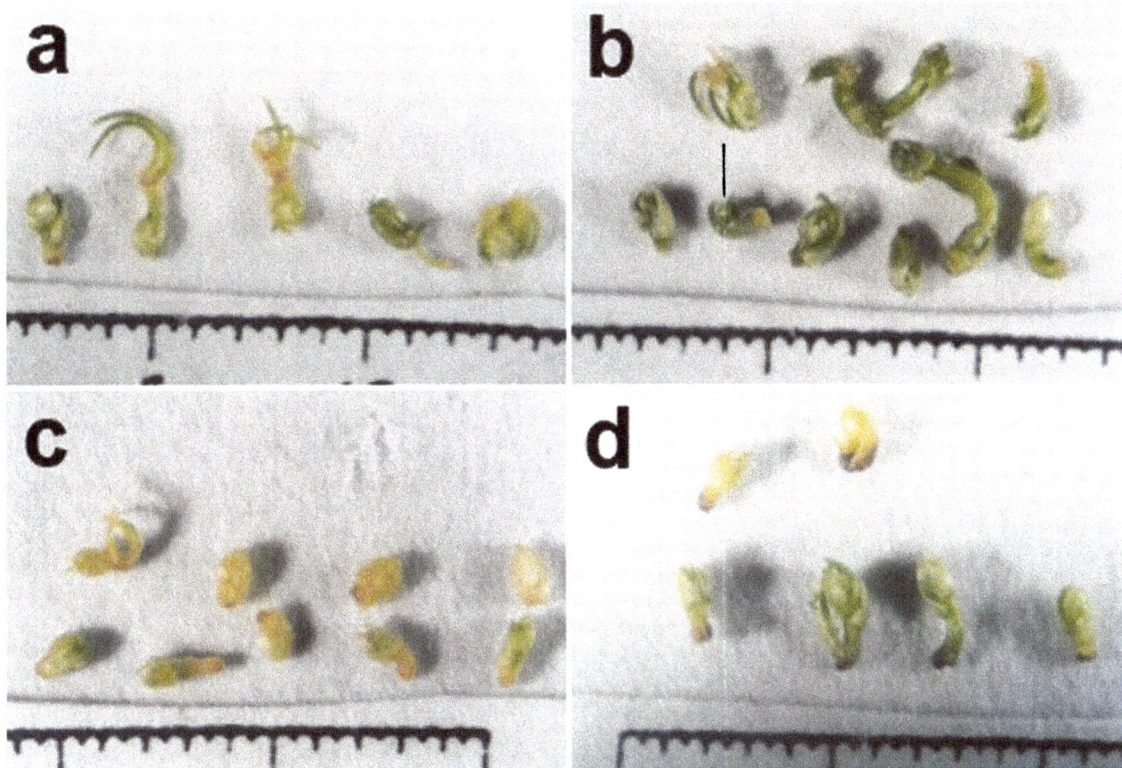

Fig. 4 Plant tissue cultures of *Asparagus officinalis* in meteorite and soil extracts, all supplemented by 5 millimol/l NH_4NO_3 and 3% sucrose. **a** Murchison CM2 meteorite. **b** DaG 476 Martian meteorite. **c** Water. **d** Hawaii lava Mars simulant. Scale divisions: small ticks 0.5 mm [9].

Tab. 2 Fertility ratings of planetary materials according to biological yields and nutrient contents, and an overall fertility rating[a].

	Algal yield	Average algal and plant yield	N nutrient	P nutrient	Fertility rating
Allende meteorite	+	++	+	+	Medium
Murchison meteorite	+	+	++	+	Medium
DaG 476 (Mars)	++	++	+++	++	High
EETA 79001 (Mars)	+++	++	+++	+++	Very High
Lunar simulant (lava ash)	0	0	0	++	Medium
Agricultural soil	++	++	0	+	High

[a] Ratings according to deviation from standard normal variate for each property: low (0), medium (+), high (++), very high (+++) yield or nutrient content [9].

The human role

Expanding life in the Solar System: serving human needs and motivations

The expansion of life in space can take dual routes: by human expansion in the Solar System, and by seeding new solar systems with microbial life. Interstellar human travel is desirable ultimately, but seems impractical with current levels of technology [15,16].

Human expansion in the Solar System will depend on physical resources, technology, and on motivation. Concerning resources, the above astroecology experiments confirmed that asteroid materials can support life, and their measured nutrient contents allow estimating the supportable human populations.

Settling the Solar System can be motivated by serving human needs. These can include: satellite solar power stations for permanent clean energy [17]; a space sunshade against global warming [18,19]; mining of asteroids metals and structural materials [11,20]; large space colonies for growth and survival [3]; high resolution lunar telescopes [21]; and lunar gene banks for saving and re-cloning endangered species, including endangered human ethnic groups [22]. The space infrastructure that develops for these purposes will then allow exponential human growth in the Solar System, and serve as a base for seeding new solar systems [23–25].

The possible scope of space colonization can be based on Solar System resources that were estimated above. The resources and populations in various stages of space settlements, up to cities of millions, were examined recently [26].

On the long term, human expansion in space can be motivated by a responsibility to secure life. By necessity, we need to be the guardians of life, because only technological humans and post-humans can secure life to realize its immense potentials in space.

Life in the Solar System: resources, biomass and populations

Accessible resources are available in C type carbonaceous chondrite asteroids, and later in comets, that contain soluble bioavailable organics, electrolyte nutrients and extractable water. The bioavailable contents can be compared with the elemental requirements of biomass. This gives the potential yield of biomass from a unit mass (kg) of resource materials

[Equation (3) and Tab. 1], with bioavailable P and N as the limiting elements [8,9]. For example, eight different carbonaceous chondrite meteorites contained an average bioavailable phosphate content of $c_{x,resource}$ = 0.0012 g/kg meteorite solids [14]. In comparison, average dry biomass contains $c_{x,biomass}$ = 15.5 g P/kg biomass [10]. Therefore the yield of biomass from this soil, $m_{x,biomass}/m_{resource}$ is 0.0012/15.5 = 7.8×10^{-5} or roughly 10^{-4} kg biomass/kg soil (Tab. 1).

On this basis, the estimated 10^{22} kg carbonaceous asteroids could yield 10^{18} kg biomass based on limiting P. If all of this was incorporated in human biomass, this could allow about 10^{16} humans.

By another estimate, in the Earth ecosystem 10^{15} kg biomass supports about 10^{10} humans, requiering 10^5 kg sustaining biomass/human. A more efficient designed ecosystem may allow 10^4 kg sustaining biomass/human. On this basis, the 10^{18} kg biomass derived from asteroid solubles could sustain a population of 10^{14} humans. In the next billion habitable years of the Solar System these would allow 10^{23} human-years and a total of 10^{27} kg-years of time-integrated biomass, a thousand times more than the estimated 10^{24} kg time-integrated biomass on Earth to date.

Human populations and the effects of wastage

Equation (2) above yields the amount of biomass that can be derived from resource materials. However, the amount of life that is produced, in terms of time-integrated biomass $BIOTA_{int}$, also depends on the duration of the biomass. This may be limited if wastage removes biomass irreversibly. This wastage decreases the amount of biomass by a fraction k_{waste} per unit time (year), and the remaining biomass after time t is then given by $m_{biomass,t} = m_{biomass,0} \exp(-k_{wastage} t)$. Here $m_{biomass,0}$ is the starting biomass and $m_{biomass,t}$ is the remaining biomass after time t [7].

From Equations (1) and (4) above follows a relation for the total time-integrated biomass $BIOTA_{int,total}$ from time zero to infinity, for a biomass $m_{biomass,0}$ formed from the resources and subject to wastage at the rate of $-k_{wastage} \times m_{biomass,t}$ (kg/year), giving $BIOTA_{int,total} = m_{biomass,0}/k_{wastage}$. This relation applies to each unit of biomass that decays exponentially, regardless of when it was formed. The total integrated biomass $BIOTA_{int,total}$ depends only on the total amount of biomass created and its decay rate, but not on the rate of formation or on the biomass that exists at any time. Therefore, the lifetime of the ecosystem, yielding a given amount of life $BIOTA_{int,total}$ can be extended by forming the biomass more slowly, and sustaining a smaller biomass longer.

A low k_{waste} = 0.01, i.e., 1% per year of the steady-state biomass will be considered. With a steady-state biomass $m_{biomass,steady-state}$ (kg), then $k_{waste} \, m_{biomass,steady-state} \, t_{ecosystem}$ biomass is wasted during the lifetime of the ecosystem.

The amount of life in the ecosystem is maximized if all the resources are used during the lifetime of the ecosystem (here 10^9 years). Considering 10^{18} kg biomass from soluble asteroid resources, the relation k_{waste} (0.01) $\times m_{biomass,steady-state}$ $\times t_{ecosystem}$ (10^9) years = 10^{18} kg allows maintaining a steady state biomass of 10^{11} kg by continually producing biomass to replace the wastage. This can support a population of 10^7 (ten million) humans, during a billion future habitable years of the Solar System. With these parameters the asteroids then yield 10^{20} kg-year time-integrated biomass supporting 10^{16} human-years in the Solar System.

If all the elemental contents of the asteroids can be used, this can produce a biomass larger by about a factor of 100. Further, the 10^{26} kg comets, assuming compositions similar to asteroids, could support populations and biomass that are larger by a factor of 10 000 than the above amounts based on the asteroids.

Energy requirements

Populations in space can use solar energy. The energy demand can be estimated based on an industrialized 1 kW power per person, plus 9 kW for the supporting biota, and a conversion efficiency of 10% of the collected solar energy to electricity, adding up to 100 kW/person (alternatively, if all the collected energy is used to support biomass, the power demand of 30W/kg biomass and 10^4 kg supporting biomass per person would require collecting 300 kW/person). The above steady-state population of 10^{11} that was based on material resources then requires a power supply of 10^{16} W, a small fraction of the 3.8×10^{26} W output of the Sun.

Considering the power demands of the biomass, the estimated 10^{15} kg terrestrial biomass is supported by 3×10^{16} W absorbed solar irradiance (240 W/m^2 absorbed irradiance $\times$ 1.3×10^{14} m^2 Earth cross section), i.e., 30 W/kg biomass. At this rate, the above 10^{18} kg biomass derived from asteroids would require a solar power supply of 3×10^{19} W, still a small fraction of the solar output. Accordingly, the biota of the Solar System is limited by materials and not by energy.

Further in the future, the total elemental contents of the 10^{26} kg materials in comets, assuming a CM2 meteorite-like composition could yield a biomass of 10^{24} kg with a solar power demand of 3×10^{25} W, about 10% of the solar output, that could be collected by a Dyson sphere [27].

The above estimates concern the upper limits of biomass and populations in the Solar System about the main sequence Sun. Beyond that, life can continue after the Sun becomes a white dwarf star, and similarly about other future stars, on cosmological time-scales as discussed below. Life-centered ethics would recommend that as much of this potential life should be realized as possible, leading to great biological, social and intellectual advancements.

The biology of human space adaptation

The new planetary and space environments may present wide ranges of atmospheres, hydrospheres, geology, pressure, temperature, chemistry, pH, lighting, radiation, and gravity. Some of these may have contributed to early life [28], and may require adaptation by future life. Further, the ultimate adaptation may be free living in open space. This may require new human features such as vacuum-tight containment; fully recycling self-contained metabolism; solar sail "wings"; asexual reproduction; hybrid algae/human organs for photosynthesis; organs for direct radio or laser communication; extended IR to UV vision; biological brains interconnected with computers; lifespans of centuries; social interactions among new life-forms; and psychological adaptations to extreme solitude or crowding.

For continuing our family of life, this transformed biota must still remain DNA/protein organic life. However, it can incorporate proteins with novel properties, using new amino acids, possibly coded by DNA that is extended with new nucleic bases [29], or other types of information-containing molecules. Moreover, to continue our genetic heritage similar to natural evolution, the genes of these human descendants should build on and incorporate human genes.

Artificial intelligence, durable robots or robot organs may be necessary in space. However, in silico "artificial life" is not life as defined commonly. Substituting ourselves with robots would eliminate, not propagate, our organic DNA/protein life-form. Even in space and through engineered evolution, the continuation of biological life is best assured through control by biological brains (as opposed to computers) with a vested interest to continue DNA/protein life.

The profound changes in biology can affect human nature itself, and even the core processes of biology. This may seem futuristic, but scientists are already developing these new biotechnologies [30]. These developments will affect the genetic heritage of all present humanity, and should be therefore subject to informed public approval.

The transformation of life will also require psychological and philosophical adaptations that promote survival. In fact, destruction (high entropy) is easier than survival, and any advanced society can self-destruct. Aggression then needs to be modified, and re-directed to motivate the expansion, not the destruction, of life. In a self-fulfilling future, the conscious pursuit of survival will be essential to secure survival.

Early life and the origins of ecology

Chemical kinetics and the probability of life

Life can exist and expand in the Solar System for its expected five billion habitable years, but this is a minute fraction of the possible future life in the universe. A main question is if life already exists in other Solar Systems, or if we are alone, with the future of life in our hands. If so, we can end our cosmic isolation by seeding new solar systems with life; and if we are alone, it is, by biotic ethics, a moral duty to do so.

The prevalence of life in the universe can be estimated by the probability that life arises and survives in favorable environments. There may be many other habitable solar systems throughout the galaxy, but the probability for life to start there may be very small, because all viable self-replicating life forms need complex interactive biological functions. The required components include all four nucleic bases and corresponding nucleosides and nucleotides, biological amino acids, ADP and ATP, membranes and mebrane transport apparatus. These components have millions of isomers and related structures, and the products of non-selective abiotic synthesis from simple compounds would have spread over millions of compounds, that would then all have too low concentrations to allow further synthetic reactions. Examples are the mixture of D and L isomers of biological and non-biological amino acids in meteorites. Similarly, if all nucleic base and amino acid analogues and isomers were present equally, their concentrations would be too low for forming more complex biomolecules.

Even the first viable proto-cell needed these complex biochemical functions and apparatus. It would have had a very small probability to self-assemble through many steps with many possible outcomes, only a few of which lead to life.

Indeed, it is remarkable that a single-cell cyanobacterium can convert a few simple compounds (water, CO_2, ammonia, phosphate) into thousands of complex molecules, which would be all required in a viable precursor. For example, the genome of cyanobacteria is in the range of 10^6–10^7 base pairs [31,32]. Assuming that in a more simple protocell sequences of only a few hundred DNA or RNA bases coded peptides and proteins, even this simple genome could have still coded tens of thousands of different peptides and proteins, each of which catalyzed the synthesis of a specific biomolecule.

In turn, in an interactive system such as a protocell, every molecule may react with another molecule, including self-reactions. Considering bimolecular reactions, n compounds can then undergo n^2 different reactions. For example, 1000 compounds can undergo 10^6 reactions, with rates determined by the pseudo-first order rate coefficients that are affected by the n reactant concentrations and other variables such as temperature, pressure, pH, ionic strength, homogenous catalysis by metal ions, heterogenous mineral catalysts, inhibitors, and IR to UV light intensities.

All of these parameters can have a continuum of values, but for simplicity, assume only ten distinct values for each. Then 10 physical parameters each with 10 discrete values allow 10^{10} different combinations affecting each of the 10^6 reaction rate coefficients, for a total of 10^{16} different chemical states (where a chemical state is a combination of the parameters that define that state). This model is still oversimplified because every bimolecular reaction may be catalyzed or inhibited by each of 1000 chemicals in the cell. These termolecular interactions increase to 10^{19} the number of possible chemical states, and decrease the probability of a spontaneous assembly of a viable protocell to 10^{-19} accordingly. Further, the protocell had to arise in a supporting environment that also has many variables, which further decreases the probability to form a viable protocell.

Moreover, this first protocell needed to arise in a survivable environment, and the microbial population had to able to adapt to changing environments, but this involves more advanced biological capabilities.

The origins of ecology

A viable first cell must emerge in an environment where it can survive, multiply and evolve. The probability $P_{biosphere}$ that a viable biosphere arises may be expressed in a truncated Drake-type Equation (5) [33], where P_{origin} is the probability that first life arises in a given environment, and $P_{survival}$ is the probability that this environment can sustain it.

$$P_{biosphere} = P_{origin} P_{survival} \qquad (5)$$

For example, by the above estimates, the probability that a viable self-propagating system arises is $P_{origin} < 10^{-19}$ in a sustaining environment where $P_{survival} < 10^{-6}$ (one of 10^6 possible states; Fig. 5), i.e., $P_{biosphere} < 10^{-19} \times 10^{-6} = 10^{-25}$.

Further, once a viable self-replicating multiplying protocell arises, its population can grow exponentially, and all the nutrients in the environment would become sequestered in this biomass. The continuing availability of nutrients and essential organics would require continuing biosynthesis (autotrophs) and recycling (heterotrophs), but such complex populations are unlikely in the first ecosystem.

Alternatively, the first cell would have to synthesize its biomolecules from simple compounds, but this already requires catalysis by complex biomolecules such as enzymes in the first place (the "chicken or eggs first" paradox).

In summary, even the first sustainable ecosystem must provide the continuing supply, production and recycling of biomass for a sustainable ecosystem. Fig. 5 summarizes this early ecology.

The active biomass can establish a steady-state involving the cycle in Fig. 5. The total mass of the system can achieve a steady state depending on the production rates of organics and on their depletion rates to form inactive end products.

Fig. 5 may be too complex for analytical solution but may be suitable for computer modelling. The amount of active biomass is controlled by six rate coefficients even in this simple scheme. If each can have only 10 distinct values, only one of which sustains life, then as in Equation (7), this ecological complexity alone would decrease the probability of a viable early ecosystem by a factor of one million.

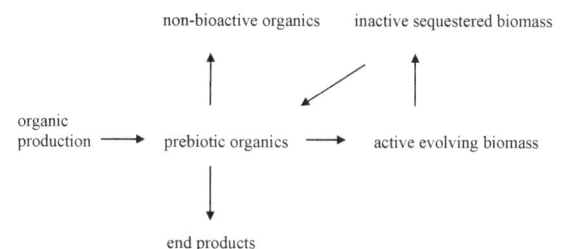

Fig. 5 Processes in an early ecosystems.

Further, once sequestered in biopolymers, these organics would not be available for further biosynthesis. Therefore, a continual supply of biomolecules is needed through photosynthesis by autotrophs and recycling by heterotrophs. A natural founding population of protocells is unlikely to be able to fulfil all of these functions, but directed seeding populations could do so. An alternative continuous source of organics may be the infall of interplanetary dust, meteorites and comets. The requirements of early ecosystems are discussed further below.

Can the probability of life be tested experimentally?

The earlier question: "can we quantify the probability of life?" can be restated: "can we compute or test experimentally the probability that life arises, given plausible source materials and environmental conditions?" A scientific model would have to explore all the chemical states of a plausible pre-biotic chemical system, i.e., all possible combination of the variables that affect the system.

For example, by the preceding estimates, a viable protosystem has to be in one of 10^{19} chemical states in an environment that can have one survivable set of parameters out of 10^6 possible sets, i.e., $P_{\text{biosphere}} = 10^{-19} \times 10^{-6} = 10^{-25}$. An experimentally verifiable model would have to test experimentally these 10^{25} states.

If one of the 10^{25} states could be tested in one day for forming a viable protocell, then the experimental testing would require 10^{25} days or about 10^{18} human-years (a billion scientists working for a billion years) to test this simplified model, which is only one of practically innumerable possible prebiotic chemical systems with various compositions and physical parameters. Even this would yield only one of many possible mechanisms for life to arise and would not identify how life actually started. Apparently, a realistic experimental model for the probability to form a viable biological ecosystem stochastically from plausible precursor chemicals may not be feasible.

In other words, the origins of complex life involve an improbable coincidence of a very large number of physical and chemical variables. It seems unlikely that such a model can be constructed, much less tested experimentally to verify proposed mechanisms. It is then not possible to quantify the probability that life arose and now exists elsewhere. Without such information, and if astronomers don't find extraterrestrial (ET) life, we may need to consider that terrestrial life may be alone (extraterrestrial means here life outside our Solar System. Microorganisms on other planets in this Solar System, that result from material exchanges among planets such as in Martian meteorites, would be irrelevant to the probability of independent origins elsewhere. To distinguish ET life in and outside the Solar System, a compact terminology for solar system may be solys, and for life outside our Solar System, extrasolys life).

Seeding the galaxy

An overview of directed panspermia

With the existence or probability of extraterrestrial life unknown, we may be alone to secure the future of life. We can then make sure that life will continue and expand, by seeding new solar systems with representatives of our organic DNA/protein life-form. In fact, our biological unity with all life, combined with our technical abilities, imply a moral responsibility to secure and expand life in space.

The essential structures and processes of our cellular organic life are present in every cell from microorganisms to humans. To carry this basic information, microorganisms can be launched to space in large numbers to start new ecosystems that can lead to new species and a galactic-scale biodiversity. There is some urgency, because we don't know how long our space-faring technology will endure. Interfering with indigenous life can be avoided by seeding newly forming young solar systems where local life could not have started yet.

Similarly, directed panspermia by another civilization could have started life on Earth after arising elsewhere [34]. Conversely, directed panspermia from Earth to space was also considered [35,36], and its scientific and ethical aspects have been developed in some detail [23–25,37–39]. From the new habitats, life may expand further by natural or directed panspermia, as an effective mechanism for the expansion of life.

Panspermia missions may aim to seed extrasolar planets, accretion disks about young stars, or star-forming zones in interstellar clouds. Each strategy has different requirements and probabilities of success.

Mature planets nearby, with liquid water, may be ready for colonizing microorganisms. These habitable extrasolar planets are small and hard to aim at accurately, but dispersing the microbial capsules in orbit can increase the probability of capture (Fig. 6) [23]. For example, for a reasonable probability of success we may send 100 capsules to each target, i.e., $n_{\text{(capsules)}} = 100$, carrying 100 000 microorganisms of 10^{-15} kg each and a biomass payload of 10^{-10} kg, delivering a total of 10^7 microorganisms to seed the planet.

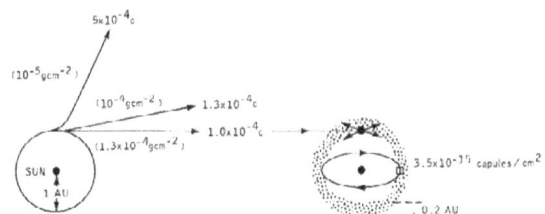

Fig. 6 Launch of microbial payloads from Earth orbit to target planetary systems by solar sail propulsion, and dispersion and capture at the extrasolar planet [23].

Seeding star-forming clouds

The preferred strategy could aim at star-forming interstellar clouds as the largest and easiest targets (Fig. 7). An important advantage is that local life, especially advanced intelligent life, would not have developed yet in these new solar systems, avoiding biological interference. Also, stars develop in clusters in these clouds and one mission can seed dozens of new solar systems [25].

One potential target is the Rho Ophiuchus cloud 520 light-years (ly) away, that contains zones with various densities as they progress to star formation [40]. The missions may target the entire cloud; within these clouds, specific dense condensation cores; within these cores, protosolar condensations that form stars; and within these condensations,

Fig. 7 Young stars in interstellar clouds (NASA Hubble Space Telecope). Panspermia missions can aim for the clouds, for dense star-forming zones within them, or for accreting planetary systems about the new stars (red dots).

accretion disks about new stars. It is desirable to target the smaller zones with high precision, where the capsules will mix with less dust and therefore a larger fraction will be then delivered to planets [25].

The probability $P_{(target)}$ that a mission will arrive at a target zone is given by Equation (6). Here A is the area of the target, δy is its positional uncertainty at the time of arrival, $r_{(target)}$ is the radius of the target zone, v is the velocity of the vehicle, α_p is the uncertainty of the angular proper motion of the target, and d is the distance to the target.

$$P_{(target)} = \frac{A_{(target)}}{\pi(\delta y)^2} = \frac{r_{(target)}^2 v^2}{\alpha_p^2 d^4} \qquad (6)$$

The probability of arrival to the target can be increased by selecting larger targets ($r_{(target)}$), increasing the velocity of travel (that decreases the positional uncertainty of the target at arrival), decreasing the uncertainty of position of the target (increasing precision of astrometry) and decreasing d (closer targets). Equation (6) yields the $P_{(target)}$ values in Tab. 3. Values of $P_{(target)} \geq 1$ means unit probability.

The microbial payload can be launched as swarms of small microbial capsules, or the capsules can be bundled and protected through the interstellar journey to the targets where they will be dispersed by collisions with dust. The panspermia capsules will mix with the dust and condense with it into frozen and shielding in asteroids, comets and interplanetary dust particles. We can predict the conditions of these objects, because the outer zones of all solar nebulae should be similar, with temperatures <50 K, and contain ice, dust and organics similar to our early solar nebula.

Panspermia payloads to accretion zones or interstellar clouds will be captured in the dust and accreting asteroids and comets solar nebulae, similar to possible early life in asteroids in our Solar System during aqueous alteration [28,41–46]. The prospects for microbial life that was seeded into or arose locally in solar nebulae [42], are discussed below.

Some of the microbial capsules stored in asteroids and comets will be delivered later by meteorites and interplanetary dust particles (IDPs) to planets that developed habitable conditions. If the capsules are mixed homogenously with the dust, the fraction of the capsules delivered to the planets will be equal to the fraction of total dust delivered to planets as IDP particles, i.e. $P_{(delivery)} = m_{(delivered\ dust)} / m_{(total\ dust)}$.

The probability $P_{(planet)}$ that a capsule originally launched to the cloud will be delivered to the planet is then given by Equation (7).

$$P_{(planet)} = P_{(target)}P_{(delivery)} = \frac{P_{(target)}m_{(delivered\ dust)}}{m_{(total\ dust)}} \qquad (7)$$

Tab. 3 The probability that panspermia missions will arrive at their target zones, and probability for eventual capture of one microbial capsule by a planet[a].

	Distance (ly)	Radius of target area (au)	Uncertainty of target position (au)	Probability of arriving in the target zone[b]	Probability of capture at a target planet	Biomass required for probable success (kg)[c]
Nearby stars with accretion disks						
Alpha PsA (Fomalhaut)	22.6	3.2	3.1	>1[b]	1.1×10^{-5}	1.0×10^{-3}
Beta Pictorius	52.8	8.7	17	0.3	2.5×10^{-6}	4.4×10^{-3}
Rho Ophiuchus star-forming cloud						
Dense fragment	520	200 000	1600	>1[b]	1×10^{-16}	1.1×10^{8}
Protostellar condensation	520	2000	80	>1[b]	1×10^{-13}	1.1×10^{5}
Early accretion disk	520	100	80	0.004	3.9×10^{-14}	2.8×10^{5}
Planetary feed zone	520	3.5	80	0.000006	4.9×10^{-11}	2.2×10^{2}

[a] Mission velocity 0.0005 c and parameters as described in [25]. [b] Probabilities greater than unity mean that arrival or capture is virtually certain. [c] Assuming the capture of 100 capsules with 0.11 microgram of microorganisms each, i.e., a total captured biomass of 11 micrograms. The required biomass is therefore given by biomass = $10^{-8}/P_{capture}$ kilograms [25].

If the probability for a launched capsule to reach a target planet is small, then sufficiently large numbers of capsules should be launched to assure a probability of landing on a planet, as given by Equation (8) will be near unity.

$$n_{(capsules)} = \frac{1}{P_{(planet)}} = \frac{m_{(total\ dust)}}{P_{(target)}m_{(delivered\ dust)}} \qquad (8)$$

Penetration to target zones with various densities within the cloud can be achieved by choosing the mass of the vehicles, since heavier projectiles can penetrate into increasingly dense zones. Each capsule can be placed in a miniature spherical reflective film solar sail that does not need attitude control.

The above factors affect the probability that the microbial capsules will reach young planets. If the probability is small, more capsules with larger biomass need to be launched to increase the probability of success. The biomass required for successful missions in Tab. 3 was derived by these considerations.

These directed panspermia programs can be realized using current-level technologies, and may be easy to implement when a space infrastructure makes launch costs affordable. Even now, with present launch costs of $10 000/kg, a few hundred tons of microbial biomass with launch costs of about $1 billion can seed dozens of new solar systems with our family of organic life to last there for eons.

Biological and ecological challenges for directed panspermia

Biological adaptation is key to life in space. The planted biota need to survive, procreate, grow and evolve in space habitats.

Fortunately, extremophile microorganisms can survive in a wide range of environments, from anaerobic to oxygen-rich conditions, from below 0°C to over 140°C, from low pressures at high altitudes to high pressures in deep seas, from basic solutions at high pH to concentrated sulfuric acid, from fresh water to concentrated brine, and also under intense radiation [47,48].

Further, at currently achievable speeds of 0.0001 c, the microorganisms need to survive cryptobiosis states for transit times up to 100 000 years or more, which may be possible at interstellar temperatures of a few K in high vacuum. Faster transit times using more advanced interstellar propulsion methods are desirable [20].

Adapting life to such diverse extreme environments can involve fundamental biological changes, down to the basic molecular levels of DNA/protein life. These transformations need to be rapid if new environments are settled rapidly.

The microbial payloads can contain natural and bioengineered species with various tolerances, including designed microorganisms with combined multiple tolerances, hardy cyanobacterium akinetes, bacterial endospores that can survive interstellar travel without nutrients, and species resistant to UV radiation, desiccation in vacuum, freezing or high temperatures, and chemicals. It is a challenge to bioengineer space-adapted microorganisms with combinations of these various tolerances.

As noted above, sustainable seeded ecosystems require both photosynthetic autotrophs and biodegrading heterotrophs. Autotrophs are needed to synthesize complex biomolecules, and heterotrophs such as fungi are needed to degrade and recycle biomass. The autotrophs could be chemotrophs but preferably photosynthetic organisms such as algae that can propagate the chlorophyll-based apparatus and the genetic mechanisms to synthesize them. The heterotrophs may include fungi in lichens that are hardy colonizers. The combination of autotrophs and heterotrophs can form self-sustaining ecosystems, and facilitate evolution by predator/pray pressures.

Microbial ecosystems can fulfil the basic objective to continue DNA/protein life. However, as humans, we may also want to induce evolution toward conscious intelligent life. For this purpose we can include, along with the first colonizer microorganisms, hardy multicellular organisms such as rotifer cysts. Rotifers have the basic body-plans of higher organisms with differentiated organs and an animal-like, but not insect-like, body-plan (Fig. 8). This will bypass a bottleneck to the evolution of multicellular organisms that took eons to develop on Earth. Rotifer eggs and cysts can possibly survive long interstellar journeys when deeply frozen. Tardigrades could also survive space travel [49].

Life in space, including directed panspermia, will encounter diverse environments that pose basic challenges for biology. Fortunately, the laws of physics and chemistry that underlie biology allow a broad range of biological transformations. The only constraints are the tests of survival that will always challenge life, whether produced naturally or by conscious designs.

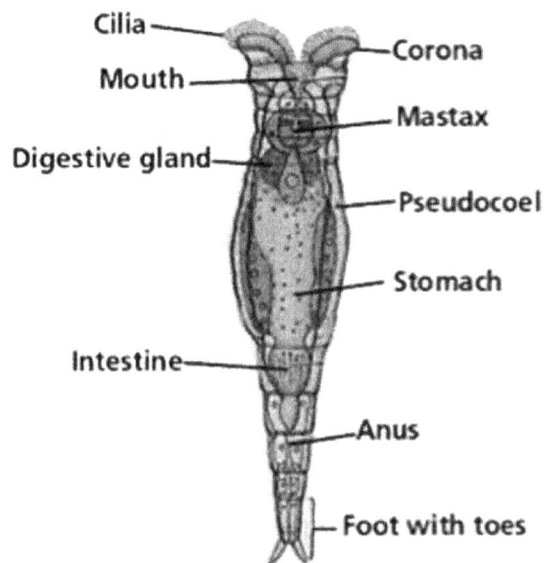

Fig. 8 Rotifers with animal body plan can jump-start higher evolution.

Prospects of directed panspermia

The technologies for directed panspermia are advancing: solar sailing and interstellar propulsion, precise astrometry, search for extrasolar planets, natural and bioengineered extremophile microorganisms. As a space infrastructure develops to fulfill human needs, launching directed panspermia missions will become possible for motivated individuals, small groups and organizations.

This program may affect the future of life, and possibly that of the physical galaxy, more than any other human activity. These fateful prospects, that affect the heritage of all, should be discussed publicly in the next decades while the technologies are advancing. The immense potentials of future life can contribute ethical incentives for space development.

Directed panspermia and panbiotic ethics

A concern about directed panspermia is that it may interfere with local life. This probability can be minimized by seeding newly forming young solar systems where local life, especially advanced life, could not have developed yet. In any case, we may at best seed a few dozen new solar systems, which would secure life but leave the vast majority of stars untouched for future exploration.

The only life known presently is our organic DNA/protein life on Earth. The existence of life outside our Solar System is unproven and its probability cannot be quantified. If our microbial messengers encounter other branches of DNA/protein life, they may be both enriched by a genetic exchange. If they meet other, different life-forms, they will not interfere with it.

If we are alone, all life in the universe will end with end of the Sun, and the immense potentials of life in space will be lost. We can secure life now to make sure that this does not happen. It would seem irresponsible to abandon the only life that we know to exist to a certain end, for concerns about extraterrestrial life that may or may not exist.

A full scope of these panbiotic ethics can be derived from life-centered principles as discussed below [4,38]. They can secure the propagation of life if these ethics themselves are always propagated.

Cosmo-ecology and the ultimate future

An overview

Once life is established in space, what is its cosmological outlook?

Life can develop in many new directions given cosmological resources and timescales. The future forms of life are unpredictable, but its amounts will be defined by the available resources. Cosmo-ecology quantifies these potentials, and life-centered panbiotic ethics aims to realize them.

Cosmo-ecology can estimate the possible amounts of future life in the galaxy and in the universe [in units of $BIOTA_{int}$ (kg-years)]. Starting with the Solar System, life can survive for 10^{20} years about the red giant Sun and then about the subsequent white dwarf Sun, and likewise about other similar stars [4,7,50].

Quantitatively, the amount of life on Earth in the last billion years has been on the order of 10^{15} kg $\times 10^9$ years = 10^{24} kg-years (assuming a constant biomass of 10^{15} kg during this period). In comparison, the energy of red and white dwarf stars can sustain 10^{46} kg-years of $BIOTA_{int}$ during trillions of eons in the galaxy [7,50]. The potential future of life is immensely greater than its past.

At the theoretical limits all matter could be incorporated into biomass, and then converted gradually into energy to power this biomass. This yields the maximum possible amounts of life in the universe. However, the actual long-term future depends on dark matter and dark energy that have not been characterized yet. The 14 billion years since the Big Bang were just a brief moment in the long-term evolution of these forces, and predicting their future, and with it the future of life and the universe, may be possible only after further observation for thousands of future eons. Speculatively, life may continue indefinitely also in other universes [51], but they are unobservable by definition.

In the known universe life has immense potentials. To realize these potentials, we can preserve, continue, secure and expand life in space. Our descendants will be here then to understand nature more deeply, and reach for eternity.

Future life in the Solar System

The following discussions assume an efficiency of 10% for collecting and converting stellar radiation to biological energy, and a power use of 10 W/kg by metabolically active biomass, altogether 100 W of stellar power/kg biomass. The energy from stellar sources in various periods is estimated [50], and future technologies may capture them all in Dyson spheres [27].

The contribution of energy sources to time-integrated biomass $BIOTA_{int}$ (kg-years) in an ecosystem is given by Equation (9), where L_{source} (W) is the luminosity of the source (energy output/time), C_{eff} is the efficiency of collection and conversion of radiation to biological energy, t_{source} (year) the life-time of the energy source, n_{source} is the number of similar energy sources in the ecosystem (here, stars in the galaxy) and $P_{biomass}$ is the power demand of the biomass [W/(kg biomass)].

$$BIOTA_{int} = \frac{L_{source}C_{eff}t_{source}n_{source}}{P_{biomass}} \quad (9)$$

After the current main sequence phase, the Sun will become a red giant and then a white dwarf star following the patterns of main sequence stars [50]. Populations can survive these transitions by moving further or closer to the Sun as its luminosity changes. Note that photosynthetic plants such as asparagus can grow under a range of light intensities, down to solar irradiation at 9 astronomical units (AU) at Saturn [9], and maybe out to 300 AU [52]. The 3.8×10^{26} W luminosity of the main sequence Sun could then sustain about 4×10^{24} kg biomass for a total of 10^{10} years, yielding $BIOTA_{int} = 4 \times 10^{34}$ kg-years, and 10^{11} similar main sequence stars in the galaxy can contribute $BIOTA_{int}$ of 4×10^{45} kg-years to the life in the galaxy [7]. Interestingly, biological resource elements from the total mass of the asteroids and comets can yield a similar time-integrated biomass.

After the Sun becomes a white dwarf, its 10^{15} W luminosity can sustain 10^{13} kg biomass, and the luminosity of the estimated 10^{12} white dwarfs can sustain 10^{25} kg biomass in the galaxy for 10^{20} years, yielding 10^{45} kg-years [7,50].

Tab. 4 shows the estimated contributions of the various types of stars to the total $BIOTA_{int}$ in the galaxy during its habitable lifetime. The eventual red giant stages of stars contribute 6×10^{44} kg-years of $BIOTA_{int}$ in the galaxy, considering

Tab. 4 Life supported by the principal resources in future periods of cosmology[a].

Location	Materials, mass (kg)	Power (W)	No. in the galaxy	Future lifetime (y)	Biomass (kg)[a,b]	$BIOTA_{int}$ (kg-y)[a,b]	$BIOTA_{int}$ in galaxy (kg-y)[a]
Earth to present				4×10^9	10^{15} c	4×10^{24} c	
Solar System	Asteroids, 10^{22}	4×10^{26}	10^{11}	5×10^9	5×10^{18} d (6×10^{20})e	3×10^{28} d (3×10^{30})e	3×10^{39} d (3×10^{41})e
Solar System	Comets, 10^{26}	4×10^{26}	10^{11}	5×10^9	5×10^{22} d (6×10^{24})e	3×10^{32} d (3×10^{34})e	3×10^{43} d (3×10^{45})e
Red giants	Comets, 10^{26}	10^{30}	10^{11}	10^9	6×10^{24} e	6×10^{33} e	6×10^{44} e
White dwarfs	Comets, 10^{26}	10^{15}	10^{12}	10^{20}	10^{13} f	10^{33} d	10^{45}
Red dwarfs		10^{23}	10^{12}	10^{13}	10^{21} f	10^{34}	10^{46}
Brown dwarfs		10^{20}	10^{12}	10^{10}	10^{18} f	10^{28}	10^{39}
Galaxy	Baryons, 10^{41}	mc^2/t		10^{37} g	$<10^{41}$		10^{48} h
Universe	Baryons, 10^{52}	mc^2/t		10^{37} g	$<10^{52}$ i		10^{59} h,i

[a] The figures are order-of-magnitude estimates and the digits shown indicate the results of the calculations but don't imply this degree of accuracy. [b] Per solar system. [c] Assuming the estimated present 10^{15} kg biomass [10] for the past 4×10^9 years, as an upper limit. [d] Biomass obtained using water-soluble elements in asteroids or comets, respectively, based on N as the limiting nutrient. [e] Biomass obtained using total elemental contents of asteroids or comets, respectively, based on N as the limiting nutrient. [f] Biomass based on power supply of 100 W/kg as the limiting factor. [g] Estimated proton decay time [50]. [h] Based on the dissipation of mass as bioavalaible energy. [i] Amount in the universe [7].

the limiting factor to be nitrogen in the resource asteroids and comets [7,8].

The limiting factor for the white, red and brown dwarf stars is assumed to be energy, and they can contribute 10^{45}, 10^{46}, and 10^{39} kg-years of time-integrated biomass, $BIOTA_{int}$ in the galaxy, respectively (Tab. 4). Interestingly, all the main long-lived types of stars except brown dwarfs contribute similar integrated energy output (luminosity × lifetime × n_{galaxy}) in the galaxy. Correspondingly, they can contribute similar $BIOTA_{int}$ of about 10^{45}–10^{46} kg-years to the potential maximum amount of life in the galaxy that can be based on stellar energy.

Life in solar nebulae

Panspermia payloads to accretion zones or interstellar clouds will be captured in solar nebulae, or microbial life may arise in or be transported naturally to asteroids during aqueous alteration [28,41,42]. Tests of meteorites showed that microorganisms can be sustained by the nutrients in these asteroid solutions [8,9].

These solutions in cavities and pores of asteroids could contain concentrated, several mol/l organic solutions as our meteorite experiments showed. The organics there can be activated by catalytic metal ions, minerals and clays [8,9], and by the radioactive decay of rocks, while trapped for 10^6 –10^7 years of aqueous alteration at temperatures from 0 to >140 C in reducing conditions under high pressure hydrogen, containing ammonia and methane. Organics in carbonaceous chondrite meteorites show that complex molecules including amino acids and nucleic bases form under these conditions.

Quantitatively, porosities of 15–25% of carbonaceous chondrites [43] allow asteroids to contain large volumes of nutrient pore solutions. In 4–6×10^6 years of aqueous alteration [44], organic reactions in solutions with half-lives of seconds or shorter can allow stepwise synthetic reactions

of thousands up to millions of steps, to produce the complex molecules required to form a microorganism.

Bioavailable nutrients in a large asteroid can then yield enough microbial populations to seed the 10^{22} kg of asteroids if the asteroid is fragmented and scattered. The microorganisms may also land on hyperbolic comets that carry them to interstellar space. Nutrients in a meteorite can yield enough microorganisms to colonize a planet, and nutrients in an asteroid belt can yield enough microorganisms to seed the galaxy.

For example, a 10 km radius, 4×10^{12} m^3 and 10^{16} kg asteroid with 20% porosity would contain 8×10^{14} l pore solutions with 10^{13} kg organic C (based on 1 g/kg soluble C). The total 10^{22} kg asteroids would contain 10^{19} kg dissolved organic C, in 0.01 kg/l (about 0.1 mol/l) solutions. During a million years of aqueous alteration, and with reaction half-lives of seconds to years, these solutions allow multi-step chemical synthesis of 10^6–10^{13} steps to build up complex proteins or RNA leading to microbial life. Similar processes can occur later in the pores of carbonaceous chondrite meteorites landed in water on planets [8,9].

If microbial life starts in an asteroid, the organics may become sequestered in microbial biomass or freeze, stopping evolution. However, these microorganisms can start evolution later when delivered by meteorites to planets.

To examine this mechanism quantitatively, a C-type asteroid with phosphorus as the limiting nutrient yield about 10^{-4} kg biomass/kg resource material. A 10 km, 10^{16} kg early asteroid could then yield a bacterial biomass of 10^{12} kg with 10^{27} microorganisms. With doubling time of one day, a single 10^{-15} kg inoculating microorganism could develop into this 10^{12} kg population of 10^{27} microorganisms in 90 days.

This microbial biomass would sequester the organics of the asteroid, and without recycling this would stop growth and evolution. However, if the 10^{27} microorganisms in the parent asteroid scatter in the asteroid belt, they can provide

10^5 inoculant microorganisms per kg asteroid (e.g., 10^{21} microorganisms for a 10 km, 10^{16} kg asteroid) to colonize all the 10^{22} kg asteroid population in the early Solar System. Similarly, microorganisms in a 10 kg meteorite with a CM2 Murchison-like composition could reach a biomass of 1 g with a population of 10^{12} microorganisms, sufficient to seed a planet. Further, soluble materials in the 10^{22} kg asteroids can yield a microbial biomass up to 10^{18} kg comprising 10^{33} microorganisms that could seed 10^{11} solar systems in the galaxy each with 10^{22} microorganisms.

Beyond asteroids, comets may also host and distribute microorganisms. Subsurface ice may melt close to perihelion [45], and its organics can yield a solution similar to those in meteorites, that support algae and fungi [8,9].

Heterotrophs, chemotrophs, lithophiles and psychrophiles could multiply in these solutions, then eject with the coma in dust particles that protect them in space. They could be also transported through interstellar space by hyperbolic comets. These mechanisms were considered for natural panspermia [46], and for directed panspermia that could also use genetically engineered microorganisms suited to these conditions [25,42,46,47].

In summary, based on measured nutrients in meteorites, nutrients in a meteorite can produce enough microorganisms to seed a planet; nutrients in an asteroid can produce enough microorganisms to seed the asteroid belt; and nutrients in one asteroid belt can produce enough microorganisms to seed all the solar systems in the galaxy.

Life in a finite universe

The theoretical upper limits of the amounts of life in the universe would be realized if all baryonic matter is converted to biomass, and a small fraction of the biomass (3.5×10^{-8} y^{-1}) is then converted relativistically gradually to energy to sustain the biomass.

Assume that the power requirement of biomass is $P_{biomass}$ (J s^{-1} kg^{-1}), and the energy yield factor of converting biomass to energy is $E_{yield,biomass} = E_{biomass,released}/m_{biomass,converted} = c^2$. If biomass is converted to energy at the rate needed to power the remaining biomass, then the biomass decays exponentially at the rate given by Equation (10).

$$(-dm_{biomass}/dt)E_{yeld,biomass} = P_{biomass}m_{biomass} \qquad (10)$$

The remaining biomass after time t is given by Equation (11):

$$m_{biomass,t} = m_{biomass,0} = \exp\left(-(P_{biomass}/E_{yield,biomass})t\right) \quad (11)$$

The maximum energy can be obtained by converting mass to energy according to the relativistic relation $E = mc^2$. In this case $E_{yield,biomass} = c^2$, and assuming a power need of $P_{biomass}$ = 100 W/kg biomass, the decay rate of the biomass is k_{use} = $100/(3 \times 10^8)^2 = 1.1 \times 10^{-15}$ s^{-1} = 3.5×10^{-8} y^{-1}.

The amount of baryonic matter that could be converted to biomass in the galaxy is on the order of 10^{41} kg. If a fraction of it was converted to energy at the rate shown in Equation (10), there would be enough biomass left for one 50 kg human in the galaxy after 2.6 billion years, and in the universe, after 3.3 billion years. The last remaining life, a single bacterium

of 10^{-15} kg, would be left after 3.7 billion years in the galaxy, and after 4.4 billion years in the universe. The total time-integrated life in the galaxy would be 10^{48} kg-years and in the universe 10^{59} kg-years. The data in last rows in Tab. 4 were derived according to these considerations.

Although these amounts of life are immense, life would become extinct rapidly compared with the estimated ultimate decay of baryonic matter in 10^{37} years [50] as the biomass is converted to the energy needed to sustain it. In order to extend life throughout these possible 10^{37} years, the 10^{41} kg baryonic matter in the galaxy could be converted to biomass and then to energy more slowly, sustaining a steady-state biomass of 3×10^{11} kg throughout this time, possibly as 10^{10} humans. The respective biomass and populations in the universe are larger than in the galaxy by a factor of 10^{11} (galaxies in the universe). Life in the galaxy and the universe would then last for 10^{37} (ten trillion trillion trillion) years through the habitable lifetime of the galaxy.

Future life is finite in a finite universe, while in an ever-expanding universe, intelligent existence (not organic life) may be extended indefinitely at an ever slowing pace [5]. The above calculations and results are theoretical upper limits that are permitted by physical law. These potential amounts of life are immense, but finite.

If life is finite, should we propagate it? Even a small fraction of the potential time-integrated biomass can secure biological life for all foreseeable time. Biotic ethics recommend to realize as much of this potential life as possible. This will allow our remote descendants to observe nature more deeply, and use it to best serve life.

Life-centered astroethics

The future of life in space depends on human actions that are governed by our philosophies and ethics. These human activities are already transforming the Earth, and in the future they can affect the Solar System and the galaxy. In this manner, ethical principles can become physical forces.

These powerful ethics can be formulated on rational science-based principles. Such a fundamental principle is our identity as living beings, united with all other known organic DNA/protein life in the complex cellular machinery that underlies all biology. All of our organic cellular life share the complex structures and mechanisms of genetics, metabolism, enzyme catalysis, membrane transport, and ATP-based energy use. These complex biological structures and processes are special in Nature, because the laws of physics just narrowly allow biology to exist.

Self-propagation is also unique and common to all life. Our basic unity with all self-propagating life therefore implies a human purpose to protect, propagate and expand life. This purpose can define the basic values of life-centered biotic ethics: acts that support life are good and acts that destroy life are evil. Most cultures observe these principles.

The human purpose to propagate life is best secured in space. Once established in many worlds, the basic patterns of life can be secure through eons. The cosmological scales of time, space and resources allow life to realize the full potentials that biology permits.

Expansion in space implies a technological society that can also self-destruct by violence, biological misengineering, or by replacing biological life with robots. Therefore in a self-fulfilling future, survival must be pursued deliberately to secure survival. This implies control by DNA/protein organic brains that have a vested self-interest to propagate biological life.

These life-centered ethics can be applied in space as panbiotic ethics that aim to secure and expand life. Human settlement of the Solar System serves this objective if all humans act in unison for the future. Directed panspermia can open even larger, galactic scale prospects for life, and it can be achieved more readily from a future space infrastructure, by groups or even individuals motivated by life-centered panbiotic ethics [4,23–25,38].

Conclusions

Expansion in space can open immense potentials for the growth of humankind and life. In turn, biological adaptation and life-centered biotic ethics are keys to this expansion [4,32,38,39,47,52–56]. To examine these potentials scientifically, it is necessary to quantify, based on experimental data, the potential scope of life in the Solar System and in the galaxy.

To this effect, experimental astroecology investigates the relationship between life and space resources, with the amount of life in an ecosystem measured as time-integrated biomass (kg-years). In these studies, meteorite models of asteroid materials showed that bacteria, algae and plants can grow on asteroid/meteorite organics and electrolytes, with fertilities similar to productive agricultural soils [4,8,9,26,57–61]. Based on the measured soluble nutrient contents, asteroid resources can yield a biomass of 10^{18} kg in the Solar System. Extrapolating to a lifetime of 10^{10} years and to 10^{11} stars in the galaxy, similar resources could support a time-integrated biomass on the order of 10^{39} kg-years during the habitable lifetime of the galaxy, vastly greater than the 10^{24} kg-years of life on Earth to date. On cosmological scales, energy from white and red dwarf stars can sustain life for 10^{20} (a hundred billion billion) years, yielding a time-integrated biomass of 10^{46} kg-years in the galaxy and 10^{57} kg-years in the universe.

This potential future life can be accessed with current-level technology. Human populations in the Solar System can use asteroid resources to achieve a population of 10^{14} (a hundred trillion) humans and their supporting biota. Considering a 1% per year irreversible waste, the Solar System can sustain 10^7 (ten million) humans and their supporting biomass of 10^{11} kg for billions of years.

These space-based human populations are vital for securing and expanding life in space. This is especially critical if we are alone [4,23–25,62,63]. To assess extra-Solar-System life, it is necessary to estimate the probability that life arises in favorable ecosystems. This probability is hard to estimate, because the vital functions of even the first viable cell would require complex processes subject to specific chemical and physical parameters, and also a sustaining early environment with restricted viable parameters [6,28,41,64]. These parameters can have many possible combinations, that cannot be tested experimentally, and only a few can support life. Therefore we may not be able to quantify the probability that life arises even in favorable environments. This leaves it possible that we are alone, and the future of life is in our hands.

We can secure this future by expansion in space. Human settlement in the Solar System can serve this objective, if all humans act in unison for the future. Directed panspermia can achieve even larger, galactic scale prospects for life, and it can be implemented more easily from a space infrastructure by groups or individuals motivated by life-centered panbiotic ethics.

This panbiotic program can allow DNA/protein organic life to access galactic scale resources, by seeding new solar systems where any local life could not have arisen yet. The payloads can include extremophiles suited to diverse environments, including autotrophs and heterotrophs to establish self-sustaining recycling ecosystems. Small hardy multicellular organisms can be also included, to jump-start higher evolution toward intelligent species that may further expand life in space.

With our powerful technologies our designs can become self-fulfilling, and to secure survival, it must be then sought purposefully. The guiding ethics can be based on scientific insights: our basic biological unity with all cellular, organic DNA/protein life; the special place of complex life in nature, that precisely allows biology to exist; and the shared drive of all life for self-propagation. Our unity with all life then implies life-centered biotic ethics that aim to propagate life, applied in space as panbiotic ethics to expand life. To secure survival in a self-designed future, these ethics themselves must be also always propagated.

With these life-centered ethics, the cosmological spans of space, time and resources can support an immense future. We can plant that future by seeding new worlds with organic DNA/protein life. A great diversity of life can then emerge, some intelligent who understand nature more deeply and seek to extend life indefinitely. In securing that future for life, our human endeavors can fulfill a cosmic purpose.

Acknowledgments
I thank the NASA/NSF "Antarctic meteorite program" for samples studied in the source papers. This work was not subject to external funding.

Competing interests
No competing interests have been declared.

References
1. Rynin NA. Interplanetary flight and communication. Volume 3, No. 7: K. E. Tsiolkovskii, life, writings, and rockets. Washington, DC: Israel Program for Scientific Translations; 1971.
2. Dyson FJ. Disturbing the universe. New York, NY: Harper and Row; 1979.

3. O'Neill GK. The colonization of space. Phys Today. 1974;27(9):32–38.

4. Mautner MN. Seeding the universe with life: securing our cosmological future: galactic ecology, astroethics and directed panspermia. Washington, DC: Legacy Books; 2000.

5. Dyson FJ. Time without end: biology and physics in an open universe. Rev Mod Phys. 1979;5:447–460.

6. Purves WK, Orians GH, Sadava D, Heller HC. Life, the science of biology. Sunderland, MA: Sinauer Associates and W.H. Freeman; 2001.

7. Mautner MN. Life in the cosmological future – resources, biomass and populations. J Br Interplanet Soc. 2005;58:167–180.

8. Mautner MN. Planetary resources and astroecology. planetary microcosm models of asteroid and meteorite interiors: electrolyte solutions and microbial growth – implications for space populations and panspermia. Astrobiology. 2002;2(1):59–76.

9. Mautner M. Planetary bioresources and astroecology. 1. Planetary microcosm bioassays of Martian and carbonaceous chondrite materials: nutrients, electrolyte solutions, and algal and plant responses. Icarus. 2002;158(1):72–86.

10. Bowen HJM. Trace elements in biochemistry. New York, NY: Academic Press; 1966.

11. Bolonkin AA. Making asteroids habitable. In: Badescu V, editor. Asteroids. Berlin: Springer; 2013. p. 561–580.

12. McKay CP, Toon OB, Kasting JF. Making Mars habitable. Nature. 1991;352(6335):489–496.

13. Fogg MJ. Terraforming: a review for environmentalists. Environmentalist. 1993;13(1):7–17.

14. Mautner MN, Sinaj S. Water-extractable and exchangeable phosphate in Martian and carbonaceous chondrite meteorites and in planetary soil analogues. Geochim Cosmochim Acta. 2002;66:3161–3174.

15. Mauldin JH. Prospects for interstellar travel. San Diego, CA: Univelt; 1992. (Science and technology series; vol 80).

16. Mallove EF. The starflight handbook: a pioneer's guide to interstellar travel. New York, NY: Wiley; 1989.

17. Glaser PE. Power from the Sun: its future. Science. 1968;162(3856):857–861.

18. Mautner MN. A space-based solar screen against climatic warming. J Br Interplanet Soc. 1991;44:135–138.

19. Bewick R, Sanchez JP, McInnes CR. Gravitationally bound geoengineering dust shade at the inner Lagrange point. Adv Space Res. 2012;50(10):1405–1410.

20. Matloff GL, Johnson L, Bangs C. Living off the land in space: green roads to the cosmos. New York, NY: Springer; 2001.

21. Sauser B. A Moon-based telescope [Internet]. MIT Technology Review. 2008 [cited 2014 Dec 20].

22. Mautner MN. Space-based genetic cryoconservation of endangered species. J Br Interplanet Soc. 1996;49:319–320.

23. Mautner MN, Matloff GL. Directed panspermia – a technical and ethical evaluation of seeding other solar systems. J Br Interplanet Soc. 1979;48:435–440.

24. Mautner MN, Matloff GL. Directed panspermia. 2. Technological advances toward seeding other solar systems, and the foundation of panbiotic ethics. J Br Interplanet Soc. 1995;48:435–440.

25. Mautner MN. Directed panspermia. 3. Strategies and motivation for seeding star-forming clouds. J Br Interplanet Soc. 1997;50:93–102.

26. Mautner MN. In situ biological resources: soluble nutrients and electrolytes in carbonaceous asteroids/meteorites. Implications for astroecology and human space populations. Planet Space Sci. 2014;104:234–243.

27. Dyson FJ. Search for artificial stellar sources of infrared radiation. Science. 1960;131(3414):1667–1668.

28. Chyba CF, McDonald GD. The origin of life in the Solar System: current issues. Annu Rev Earth Planet Sci. 1995;23(1):215–249.

29. Lynch SR, Liu H, Gao J, Kool ET. Toward a designed, functioning genetic system with expanded-size base pairs: solution structure of the 8-base xDNA double helix. J Am Chem Soc. 2006;128(45):14704–14711.

30. Pinheiro VB, Taylor AI, Cozens C, Abramov M, Renders M, Zhang S, et al. Synthetic genetic polymers capable of heredity and evolution. Science. 2012;336(6079):341–344.

31. Herdman M, Janvier M, Rippka R, Stanier RY. Genome size of cyanobacteria. J Gen Microbiol. 1979;111(1):73–85.

32. Jacobsen JH. Genetic engineering of cyanobacteria [PhD thesis]. Copenhagen: University of Copenhagen; 2012.

33. Maccone C. The statistical Drake equation. Acta Astronaut. 2010;67(11–12):1366–1383.

34. Haldane JBS. The origins of life. New Biol. 1954;16:12–27.

35. Shklovskii IS, Sagan C. Intelligent life in the Universe. San Francisco, CA: Holden-Day; 1996.

36. Crick FHC, Orgel LE. Directed panspermia. Icarus. 1973;19(3):341–346.

37. Zuckerman B. Space telescopes, interstellar probes and directed panspermia. J Br Interplanet Soc. 1981;34:367–370.

38. Mautner MN. Life-centered ethics, and the human future in space. Bioethics. 2009;23(8):433–440.

39. Makukov MA, shCherbak VI. Space ethics to test directed panspermia. Life Sci Space Res Amst. 2014;3:10–17.

40. Mezger PG. The search for protostars using millimeter/submillimeter dust emission as a tracer. In: Burke BF, Rahe JH, Roettger EE, editors. Planetary systems: formation, evolution, and detection. Dordrecht: Springer; 1994. p. 197–214.

41. Pizzarello S. The chemistry of life's origin: a carbonaceous meteorite perspective. Acc Chem Res. 2006;39(4):231–237.

42. Mautner MN, Ibrahim Y, El-Shall MS. Organic synthesis and potential microbiology in the Solar Nebula: are early Solar Systems nurseries for microorganisms? Int J Astrobiol. 2004;3(1 suppl):101.

43. Macke RJ. Survey of meteorite physical properties: density, porosity and magnetic susceptibility [PhD thesis]. Orlando, FL: University of Central Florida; 2010.

44. Guo W, Eiler JM. Temperatures of aqueous alteration and evidence for methane generation on the parent bodies of the CM chondrites. Geochim Cosmochim Acta. 2007;71(22):5565–5575.

45. Lewis JS. Physics and chemistry of the solar system. New York, NY: Academic Press; 1997.

46. Hoyle F. Lifecloud: the origin of life in the universe. London: J. M. Dent; 1978.

47. Montague M, McArthur GH, Cockell CS, Held J, Marshall W, Sherman LA, et al. The role of synthetic biology for in situ resource utilization (ISRU). Astrobiology. 2012;12(12):1135–1142.

48. Horikoshi K, Grant WD, editors. Extremophiles: microbial life in extreme environments. New York, NY: Wiley; 1998.

49. Jönsson KI, Rabbow E, Schill RO, Harms-Ringdahl M, Rettberg P. Tardigrades survive exposure to space in low Earth orbit. Curr Biol. 2008;18(17):R729–R731.

50. Adams F, Laughlin G. The five ages of the universe: inside the physics of eternity. New York, NY: Touchstone; 1999.

51. Bousso R, Susskind L. The multiverse interpretation of quantum mechanics. Phys Rev Part Fields. 2012;85(4).

52. Raven JA, Kübler JE, Beardall J. Put out the light, and then put out the light. J Mar Biol Assoc UK. 2000;80(01):1–25.

53. Chyba C, Sagan C. Endogenous production, exogenous delivery

and impact-shock synthesis of organic molecules: an inventory for the origins of life. Nature. 1992;355(6356):125–132.

54. Olsson-Francis K, de la Torre R, Towner MC, Cockell CS. Survival of akinetes (resting-state cells of cyanobacteria) in low earth orbit and simulated extraterrestrial conditions. Orig Life Evol Biosph. 2009;39(6):565–579.

55. Hart MH. Interstellar migration, the biological revolution, and the future of the galaxy. In: Finney BR, Jones EM, editors. Interstellar migration and the human experience. Berkeley, CA: University of California Press; 1986. p. 278–291.

56. Fukuyama F. Our posthuman future: consequences of the biotechnology revolution. New York, NY: Farrar Straus & Giroux; 2002.

57. Mautner MN, Leonard RL, Deamer DW. Meteorite organics in planetary environments: hydrothermal release, surface activity, and microbial utilization. Planet Space Sci. 1995;43(1-2):139–147.

58. Mautner MN, Conner AJ, Killham K, Deamer DW. Biological potential of extraterrestrial materials. 2. Microbial and plant responses to nutrients in the Murchison carbonaceous meteorite. Icarus. 1997;129:245–253.

59. Mautner MN. Biological potential of extraterrestrial materials. I. Nutrients in carbonaceous meteorites, and effects on biological growth. Planet Space Sci. 1997;45(6):653–664.

60. Kennedy J, Mautner MN, Barry B, Markwitz A. Microprobe analysis of brine shrimp grown on meteorite extracts. Nucl Instrum Methods Phys Res B. 2007;260(1):184–189.

61. Marcano V, Matheus P, Cedeño C, Falcon N, Palacios-Prü E. Effects of non-carbonaceous meteoritic extracts on the germination, growth and chlorophyll content of edible plants. Planet Space Sci. 2005;53(12):1263–1279.

62. O'Neill GK. The high frontier. New York, NY: William Morrow; 1977.

63. Davies P. The eerie silence. Boston, MA: Houghton Mifflin Harcourt; 2010.

64. Yockey HP. Origin of life on Earth and Shannon's theory of communication. Comput Chem. 2000;24(1):105–123. http://dx.doi.org/10.1016/S0097-8485(00)80010-8

Permissions

List of Contributors

Janet María León Morales
Department of Biotechnology, Center of Biotic Products Development, National Polytechnic Institute, CeProBi 8, San Isidro, Yautepec, Morelos 62731, México

Mario Rodríguez-Monroy
Department of Biotechnology, Center of Biotic Products Development, National Polytechnic Institute, CeProBi 8, San Isidro, Yautepec, Morelos 62731, México

Gabriela Sepúlveda-Jiménez
Department of Biotechnology, Center of Biotic Products Development, National Polytechnic Institute, CeProBi 8, San Isidro, Yautepec, Morelos 62731, México

Mateusz Labudda
Department of Biochemistry, Warsaw University of Life Sciences – SGGW, Nowoursynowska 159, 02-776 Warsaw, Poland

Fardous Mohammad Safiul Azam
Department of Biotechnology and Genetic Engineering, University of Development Alternative, 80 Satmasjid Road, Dhanmondi R/A, Dhaka-1209, Bangladesh

Wei Zhao
School of Bioscience and Bioengineering, South China University of Technology, Guangzhou 510006, People's Republic of China
Department of Pharmacy, General Hospital of Guangzhou Military Command, Guangzhou 510010, People's Republic of China

Shujing Sheng
Department of Pharmacy, General Hospital of Guangzhou Military Command, Guangzhou 510010, People's Republic of China

Zhongyu Liu
Department of Pharmacy, General Hospital of Guangzhou Military Command, Guangzhou 510010, People's Republic of China

Di Lu
Department of Pharmacy, General Hospital of Guangzhou Military Command, Guangzhou 510010, People's Republic of China

Kuanpeng Zhu
School of Bioscience and Bioengineering, South China University of Technology, Guangzhou 510006, People's Republic of China

Department of Pharmacy, General Hospital of Guangzhou Military Command, Guangzhou 510010, People's Republic of China

Xiaoze Li
School of Bioscience and Bioengineering, South China University of Technology, Guangzhou 510006, People's Republic of China
Department of Pharmacy, General Hospital of Guangzhou Military Command, Guangzhou 510010, People's Republic of China

Shujin Zhao
Department of Pharmacy, General Hospital of Guangzhou Military Command, Guangzhou 510010, People's Republic of China

Yan Yao
School of Life Sciences, Guangzhou University, Guangzhou 510006, People's Republic of China

Sun Cheng-Zhen
State Key Laboratory of Crop Genetics & Germplasm Enhancement, Horticultural College, Nanjing Agricultural University, Nanjing, 210095, China

Li Ying
State Key Laboratory of Crop Genetics & Germplasm Enhancement, Horticultural College, Nanjing Agricultural University, Nanjing, 210095, China

Zhang Shu-Ning
State Key Laboratory of Crop Genetics & Germplasm Enhancement, Horticultural College, Nanjing Agricultural University, Nanjing, 210095, China

Zheng Jin-Shuang
State Key Laboratory of Crop Genetics & Germplasm Enhancement, Horticultural College, Nanjing Agricultural University, Nanjing, 210095, China

Przemysław Jagodzik
Institute of Experimental Biology, Adam Mickiewicz University, Umultowska 89, 61-614 Poznań; Poland

Małgorzata Adamiec
Institute of Experimental Biology, Adam Mickiewicz University, Umultowska 89, 61-614 Poznań; Poland

Grzegorz Jackowski
Institute of Experimental Biology, Adam Mickiewicz University, Umultowska 89, 61-614 Poznań; Poland

Jagna Chmielowska-Bąk
Department of Plant Ecophysiology, Institute of Experimental Biology, Faculty of Biology, Adam Mickiewicz University, Umultowska 89, 61-614 Poznań, Poland

Isabelle Lefèvre
Groupe de Recherche en Physiologie végétale (GRPV), Earth and Life Institute, Université catholique de Louvain, Croix du Sud, 4-5, bte L7.07.13, 1348 Louvain-la-Neuve, Belgium

Stanley Lutts
Groupe de Recherche en Physiologie végétale (GRPV), Earth and Life Institute, Université catholique de Louvain, Croix du Sud, 4-5, bte L7.07.13, 1348 Louvain-la-Neuve, Belgium

Agata Kulik
Department of Plant Ecophysiology, Institute of Experimental Biology, Faculty of Biology, Adam Mickiewicz University, Umultowska 89, 61-614 Poznań, Poland

Joanna Deckert
Department of Plant Ecophysiology, Institute of Experimental Biology, Faculty of Biology, Adam Mickiewicz University, Umultowska 89, 61-614 Poznań, Poland

Paulina Mistrzak
Department of Pharmaceutical Biology and Medicinal Plant Biotechnology, Medical University of Warsaw, Banacha 1, 02-097 Warsaw, Poland
Department of Pharmacognosy and Molecular Basis of Phytotherapy, Medical University of Warsaw, Banacha 1, 02-097 Warsaw, Poland

Hanna Celejewska-Marciniak
Department of Pharmaceutical Biology and Medicinal Plant Biotechnology, Medical University of Warsaw, Banacha 1, 02-097 Warsaw, Poland

Wojciech J. Szypuła
Department of Pharmaceutical Biology and Medicinal Plant Biotechnology, Medical University of Warsaw, Banacha 1, 02-097 Warsaw, Poland

Olga Olszowska
Department of Pharmaceutical Biology and Medicinal Plant Biotechnology, Medical University of Warsaw, Banacha 1, 02-097 Warsaw, Poland

Anna K. Kiss
Department of Pharmacognosy and Molecular Basis of Phytotherapy, Medical University of Warsaw, Banacha 1, 02-097 Warsaw, Poland

Junjun Liang
Chengdu Institute of Biology, Chinese Academy of Sciences, No. 9 Section 4, Renmin South Road, Chengdu 610041, China

Xin Chen
Chengdu Institute of Biology, Chinese Academy of Sciences, No. 9 Section 4, Renmin South Road, Chengdu 610041, China
College of Life Sciences, Sichuan University, No. 24 South Section 1, Yihuan Road, Chengdu 610065, China

Huanhuan Zhao
Chengdu Institute of Biology, Chinese Academy of Sciences, No. 9 Section 4, Renmin South Road, Chengdu 610041, China

Shuiyang Yu
Chengdu Institute of Biology, Chinese Academy of Sciences, No. 9 Section 4, Renmin South Road, Chengdu 610041, China

Hai Long
Chengdu Institute of Biology, Chinese Academy of Sciences, No. 9 Section 4, Renmin South Road, Chengdu 610041, China

Guangbing Deng
Chengdu Institute of Biology, Chinese Academy of Sciences, No. 9 Section 4, Renmin South Road, Chengdu 610041, China

Zhifen Pan
Chengdu Institute of Biology, Chinese Academy of Sciences, No. 9 Section 4, Renmin South Road, Chengdu 610041, China

Maoqun Yu
Chengdu Institute of Biology, Chinese Academy of Sciences, No. 9 Section 4, Renmin South Road, Chengdu 610041, China

Chaoqiang Jiang
Tobacco Research Institute, Anhui Academy of Agricultural Sciences, Nongkenan Road 40, Hefei, Anhui, China

Chaolong Zu
Tobacco Research Institute, Anhui Academy of Agricultural Sciences, Nongkenan Road 40, Hefei, Anhui, China

Jia Shen
Tobacco Research Institute, Anhui Academy of Agricultural Sciences, Nongkenan Road 40, Hefei, Anhui, China

Fuwen Shao
Tobacco Research Institute, Anhui Academy of Agricultural Sciences, Nongkenan Road 40, Hefei, Anhui, China

Tian Li
Tobacco Research Institute, Anhui Academy of Agricultural Sciences, Nongkenan Road 40, Hefei, Anhui, China

Cai-Feng Jia
School of Life Sciences, East China Normal University, 500 Dongchuan Rd, Shanghai 200241, China

Wan-Hong Hu
School of Life Sciences, East China Normal University, 500 Dongchuan Rd, Shanghai 200241, China

Zhong-yi Chang
School of Life Sciences, East China Normal University, 500 Dongchuan Rd, Shanghai 200241, China

Hong-Liang Gao
School of Life Sciences, East China Normal University, 500 Dongchuan Rd, Shanghai 200241, China

Jolanta Białek
Department of Biophysics and Plant Morphogenesis, University of Silesia in Katowice, Jagiellońska 28, 40-032 Katowice, Poland

Izabela Potocka
Laboratory of Cell Biology, University of Silesia in Katowice, Jagiellońska 28, 40-032 Katowice, Poland

Joanna Maria Szymanowska-Pułka
Department of Biophysics and Plant Morphogenesis, University of Silesia in Katowice, Jagiellońska 28, 40-032 Katowice, Poland

Julian David Janna Olmos
Faculty of Biology, University of Warsaw, Miecznikowa 1, 02-096 Warsaw, Poland

Joanna Kargul
Centre of New Technologies, University of Warsaw, Banacha 2c, 02-097 Warsaw, Poland

Joanna Simińska
Department of Biochemistry, Faculty of Agriculture and Biology, Warsaw University of Life Sciences – SGGW, Nowoursynowska 159, 02-776 Warsaw, Poland

Wiesław Bielawski
Department of Biochemistry, Faculty of Agriculture and Biology, Warsaw University of Life Sciences – SGGW, Nowoursynowska 159, 02-776 Warsaw, Poland

Takuya Okabe
Graduate School of Integrated Science and Technology, Shizuoka University, 3-5-1 Johoku, Naka-ku, 432-8561 Hamamatsu, Japan

Justyna Fidler
Department of Biochemistry, Warsaw University of Life Sciences – SGGW, Nowoursynowska 159, 02-776 Warsaw, Poland

Edyta Zdunek-Zastocka
Department of Biochemistry, Warsaw University of Life Sciences – SGGW, Nowoursynowska 159, 02-776 Warsaw, Poland

Wiesław Bielawski
Department of Biochemistry, Warsaw University of Life Sciences – SGGW, Nowoursynowska 159, 02-776 Warsaw, Poland

Rainer W. Bussmann
Missouri Botanical Garden, Box 299, St. Louis, Missouri 63166-0299, USA

Narel Y. Paniagua Zambrana
Instituto de Ecología, Universidad Mayor de San Andrés, Box 10077, Correo Central, La Paz, Bolivia

Eva C.M. Nowack
Department of Biology, Heinrich-Heine-Universität Düsseldorf, Universitätsstrasse 1, 40225 Düsseldorf, Germany

Philippe Deschamps
Unité d'Ecologie, Systématique et Evolution, CNRS UMR 8079, Université Paris-Sud, 91405 Orsay, France

Alicja Banasiak
Department of Developmental Plant Biology, Institute of Experimental Biology, University of Wrocław, Kanonia 6/8, 50-328 Wrocław, Poland

Takuro Nakayama
Center for Computational Sciences, University of Tsukuba, 1-1-1 Tennoudai, Tsukuba, Ibaraki 305-8577, Japan

Yuji Inagaki
Center for Computational Sciences, University of Tsukuba, 1-1-1 Tennoudai, Tsukuba, Ibaraki 305-8577, Japan
Graduate School of Life and Environmental Sciences, University of Tsukuba, 1-1-1 Tennoudai, Tsukuba, Ibaraki 305-8572, Japan

Przemysław Gagat
Department of Genomics, Faculty of Biotechnology, University of Wrocław, Fryderyka Joliot-Curie 14a, 50-383 Wrocław, Poland

Paweł Mackiewicz
Department of Genomics, Faculty of Biotechnology, University of Wrocław, Fryderyka Joliot-Curie 14a, 50-383 Wrocław, Poland

Romana Petrželková
Department of Biology and Ecology, Faculty of Science, University of Ostrava, Chittussiho 10, 710 00 Ostrava, Czech Republic

Marek Eliáš
Department of Biology and Ecology, Faculty of Science, University of Ostrava, Chittussiho 10, 710 00 Ostrava, Czech Republic

Björn Krenz
Lehrstuhl für Biochemie, Department Biologie, Staudtstr. 5, 91058 Erlangen, Germany

Tai Wei Guo
Department of Plant Pathology and Plant-Microbe Biology, Cornell University, Ithaca NY, USA

Tatjana Kleinow
Institut für Biomaterialien und biomolekulare Systeme, Abteilung für Molekularbiologie und Virologie der Pflanzen, Universität Stuttgart, Pfaffenwaldring 57, 70550 Stuttgart, Germany

Michael N. Mautner
Department of Chemistry, Virginia Commonwealth University, 1001 West Main Street, Richmond, VA 23284-2006, USA